T0212429

Lecture Notes in Computer Science 12097

More information about this series at http://www.springer.com/series/7407

Anna Maria Bigatti · Jacques Carette ·
James H. Davenport · Michael Joswig ·
Timo de Wolff (Eds.)

Mathematical Software – ICMS 2020

7th International Conference
Braunschweig, Germany, July 13–16, 2020
Proceedings

 Springer

Editors
Anna Maria Bigatti (ID)
Dipartimento di Matematica
Università degli Studi di Genova
Genova, Genova, Italy

Jacques Carette (ID)
Computing and Software
McMaster University
Hamilton, ON, Canada

James H. Davenport (ID)
Department of Computer Science
University of Bath
Bath, UK

Michael Joswig (ID)
Institut für Mathematik, MA 6-2
TU Berlin
Berlin, Germany

Timo de Wolff (ID)
Technische Universität Braunschweig
Braunschweig, Germany

ISSN 0302-9743   ISSN 1611-3349   (electronic)
Lecture Notes in Computer Science
ISBN 978-3-030-52199-8   ISBN 978-3-030-52200-1   (eBook)
https://doi.org/10.1007/978-3-030-52200-1

LNCS Sublibrary: SL1 – Theoretical Computer Science and General Issues

This Springer imprint is published by the registered company Springer Nature Switzerland AG
The registered company address is: Gewerbestrasse 11, 6330 Cham, Switzerland

Preface

These are the proceedings of the 7th International Congress on Mathematical Software, which was held during July 13–16, 2020, and hosted by the Technische Universität Braunschweig, Germany. In fact, in the middle of the pandemic caused by the Sars-Cov-2 virus, the conference took place online, and the video recordings of the talks are available on the conference website to accompany this book.

The ICMS community believes that the appearance of mathematical software is one of the most important current developments in mathematics, and this phenomenon should be studied as a coherent whole. We hope this conference can serve as the main forum for mathematicians, scientists, and programmers who are interested in development of mathematical software. The program of the 2020 meeting consisted of 14 topical sessions, which made up the core of the program, consisting of more than 120 contributed talks. Session contributors were given the option to submit their work for publication in these proceedings, and 48 papers were selected through a peer-review process. For the first time this ICMS featured a special session with parallel software demonstrations.

The conference also featured three invited talks. Erika Ábrahám gave a talk on "Solving Real-Algebraic Formulas with SMT-RAT," Alan Edelman on "Julia—The Power of Language," and Victor Shoup on "NTL: a Library for Doing Number Theory." Short abstracts of these talks also appear in these proceedings. We thank the invited speakers for accepting our invitations to speak at ICMS 2020. We also thank all the contributors, session organizers, PC members, as well the local arrangement team and the members of the Advisory Board for helping to make this conference a success. Finally, we thank our sponsors, listed on the following pages, for the financial support of the event. Sebastian Gutsche and Amazon kindly provided technical support.

July 2020

Anna Maria Bigatti
Jacques Carette
James H. Davenport
Michael Joswig
Timo de Wolff

Organization

General Chair

Michael Joswig   TU Berlin and MPI for Mathematics in the Sciences, Germany

Program Committee Chairs

Anna Maria Bigatti   Università degli Studi di Genova, Italy
Folkmar Bornemann   TU München, Germany
Jacques Carette   McMaster University, Canada

Program Committee

Erika Ábrahám   RWTH Aachen, Germany
Carlo Angiuli   Carnegie Mellon University, USA
Katja Berčič   FAU Erlangen-Nürnberg, Germany
Anton Betten   Colorado State University, USA
Gavin Brown   University of Warwick, UK
Taylor Brysiewicz   Texas A&M University, USA
Changbo Chen   Chinese Academy of Sciences, China
Tom Coates   Imperial College London, UK
Wolfgang Dalitz   Zuse Institute Berlin, Germany
James H. Davenport   University of Bath, UK
Bettina Eick   TU Braunschweig, Germany
Matthew England   Coventry University, UK
Claus Fieker   TU Kaiserslautern, Germany
Florian Hess   Oldenburg University, Germany
Alexander Kasprzyk   Nottingham University, UK
Michael Kohlhase   FAU Erlangen-Nürnberg, Germany
Christoph Koutschan   RICAM Linz, Austria
Viktor Levandovskyy   RWTH Aachen, Germany
Anders Mörtberg   Stockholm University, Sweden
Yue Ren   Swansea University, UK
Anna-Laura Sattelberger   MPI for Mathematics in the Sciences, Germany
Benjamin Schröter   Binghamton University, USA
Moritz Schubotz   FIZ Karlsruhe, Germany
Emre Can Sertöz   Leibniz Universität Hannover, Germany
Nicolas M. Thiery   Université Paris-Saclay, France
Rebecca Waldecker   Martin-Luther-University Halle-Wittenberg, Germany
Timo de Wolff   TU Braunschweig, Germany

Advisory Board

James Davenport (Chair)	University of Bath, UK
Manuel Kauers	University of Linz, Austria
George Labahn	University of Waterloo, Canada
Josef Urban	Czech Technical University in Prague, Czech Republic
Jonathan Hauenstein	University of Notre Dame, USA
Peter Paule	RISC Linz, Austria
Andrew Sommese	University of Notre Dame, USA
Thorsten Koch	TU Berlin, Germany

Local Organizers

Timo de Wolff (Chair)
Alexander Cant
Bettina Eick
Janin Heuer
Birgit Komander
Khazhgali Kozhasov
Dirk Lorenz
Tobias Moede
Matthias Neumann-Brosig
Silke Thiel
Oğuzhan Yürük

Additional Reviewers

Christoph Brauer
Lars Kastner
Benjamin Lorenz
Dennis Müller
Tom Wiesing

Sponsors

Technische Universität Braunschweig
Emmy Noether-Programm der DFG
Einstein Stiftung Berlin
Berlin Mathematics Research Center MATH$^+$
SFB/TRR 195: Symbolic Tools in Mathematics and their Application
Amazon
Maplesoft

Abstracts of Invited Talks

Solving Real-Algebraic Formulas with SMT-RAT

Erika Ábrahám

RWTH Aachen, Germany
abraham@cs.rwth-aachen.de

In this talk we present our SMT solver named SMT-RAT, a tool for the automated check of quantifier-free real and integer arithmetic formulas for satisfiability. As a distinguishing feature, SMT-RAT provides a set of decision procedures and supports their strategic combination. We describe our CArL C++ library for arithmetic computations, the available modules implemented on top of CArL, and how modules can be combined to satisfiability-modulo-theories (SMT) solvers. Besides the traditional SMT approach, some new modules support also the recently proposed and highly promising model-constructing satisfiability calculus approach.

Julia—The Power of Language

Alan Edelman

MIT, USA
edelman@mit.edu

We often think of a programming language as a means to implement and check mathematics or as a sideshow to raise conjectures only to be hidden when the paper comes out. We argue that while all this is true, elegant open source software has the power to create much more. In this talk, we will provide examples, and suggest how the traditional academic view of paper first, computation secondary needs to be more flexible.

NTL: A Library for Doing Number Theory

Victor Shoup

Courant Institute, USA
shoup@cs.nyu.edu

NTL is a high-performance C++ library for doing arithmetic on polynomial over various rings (integers and finite fields), as well as a number of other algebraic structures. I will discuss the history of NTL, as well as some of the basic elements of its design and the algorithms it employs. I will also discuss recent work on making NTL exploit multicore and SIMD computer architectures, as well as NTL's use in implementing fully homomorphic encryption schemes.

Topical Sessions

Gröbner Bases in Theory and Practice

Session Chair

Viktor Levandovskyy — RWTH Aachen, Germany

Real Algebraic Geometry

Session Chairs

Erika Ábrahám — RWTH Aachen, Germany
James Davenport — University of Bath, UK
Matthew England — Coventry University, UK

Algebraic Geometry via Numerical Computation

Session Chairs

Taylor Brysiewicz — Texas A&M University, USA
Emre Sertöz — Leibniz Universität Hannover, Germany

Computational Algebraic Analysis

Session Chairs

Christoph Koutschan — RICAM, Austria
Anna-Laura Sattelberger — MPI for Mathematics in the Sciences, Germany

Software for Number Theory and Arithmetic Geometry

Session Chairs

Claus Fieker — TU Kaiserslautern, Germany
Florian Hess — Oldenburg University, Germany

Groups and Group Actions

Session Chairs

Bettina Eick — TU Braunschweig, Germany
Rebecca Waldecker — Martin-Luther-University Halle-Wittenberg, Germany

The Classification Problem in Geometry

Session Chair

Anton Betten — Colorado State University, USA

Polyhedral Methods in Geometry and Optimization

Session Chairs

Yue Ren	Swansea University, UK
Benjamin Schröter	Binghamton University, USA

Univalent Mathematics: Theory and Implementation

Session Chairs

Carlo Angiuli	Carnegie Mellon University, USA
Anders Mörtberg	Stockholm University, Sweden

Artificial Intelligence and Mathematical Software

Session Chairs

Changbo Chen	Chinese Academy of Sciences, China
Matthew England	Coventry University, UK

Databases in Mathematics

Session Chairs

Gavin Brown	University of Warwick, UK
Tom Coates	Imperial College London, UK
Alexander Kasprzyk	Nottingham University, UK

Accelerating Innovation Speed in Mathematics by Trading Mathematical Research Data

Session Chairs

Katja Berčič	FAU Erlangen-Nürnberg, Germany
Wolfgang Dalitz	Zuse Institute Berlin, Germany
Moritz Schubotz	FIZ Karlsruhe, Germany

The Jupyter Environment for Computational Mathematics

Session Chair

Nicolas M. Thiéry	Paris-Sud University, France

General Session

Session Chair

Michael Joswig	TU Berlin and MPI for Mathematics in the Sciences, Germany

Software Demonstrations

- The Quantifier Elimination Package in Maple for QE and Real Algebraic Geometry
  Zak Tonks

- HomotopyContinuation.jl - a package for solving systems of polynomials in Julia
  Sascha Timme and Paul Breiding

- A variant of van Hoeij's algorithm for computing hypergeometric term solutions of holonomic recurrence equations
  Bertrand Teguia Tabuguia

- Classifying Cubic Surfaces over Finite Fields with Orbiter
  Anton Betten and Fatma Karaoglu

- Software for Generating Points Uniformly in a Convex Polytope
  Mark Korenblit and Efraim Shmerling

Contents

The Jupyter Environment for Computational Mathematics

General Session

Gröbner Bases in Theory and Practice

A Design and an Implementation of an Inverse Kinematics Computation in Robotics Using Gröbner Bases

Noriyuki Horigome[1,2P], Akira Terui[1(✉)] ⓘ, and Masahiko Mikawa[1] ⓘ

[1] University of Tsukuba, Tsukuba, Japan
{horigome,terui}@math.tsukuba.ac.jp, mikawa@slis.tsukuba.ac.jp
[2] Internet Initiative Japan Inc., Tokyo, Japan
https://researchmap.jp/aterui
https://www.iij.ad.jp/en/

Abstract. The solution and a portable implementation of the inverse kinematics computation of a 3 degree-of-freedom (DOF) robot manipulator using Gröbner bases are presented. The main system was written Python with computer algebra system SymPy. Gröbner bases are computed with computer algebra system Risa/Asir, called from Python via OpenXM infrastructure for communicating mathematical software systems. For solving a system of algebraic equations, several solvers (both symbolic and numerical) are used from Python, and their performance has been compared. Experimental results with different solvers for solving a system of algebraic equations are shown.

Keywords: Gröbner bases · Robotics · Inverse kinemetics

1 Introduction

In robotics, inverse kinematics computation is one of the central topics in motion planning [17]. In the field of computer algebra, methods of the inverse kinematics computation using Gröbner bases have been proposed ([2,7,22,23], and the references therein). After formulating the forward kinematics problem using the Denavit–Hartenberg convention, the inverse kinematics problem is derived as a system of algebraic equations by conversion of trigonometric expressions to polynomials. Then, the system is triangularized by computing the Gröbner basis with respect to the lexicographic ordering and solved by appropriate solvers.

Since methods using Gröbner bases solve the inverse kinematics problem directly, these methods have advantages that one can verify if the given inverse kinematics problem is solvable (with the certification of the solution if needed), and if it is solvable, one can obtain the configuration of parameters for the desired motion of the robot directly, before the actual motion. On the other hand, the computational cost of methods using Gröbner bases tends to be high compared to that of numerical methods, thus it is desired to decrease computational cost for

© Springer Nature Switzerland AG 2020
A. M. Bigatti et al. (Eds.): ICMS 2020, LNCS 12097, pp. 3–13, 2020.
https://doi.org/10.1007/978-3-030-52200-1_1

effective computation of solving the inverse kinematics problem using Gröbner bases. Furthermore, for the use of these methods in robotics simulators such as the Robot Operating System (ROS) [9], an implementation that can easily be integrated with these simulators is needed.

In this paper, we present the solution and a portable implementation of the inverse kinematics computation of a 3 degree-of-freedom (DOF) robot manipulator using Gröbner bases. For portable implementation and rapid development, the main program is developed in Python with computer algebra system SymPy [12]. Gröbner bases are computed effectively using computer algebra system Risa/Asir [14], connected to Python with OpenXM infrastructure for communicating mathematical software systems [10]. Then, the system of algebraic equations is solved using an appropriate solver called within Python. As the main focus of our paper, several solvers for solving a system of algebraic equations have been compared: an exact solver included in SymPy, a multivariate numerical solver using the secant method, and a univariate numerical solver with successive substitutions.

The rest of the paper is organized as follows. In Sect. 2, the method of inverse kinematics computation of a 3 DOF manipulator using Gröbner bases is explained. In Sect. 3, the description of the proposed system for solving the inverse kinematics problem is presented. In Sect. 4, the result of experiments is presented.

2 Inverse Kinematics of a 3 Degree-of-Freedom (DOF) Robot Manipulator

In this paper, an example of 3 degree-of-freedom (DOF) manipulator has been built using LEGO® MINDSTORMS® EV3 Education[1] (henceforth abbreviated to EV3) (Fig. 1). An EV3 set has a computer (which is called "EV3 Intelligent Brick"), servo motors and sensors (gyro, ultrasonic, color and touch sensors), along with bricks used for constructing building blocks of robots. While a GUI-based programming environment for controlling the robot is available, several programming languages such as Python, Ruby, C, and Java can also be used on the top of other programming environments.

We first solve the forward kinematics problem. Components of the manipulator are defined as shown in Fig. 2. Segments (links) are called Segment i ($i = 1, 2, 3, 4$) from the one fixed on the ground towards the end effector, and a joint connecting Segment i and $i + 1$ is called Joint i. For Joint i, the coordinate system Σ_i, with the x_i, y_i and z_i axes and the origin at Joint i, is defined according to the Denavit–Hartenberg convention [18] (Fig. 2). Note that, since the present coordinate system is right-handed, the positive axis pointing upwards and downwards is denoted by "⊙" and "⊗", respectively. Furthermore, let Σ_0 be the coordinate system satisfying that the origin is placed at the perpendicular foot from the origin of Σ_1, and the direction of axes x_0, y_0 and z_0 are the same

---

[1] LEGO and MINDSTORMS are trademarks of the LEGO Group.

Fig. 1. A 3 DOF manipulator built with EV3.

as that of axes x_1, y_1 and z_1, respectively. Also, let Joint 5 be the end effector, and Σ_5 be the coordinate system with the origin placed at the position of Joint 5 and that the axes x_5, y_5 and z_5 have the same direction as the axes x_4, y_4 and z_4, respectively.

Let a_i be the distance between axes z_{i-1} and z_i, α_i the angle between axes z_{i-1} and z_i with respect to x_i axis, d_i the distance between axes x_{i-1} and x_i, and θ_i be the angle between axes x_{i-1} and x_i with respect to z_i axis. Then, the coordinate transformation matrix $^{i-1}T_i$ from the coordinate system Σ_{i-1} to Σ_i is expressed as

$$
^{i-1}T_i = \begin{pmatrix} 1 & 0 & 0 & a_i \\ 0 & 1 & 0 & 0 \\ 0 & 0 & 1 & 0 \\ 0 & 0 & 0 & 1 \end{pmatrix} \begin{pmatrix} 1 & 0 & 0 & 0 \\ 0 & \cos\alpha_i & -\sin\alpha_i & 0 \\ 0 & \sin\alpha_i & \cos\alpha_i & 0 \\ 0 & 0 & 0 & 1 \end{pmatrix} \begin{pmatrix} 1 & 0 & 0 & 0 \\ 0 & 1 & 0 & 0 \\ 0 & 0 & 1 & d_i \\ 0 & 0 & 0 & 1 \end{pmatrix} \begin{pmatrix} \cos\theta_i & -\sin\theta_i & 0 & 0 \\ \sin\theta_i & \cos\theta_i & 0 & 0 \\ 0 & 0 & 1 & 0 \\ 0 & 0 & 0 & 1 \end{pmatrix}
$$

$$
= \begin{pmatrix} \cos\theta_i & -\sin\theta_i & 0 & a_i \\ \cos\alpha_i \sin\theta_i & \cos\alpha_i \cos\theta_i & -\sin\alpha_i & -d_i \sin\alpha_i \\ \sin\alpha_i \sin\theta_i & \sin\alpha_i \cos\theta_i & \cos\alpha_i & d_i \cos\alpha_i \\ 0 & 0 & 0 & 1 \end{pmatrix}.
$$

Since the joint parameters a_i, α_i, d_i and θ_i for EV3 are given as shown in Table 1 (note that the unit of a_i and d_i is centimeters) and the transformation matrix T from the coordinate system Σ_5 to Σ_0 is calculated as $T = {}^0T_1{}^1T_2{}^2T_3{}^3T_4{}^4T_5$, the position ${}^t(x, y, z)$ of the end effector with respect to the coordinate system Σ_0 is expressed as

$$
\begin{pmatrix} x \\ y \\ z \end{pmatrix} = \begin{pmatrix} 8(2\cos(\theta_2 + \pi/4) + 2\cos(\theta_2 + \theta_3 + \pi/4) + \sqrt{2})\cos\theta_1 \\ 8(2\cos(\theta_2 + \pi/4) + 2\cos(\theta_2 + \theta_3 + \pi/4) + \sqrt{2})\sin\theta_1 \\ 16\sin(\theta_2 + \theta_3 + \pi/4) + 8 + 8\sqrt{2} \end{pmatrix}. \tag{1}
$$

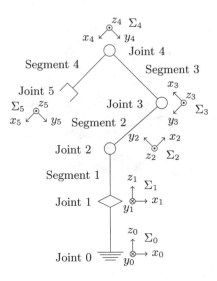

Fig. 2. Components and the coordinate systems of the manipulator.

Table 1. Joint parameters for EV3.

i	a_i (cm)	α_i	d_i (cm)	θ_i
1	0	0	8	θ_1
2	0	$\pi/2$	0	$\pi/4$
3	16	0	0	θ_2
4	16	0	0	θ_3
5	16	0	0	0

Next, for solving the inverse kinematics problem, we solve Eq. (1) with respect to $\theta_1, \theta_2, \theta_3$. With the expansion of trigonometric functions using the trigonometric addition formulas and transformation of trigonometric functions defined as

$$c_i = \cos\theta_i, \quad s_i = \sin\theta_i,$$

subject to $c_i^2 + s_i^2 = 1$, Eq. (1) is transferred to a system of algebraic equations:

$$
\begin{aligned}
f_1 &= 8\sqrt{2}c_1(c_2 + c_3(c_2 - s_2)) - s_2 - s_3(c_2 + s_2) + 1) - x = 0, \\
f_2 &= 8\sqrt{2}s_1(c_2 + c_3(c_2 - s_2)) - s_2 - s_3(c_2 + s_2) + 1) - y = 0, \\
f_3 &= 8\sqrt{2}(c_2 + c_3(c_2 + s_2)) + s_2 + s_3(c_2 - s_2) + 1) + 8 - z = 0, \\
f_4 &= s_1^2 + c_1^2 - 1 = 0, \quad f_5 = s_2^2 + c_2^2 - 1 = 0, \quad f_6 = s_3^2 + c_3^2 - 1 = 0.
\end{aligned}
\tag{2}
$$

Then, by putting the coordinate $^t(x, y, z)$ into Eq. (2) and computing Gröbner basis of an ideal $\langle f_1, \ldots, f_6 \rangle$ with respect to the lexicographic (lex) ordering, a "triangularized" system of equations is obtained. By solving the triangularized system of equations, configuration of the joint angles $\theta_1, \theta_2, \theta_3$ are obtained.

Remark 1. We see that the ideal $\langle f_1, \ldots, f_6 \rangle$ is zero-dimensional for generic values of x, y, z, as follows. By computing the comprehensive Gröbner system [8] of $\langle f_1, \ldots, f_6 \rangle$ with parameters x, y, z and variables $c_1, s_1, c_2, s_2, c_3, s_3$ in $\mathbb{R}[x, y, z, c_1, s_1, c_2, s_2, c_3, s_3]$ with respect to lex order with $c_1 \succ s_1 \succ c_2 \succ s_2 \succ c_3 \succ s_3$, we have $\{g_1, g_2, g_3, g_4, g_5, g_6\}$ as the Gröbner basis for the generic case[2]. In the generic case, we have $\mathrm{LM}(g_1) = s_3^4$, $\mathrm{LM}(g_2) = c_3$, $\mathrm{LM}(g_3) = s_2$, $\mathrm{LM}(g_4) = c_2$, $\mathrm{LM}(g_5) = s_1$, $\mathrm{LM}(g_6) = c_1$, where $\mathrm{LM}(g)$ denotes the leading monomial of g. This shows that the ideal is zero-dimensional for the generic case [1].

3   Implementation

We have implemented a system for the inverse kinematics computation of the manipulator in SymPy [12] on the top of Python, connecting with the computer algebra system Risa/Asir [14] via OpenXM infrastructure for communicating mathematical software systems [10].

Python (and SymPy) has been chosen for rapid development including the use of the library for solving algebraic equations, and interoperability with the Robot Operating System (ROS) [9] for embedding our present implementation as a simulation environment or an inverse kinematics solver in the future.

OpenXM (which stands for "Open message eXchange protocol for Mathematics") consists of definitions of protocols and data formats for communication and/or interchange of mathematical information among mathematical software systems. It also includes implementation of interface for various mathematical softwares including Risa/Asir, Kan/sm1 [20], Maple [11], Mathematica [26], MixedVol [3], NTL [16], PHC Pack [25] and TiGERS [4].

Risa/Asir is used for effective computation of Gröbner bases. After receiving input polynomials from Python/SymPy, it first computes the Gröbner basis with respect to the graded reverse lexicographic (grlex) ordering. Then, it converts the basis to the one with respect to lex ordering (with a modular FGLM algorithm [14]) before returning the final result to the host program. Risa/Asir can be called from Python easily using `ctypes` library [15] with the OpenXM interface library for Risa/Asir.

4   Experiments

We have tested our implementation for inverse kinematics computation for randomly given points within the feasible region.

For solving a system of algebraic equations, the following solvers have been used:

1. a built-in exact solver in SymPy (`sympy.solvers.solvers.solve`),

---

[2] We have computed the comprehensive Gröbner system on Risa/Asir using an implementation by Nabeshima [13].

2. a numerical solver in Python's npmath library [6] (`mpmath.findroot`) using multivariate secant method[3] (called by `sympy.solvers.solvers.nsolve` function in SymPy package),
3. a numerical solver in Python's NumPy package [24] (`numpy.roots`) solving univariate algebraic equations with successive substitutions.

For each solver, 10 sets of experiments were conducted with 100 random points given in each set of experiments (thus 1000 random points were given in total).

The computing environment is as follows.

Host environment. Intel Core i5-7360U 2.3GHz, RAM 8GB, macOS 10.15.1, Parallels Desktop Lite 1.4.0.

Guest environment. RAM 2GB, Linux 4.15.0, Python 2.7.12, SymPy 1.4, mpmath 1.1.0, numpy 1.11.0, OpenXM 1.3.3-1, Asir 20191101.

Remark 2. As shown in Remark 1, if the given point is within the feasible region, the system of algebraic equations $g_1 = \cdots = g_6 = 0$ has real solution(s) and can be solved rigorously, where $\{g_1, \ldots, g_6\}$ is the Gröbner basis of the ideal $\langle f_1, \ldots, f_6 \rangle$ with respect to lex order.

Remark 3. For the exact computation of Gröbner bases, the coordinates of the sample points are given as rational numbers with the magnitude of the denominator is less than 100.

4.1   The Result with an Exact Solver (`solve`)

Table 2 shows the result of experiments with the exact solver (`sympy.solvers.solvers.solve`) [19]. For a system of algebraic equations, the solver computes a Gröbner basis with lex order, solve univariate equations and substitute the roots in the other equations to obtain the other coordinate[4].

In each test, T_{GB} is the average of computing time of Gröbner basis, T_{Solve} is the average of computing time for solving the system of algebraic equations, T_{Total} is the average of total computing time for inverse kinematics computation, and 'Error' is the average of absolute error of the position of the end effector with the configuration of joint angles $\theta_1, \theta_2, \theta_3$, computed by solving the inverse kinematics problem, from the randomly given position. Note that T_{Total} includes the time for synchronizing received data of Gröbner basis from Risa/Asir ($\simeq 1.5\,\text{s}$) and the time for evaluation of Error. The bottom row 'Average' shows the average of the values in each column of the 10 test sets.

In all the test cases, the system of algebraic equations has been rigorously solved with finding appropriate real roots, although it took much time for finding the roots. For finding joint angles $\theta_1, \theta_2, \theta_3$, the solutions $s_1, c_1, s_2, c_2, s_3, c_3$ have been approximated by double precision floating-point numbers for efficient use of arctan function. We see that, though the approximation of the solution, the position of the end effector has been computed with high precision, compared to the length of the segments (Table 1).

---

[3] As the initial values, $(c_1, s_1, c_2, s_2, c_3, s_3) = (1, 1, 1, 1, 1, 1)$ were given.

[4] The solver may not need a Gröbner basis of lex order as an input, but it might be better to compute beforehand for faster computation.

Table 2. A result of the inverse kinematics computation with the exact solver (`nsolve`).

Test	T_{GB} (s)	T_{solve} (s)	T_{Total} (s)	Error (cm)
1	0.511	37.410	58.771	4.775×10^{-11}
2	0.244	38.268	60.528	4.747×10^{-11}
3	0.378	37.890	59.970	5.015×10^{-11}
4	0.311	38.123	57.955	4.951×10^{-11}
5	0.234	38.615	61.405	4.633×10^{-11}
6	0.471	37.599	59.809	4.525×10^{-11}
7	0.292	37.839	59.784	4.884×10^{-11}
8	0.166	22.145	36.476	4.976×10^{-11}
9	0.254	21.568	35.650	4.794×10^{-11}
10	0.267	21.501	34.217	4.960×10^{-11}
Average	0.312	33.092	52.457	4.826×10^{-11}

4.2 The Result with the Multivariate Numerical Solver (`nsolve`)

Table 3 shows the result of experiments with the multivariate numerical solver (`mpmath.findroot`) [5]. The columns 'Test', T_{GB}, T_{Solve}, T_{Total}, 'Error', and the bottom row 'Average' are the same as those in Table 2. In these sets of experiments, we have observed that the method did not converge in many cases, so the column '#Fail' shows the number of tests in which the method did not converge in each test set. Note that the data in T_{GB}, T_{Solve}, T_{Total}, 'Error' and 'Average' have been taken for the tests in which the method has successfully converged. The number '(564)' in the 'Average' row and the '#Fail' column shows the total number of tests in which the method did not converge[5].

The result shows that the method did not converge in approximately half of the test cases, while, in the cases that the method converged, the method is more efficient than the exact root-finding method. It also shows that, in the cases that the method converged, the magnitude of the absolute error of the solution is approximately 10 times larger than those in the case of the exact method, which is sufficiently small for practical use [6].

---

[5] We have tested the method with other initial values. With the initial values $(1, 0, 1, 0, 1, 0)$, the number of test cases in which approximate roots do not converge was the same as the test cases with initial values $(1, 1, 1, 1, 1, 1)$. With initial values $(0, 1, 0, 1, 0, 1)$, the approximate roots have never converged to the roots.

[6] We have also applied the multivariate numerical solver to the original system of equations with initial values $(1, 1, 1, 1, 1, 1)$, $(0, 0, 0, 0, 0, 0)$, $(1, 0, 1, 0, 1, 0)$ and $(0, 1, 0, 1, 0, 1)$, and found that none of the initial values converge to true roots in all the test cases.

Table 3. A result of the inverse kinematics computation with the multivariate numerical solver (`nsolve`).

Test	T_{GB} (s)	T_{Solve} (s)	T_{Total} (s)	Error (cm)	#Fail
1	0.249	0.261	4.096	2.689×10^{-10}	54
2	0.248	0.256	4.086	2.819×10^{-10}	49
3	0.274	0.275	4.140	2.648×10^{-10}	55
4	0.264	0.268	4.112	3.182×10^{-10}	60
5	0.264	0.271	4.121	2.375×10^{-10}	62
6	0.264	0.269	4.119	2.618×10^{-10}	68
7	0.268	0.276	4.126	2.612×10^{-10}	52
8	0.272	0.274	4.133	2.770×10^{-10}	50
9	0.270	0.265	4.119	2.995×10^{-10}	50
10	0.290	0.254	4.118	2.705×10^{-10}	64
Average	0.266	0.267	4.117	2.741×10^{-10}	(564)

Table 4. The result of the inverse kinematics computation with the univariate numerical solver (`roots`) with successive substitutions.

Test	T_{GB} (s)	T_{Solve} (s)	T_{Total} (s)	Error (cm)
1	0.245	0.309	4.181	2.720×10^{-10}
2	0.246	0.306	4.179	2.659×10^{-10}
3	0.246	0.306	4.179	2.639×10^{-10}
4	0.247	0.307	4.181	2.741×10^{-10}
5	0.251	0.313	4.187	2.757×10^{-10}
6	0.250	0.316	4.197	2.720×10^{-10}
7	0.254	0.314	4.198	2.588×10^{-10}
8	0.252	0.307	4.183	2.653×10^{-10}
9	0.254	0.309	4.186	2.773×10^{-10}
10	0.253	0.307	4.181	2.755×10^{-10}
Average	0.250	0.309	4.185	2.700×10^{-10}

4.3 The Result with the Univariate Numerical Solver (roots) with successive substitutions

Table 4 shows the result of experiments with the univariate numerical solver (`numpy.roots`) [21]. The columns 'Test', T_{GB}, T_{Solve}, T_{Total}, 'Error', and the bottom row 'Average' are the same as those in Tables 2 and 3. The result shows that the method successfully converged and found appropriate solutions with sufficiently small errors in all the tests.

5   Concluding Remarks

In this paper, we have presented a portable implementation of the inverse kinematics computation of a 3 DOF robot manipulator using Gröbner bases. The implementation is made on the top of Python and SymPy with using Risa/Asir for efficient computation of Gröbner bases and symbolic and/or numerical solvers called within Python for solving a system of algebraic equations. Risa/Asir can easily be called from Python via OpenXM infrastructure.

The experiments have shown the following features of solvers used in solving the system of algebraic equations used in the present computation:

1. Symbolic solver can be used for inverse kinematics computation with high accuracy with the certification of real roots, although the computing time increases.
2. Multivariate numerical solver is often unstable, although it can be used to solve the inverse kinematics problem with high efficiency and accuracy in stable cases.
3. Univariate numerical solver with successive substitutions is stable with high efficiency and accuracy for all the tests in the present paper.

Thus, we see that univariate numerical solver with successive substitutions is effective for solving the inverse kinematics problem in the present case, although certification of real roots may be needed.

For future research, we need to improve implementation for calling Risa/Asir from Python via OpenXM in a more sophisticated way for more efficient computation.[7] Furthermore, we intend to extend our implementation for embedding our solver in robotics simulators such as ROS and/or controlling the actual EV3 manipulators including the one we have built in the present paper.

References

1. Cox, D.A., Little, J., O'Shea, D.: Using algebraic geometry (2005). https://doi.org/10.1007/978-1-4757-6911-1
2. Faugère, J.C., Merlet, J.P., Rouillier, F.: On solving the direct kinematics problem for parallel robots. Research report RR-5923, INRIA (2006). https://hal.inria.fr/inria-00072366
3. Gao, T., Li, T.Y., Wu, M.: Algorithm 846: mixedvol: a software package for mixed-volume computation. ACM Trans. Math. Softw. **31**(4), 555–560 (2005). https://doi.org/10.1145/1114268.1114274
4. Huber, B., Thomas, R.R.: Computing Gröbner fans of toric ideals. Exp. Math. **9**(3), 321–331 (2000). https://doi.org/10.1080/10586458.2000.10504409

---

[7] At this time Risa/Asir is invoked by sending the command in the text form, with the waiting time (approximately 1.5 s) for synchronizing output data is set. We expect that this process becomes more efficient by using appropriate API for sending/receiving commands and data.

5. Johansson, F.: mpmath developers: mpmath 1.1.0 documentation: Root-finding and optimization (2018). http://mpmath.org/doc/current/calculus/optimization. html

6. Johansson, F.: mpmath developers: mpmath: a Python library for arbitrary-precision floating-point arithmetic (version 1.1.0) (2018). http://mpmath.org/. Accessed 20 Mar 2020

7. Kalker-Kalkman, C.M.: An implementation of Buchbergers' algorithm with applications to robotics. Mech. Mach. Theory **28**(4), 523–537 (1993). https://doi.org/10.1016/0094-114X(93)90033-R

8. Kapur, D., Sun, Y., Wang, D.: An efficient method for computing comprehensive gröbner bases. J. Symbolic Comput. **52**, 124–142 (2013). https://doi.org/10.1016/j.jsc.2012.05.015

9. Koubaa, A. (ed.): Robot Operating System (ROS). SCI, vol. 625. Springer, Cham (2016). https://doi.org/10.1007/978-3-319-26054-9

10. Maekawa, M., Noro, M., Ohara, K., Takayama, N., Tamura, K.: The design and implementation of OpenXM-RFC 100 and 101. In: Shirayanagi, K., Yokoyama, K. (eds.) Computer Mathematics: Proceedings of the Fifth Asian Symposium on Computer Mathematics (ASCM 2001), pp. 102–111. World Scientific (2001). https://doi.org/10.1142/9789812799661_0011

11. Maplesoft, a division of Waterloo Maple Inc.: Maple 2020 [computer software] (2020). https://www.maplesoft.com/products/maple/. Accessed 17 Mar 2020

12. Meurer, A., et al.: SymPy:symbolic computing in Python. PeerJ Comput. Sci. **3** (2017). https://doi.org/10.7717/peerj-cs.103

13. Nabeshima, K.: An implementation of GCS algorithm for Risa/Asir. Private communication (2012)

14. Noro, M.: A computer algebra system: Risa/Asir. In: Joswig, M., Takayama, N. (eds.) Algebra, Geometry and Software Systems, pp. 147–162. Springer, Heidelberg (2003). https://doi.org/10.1007/978-3-662-05148-1_8

15. Python Software Foundation: The Python Standard Library. Python Software Foundation (2020). https://docs.python.org/3/library/. Accessed 15 Mar 2020

16. Shoup, V.: NTL: a library for doing number theory [computer software] (version 11.4.3) (2020). http://www.shoup.net/ntl/. Accessed 18 Mar 2020

17. Siciliano, B., Khatib, O. (eds.): Springer Handbook of Robotics. SHB. Springer, Cham (2016). https://doi.org/10.1007/978-3-319-32552-1

18. Siciliano, B., Sciavicco, L., Villani, L., Oriolo, G.: Springer. Robotics: Modelling, Planning and Control (2008). https://doi.org/10.1007/978-1-84628-642-1

19. SymPy Development Team: Sympy 1.5.1 documentation: Solvers (2019). https://docs.sympy.org/latest/modules/solvers/solvers.html

20. Takayama, N.: Kan/sm1: a system for computing in the ring of differential operators [computer software] (1991–2003), http://www.math.kobe-u.ac.jp/KAN/. Accessed 17 Mar 2020

21. The SciPy community: Numpy v1.18 manual: numpy.roots (2019). https://numpy.org/doc/stable/reference/generated/numpy.roots.html

22. Uchida, T., McPhee, J.: Triangularizing kinematic constraint equations using Gröbner bases for real-time dynamic simulation. Multibody Syst. Dyn. **25**, 335–356 (2011). https://doi.org/10.1007/s11044-010-9241-8

23. Uchida, T., McPhee, J.: Using Gröbner bases to generate efficient kinematic solutions for the dynamic simulation of multi-loop mechanisms. Mech. Mach. Theory **52**, 144–157 (2012). https://doi.org/10.1016/j.mechmachtheory.2012.01.015

24. van der Walt, S., Colbert, S.C., Varoquaux, G.: The numpy array: a structure for efficient numerical computation. Comput. Sci. Eng. **13**(2), 22–30 (2011). https://doi.org/10.1109/MCSE.2011.37

25. Verschelde, J.: Algorithm 795: PHCpack: a general-purpose solver for polynomial systems by homotopy continuation. ACM Trans. Math. Softw. **25**(2), 251–276 (1999). https://doi.org/10.1145/317275.317286

26. Wolfram Research Inc: Mathematica, Version 12.1 [computer software] (2020). https://www.wolfram.com/mathematica. Accessed 17 Mar 2020

Real Algebraic Geometry

Curtains in CAD: Why Are They a Problem and How Do We Fix Them?

Akshar Nair$^{(\boxtimes)}$ ⓘ, James Davenport ⓘ, and Gregory Sankaran ⓘ

University of Bath, Claverton Down, Bath BA2 7AY, UK
{asn42,masjhd,masgks}@bath.ac.uk

Abstract. This paper is part of our ongoing research on the adaptation of Lazard's CAD to benefit from equational constraints in formulae. In earlier work we combined the CAD methods of McCallum and Lazard so as to produce an efficient algorithm for decomposing a hypersurface rather than the whole of $\mathbb{R}^n$ (exploiting an equational constraint $f = 0$). That method, however, fails if f is nullified (in McCallum's terminology): we call the set where this happens a curtain. Here we provide a further modification which, at the cost of a trade off in terms of complexity, is valid for any hypersurface, including one containing curtains.

1 Introduction

A Cylindrical Algebraic Decomposition (CAD) is a decomposition of a semi-algebraic set $S \subseteq \mathbb{R}^n$ (for any n) into semi-algebraic sets (also known as cells) homeomorphic to $\mathbb{R}^m$, where $0 \leq m \leq n$, such that the projection of any two cells onto the first k coordinates is either the same or disjoint. We generally want the cells to have some property relative to some given set of input polynomials, often used to form constraints using sign conditions. For example, we might require sign-invariance, i.e. the sign of each input polynomial is constant on each cell, as in the original algorithm of [3].

CAD algorithms have many applications: epidemic modelling [1], artificial intelligence to pass exams [13], financial analysis [10], and many more, so efficient algorithms are of more than theoretical interest.

If S is contained in a subvariety of $\mathbb{R}^n$ it is clearly wasteful to compute a decomposition of $\mathbb{R}^n$. McCallum, in [7], adapts his earlier algorithm [6] to this situation. To explain this idea more precisely, we need some terminology.

Definition 1. *A Quantifier Free Tarski Formula (QFF) is made up of atoms connected by the standard boolean operators $\wedge, \vee$ and $\neg$. The atoms are statements about signs of polynomials $f \in \mathbb{R}[x_1, \ldots, x_n]$, of the form $f * 0$ where $* \in \{=, <, >\}$ (and by combination also $\{\geq, \leq, \neq\}$).*

Strictly speaking we need only the relation $<$, but this form is more convenient because of the next definition.

Supported by University of Bath and EPSRC.

A. M. Bigatti et al. (Eds.): ICMS 2020, LNCS 12097, pp. 17–26, 2020.
https://doi.org/10.1007/978-3-030-52200-1_2

Definition 2. *[4] An Equational Constraint (EC) is a polynomial equation logically implied by a QFF. If it is an atom of the formula, it is said to be* explicit; *if not, then it is* implicit. *If the constraint is visibly an equality one from the formula, i.e. the formula Φ is $f = 0 \wedge \Phi'$, we say the constraint is* syntactically explicit.

Although implicit and explicit ECs have the same logical status, in practice only the syntactically explicit ECs will be known to us and therefore be available to be exploited.

Example 1. [4] Let f and g be two real polynomials.

1. The formula $f = 0 \wedge g > 0$ has an explicit EC, $f = 0$.
2. The formula $f = 0 \vee g = 0$ has no explicit EC, but the equation $fg = 0$ is an implicit EC.
3. The formula $f^2 + g^2 \leq 0$ also has no explicit EC, but it has two implicit ECs: $f = 0$ and $g = 0$.
4. The formula $f = 0 \vee f^2 + g^2 \leq 0$ logically implies $f = 0$, and the equation is an atom of the formula which makes it an explicit EC according to the definition. However, since this deduction is semantic rather than syntactic, it is more like an implicit EC rather than an explicit EC.

Definition 3. *Let A be a set of polynomials in $\mathbb{R}[x_1, \ldots, x_n]$ and $P\colon \mathbb{R}[x_1, \ldots, x_n] \times \mathbb{R}^n \to \Sigma$ a function to some set Σ. If $C \subset \mathbb{R}^n$ is a cell and the value of $P(f, \alpha)$ is independent of $\alpha \in C$ for every $f \in A$, then A is called P-invariant over that cell. If this is true for all the cells of a decomposition, we say the decomposition is P-invariant.*

Much work has been done on sign-invariant CADs, but we first focus our attention on the algorithm for lex-least invariant CADs introduced by Lazard [5] (for a validity proof, see [9]). Unlike the algorithm in [6] it works in the presence of curtains (see Definition 11 below), and has some complexity advantages also. In [12] we adapted [5] to the case of an EC, analogously to the adaptation of [6] in [7], but in doing so we reintroduced the problem of curtains. This paper revisits [12] and provides a hybrid algorithm, which we believe to be the first one that directly gives a CAD of the variety defined by an EC and yet is valid even on curtains. This process avoids going via a CAD of $\mathbb{R}^n$ ($\mathbb{C}^n$ for regular chains [2]).

In Sect. 3 we analyse the reasons why curtains are a problem, and explain how the previous literature has concentrated on valuations rather than curtains themselves. Section 4 consists of the complexity analysis of our algorithm. As in [12], this algorithm cannot be used recursively because the projection operator used in the first stage of projection would output a partial CAD which is a hybrid between sign-invariant and lex-least invariant on curtains of the equational constraint.

2   Lex-Least Valuation and Its Applications in CAD

In order to understand lex-least valuation, let us recall *lexicographic order* $\geq_{\text{lex}}$ on $\mathbb{N}^n$, where $n \geq 1$.

Definition 4. *We say that $v = (v_1, \ldots, v_n) \geq_{\text{lex}} (w_1, \ldots, w_n) = w$ if and only if either $v = w$ or there exists an $i \leq n$ such that $v_i > w_i$ and $v_k = w_k$ for all k in the range $1 \leq k < i$.*

Definition 5. *[9, Definition 2.4] Let $n \geq 1$ and suppose that $f \in \mathbb{R}[x_1, \ldots, x_n]$ is non-zero and $\alpha = (\alpha_1, \ldots, \alpha_n) \in \mathbb{R}^n$. The lex-least valuation $\nu_\alpha(f)$ at α is the least (with respect to $\geq_{\text{lex}}$) element $v = (v_1, \ldots, v_n) \in \mathbb{N}^n$ such that f expanded about α has the term*

$$c(x_1 - \alpha_1)^{v_1} \cdots (x_n - \alpha_n)^{v_n},$$

where $c \neq 0$.

Note that $\nu_\alpha(f) = (0, \ldots, 0)$ if and only if $f(\alpha) \neq 0$. The lex-least valuation is referred to as the Lazard valuation in [9].

Example 2. If $n = 1$ and $f(x_1) = x_1^3 - 2x_1^2 + x_1$, then $\nu_0(f) = 1$ and $\nu_1(f) = 2$. If $n = 2$ and $f(x_1, x_2) = x_1(x_2 - 1)^2$, then $\nu_{(0,0)}(f) = (1, 0)$, $\nu_{(2,1)}(f) = (0, 2)$ and $\nu_{(0,1)}(f) = (1, 2)$.

Definition 6. *[9] Let $n \geq 2$, and suppose that $f \in \mathbb{R}[x_1, \ldots, x_n]$ is non-zero and that $\beta \in \mathbb{R}^{n-1}$. The Lazard residue $f_\beta \in \mathbb{R}[x_n]$ of f at β, and the lex-least valuation $N_\beta(f) = (\nu, \ldots, \nu_{n-1})$ of f above β are defined to be the result of Algorithm 1.*

Algorithm 1. Lazard residue

Input: $f \in \mathbb{R}[x_1, \ldots, x_n]$ and $\beta \in \mathbb{R}^{n-1}$.
Output: Lazard residue f_β and Lex-least valuation of f above β.
1: $f_\beta \leftarrow f$
2: **for** $i \leftarrow 1$ to $n - 1$ **do**
3:     $\nu_i \leftarrow$ greatest integer ν such that $(x_i - \beta_i)^\nu | f_\beta$.
4:     $f_\beta \leftarrow f_\beta / (x_i - \beta_i)^{\nu_i}$.
5:     $f_\beta \leftarrow f_\beta(\beta_i, x_{i+1}, \ldots, x_n)$
6: **end for**
7: **return** $f_\beta, (\nu_1, \ldots, \nu_{n-1})$

Do not confuse the lex-least valuation $N_\beta(f) \in \mathbb{Z}^{n-1}$ of f above $\beta \in \mathbb{R}^{n-1}$ with the lex-least valuation $\nu_\alpha(f) \in \mathbb{Z}^n$ at $\alpha \in \mathbb{R}^n$, defined in Definition 5. Notice that if $\alpha = (\beta, b_n) \in \mathbb{R}^n$ then $\nu_\alpha(f) = (N_\beta(f), \nu_n)$ for some integer ν_n: in other words, $N_\beta(f)$ consists of the first $n - 1$ coordinates of the valuation of f at any point above β.

Definition 7. *[9, Definition 2.10] Let $S \subseteq \mathbb{R}^{n-1}$ and $f \in \mathbb{R}[x_1, \ldots, x_n]$. We say that f is Lazard delineable on S if:*

i) *The lex-least valuation of f above β is the same for each point $\beta \in S$.*
ii) *There exist finitely many continuous functions $\theta_i \colon S \to \mathbb{R}$, such that $\theta_1 < \ldots < \theta_k$ if $k > 0$ and such that for all $\beta \in S$, the set of real roots of f_β is $\{\theta_1(\beta), \ldots, \theta_k(\beta)\}$.*
iii) *If $k = 0$, then the hypersurface $f = 0$ does not pass over S. If $k \geq 1$, then there exist positive integers $m_1, \ldots, m_k$ such that, for all $\beta \in S$ and for all $1 \leq i \leq k$, m_i is the multiplicity of $\theta_i(\beta)$ as a root of f_β.*

Definition 8. *[9, Definition 2.10] Let f be Lazard delineable on $S \subseteq \mathbb{R}^{n-1}$. Then*

i) *The graphs θ_i are called Lazard sections and m_i is the associated multiplicity of these sections.*
ii) *The regions between consecutive Lazard sections[1] are called Lazard sectors.*

Remark 1. If f is Lazard delineable on S and the valuation of f above any point in S is the zero vector, then the Lazard sections of f are the same as the sections of f defined as in [3] and [7].

Remark 2. We can use Algorithm 1 to compute the lex-least valuation of f at $\alpha \in \mathbb{R}^n$. After the loop is finished, we proceed to the first step of the loop and perform it for $i = n$ and the n-tuple $\nu_1, \ldots, \nu_n$ is the required valuation.

Definition 9. *[12] Let $A \subset \mathbb{R}[x_1, \ldots, x_n]$ be a set of polynomials. Let $E \subseteq A$, and define the projection operator $PL_E(A)$ as follows*

$$PL_E(A) = \mathrm{ldcf}(E) \cup \mathrm{trcf}(E) \cup \mathrm{disc}(E) \cup \{\mathrm{res}_{x_n}(f, g) \mid f \in E, \, g \in A \setminus E\}.$$

Here $\mathrm{ldcf}(E)$ and $\mathrm{trcf}(E)$ are the sets of leading and trailing coefficients of elements of E. The set $\mathrm{disc}(E)$ is the set of all discriminants of E and $\mathrm{res}(E) = \{\mathrm{res}(f, g) : f, g \in E\}$ the set of cross-resultants. We will be comparing this to Lazard's projection operator $PL(A)$ defined in [9]. Note that in the practical use of the operator the set E corresponds to equational constraints.

Theorem 1. *[9] Let $f(x, x_n) \in \mathbb{R}[x, x_n]$ have positive degree d in x_n, where $x = (x_1, \ldots, x_{n-1})$. Let $D(x)$, $l(x)$ and $t(x)$ denote the discriminant, leading coefficient and trailing coefficient (with respect to x_n) of f, respectively, and suppose that each of these polynomials is non-zero (as an element of $\mathbb{R}[x]$). Let S be a connected analytic subset of $\mathbb{R}^{n-1}$ in which $D(x)$, $l(x)$ and $t(x)$ are all lex-least invariant. Then f is Lazard delineable on S, and hence f is lex-least invariant in every Lazard section and sector over S. Moreover, the same conclusion holds for the polynomial $f^*(x, x_n) = x_n f(x, x_n)$.*

---

[1] Including $\theta_0 = -\infty$ and $\theta_{k+1} = +\infty$.

Theorem 2. *[12] Let $n \geq 2$ and let $f, g \in \mathbb{R}[x_1, \ldots, x_n]$ be of positive degrees in the main variable x_n. Suppose that f is Lazard delineable on a connected subset $S \subset \mathbb{R}^{n-1}$, in which $R = \mathrm{res}_{x_n}(f, g)$ is lex-least invariant, and f does not have a curtain on S (see Definition 12 below). Then g is sign-invariant in each section of f over S.*

Definition 10. *Let A be a set of polynomials and D be the lex-least invariant CAD of A. Let C be a cell of D. The valuation of cell C with respect to a polynomial $f \in A$ is the lex-least valuation f at any point $\alpha \in C$.*

3 Implications of Curtains on CAD

We propose some geometric terminology to describe the conditions under which a polynomial is nullified in the terminology of [7].

Definition 11. *A variety $C \subseteq \mathbb{R}^n$ is called a curtain if, whenever $(x, x_n) \in C$, then $(x, y) \in C$ for all $y \in \mathbb{R}$.*

Definition 12. *Suppose $f \in \mathbb{R}[x_1, \ldots, x_n]$ and $W \subseteq \mathbb{R}^{n-1}$. We say that f has a curtain at W if for all $\underline{x} \in W$ and $y \in \mathbb{R}$ we have $f(\underline{x}, y) = 0$.*

Remark 3. Lazard delineability differs from delineability as in [3] and [7] in two important ways. First, we require lex-least invariance on the sections. Second, delineability is not defined on curtains, but Lazard delineability is because the Lazard sections are of f_β rather than f.

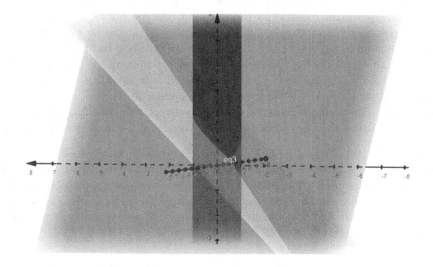

Fig. 1. Example illustrating the problems with curtains.

The algorithms in [7] and [8] exploit ECs but both rely on order-invariance. Because order is not defined on curtains, these algorithms fail there. It is therefore natural to try to use Lazard's algorithm, which does not have this issue when used to construct a full CAD of $\mathbb{R}^n$. However, when we try to exploit an EC, a difficulty does still arise if the EC has a curtain (Fig. 1).

Example 3. Let $f = x^2 + y^2 - 1$ (which we assume to be an EC), $g_1 = z - x - 1$ and $g_2 = z - y - 1$ (which we assume are not ECs). Then $\text{res}(f, g_1) = x^2 + y^2 - 1 = \text{res}(f, g_2)$, and this gives us no information about $\text{res}(g_1, g_2)$. In such cases, when the EC has a curtain, it becomes impossible to use PL_E to detect the intersections of the other constraints on that curtain. Because of this, we must resort to a general-purpose projection such as the one in [3], which includes these resultants, when we are on a curtain contained in the hypersurface defined by the EC.

In one common case, where W in Definition 12 is a single point, we can avoid this complication.

Definition 13. *We say that $f \in \mathbb{R}[x_1, \ldots, x_n]$ has a point curtain at $\alpha \in \mathbb{R}^{n-1}$ if $N_\alpha(f) \neq (0, \ldots, 0)$ and there exists a neighbourhood U of α such that $N_{\alpha'}(f) = (0, \ldots, 0)$ for all $\alpha' \in U \setminus \{\alpha\}$.*

In this case we do not need to consider the resultants between the non-equational constraints $A \setminus \{f\}$ when projecting.

Theorem 3. *Let $f \in \mathbb{R}[x_1, \ldots, x_n]$ and let $\alpha \in \mathbb{R}^{n-1}$. If f is an equational constraint and has a point curtain at α, then PL_E is sufficient to obtain a sign-invariant CAD.*

Indeed, if f has a curtain on a set S of positive dimension, we need the resultants of the non-equational constraints to determine sample points in $\mathbb{R}^{n-1}$ above which two such constraints meet, as in Example 3. For a point curtain, there is only one sample point in $\mathbb{R}^{n-1}$, namely α. We calculate the roots of Lazard residues of all constraints to determine the sample points on the curtain (which is the fibre above α), and nothing more is needed: the algorithm still produces a lex-least invariant CAD. Further details will appear in [11].

For this to be useful we need to be able to detect and classify curtains (i.e. tell whether or not they are point curtains), Algorithm 2 describes this process.

Remark 4. In steps 5 and 6 of Algorithm 2 we explicitly are looking at the neighbouring cells of the sample points rather than just checking the parity of their indices. This measure is taken to avoid cases where a sample point might seem to be a point curtain, but is actually a part of a larger curtain. This can be seen in Example 3, where the curtains above $(1, 0)$ and $(-1, 0)$ are part of a larger curtain.

Remark 5. Algorithm 3 is called once a CAD is computed using the reduced Lazard projection operator. This algorithm focuses only on decomposing the curtains of the equational constraint.

Algorithm 2. Detecting and Classifying Curtains

$(B) \leftarrow PC(f, I, S, n)$

Input = Set of indices I, set of sample points S with respect to the indices I which correspond to the CAD cells, equational constraint $f \in \mathbb{R}[x_1, \ldots, x_n]$.

Output = B, B', where B is the set of sample points that are point curtains and B' is the set of sample points that are curtains (but not point curtains).

1: $B \leftarrow$ Empty List
2: $B' \leftarrow$ Empty List
3: **for** $\alpha \in S$ **do**
4:     **if** $\nu_\alpha(f) \neq 0$ **then**
5:         Check if the nearest 1-cell neighbours have zero valuation.
6:         If all neighbours are zero valuation add α to B otherwise add it to B'
7:     **end if**
8: **end for**
9: **return** (B, B')

Algorithm 3. Partial CAD for Curtains

$(I', S') \leftarrow DCBS(A, I, S, C, n)$

Input = Set of indices I, set of sample points S with respect to the indices I, set of polynomials A, C is a list of tuples of sample points that are non-point curtains and their respective indices and n the dimension of our space.

Output = I', S' list of sample points and their indices for sections of the equational constraints that are curtains.

1: If $n \geq 2$ then go to step 5.
2: **for** each $(c, i) \in C$ **do**
3:     Isolate the real roots of the irreducible factors of the non-zero elements of A between the neighbours of c. Construct cell indices I' and sample points S' from the real roots. Exit.
4: **end for**
5: $B \leftarrow$ the square free basis of the primitive parts of all elements of A.
6: $P \leftarrow \text{cont}(A) \cup PL(B)$.
7: $(I'', S'') \leftarrow DCBS(P, I, S, C, n - 1)$.
8: $(I', S') \leftarrow$ (empty list,empty list).
9: **for** each $\alpha \in S''$ **do**
10:     $f^* \leftarrow \{f_\alpha \mid f \in B\}$.
11:     **for** each $(c, i) \in C$ **do**
12:         Isolate the real roots of all the polynomials in f^* between the neighbours of c (looking at the nth coordinate).
13:         Construct cell indices and sample points for Lazard sections and sectors of elements of B from i, α and the isolated real roots of f^*.
14:         Add the sample points to I' and S'.
15:     **end for**
16:     **return** (I', S').
17: **end for**

Example 4. Suppose Fig. 2 describes the sample points obtained using the CAD algorithm with the reduced Lazard projection operator. In this example, sample points $S_{3,3,1}$, $S_{3,3,2}$ and $S_{3,3,3}$ describe the cells that are part of the curtain of the equational constraint. The sample points circled in blue represent the sample points that lead to $S_{3,3,1}$, $S_{3,3,2}$ and $S_{3,3,3}$ during lifting. The following steps describe Algorithm 3. First, we project all non-equational constraints using Lazard's original projection operator. When computing sample points, we only consider those which lie within the neighbours of the curtain sample point. For example in our case, at level x_1, we calculate all sample points between S_2 and S_4.

Since we are only decomposing the curtain part of the equational constraint, we only need to look at the cell described by S_3. This results in computing the roots between S_2 and S_4 (and their respective sample points as well). We lift these sample points to level x_2. When computing at level x_2, we consider the roots (and their respective sample points) between $S_{3,2}$ and $S_{3,4}$, by the same argument. If at any level the index of the sample point is even, we do not need to calculate roots, but consider that value as our sample point for lifting. Once we lift to x_3, we have successfully decomposed our curtain.

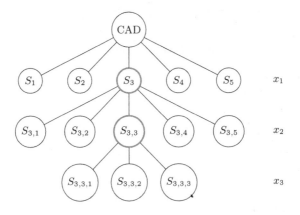

Fig. 2. Example of root finding in the lifting modification

4   Complexity Analysis of Curtain Solving Method

In this section we look at the complexity of the number of cells produced by Algorithm 3. Note that Algorithm 3 works so to speak on top of the algorithm described in [12]. The idea is to do a second decomposition on the sections of the equational constraint that contain curtains.

Theorem 4. *Given a set A of m polynomials in n variables with maximum degree d in any one variable, Algorithm 3 outputs a partial CAD (i.e. a CAD of $\{f = 0\}$ only) with the number of cells being at most*

$$2^{2^n-1}(m(3m+1)^{2^{n-1}-1} + (m-1)m^{2^{n-1}-1})d^{2^n-1}. \tag{1}$$

This complexity is an improvement on the existing method of using [5] in full without exploiting the EC, and unlike [12] and [7] the algorithm is valid on curtains.

Proof. The first step, using a single equational constraint, has the same complexity as [7]. Further to this, if the equational constraint has curtains, we need to re-project the non-equational constraints, thus performing a CAD of $m - 1$ polynomials of maximum total degree d. Combining these two gives the bound in (1).

The following verifies that the modified algorithm is better than [5] in terms of complexity. It has worse complexity than [12] but is valid for equational constraints with curtains.

The complexity of [5] is $2^{2^n-1}(m+1)^{2^n-2}md^{2^n-1}$ so after removing a factor of $2^{2^n-1}md^{2^n-1}$ the claim is that

$$(3m+1)^{2^{n-1}-1} + (m-1)m^{2^{n-1}-2} < (m+1)^{2^n-2}.$$

We may assume $m \geq 2$ and $n \geq 2$ (otherwise there is nothing to do) so

$$\begin{aligned}
(3m+1)^{2^{n-1}-1} + (m-1)m^{2^{n-1}-2} &< 3^{2^{n-1}-1}(m+1)^{2^{n-1}-1} + (m+1)^{2^{n-1}-1}\\
&= (3^{2^{n-1}-1}+1)(m+1)^{2^{n-1}-1}\\
&= (m+1)^{2^n-2}\frac{3^{2^{n-1}-1}+1}{(m+1)^{2^n-2^{n-1}-1}}\\
&\leq (m+1)^{2^n-2}\frac{3^{2^{n-1}-1}+1}{3^{2^n-2^{n-1}-1}}\\
&= (m+1)^{2^n-2}(3^{-2^n}+3^{-2^n+2^{n-1}+1})\\
&< (m+1)^{2^n-2}
\end{aligned}$$

since $3^{-2^n} + 3^{-2^n+2^{n-1}+1} \leq 3^{-4} + 3^{-1} = \frac{28}{81} < 1$.

5   Conclusion and Further Research

Algorithm 3 is the first partial CAD algorithm which is a hybrid between sign-invariant and lex-least invariant CAD algorithms. It allows us to exploit equational constraints unconditionally. The novelty lies in performing a second decomposition of the curtain sections of the equational constraints: the worst case for the complexity analysis is when the entire equational constraint is a curtain. More analysis needs to done, as better complexity should be achievable, and we hope to do this the near future. We are currently looking into extending this approach so as to produce as output a partial CAD that has a hybrid of order invariance and lex-least invariance. Note that lex-least invariance over a given region does not imply order invariance, but an order invariant CAD is also lex-least invariant. The desire to produce lex-least invariance in future work is driven by the aim of a recursive result.

Acknowledgements. We are grateful for many conversations with Matthew England, Scott McCallum and Zak Tonks.

References

1. Brown, C.W., El Kahoui, M., Novotni, D., Weber, A.: Algorithmic methods for investigating equilibria in epidemic modeling. J. Symbolic Comp. **41**, 1157–1173 (2006)
2. Chen, C., Moreno Maza, M.: Cylindrical algebraic decomposition in the RegularChains library. In: Hong, H., Yap, C. (eds.) ICMS 2014. LNCS, vol. 8592, pp. 425–433. Springer, Heidelberg (2014). https://doi.org/10.1007/978-3-662-44199-2_65
3. Collins, G.E.: Quantifier elimination for real closed fields by cylindrical algebraic decomposition. In: Proceedings 2nd. GI Conference Automata Theory & Formal Languages, pp. 134–183 (1975)
4. England, M., Bradford, R.J., Davenport, J.H.: Cylindrical Algebraic Decomposition with equational constraints. In: Davenport, J.H., England, M., Griggo, A., Sturm, T., Tinelli, C. (eds.) Symbolic Computation and Satisfiability Checking, vol. 100, pp. 38–71 (2020). Journal of Symbolic Computation
5. Lazard, D.: An improved projection operator for cylindrical algebraic decomposition. In: Bajaj, C.L. (ed.) Proceedings Algebraic Geometry and Its Applications: Collections of Papers from Shreeram S. Abhyankar's 60th Birthday Conference, pp. 467–476 (1994)
6. McCallum, S.: An improved projection operation for cylindrical algebraic decomposition. Ph.D. thesis, University of Wisconsin-Madison Computer Science (1984)
7. McCallum, S.: On projection in CAD-based quantifier elimination with equational constraints. In: Dooley, S. (ed.) Proceedings ISSAC 1999, pp. 145–149 (1999)
8. McCallum, S.: On propagation of equational constraints in CAD-based quantifier elimination. In: Mourrain, B. (ed.) Proceedings ISSAC 2001, pp. 223–230 (2001)
9. McCallum, S., Parusiński, A., Paunescu, L.: Validity proof of Lazard's method for CAD construction. J. Symbolic Comp. **92**, 52–69 (2019)
10. Mulligan, C.B., Davenport, J.H., England, M.: TheoryGuru: a mathematica package to apply quantifier elimination technology to economics. In: Davenport, J.H., Kauers, M., Labahn, G., Urban, J. (eds.) ICMS 2018. LNCS, vol. 10931, pp. 369–378. Springer, Cham (2018). https://doi.org/10.1007/978-3-319-96418-8_44
11. Nair, A.S.: Exploiting equational constraints to improve the algorithms for computing cylindrical algebraic decompositions. Ph.D. thesis, University of Bath (2021)
12. Nair, A.S., Davenport, J.H., Sankaran, G.K.: On benefits of equality constraints in lex-least invariant CAD (extended abstract). In: Proceedings SC2 2019, September 2019
13. Wada, Y., Matsuzaki, T., Terui, A., Arai, N.H.: An automated deduction and its implementation for solving problem of sequence at university entrance examination. In: Greuel, G.-M., Koch, T., Paule, P., Sommese, A. (eds.) ICMS 2016. LNCS, vol. 9725, pp. 82–89. Springer, Cham (2016). https://doi.org/10.1007/978-3-319-42432-3_11

Chordality Preserving Incremental Triangular Decomposition and Its Implementation

Changbo Chen[1,2(✉)]

[1] Chongqing Key Laboratory of Automated Reasoning and Cognition,
Chongqing Institute of Green and Intelligent Technology,
Chinese Academy of Sciences, Chongqing, China
chenchangbo@cigit.ac.cn
[2] University of Chinese Academy of Sciences, Beijing, China
http://www.arcnl.org/cchen

Abstract. In this paper, we first prove that the incremental algorithm for computing triangular decompositions proposed by Chen and Moreno Maza in ISSAC' 2011 in its original form preserves chordality, which is an important property on sparsity of variables. On the other hand, we find that the current implementation in Triangularize command of the RegularChains library in Maple may not always respect chordality due to the use of some simplification operations. Experimentation show that modifying these operations, together with some other optimizations, brings significant speedups for some super sparse polynomial systems.

Keywords: Triangular decomposition · Chordal graph · Incremental algorithm · Regular chain

1 Introduction

The method of triangular decomposition pioneered by Ritt [19] and Wu [23] has become a basic tool for computing the solutions of polynomial systems over an algebraically closed field. Given a finite set of polynomials F, this method decomposes F into finitely many systems of triangular shape such that the union of their zero sets is the same as that of F. With such decomposition in hand, many information on the solution set, such as emptiness, dimension, cardinality, etc., can be easily obtained. Triangular decomposition has been studied and gradually improved by many others in both theory [1,2,14,15,25] and algorithms [7,10,12,13,16,17,22,24]. Efficient implementations exist in a Maple built-in package RegularChains as well as many other libraries and softwares, such as Epsilon, Wsolve, Magma, and so on.

Nowadays, triangular decomposition has also become an important back-end engine for several algorithms in studying semi-algebraic sets, such as real root classification [24] and comprehensive triangular decomposition of parametric semi-algebraic sets [5], computing sample points of semi-algebraic sets [4],

© Springer Nature Switzerland AG 2020
A. M. Bigatti et al. (Eds.): ICMS 2020, LNCS 12097, pp. 27–36, 2020.
https://doi.org/10.1007/978-3-030-52200-1_3

describing semi-algebraic sets [4], as well as computing cylindrical algebraic decompositions and performing quantifier elimination [3,8,9]. These algorithms and their implementations make triangular decomposition become an efficient tool in many applications, such as theorem proving, program verification, stability analysis of biological systems, and so on.

To further improve the efficiency of triangular decomposition, one important direction is to explore the structure of input systems, such as symmetry and sparsity. The work of [11] brings the concept of variable sparsity to the world of triangular decomposition by virtue of the chordal graph of polynomial systems. Chordal graph already exists in many other contexts, such as Gauss elimination [20] and semidefinite optimization [21].

It has already been shown that some top-down algorithms [22] (up to minor modification of its original form) preserve chordality [18]. Incremental algorithms [7,16,17] are another important class, which compute triangular decompositions by induction on the number of polynomials. It is a natural question to ask if the incremental algorithms can also preserve chordality. In this paper, we provide an affirmative answer to this question for the incremental algorithm proposed in [6,7]. On the other hand, we find that the current implementation of this algorithm in Triangularize command of the RegularChains library in Maple may not always respect chordality. After a careful examination of the implementation, we point out this is due to the use of some simplification operations. Finally, we show by experimentation that modifying these operations, together with some other optimizations, bring significant speedups for Triangularize on some super sparse polynomial systems.

2   Basic Lemmas

Definition 1 (Graph). *Let* $\mathbf{x} = x_1, \ldots, x_n$ *and* $F \subset \mathbf{k}[\mathbf{x}]$. *The (associated) graph* $\mathfrak{G}(F)$ *of* F *is an undirected graph defined as follows:*

- *The set* V *of vertices of* $\mathfrak{G}(F)$ *is the set of variables appearing in* F.
- *The set* E *of edges of* $\mathfrak{G}(F)$ *is the set of* (x_i, x_j), $i \neq j$, *where* x_i *and* x_j *simultaneously appear in some* $f \in F$.

Denote by $\mathfrak{v}(\mathfrak{G}(F))$ *an operation which returns* V.

Definition 2 (Perfect elimination ordering). *Let* $\mathfrak{G} = (V, E)$ *be a graph with vertices* $V = \{x_1, \ldots, x_n\}$. *An ordering* $\mathbf{x} = x_{i_1} > \cdots > x_{i_n}$ *is a perfect elimination ordering for* $\mathfrak{G}$ *if for any* x_{i_j}, *the induced subgraph on the set of vertices* $V_{i_j} = \{x_{i_j}\} \cup \{x_{i_k} \mid x_{i_k} < x_{i_j} \text{ and } (x_{i_j}, x_{i_k}) \in E\}$ *is a clique. If a perfect elimination ordering* $\mathbf{x}$ *exists for* $\mathfrak{G}$, *we say* $\mathfrak{G}$ *is chordal (w.r.t.* $\mathbf{x}$). *We say that a graph* $\overline{\mathfrak{G}}$ *with vertices* $V = \{x_1, \ldots, x_n\}$ *is a chordal completion of* $\mathfrak{G}$, *if* $\overline{\mathfrak{G}}$ *is chordal and* $\mathfrak{G}$ *is a subgraph of* $\overline{\mathfrak{G}}$.

Example 1. *Let* $F := \{x_1^2 - x_2^2 + x_2 x_4, x_2^2 - x_3 x_4\}$. *Then* $\mathfrak{G}(F)$ *is illustrated in Fig. 1, where the vertex* x_i *is renamed as* i *for short. The ordering* $x_1 > x_2 > x_3 > x_4$ *is a perfect elimination ordering and* $\mathfrak{G}$ *is chordal w.r.t. this ordering. Another ordering* $x_2 > x_1 > x_3 > x_4$ *is not a perfect elimination ordering.*

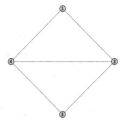

Fig. 1. Chordal graph.

Definition 3. *Let* $\mathbf{x}$ *be a given ordering and* $F \subset \mathbf{k}[\mathbf{x}]$. *Let* $M : 2^{\mathbf{k}[\mathbf{x}]} \rightarrow 2^{2^{\mathbf{k}[\mathbf{x}]}}$. *Let* $\mathfrak{G}(F)$ *be any chordal completion of* $\mathfrak{G}(F)$ *w.r.t. the ordering* $\mathbf{x}$ *(that is* $\mathbf{x}$ *is a perfect elimination ordering for* $\mathfrak{G}(F)$*). We say that* M *preserves chordality in* $\mathfrak{G}(F)$ *(resp.* $\overline{\mathfrak{G}(F)}$*) if for any* $S \in M(F)$*, we have* $\mathfrak{G}(S) \subseteq \mathfrak{G}(F)$ *(resp.* $\mathfrak{G}(S) \subseteq \overline{\mathfrak{G}(F)}$*). In the former case, we say that* M *strongly preserves chordality (for* F*). In the latter case, that is* M *preserves chordality in any chordal completion of* $\mathfrak{G}(F)$ *w.r.t.* $\mathbf{x}$*, we say that* M *preserves chordality (for* F*).*

Remark 1. *Let* $O : \mathbf{k}[\mathbf{x}] \rightarrow 2^{\mathbf{k}[\mathbf{x}]}$ *be a unary operation which maps* $f \in \mathbf{k}[\mathbf{x}]$ *to* $O(f) \subset \mathbf{k}[\mathbf{x}]$*. Then it induces a map* M *which maps* $\{f\}$ *to* $\{O(f)\}$*. Let* $O : \mathbf{k}[\mathbf{x}] \times \mathbf{k}[\mathbf{x}] \rightarrow 2^{\mathbf{k}[\mathbf{x}]}$ *be a binary operation which maps* (f, g) *to* $O(f, g)$*. Then it induces a map* M *which maps* $\{f, g\}$ *to* $\{O(f, g)\}$*. For both cases, if* M *(strongly) preserves chordality, we say that the operation* O *(strongly) preserves chordality.*

Example 2. *Consider* $F := \{x_1{}^3 - 1, x_2{}^3 - 1, x_3{}^3 - 1, x_4{}^3 - 1, x_1{}^2 + x_1 x_2 + x_2{}^2, x_2{}^2 + x_2 x_3 + x_3{}^2, x_3{}^2 + x_3 x_4 + x_4{}^2, x_1{}^2 + x_1 x_4 + x_4{}^2\}$. *Figure 1 depicts a chordal completion* $\mathfrak{G}(F)$ *of* $\mathfrak{G}(F)$ *w.r.t. the ordering* $x_1 > x_2 > x_3 > x_4$*. Let* M *be the* Triangularize *command in Maple 2019, which computes a set of regular chains as follows:*

$$\{\{x_1 + x_4 + 1, x_2 - x_4, x_3 + x_4 + 1, x_4{}^2 + x_4 + 1\},$$
$$\{x_1 - 1, x_2 - x_4, x_3 + x_4 + 1, x_4{}^2 + x_4 + 1\},$$
$$\{(x_4 - 1)x_1 - x_4 - 2, x_2 - 1, x_3 + x_4 + 1, x_4{}^2 + x_4 + 1\},$$
$$\{x_1 - 1, x_2{}^2 + x_2 + 1, x_3 - 1, x_4{}^2 + x_4 + 1\},$$
$$\{x_1 + x_4 + 1, (x_4 + 2)x_2 - x_4 + 1, x_3 - 1, x_4{}^2 + x_4 + 1\},$$
$$\{(x_3 + 2)x_1 - x_3 + 1, x_2 + x_3 + 1, x_3{}^2 + x_3 + 1, x_4 - 1\},$$
$$\{x_1{}^2 + x_1 + 1, x_2 - 1, x_3{}^2 + x_3 + 1, x_4 - 1\}\}.$$

As we see, it does not preserve chordality since the graph of the fifth regular chain is not a subgraph of $\overline{\mathfrak{G}(F)}$*.*

Definition 4. *Let* $O : K[\mathbf{x}] \rightarrow 2^{K[\mathbf{x}]}$ *be a unary operation. We say that it respects the (elimination) ordering* $\mathbf{x}$ *for* $f \in \mathbf{k}[\mathbf{x}]$ *if* $\mathfrak{v}(O(f)) \subseteq \mathfrak{v}(f)$*.*

Lemma 1. *If a unary operation* O *respects the ordering, then it strongly preserves chordality.*

Proof. It trivially holds since $\mathfrak{G}(f)$ is a clique for any $f \in \mathbf{k}[\mathbf{x}]$.

Definition 5. *Given $f \in \mathbf{k}[\mathbf{x}]$, let $\mathrm{mvar}(f)$ be the largest variable appearing in f. Let $O : \mathbf{k}[\mathbf{x}] \times \mathbf{k}[\mathbf{x}] \to 2^{\mathbf{k}[\mathbf{x}]}$ be a binary operation. Let $S := O(f, g)$. We say that O respects the ordering $\mathbf{x}$ (for f and g) if the following are satisfied:*

- *If $\mathrm{mvar}(f) = \mathrm{mvar}(g)$, we have $\mathfrak{v}(S) \subseteq \mathfrak{v}(\{f, g\})$.*
- *If $\mathrm{mvar}(f) \neq \mathrm{mvar}(g)$, we have $\mathfrak{v}(S) \subseteq \mathfrak{v}(f)$ or $\mathfrak{v}(S) \subseteq \mathfrak{v}(g)$.*

Lemma 2. *Suppose that O respects the ordering $\mathbf{x}$, then we have the following.*

- *If $\mathrm{mvar}(f) \neq \mathrm{mvar}(g)$, then O strongly preserves chordality.*
- *If $\mathrm{mvar}(f) = \mathrm{mvar}(g)$, then O preserves chordality.*

Proof. Let $\overline{\mathfrak{G}}$ be any chordal completion of $\mathfrak{G}(\{f, g\})$ and x_i be the common main variable of f and g. By the definition of chordal graph, the subgraph of $\overline{\mathfrak{G}}$ induced by the set of vertices $\mathfrak{v}(\overline{\mathfrak{G}}) \setminus \{x_i\}$ is a clique. So the conclusion holds.

Corollary 1. *The irreducible factorization* Factor *strongly preserves chordality. If the two input polynomials have the same main variable, then the subresultant chain* SubRes, *the resultant* res, *the pseudo remainder* prem *and the pseudo quotient* pquo *operations w.r.t. the main variable preserve chordality.*

If the input two polynomials do not have the same main variable, then these operations may destroy chordality, see Example 3.

Example 3. *Consider the system $F := \{f_1, f_2\}$ again from Example 1 and the ordering $x_1 > x_2 > x_3 > x_4$. We have $\mathrm{mvar}(f_1) = x_1$ and $\mathrm{mvar}(f_2) = x_2$. Then* prem$(f_1, f_2, x_2) = x_1{}^2 + x_2 x_4 - x_3 x_4$. *Clearly* prem *does not preserve chordality for f_1 and f_2.*

Lemma 3. *Let $F \subset \mathbb{Q}[\mathbf{x}]$. Let p be a polynomial in F. Assume that F is chordal w.r.t. the variable order $\mathbf{x}$. Let $F' := F \setminus \{p\}$. Then there exists a chordal completion $\overline{G(F')}$ of $G(F')$ (with the same set of vertices) w.r.t. the order $\mathbf{x}$ such that $\overline{G(F')} \subseteq \overline{G(F)}$.*

Proof. Let $\overline{G(F')} := \{(u, v) \mid (u, v) \in \overline{G(F)}, v < u, u \in G(F') \text{ and } v \in G(F')\}$. For any u in $\mathfrak{v}(F')$, the set $\{(u, v), v < u, v \in \mathfrak{v}(F')\}$ is a also clique since $\{(u, v) \mid (u, v) \in \overline{G(F)}, v < u, v \in \mathfrak{v}(F)\}$ is a clique by the assumption that F is chordal. Thus $\overline{G(F')}$ is a chordal completion of $G(F')$.

3   The Incremental Algorithm Preserves Chordality

In this section, we prove that the incremental algorithm, namely Algorithm 1, proposed in [7] preserves chordality. The main subroutine of Algorithm 1 is the Intersect operation, which takes a polynomial p and a regular chain T as input, and returns a sequence of regular chains $T_1, \ldots, T_r$ such that

$$V(p) \cap W(T) \subseteq \cup_{i=1}^{r} W(T_i) \subseteq V(p) \cap \overline{W(T)}, \tag{1}$$

where $V(p)$ is the variety of p, $W(T)$ is the quasi-component of T and $\overline{W(T)}$ is the Zariski closure of $W(T)$. Due to limited space, we refer the reader to [7] for a precise definition of these concepts and a detailed description of the algorithm Intersect and its subroutines Extend, IntersectAlgebraic, IntersectFree, CleanChain, and RegularGcd.

Algorithm 1: Triangularize(F, R)

---

1 **if** $F = \{\ \}$ **then** return $\{\varnothing\}$;
2 Choose a polynomial $p \in F$;
3 **for** $T \in$ Triangularize$(F \setminus \{p\}, R)$ **do** output Intersect(p, T, R);

Lemma 4. *One can transform Algorithm* Triangularize *into an equivalent one with the original flow graph illustrated by the left subgraph of Fig. 2 replaced by the one depicted in the right subgraph of Fig. 2.*

Proof. The transformation can be done in two steps. Firstly one can easily replace the direct recursions in Extend, Triangularize and IntersectAlgebraic by iterations. Secondly one can make the function calls to Extend, IntersectAlgebraic, IntersectFree, CleanChain, RegularGcd inline. As a consequence, one obtains an equivalent form of Triangularize with the flow graph of function calls depicted in the right subfig of Fig. 2.

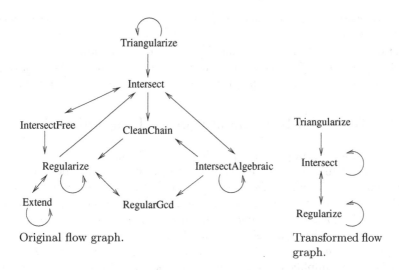

Original flow graph.

Transformed flow graph.

Fig. 2. Transform flow graph of the algorithm.

Lemma 5. *For each algorithm in Fig. 2, any polynomial in it, if appearing in the output of the algorithm or appearing as the output or input of subroutines in Fig. 2 of the algorithm, is obtained through chordality preserving operations. Moreover, if we remove step 1 in* Intersect, *which calls* prem *to test if p belongs to the saturated ideal of T (the algorithm is still correct), all basic operations appearing in the algorithm preserve chordality.*

Lemma 6. *Let $F \subset \mathbb{Q}[\mathbf{x}]$ be finite and assume that F is chordal. Let p be a polynomial and T be a regular chain of $\mathbb{Q}[\mathbf{x}]$ such that $\mathfrak{G}(p) \subseteq \mathfrak{G}(F)$ and $\mathfrak{G}(T) \subseteq \mathfrak{G}(F)$. Let $(p_1, T_1), \ldots, (p_e, T_e)$ be the processes in the output of* Intersect *or* Regularize, *then we have $\mathfrak{G}(p_i, T_i) \subseteq \mathfrak{G}(F)$. (The output T_i is treated as a special process $(0, T_i)$.) If this is true, we say that* Intersect *and* Regularize *preserves chordality (w.r.t. F).*

Proof. We prove this by induction on the rank of the process (p, T).

– *Base:* For each returned process (p_i, T_i), if it is obtained without relying on the output of recursive calls to Intersect or Regularize, then $\mathfrak{G}(\{p_i\} \cup T_i) \subseteq \mathfrak{G}(\{p\} \cup T)$ holds.
– *Induction:* For each recursive calls to Intersect or Regularize, the rank of input process is less than that of (p, T). By the induction hypothesis, the recursive calls preserve chordality. Moreover, by Lemma 5, we notice that the input process for each recursive call is obtained by chordality preserving operations as well as the output process of each recursive call is processed by chordality preserving operations. Thus the lemma holds.

Theorem 1. *Algorithm* Triangularize *preserves chordality.*

Proof. We prove it by induction on the number of elements of F. The base case trivially holds. Let $p \in F$ and $F' := F \setminus \{p\}$. By Lemma 3, there exists a chordal completion $\overline{G(F')}$ of $G(F')$ such that $\overline{G(F')} \subseteq G(F)$. By induction hypothesis, for each regular chain T in the output of the recursive call to Triangularize, we have $G(T) \subseteq \overline{G(F')} \subseteq G(F)$. Then the conclusion follows from Lemma 6.

4   Modifying the Implementation of Triangularize

Algorithm 1 has been implemented in the RegularChains library with the same name. After carefully tracing the code, we find that the only operation in the implementation of Triangularize that may destroy the chordality is the pseudo-reminder operation prem. As illustrated by Example 3 in Sect. 2, if two input polynomials do not have the same main variable, prem can destroy chordality. In particular, the operation prem was empolyed in several places for performing the simplification $q := \mathsf{prem}(p, T)$, where p, q are two polynomials and T is a regular chain. Such simplification does not hurt the correctness of the algorithm, although it may affect the efficiency and should be empolyed with caution. For instance, it was empolyed as a preprocessing step for Intersect but only if the initials of polynomials in T are constants, which may reduce the degree of the polynomial. Note that we have $V(p) \cap W(T) = V(q) \cap W(T)$, thus the correctness of Intersect is not hurt by Eq. (1). However, if q is involved in producing polynomials as input or output of some algorithms in Fig. 2, then Triangularize may not preserve chordality. Thus, for all these places, we simply do not call prem to preserve chordality. Note that $q = 0$ if and only if p is contained in the saturated ideal of T (membership testing) [6]. If prem is only used for membership testing, we do not suspend the call to prem as it does not affect chordality. There are some other operations, such as iterated resultant, which may not preserve the chordality either. But since they do not produce polynomials as input or output of algorithms in Fig. 2, we keep the calls to them unchanged.

There are several other changes we made to improve the efficiency of the code for chordal input. One is to control the generation of redundant regular chains in Algorithm 1 after each recursive call. Another is to change the order of polynomials in F to solve in Algorithm 1. In the current implementation of Triangularize, one first solves polynomials with smaller rank (in particular with smaller main variables). But this strategy seems to increase the chance of calling operations not preserving chordality. So we instead now first solve polynomials with larger rank. As an example, for lattice-r-10 in Table 1, the two different strategies respectively lead to calling $\mathsf{prem}(p, T)$ 2659 and 964 times.

The implementation preserving chordality is available in Triangularize in the updated RegularChains library (downloadable from http://www.arcnl.org/cchen/software/chordal) through option chordal = true.

5   Experiments

Table 2 compares its performance with Triangularize in Maple 2019. We also include the performance of the regser command of the Epsilon library as a reference. In Table 1, the examples minor-k, lattice-k and coloring-k are chosen from [11]. The examples minor-r-k (resp. lattice-r-k) is a slight modification of minor-k (resp. lattice-k), but have the same associated graph as minor-k (resp. lattice-k).

Table 1. Benchmark examples.

minor-k	$\{x_{2i-1}x_{2i+2} - x_{2i}x_{2i+1} \mid i = 1, \ldots, k\}$
minor-r-k	$\{x_{2i-1}x_{2i+2} - x_{2i}x_{2i+1} + x_{2i} + x_{2i+1} \mid i = 1, \ldots, k\}$
lattice-k	$\{x_i x_{i+3} - x_{i+1}x_{i+2} \mid i = 1, \ldots, k\}$
lattice-r-k	$\{x_i x_{i+3} - x_{i+1}x_{i+2} + x_{i+3}^2 \mid i = 1, \ldots, k\}$
coloring-k	$\{x_i^3 - 1 \mid i = 1, \ldots, k\} \cup \{\sum_{j=0}^2 x_i^{2-j} x_{(i \bmod k)+1}^j \mid i = 1, \ldots, k\}$

In Table 2, we write Triangularize for short as Tri and Triangularize with chordal = true as Tri-C. Denote by K and L respectively the Kalkbrener and Lazard triangular decomposition. For each system F, let n be the number of variables, m be the number of polynomials in F, d be the maximum degree of polynomials in F, t be the computation time (in seconds), c be the number of components in the output. The timeout $(-)$ is set as one hour. As we can see from the table, for these particular sparse systems, Tri-C significantly outperforms Tri. Meanwhile, Tri-C and regser each have their own favorite examples.

Table 2. Benchmark.

Sys	n	m	d	regser		Tri (K)		Tri (L)		Tri-C (K)		Tri-C (L)	
				Time	c	Time	c	Time	c	Time	c	Time	c
minor-10	22	10	2	2.334	455	27.08	89	467.6	875	6.738	89	24.75	767
minor-15	32	15	2	86.13	8236	1701.9	987	–	–	1147.7	987	–	–
minor-18	38	18	2	1402.1	46810	–	–	–	–	–	–	–	–
minor-20	42	20	2	–	–	–	–	–	–	–	–	–	–
minor-r-10	22	10	2	32.05	2214	10.72	1	227.0	498	1.700	1	11.98	351
minor-r-12	26	12	2	336.7	11667	41.30	1	1859.6	1713	5.366	1	75.51	1081
minor-r-14	30	14	2	–	–	153.4	1	–	–	17.87	1	584.8	3329
minor-r-15	32	15	2	–	–	–	–	–	–	33.23	1	1762.4	5842
lattice-10	13	10	2	0.140	18	7.065	15	7.6	15	0.417	24	0.417	24
lattice-20	23	20	2	1.640	154	–	–	–	–	8.93	187	8.913	187
lattice-30	33	30	2	19.47	1285	–	–	–	–	409.4	1549	406.2	1549
lattice-40	43	40	2	259.6	10733	–	–	–	–	–	–	–	–
lattice-r-10	13	10	2	0.459	13	21.41	1	26.27	13	0.436	1	0.506	13
lattice-r-15	18	15	2	27.19	18	–	–	–	–	2.134	1	2.297	18
lattice-r-18	21	18	2	–	–	–	–	–	–	150.2	1	152.7	21
lattice-r-20	23	20	2	–	–	–	–	–	–	–	–	–	–
coloring-10	10	20	3	6.45	123	13.45	123	9.89	123	4.916	102	4.636	102
coloring-12	12	24	3	56.87	322	92.67	322	56.96	322	23.06	267	22.79	267
coloring-14	14	28	3	986.5	843	667.7	843	380.1	843	128.8	699	130.2	699
coloring-15	15	30	3	–	–	–	–	–	–	315.7	1131	312.7	1131

6   Conclusion

In this paper, we first proved that the incremental algorithm for computing triangular decompositions proposed in [7] preserves chordality. Then we pointed out that some simplification operations used in the implementation may destroy chordality. We resolve this problem by carefully modifying the implementation in Triangularize and the experimentation shows that significant speedups are obtained for some very sparse polynomial systems. Finally, we remark that more extensive experimentations on diverse polynomial systems are needed to decide the best use of these simplifications with the guidance of theory and possibly the help of artificial intelligence rather than simply relying on experience.

Acknowledgments. The authors would like to thank anonymous referees for helpful comments. This research was supported by NSFC (11771421, 11671377, 61572024), CAS "Light of West China" Program, the Key Research Program of Frontier Sciences of CAS (QYZDB-SSW-SYS026), and cstc2018jcyj-yszxX0002 of Chongqing.

References

1. Aubry, P., Lazard, D., Moreno Maza, M.: On the theories of triangular sets. J. Symb. Comput. **28**(1–2), 105–124 (1999)
2. Boulier, F., Lemaire, F., Moreno Maza, M.: Well known theorems on triangular systems and the D5 principle. In: Proceedings of Transgressive Computing 2006, Granada, Spain (2006)
3. Chen, C., Moreno Maza, M.: An incremental algorithm for computing cylindrical algebraic decompositions. In: Computer Mathematics: Proceedings of ASCM 2012, pp. 199–222 (2014)
4. Chen, C., Davenport, J.H., May, J.P., Moreno Maza, M., Xia, B., Xiao, R.: Triangular decomposition of semi-algebraic systems. J. Symb. Comput. **49**, 3–26 (2013)
5. Chen, C., Golubitsky, O., Lemaire, F., Maza, M.M., Pan, W.: Comprehensive triangular decomposition. In: Ganzha, V.G., Mayr, E.W., Vorozhtsov, E.V. (eds.) CASC 2007. LNCS, vol. 4770, pp. 73–101. Springer, Heidelberg (2007). https://doi.org/10.1007/978-3-540-75187-8_7
6. Chen, C., Moreno Maza, M.: Algorithms for computing triangular decompositions of polynomial systems. In: Proceedings of ISSAC, pp. 83–90 (2011)
7. Chen, C., Moreno Maza, M.: Algorithms for computing triangular decomposition of polynomial systems. J. Symb. Comput. **47**(6), 610–642 (2012)
8. Chen, C., Moreno Maza, M.: Quantifier elimination by cylindrical algebraic decomposition based on regular chains. J. Symb. Comput. **75**, 74–93 (2016)
9. Chen, C., Moreno Maza, M., Xia, B., Yang, L.: Computing cylindrical algebraic decomposition via triangular decomposition. In: Proceedings of ISSAC 2009, pp. 95–102 (2009)
10. Chen, X.F., Wang, D.K.: The projection of quasi variety and its application on geometric theorem proving and formula deduction. In: Winkler, F. (ed.) ADG 2002. LNCS (LNAI), vol. 2930, pp. 21–30. Springer, Heidelberg (2004). https://doi.org/10.1007/978-3-540-24616-9_2
11. Cifuentes, D., Parrilo, P.A.: Chordal networks of polynomial ideals. SIAM J. Appl. Algebra Geom. **1**(1), 73–110 (2017)
12. Dahan, X., Moreno Maza, M., Schost, E., Wu, W., Xie, Y.: Lifting techniques for triangular decompositions. In: Proceedings of ISSC, pp. 108–115 (2005)
13. Gao, X.S., Chou, S.C.: Computations with parametic equations. In: Proceedings of ISSAC, pp. 122–127 (1991)
14. Hubert, E.: Notes on triangular sets and triangulation-decomposition algorithms I: polynomial systems. In: Winkler, F., Langer, U. (eds.) SNSC 2001. LNCS, vol. 2630, pp. 1–39. Springer, Heidelberg (2003). https://doi.org/10.1007/3-540-45084-X_1
15. Kalkbrener, M.: A generalized euclidean algorithm for computing triangular representations of algebraic varieties. J. Symb. Comput. **15**(2), 143–167 (1993)
16. Lazard, D.: A new method for solving algebraic systems of positive dimension. Discrete Appl. Math. **33**(1–3), 147–160 (1991)
17. Moreno Maza, M.: On triangular decompositions of algebraic varieties. Technical report, TR 4/99, NAG Ltd., Oxford, UK (1999). Presented at MEGA-2000

18. Mou, C., Bai, Y.: On the chordality of polynomial sets in triangular decomposition in top-down style. In: Proceedings of ISSAC, pp. 287–294 (2018)
19. Ritt, J.F.: Differential equations from the algebraic standpoint, vol. 14. American Mathematical Society (1932)
20. Rose, D.J.: Triangulated graphs and the elimination process. J. Math. Anal. Appl. **32**(3), 597–609 (1970)
21. Vandenberghe, L., Andersen, M.S.: Chordal graphs and semidefinite optimization. Found. Trends Optim. **1**(4), 241–433 (2015)
22. Wang, D.: Elimination Methods. Springer, Vienna (2001). https://doi.org/10.1007/978-3-7091-6202-6
23. Wu, W.T.: Basic principles of mechanical theorem proving in elementary geometries. J. Auto. Reasoning **2**(3), 221–252 (1986)
24. Yang, L., Hou, X., Xia, B.: A complete algorithm for automated discovering of a class of inequality-type theorems. Sci. China Seri. F Inf. Sci. **44**(1), 33–49 (2001)
25. Yang, L., Zhang, J.: Searching dependency between algebraic equations: an algorithm applied to automated reasoning. Technical report, International Centre for Theoretical Physics (1990)

Algebraic Geometry via Numerical Computation

Q($\sqrt{-3}$)-Integral Points on a Mordell Curve

Francesca Bianchi[✉][iD]

Bernoulli Institute for Mathematics, Computer Science and Artificial Intelligence,
University of Groningen, Groningen, The Netherlands
francesca.bianchi@rug.nl

Abstract. We use an extension of quadratic Chabauty to number fields, recently developed by the author with Balakrishnan, Besser and Müller, combined with a sieving technique, to determine the integral points over Q($\sqrt{-3}$) on the Mordell curve $y^2 = x^3 - 4$.

Keywords: Elliptic curves · Quadratic Chabauty · Integral points

1 Introduction

Let E be an elliptic curve over a number field K, described by a Weierstrass equation

$$E\colon y^2 + a_1 xy + a_3 y = x^3 + a_2 x^2 + a_4 x + a_6, \qquad a_i \in \mathcal{O}_K, \tag{1}$$

where $\mathcal{O}_K$ is the ring of integers of K. By the set of (K-)integral points on E we mean the subset of solutions (x, y) to (1) with $x, y \in \mathcal{O}_K$. Such a set is finite by Siegel's Theorem [19].

Assume for now that $K = \mathbb{Q}$. Different solutions to the problem of effectively determining the set of integral points for a fixed elliptic curve have been given. The most notable include reducing the problem to solving some Thue equations, and using elliptic logarithms. See [24] for an overview.

An alternative more recent approach to an effective version of Siegel's Theorem for elliptic curves comes from a very special instance of Kim's non-abelian Chabauty programme. At present, explicit versions of this are known only for elliptic curves of Mordell–Weil rank at most 1 [5,6,11,16] and can be understood in terms of the theory of p-adic heights and formal group logarithms, where p is some fixed prime. Algorithmically, this method, often referred to as "quadratic Chabauty" in the literature, outputs a finite set of (approximations of) p-adic points on E, containing the integral points. When the rank is equal to one, this set typically also contains some points that we cannot *recognise* as rational, or even as algebraic. In order to determine the integral points, we need to be able to *prove* that such points are not p-adic approximations of rational points.

A. M. Bigatti et al. (Eds.): ICMS 2020, LNCS 12097, pp. 39–50, 2020.
https://doi.org/10.1007/978-3-030-52200-1_4

Unfortunately, there is no known method to achieve this, which might place this approach at a disadvantage compared to the previously mentioned ones. However, the idea of using p-adic heights and linear $\mathbb{Q}_p$-valued functionals on $E(\mathbb{Q})$ and $E(\mathbb{Q}_p)$ to study the integral points turned out to be amenable to extensions to some curves of higher genus [2,3], where the Mordell–Weil sieve is an effective tool used to address the problem of eliminating spurious p-adic points.

Generalisations to higher genus curves are very interesting from a Diophantine point of view. Nevertheless, we are left with the somewhat unsatisfactory impression that most quadratic Chabauty computations for integral points on elliptic curves appearing in the literature are carried out more as a "proof of concept" than as an actual way to determine integral points (but see [3, Appendix A] and [10, Appendix 4.A] for some examples of full computations of integral points).

In recent work with Balakrishnan, Besser and Müller [1], using restriction of scalars, we extended these explicit quadratic Chabauty techniques for integral points on elliptic curves, and, more generally, odd degree hyperelliptic curves, to arbitrary number fields. Combined with the Mordell–Weil sieve, the method successfully determined the integral points over an imaginary quadratic field on a genus 2 curve, as well as the rational points over some quadratic fields on some genus 2 bielliptic curves. Once again, though, examples of quadratic Chabauty computations for elliptic curves over number fields were also presented, but not turned into a provable determination of the set of integral points.

In this note we use the techniques of [1] to determine the set of $\mathbb{Q}(\sqrt{-3})$-integral points on the Mordell curve

$$E\colon y^2 = x^3 - 4. \qquad (2)$$

The goal is twofold. First, we want to show that quadratic Chabauty can indeed be a tool to determine the set of integral points on an elliptic curve (even over number fields), and that it is not only a test case for more general techniques that apply to higher genus curves. In order to achieve this, we need to substitute the Mordell–Weil sieve step with an analogous sieve for elliptic curves: we do this by extending to number fields the technique of [3, Appendix A]. Secondly, presenting this particular computation allows us to overcome a lot of the technicalities and notational complexity of [1], while still conveying the general strategy.

A few words about our choices of curve and field. The techniques for the computation of integral points of elliptic curves over $\mathbb{Q}$ were extended to arbitrary number fields by Smart and Stephens [23]. However, it appears that they are currently implemented in Magma [12] only over totally real fields; hence the choice of an imaginary quadratic field. In order to get rid of presumed non-integral points in the quadratic Chabauty computation, it is convenient to compare quadratic Chabauty outputs at different primes. For our choice of curve, there are two prime numbers, namely $p_1 = 7$ and $p_2 = 13$, which satisfy $\#E(\mathbb{F}_{p_i}) \equiv 0 \bmod p_j$ for $i \neq j$, making the comparison of the respective quadratic Chabauty outputs easier. In fact, this was used in [3, Appendix A] to compute the integral points over $\mathbb{Q}$, since E has rank 1 over $\mathbb{Q}$ and hence "classical" quadratic Chabauty is applicable. Finally, E attains rank 2 over $\mathbb{Q}(\sqrt{-3})$, both p_1 and p_2 split in $\mathbb{Q}(\sqrt{-3})$, and the latter is also the field of complex multiplication of E, making it a nice example to which our techniques can be applied.

The code to perform quadratic Chabauty for an elliptic curve over $\mathbb{Q}$ base-changed to an imaginary quadratic field, as well as for the sieving routines of this example, is available at [9].

2   Quadratic Chabauty for $E/\mathbb{Q}(\sqrt{-3})$

Let K be a number field of degree d and let $\mathbb{A}_{K,f}^{\times}$ be the group of finite ideles of K; let p be an odd prime, unramified in K. For ease of exposition, we assume that K has class number 1 and for every prime $\mathfrak{q}$ of K we fix a generator $\xi_{\mathfrak{q}}$ for $\mathfrak{q}$. Furthermore, we let $|\cdot|_{\mathfrak{q}}$ be the absolute value on $K_{\mathfrak{q}}$, normalised so that $|q|_{\mathfrak{q}} = q^{-1}$, if q is the rational prime below $\mathfrak{q}$. Thus $|\xi_{\mathfrak{q}}| = q^{-1/r_{\mathfrak{q}}}$, for some positive integer $r_{\mathfrak{q}}$ which divides the degree $n_{\mathfrak{q}}$ of the extension $K_{\mathfrak{q}}/\mathbb{Q}_q$.

An idele class character is a continuous homomorphism

$$\chi = \sum_{\mathfrak{q}} \chi_{\mathfrak{q}} : \mathbb{A}_{K,f}^{\times}/K^{\times} \to \mathbb{Q}_p.$$

In principle, the idele class character χ is determined by the value, at each prime $\mathfrak{q}$, of $\chi_{\mathfrak{q}}$ on $\xi_{\mathfrak{q}}$ and on the units $\mathcal{O}_{\mathfrak{q}}^{\times}$ of the ring of integers of $K_{\mathfrak{q}}$. However, the incompatibility between the $\mathfrak{q}$-adic and p-adic topology for $\mathfrak{q} \neq p$ and the required vanishing of χ on $K^{\times}$ have two strong consequences:

- The restrictions of $\chi_{\mathfrak{p}}$ to $\mathcal{O}_{\mathfrak{p}}^{\times}$ at the primes $\mathfrak{p} \mid p$ uniquely determine χ. Indeed, $\chi_{\mathfrak{q}}$ vanishes on $\mathcal{O}_{\mathfrak{q}}^{\times}$ if $\mathfrak{q} \nmid p$, and for every prime $\mathfrak{q}$ we have the formula

$$\chi_{\mathfrak{q}}(\xi_{\mathfrak{q}}) = -\sum_{\substack{\mathfrak{p} \mid p \\ \mathfrak{p} \neq \mathfrak{q}}} \chi_{\mathfrak{p}}(\xi_{\mathfrak{q}}). \tag{3}$$

- Each fundamental unit for K imposes an additional constraint on the characters $\chi_{\mathfrak{p}}$. In particular, the idele class characters for K form a $\mathbb{Q}_p$-vector space V of dimension greater than or equal to $d - \mathrm{rank}(\mathcal{O}_K^{\times})$.

Computing a basis for V amounts to doing some linear algebra.

Example 1. In our situation of interest, K is an imaginary quadratic field, in which p splits, say $p = \mathfrak{p}_1\mathfrak{p}_2$, and we may fix isomorphisms $K_{\mathfrak{p}_i} \simeq \mathbb{Q}_p$. Then V is spanned by two characters: the *cyclotomic* character χ^{cyc} and the *anticyclotomic* character χ^{anti}. The cyclotomic character is the unique character satisfying

$$\chi_{\mathfrak{p}_i}^{\mathrm{cyc}}(x) = \log(x) \qquad \text{for all } x \in \mathcal{O}_{\mathfrak{p}_i}^{\times},$$

where $\log\colon \mathbb{Z}_p^\times \to \mathbb{Q}_p$ is the p-adic logarithm, and we view x as an element of $\mathbb{Z}_p^\times$ via our fixed isomorphism. The anticyclotomic character (depending on our choice of ordering of the primes above p) is determined by

$$\chi_{\mathfrak{p}_1}^{\mathrm{anti}}(x) = \log(x) \quad \text{for all } x \in \mathcal{O}_{\mathfrak{p}_1}^\times; \qquad \chi_{\mathfrak{p}_2}^{\mathrm{anti}}(x) = -\log(x) \quad \text{for all } x \in \mathcal{O}_{\mathfrak{p}_2}^\times.$$

The p-adic logarithm comes in the picture because we are considering homomorphisms from the multiplicative $\mathcal{O}_{\mathfrak{p}_i}^\times$ to the additive $\mathbb{Q}_p$.

Formula (3) gives

$$\chi_{\mathfrak{q}}^{\mathrm{cyc}}(q) = r_{\mathfrak{q}}\chi_{\mathfrak{q}}^{\mathrm{cyc}}(\xi_{\mathfrak{q}}) = -n_{\mathfrak{q}}\log(q) \qquad \text{for all } q \neq p; \tag{4}$$

$$\chi_{\mathfrak{q}}^{\mathrm{anti}}(q) = r_{\mathfrak{q}}\chi_{\mathfrak{q}}^{\mathrm{cyc}}(\xi_{\mathfrak{q}}) = 0 \qquad \text{for all } q \neq p \text{ such that } n_{\mathfrak{q}} = 2. \tag{5}$$

Let E/K now be an elliptic curve with good reduction at all primes above p, described by an equation of the form (2), which, for simplicity, we further assume to be minimal at all primes. Let S be the set of primes of bad reduction for E, and for each prime $\mathfrak{q}$ of K, let $c_{\mathfrak{q}}$ be the Tamagawa number of E at $\mathfrak{q}$ (cf. [18, Theorem 4.11]).

If none of the primes above p is in S, by work of Bernardi [7], Mazur–Tate [17], Coleman–Gross [14] and others, to every idele class character $\chi \in V$ we can attach a quadratic function

$$h_p^\chi\colon E(K) \to \mathbb{Q}_p \qquad \text{by} \qquad h_p^\chi(P) = \begin{cases} (\prod_{\mathfrak{q}\in S} c_{\mathfrak{q}})^{-1}\chi(\iota(P)) & \text{if } P \neq O, \\ 0 & \text{otherwise,} \end{cases}$$

for some suitably chosen $\iota(P) \in \mathbb{A}_{K,f}^\times$, which is independent of χ and whose $\mathfrak{p}$-adic component at a prime $\mathfrak{p} \mid p$ depends on the choice of a subspace of $H_{\mathrm{dR}}^1(E/K_{\mathfrak{p}})$, complementary to the space of holomorphic forms.

We call h_p^χ the *(global) p-adic height* attached to χ, where we should not forget, however, that we have made some preliminary choices that are not reflected in our notation.

More generally, at every prime $\mathfrak{q}$, given $P_{\mathfrak{q}} \in E(K_{\mathfrak{q}}) \setminus \{O\}$, there exists $\iota_{\mathfrak{q}}(P_{\mathfrak{q}}) \in K_{\mathfrak{q}}^\times$ and a *local p-adic height* function

$$\lambda_{\mathfrak{q}}^\chi\colon E(K_{\mathfrak{q}}) \setminus \{O\} \to \mathbb{Q}_p, \qquad \lambda_{\mathfrak{q}}^\chi(P_{\mathfrak{q}}) = c_{\mathfrak{q}}^{-1}\chi_{\mathfrak{q}}(\iota_{\mathfrak{q}}(P_{\mathfrak{q}})),$$

so that $\iota(P)_{\mathfrak{q}} = \iota_{\mathfrak{q}}(P)^{(\prod_{\mathfrak{r}\in S\setminus\mathfrak{q}} c_{\mathfrak{r}})}$ and hence

$$h_p^\chi(P) = \sum_{\mathfrak{q}} \lambda_{\mathfrak{q}}^\chi(P) \qquad \text{if } P \in E(K) \setminus \{O\}.$$

The failure of each $\lambda_{\mathfrak{q}}^\chi$ to be a quadratic function is absorbed into the vanishing of χ on $K^\times$, i.e. we have

$$h_p^\chi(mP) = m^2 h_p^\chi(P) \qquad \text{for every } m \in \mathbb{Z} \text{ and } P \in E(K). \tag{6}$$

We do not go into the details here of how to define $\iota_{\mathfrak{q}}(P_{\mathfrak{q}})$. We just note that the definition of h_p^χ mimics Néron's definition of the real canonical height (for which see, for instance, [21, Chapter VI]), and we record the following properties.

Proposition 1. *1. At a prime $\mathfrak{p} \mid p$, the function $\lambda_\mathfrak{p}^\chi$ restricts to a locally analytic function on $E(\mathcal{O}_\mathfrak{p})$, whose power series expansion in a local coordinate t in each residue disc[1] can be computed up to any desired $\mathfrak{p}$-adic and t-adic precision.*
2. At a prime $\mathfrak{q} \nmid p$, $\iota_\mathfrak{q}(P)$ is independent of p. We have
   – If P reduces to a non-singular point modulo $\mathfrak{q}$, then

$$\iota_\mathfrak{q}(P) = \max\{1, |x(P)|_\mathfrak{q}\}^{-c_\mathfrak{q}}.$$

   – If P reduces to a singular point, then $\iota_\mathfrak{q}(P)^{r_\mathfrak{q}} \in W_\mathfrak{q}$, where $W_\mathfrak{q} \subset \mathbb{Q}^\times$ is a finite set which can be deduced from knowledge of the Tamagawa number and Kodaira symbol of E at $\mathfrak{q}$.

Corollary 1. *Let S be the set of primes of K at which E has bad reduction. For every $\mathfrak{q} \in S$, there exists a finite set $T_\mathfrak{q} \subset \mathbb{Q}^\times$ such that for every $P \in E(\mathcal{O}_K)$ there exists $t_\mathfrak{q} \in T_\mathfrak{q}$ with*

$$h_p^\chi(P) = \sum_{\mathfrak{p} \mid p} \lambda_\mathfrak{p}^\chi(P) + \sum_{\mathfrak{q} \in S} c_\mathfrak{q}^{-1} r_\mathfrak{q}^{-1} \chi_\mathfrak{q}(t_\mathfrak{q}).$$

Moreover, $T_\mathfrak{q}$ and $t_\mathfrak{q}$ are independent of p and χ.

Corollary 1 and Eq. (6) are the key ingredients for our method. In fact, a point $P \in E(\mathcal{O}_K)$ carries global information by its being a point in $E(K)$. This is encoded in (6), but alone would not suffice to give it a finite characterisation inside $\prod_{\mathfrak{p}|p} E(K_\mathfrak{p})$, since $E(K)$ could be infinite. Thus, we also need to exploit the local information that it carries by its being integral at every prime and this is where Corollary 1 comes into play.

In order to use (6) effectively, we need to express the p-adic height h_p^χ as the restriction of a locally analytic function $\prod_{\mathfrak{p}|p} E(K_\mathfrak{p}) \to \mathbb{Q}_p$. This is where we are forced to impose some conditions on the rank of $E(K)$. We will focus now on our example of interest; see [1] for the general treatment of the quadratic Chabauty steps. The sieving method that we describe is an extension to number fields of [3, Appendix A]; while for the moment we restrict the exposition to this example, the method would apply more generally: see Sect. 3.

Example 2. Let E be the curve over $K = \mathbb{Q}(\sqrt{-3})$ defined by (2). Let

$$E^{-3}: y^2 = x^3 + 108 \tag{7}$$

be the quadratic twist of E by -3. By computing generators for $E(\mathbb{Q})$ and $E^{-3}(\mathbb{Q})$ with SageMath [25], we deduce that

$$E(K) = \langle Q, R \rangle \simeq \mathbb{Z}^2,$$

---

[1] A residue disc is a fibre of the reduction map $E(K_\mathfrak{p}) \to E(\mathbb{F}_\mathfrak{p})$ where $\mathbb{F}_\mathfrak{p}$ is the residue field of $K_\mathfrak{p}$.

where
$$Q = (2, 2), \qquad R = (-\sqrt{-3} + 1, 2\sqrt{-3}).$$

The set of primes at which E/K has bad reduction is

$$S = \{(2), (\sqrt{-3})\}, \qquad \text{with } c_{(2)} = 4, \quad c_{(\sqrt{-3})} = 3.$$

Let $p \in \{7, 13\}$. The prime p splits completely in K and we are in the situation of Example 1, so we have two global p-adic heights

$$h_p^{\mathrm{cyc}} := h_p^{\chi^{\mathrm{cyc}}}, h_p^{\mathrm{anti}} := h_p^{\chi^{\mathrm{anti}}} \colon E(K) \to \mathbb{Q}_p,$$

where we have chosen the local heights at $\mathfrak{p}_i$ to be the ones of Mazur–Tate (which coincide with Bernardi's in this case). For $i \in \{1, 2\}$, we may also consider

$$f_i \colon E(K) \hookrightarrow E(K_{\mathfrak{p}_i}) \xrightarrow{\mathrm{Log}_i} \mathbb{Q}_p,$$

where Log_i is the unique homomorphism of abelian groups which restricts to the formal group logarithm (composed with $K_{\mathfrak{p}_i} \simeq \mathbb{Q}_p$) on the formal group. By construction, the functions Log_1 and Log_2 are locally analytic, and Lemma 6.4 of [1] shows that f_1 and f_2 (or, more precisely, their extensions to $E(K) \otimes \mathbb{Q}_p$) are linearly independent. Since the rank of $E(K)$ is 2 by (2), it follows that any $\mathbb{Q}_p$-valued quadratic function on it must be a $\mathbb{Q}_p$-linear combination of

$$f_1^2, \quad f_1 f_2, \quad f_2^2.$$

In particular, there exist $\alpha_i^{\mathrm{cyc}}, \alpha_i^{\mathrm{anti}} \in \mathbb{Q}_p$ such that

$$h_p^{\mathrm{cyc}} = \alpha_1^{\mathrm{cyc}} f_1^2 + \alpha_2^{\mathrm{cyc}} f_1 f_2 + \alpha_3^{\mathrm{cyc}} f_2^2, \qquad h_p^{\mathrm{anti}} = \alpha_1^{\mathrm{anti}} f_1^2 + \alpha_2^{\mathrm{anti}} f_1 f_2 + \alpha_3^{\mathrm{anti}} f_2^2. \tag{8}$$

Properties of the cyclotomic and anticyclotomic characters show that $\alpha_1^{\mathrm{cyc}} = \alpha_3^{\mathrm{cyc}}$; $\alpha_1^{\mathrm{anti}} = -\alpha_3^{\mathrm{anti}}$ and $\alpha_2^{\mathrm{anti}} = 0$ (cf. [1, §6.2]). Furthermore, since h_p^{χ} is invariant under any automorphism of E/K (while f_1^2 and f_2^2 are only invariant under ± 1), in this case all the constants are identically zero, except for α_2^{cyc}, which is non-zero by [8]. We can compute it up to our desired precision, by comparing the values of the p-adic height and of $f_1 f_2$ on any point of infinite order of $E(K)$.

To make (8) sensitive to the difference between K-rational and K-integral points, we use Corollary 1. By work of Cremona–Prickett–Siksek [15] (see also [11, Proposition 2.4]), we may take

$$T_{(2)} = \{1, 2^4\}, \qquad T_{(\sqrt{-3})} = \{1, 3^2\}.$$

Consider the locally analytic functions

$$\rho_p^{\mathrm{cyc}} \colon E(\mathcal{O}_{\mathfrak{p}_1}) \times E(\mathcal{O}_{\mathfrak{p}_2}) \to \mathbb{Q}_p, \qquad \rho_p^{\mathrm{anti}} \colon E(\mathcal{O}_{\mathfrak{p}_1}) \times E(\mathcal{O}_{\mathfrak{p}_2}) \to \mathbb{Q}_p,$$

defined by

$$\rho_p^{\star}(P_1, P_2) = \lambda_{\mathfrak{p}_1}^{\star}(P_1) + \lambda_{\mathfrak{p}_2}^{\star}(P_2) - \alpha_2^{\star} \mathrm{Log}_1(P_1) \mathrm{Log}_2(P_2).$$

By Corollary 1 and (4)-(5), for every $P \in E(\mathcal{O}_K)$, there exists $t_{(2)} \in T_{(2)}$ and $t_{(\sqrt{-3})} \in T_{(\sqrt{-3})}$ such that

$$\rho_p^{\mathrm{cyc}}(\psi_p(P)) - \frac{1}{2} \log t_{(2)} - \frac{1}{3} \log t_{(\sqrt{-3})} = 0 = \rho_p^{\mathrm{anti}}(\psi_p(P)),$$

where $\psi_p \colon E(\mathcal{O}_K) \hookrightarrow E(\mathcal{O}_{\mathfrak{p}_1}) \times E(\mathcal{O}_{\mathfrak{p}_2})$ is the map induced by the completion maps. The strategy is then as follows.

1. Fix a positive integer B and compute[2]

$$E(\mathcal{O}_K)_{\mathrm{known}} = \{nQ + mR : |n|, |m| \leq B \quad \text{and} \quad nQ + mR \in E(\mathcal{O}_K)\}.$$

   By picking B sufficiently large, we might expect that $E(\mathcal{O}_K)_{\mathrm{known}} = E(\mathcal{O}_K)$ and the following steps try to prove the equality.
2. Fix $(t_{(2)}, t_{(\sqrt{-3})}) \in T_{(2)} \times T_{(\sqrt{-3})}$.
3. Let $p = 7$. Compute, modulo some fixed p-adic precision, all $(P_1, P_2) \in E(\mathcal{O}_{\mathfrak{p}_1}) \times E(\mathcal{O}_{\mathfrak{p}_2})$ such that

$$\rho_7^{\mathrm{cyc}}(P_1, P_2) - \frac{1}{2} \log t_{(2)} - \frac{1}{3} \log t_{(\sqrt{-3})} = 0 = \rho_7^{\mathrm{anti}}(P_1, P_2). \tag{9}$$

   If $\varphi \in \mathrm{Aut}(E/K)$ with induced $\varphi_i \in \mathrm{Aut}(E/K_{\mathfrak{p}_i})$, then $(\varphi_1(P_1), \varphi_2(P_2))$ is a solution to (9) if (P_1, P_2) is. Furthermore, if (P_1, P_2) is a solution, then so is (P_2, P_1) where we are abusing notation, in view of the isomorphisms $K_{\mathfrak{p}_1} \simeq \mathbb{Q}_p \simeq K_{\mathfrak{p}_2}$.
4. For every (P_1, P_2) computed in step 3, check if there exists $P \in E(\mathcal{O}_K)_{\mathrm{known}}$ with $\psi_7(P) \equiv (P_1, P_2)$. If such P does not exist, suppose there exists $T \in E(\mathcal{O}_K) \setminus E(\mathcal{O}_K)_{\mathrm{known}}$ with $\psi_7(T) = (P_1, P_2)$. Then $T = nQ + mR$ for some $n, m \in \mathbb{Z}$ satisfying

$$\begin{pmatrix} n \\ m \end{pmatrix} = \begin{pmatrix} f_1(Q) & f_1(R) \\ f_2(Q) & f_2(R) \end{pmatrix}^{-1} \begin{pmatrix} \mathrm{Log}_1(P_1) \\ \mathrm{Log}_2(P_2) \end{pmatrix} \mod 7.$$

   Furthermore we can compute

$$\begin{pmatrix} n \\ m \end{pmatrix} \mod 13,$$

   by considering the images of P_1, P_2, Q, R in $E(\mathbb{F}_{\mathfrak{p}_1})$ and $E(\mathbb{F}_{\mathfrak{p}_2})$, since $\#E(\mathbb{F}_{\mathfrak{p}_i}) = \#E(\mathbb{F}_7) = 13$. (Note that it is straightforward to verify that there can be at most one such possibility).
5. The solution set of

$$\rho_{13}^{\mathrm{cyc}}(P_1, P_2) - \frac{1}{2} \log t_{(2)} - \frac{1}{3} \log t_{(\sqrt{-3})} = 0 = \rho_{13}^{\mathrm{anti}}(P_1, P_2)$$

---

[2] This step could be skipped; assuming this computation just simplifies the exposition of the other steps.

must contain $\psi_{13}(T)$. Since $\#E(\mathbb{F}_{13}) = 3 \cdot 7$, we can, similarly to steps 3 and 4, deduce another list of possibilities for the pair

$$\binom{n}{m} \bmod 13, \qquad \binom{n}{m} \bmod 7.$$

If none of these pairs matches the one of step 4, then we have shown that (P_1, P_2) is not in the image of ψ_7.

6. We repeat steps 3–5 for each element of $T_{(2)} \times T_{(\sqrt{-3})}$.

Remark 1. The computations of the zero sets in steps 3 and 5 are carried out locally, i.e. for every pair $(\overline{P_1}, \overline{P_2}) \in E(\mathbb{F}_{\mathfrak{p}_1}) \times E(\mathbb{F}_{\mathfrak{p}_2})$, we can give a two-variable parametrisation of the points in $E(K_{\mathfrak{p}_1}) \times E(K_{\mathfrak{p}_2})$ reducing to $(\overline{P_1}, \overline{P_2})$. The task is then reduced to computing zero sets of systems of two-variable equations in two variables. This can be done either naively, or, when the hypotheses apply, using a two-variable version of Hensel's lemma [1, Appendix A]. For the strategy to give a provable determination of the integral points, it is necessary to verify that the approximations of the zeros corresponding to points in $E(\mathcal{O}_K)$ lift to unique zeros. It is on the other hand not necessary in general to show that the other zeros lift uniquely.

We implement this technique in SageMath and run it with $B = 10$. As we vary $t_{(2)}$ and $t_{(\sqrt{-3})}$, this gives 426 solutions to (9) which do not come from points in $E(\mathcal{O}_K)_{\text{known}}$. Out of these 426, there are only 4 up to automorphisms and conjugation which survive our sieve.

We need to show these 4 solutions do not come from points in $E(\mathcal{O}_K)$. We first note that they are p-adically isolated. Then we show that, if one of them did come from a point T in $E(\mathcal{O}_K)$, then, up to automorphism, it would come from a point in $E(\mathbb{Z})$ or a point in $E^{-3}(\mathbb{Z})$ where we take the minimal model (7) (this amounts to showing that in the automorphism class of the solution, there is a point of the form (P_1, P_1) or $(P_1, -P_1)$, where we are using our isomorphisms $K_{\mathfrak{p}_i} \simeq \mathbb{Q}_p$). At this point, we could invoke existing implementations to compute $E(\mathbb{Z})$ and $E^{-3}(\mathbb{Z})$, to show that all the points in $E(\mathcal{O}_K)$ coming from points in these sets are in $E(\mathcal{O}_K)_{\text{known}}$. Alternatively, we observe that being in the image of $E(\mathbb{Z})$ or $E^{-3}(\mathbb{Z})$ imposes additional contraints on the local heights away from p of P, which allow us to eliminate at once 3 of the automorphisms classes. The remaining one must come from $E^{-3}(\mathbb{Z})$ and hence must be of the form mR. There is some information that we have not used yet: since $3 \mid \#E(\mathbb{F}_{13})$, we also know $m \bmod 3$. In particular, $m \equiv 0 \bmod 3$. But then mR is a multiple of $3R$, which is in the formal group at $(\sqrt{-3})$: thus mR cannot be integral. This shows that

Theorem 1. $E(\mathcal{O}_K) = \{\varphi(P), \varphi(P^c) : P \in G, \varphi \in \text{Aut}(E/K)\}$, *where P^c is the Galois conjugate of P and*

$$G = \{(2, 2), (5, 11), (-2, 2\sqrt{-3}), (1, \sqrt{-3}),$$
$$(-122, 778\sqrt{-3}), (3\sqrt{-3} - 5, 4\sqrt{-3} + 18)\}.$$

3   A More General Approach

We end this article with an outline of the strategy that was implicitly used in Example 2, in the hope that this discussion will convince the reader that similar methods could be used to determine the integral points of other elliptic curves over number fields.

Let E be an elliptic curve over an imaginary quadratic field K. Suppose further that existing algorithms - e.g. as implemented in [12] - succeed in the computation of the rank r of $E(K)$. If $r \geq 3$, our methods are not applicable. If $r = 0$, the computation of $E(\mathcal{O}_K)$ is trivial; if $r = 1$, minor modifications to the quadratic Chabauty method over $\mathbb{Q}$ are sufficient [4].

Thus, we may assume that r is exactly equal to 2, and that we can compute generators for the Mordell–Weil group $E(K)$. For simplicity, we further assume that the torsion subgroup of $E(K)$ is trivial. Hence,

$$E(K) = \langle Q, R \rangle$$

for some points of infinite order Q and R. Our goal is the computation of the integral points $E(\mathcal{O}_K)$.

It is reasonable to first compute the integral points up to a certain height bound and we shall denote the resulting set by $E(\mathcal{O}_K)_{\text{known}}$. Indeed, for instance, if Q and R are non-integral at the same prime $\mathfrak{q}$, then $E(\mathcal{O}_K)$ is empty and no further step should be taken (see [22, Chapter VII]).

Let p be an odd prime such that E has good reduction at every $\mathfrak{p} \mid p$. Current restrictions in the implementation of some of the techniques that we need also require us to assume that p is split in K, say $p\mathcal{O}_K = \mathfrak{p}_1\mathfrak{p}_2$ for distinct primes $\mathfrak{p}_1$ and $\mathfrak{p}_2$ of $\mathcal{O}_K$. We have homomorphisms

$$f_i \colon E(K) \to \mathbb{Q}_p,$$

obtained by composing the completion map $E(K) \hookrightarrow E(K_{\mathfrak{p}_i})$ with $\mathfrak{p}_i$-adic abelian integration and with the isomorphism $K_{\mathfrak{p}_i} \cong \mathbb{Q}_p$. If f_1 and f_2 are linearly independent, then we can carry out quadratic Chabauty for the prime p on E/K. Linear independence is guaranteed by [1, Lemma 6.4] if E is the base-change of an elliptic curve over $\mathbb{Q}$ of rank 1, and we also expect it to hold if E does not descend to $\mathbb{Q}$.

In the belief that the exposition of the quadratic Chabauty computation in Sect. 2 was, when not already in full generality, easily generalisable, we do not elaborate on that further. The upshot is that we can compute a set U_p of points in $E(\mathcal{O}_{\mathfrak{p}_1}) \times E(\mathcal{O}_{\mathfrak{p}_2})$ such that if $P \in E(\mathcal{O}_K) \setminus E(\mathcal{O}_K)_{\text{known}}$ then the image of P under the completion maps lies in U_p. By "computing", we mean that given a large enough integer n we can find the finite set of approximations modulo[3] p^n of the points in U_p. This is provided that we can show that, modulo p^n, there are no points in U_p that have the same reduction as any point in $E(\mathcal{O}_K)_{\text{known}}$. Based on experimental data, we can say that the latter task can generally be addressed

---

[3] Given that $K_{\mathfrak{p}_i} \cong \mathbb{Q}_p$, the phrasing "modulo p^n" has the obvious meaning.

using the multivariable Hensel's lemma, but occasionally needs support from theoretical arguments [1, Appendix A] (see also Remark 1).

If U_p is empty, we deduce that $E(\mathcal{O}_K) = E(\mathcal{O}_K)_{\text{known}}$. Otherwise, we can try to extrapolate more information from U_p, or compute U_p for more than one choice of p. If the primes are chosen wisely, it is possible to compare the outputs U_p. We explain here a good choice; many others could be possible.

For $i \in \{1, 2\}$, let $k_{\mathfrak{p}_i}$ be the greatest common divisor of the orders in $E(\mathbb{F}_{\mathfrak{p}_i})$ of the reduction of Q and R. Assume that $(P_{\mathfrak{p}_1}, P_{\mathfrak{p}_2}) \in U_p$ is the localisation of a point $P \in E(\mathcal{O}_K) \setminus E(\mathcal{O}_K)_{\text{known}}$. Then there must exist $n, m \in \mathbb{Z}$ such that

$$P = nQ + mR.$$

Using $\mathfrak{p}_i$-adic abelian integration, we may compute a unique pair $(n_p, m_p) \in (\mathbb{Z}/p\mathbb{Z})^2$ such that

$$(n, m) \equiv (n_p, m_p) \bmod p. \tag{10}$$

Furthermore, for every i we can compute a list of possibilities for

$$(n, m) \bmod k_{\mathfrak{p}_i}, \tag{11}$$

by considering the images of Q, R and $P_{\mathfrak{p}_i}$ in $E(\mathbb{F}_{\mathfrak{p}_i})$.

If $p \mid k_{\mathfrak{p}_i}$ for at least one i, then (10) and (11) can sometimes be used to rule out points in U_p from being localisations of points in $E(\mathcal{O}_K)$.

If this is not sufficient to prove that no point in U_p can correspond to a point in $E(\mathcal{O}_K)$, or if p is coprime with each $k_{\mathfrak{p}_i}$, then we may look for a split odd prime q of good reduction such that $q \mid k_{\mathfrak{p}_i}$ and $p \mid k_{\mathfrak{q}_j}$ for at least one (i, j), if $q\mathcal{O}_K = \mathfrak{q}_1\mathfrak{q}_2$. If the element $(P_{\mathfrak{p}_1}, P_{\mathfrak{p}_2}) \in U_p$ corresponds to $P \in E(\mathcal{O}_K)$, then P must map to some $(P_{\mathfrak{q}_1}, P_{\mathfrak{q}_2}) \in U_q$ under the $(\mathfrak{q}_1, \mathfrak{q}_2)$-completion maps. By running through U_q, we obtain a list of possibilities for the pair

$$(n, m) \bmod q \qquad \text{and} \qquad (n, m) \bmod k_{\mathfrak{q}_j},$$

which we can compare with (10) and (11).

If this is still not sufficient to show that $E(\mathcal{O}_K) = E(\mathcal{O}_K)_{\text{known}}$, we can look for sequences of primes with similar patterns.

Elliptic curves E over $\mathbb{Q}$ admitting a pair of primes of good reduction (p, q) satisfying $\#E(\mathbb{F}_p) = q$ and $\#E(\mathbb{F}_q) = p$ were studied by Silverman–Stange [20]. Assuming that there are infinitely many primes ℓ for which $\#E(\mathbb{F}_\ell)$ is prime, the authors conjectured that the number of such pairs (p, q) with $p < q$ and $p \le X$ should grow like a multiple of $X/(\log X)^2$ if E has CM, and of $\sqrt{X}/(\log X)^2$ otherwise. More generally, they considered sequences of primes $(p_1, \ldots, p_m)$ of arbitrary length m such that $\#E(\mathbb{F}_{p_i}) = p_{i+1}$, where i is taken modulo m.

This study is relevant for us if our elliptic curve over K is the base-change of an elliptic curve over $\mathbb{Q}$, as the one of Example 2. For more general elliptic curves over number fields, see [13].

Finally, the discussion of this section is readily generalisable to number fields other than imaginary quadratic ones, after deriving the analogue of Example 1 and suitably replacing the condition $r = 2$ (see [1]).

Acknowledgements. The author would like to thank Jennifer Balakrishnan and Steffen Müller for providing feedback on an earlier draft and for useful conversations. She is grateful to the anonymous referees for their comments and suggestions. She is supported by an NWO Vidi grant.

References

1. Balakrishnan, J.S., Besser, A., Bianchi, F., Müller, J.S.: Explicit quadratic Chabauty over number fields. Isr. J. Math., to appear (2019). (arXiv:1910.04653)
2. Balakrishnan, J.S., Besser, A., Müller, J.S.: Quadratic Chabauty: p-adic heights and integral points on hyperelliptic curves. J. Reine Angew. Math. **720**, 51–79 (2016). https://doi.org/10.1515/crelle-2014-0048
3. Balakrishnan, J.S., Besser, A., Müller, J.S.: Computing integral points on hyperelliptic curves using quadratic Chabauty. Math. Comp. **86**(305), 1403–1434 (2017). https://doi.org/10.1090/mcom/3130
4. Balakrishnan, J.S., Dogra, N.: Quadratic Chabauty and rational points, I: p-adic heights. Duke Math. J. **167**(11), 1981–2038 (2018). https://doi.org/10.1215/00127094-2018-0013. With an appendix by J.S. Müller
5. Balakrishnan, J.S., Kedlaya, K.S., Kim, M.: Appendix and erratum to "Massey products for elliptic curves of rank 1". J. Amer. Math. Soc. **24**(1), 281–291 (2011)
6. Balakrishnan, J.S., Dan-Cohen, I., Kim, M., Wewers, S.: A non-abelian conjecture of Tate-Shafarevich type for hyperbolic curves. Math. Ann. **372**(1–2), 369–428 (2018). https://doi.org/10.1007/s00208-018-1684-x
7. Bernardi, D.: Hauteur p-adique sur les courbes elliptiques, Séminaire de Théorie des Nombres Paris, 1979–80. Progr. Math. **12**, 1–14 (1981). Birkhäuser, Boston, Mass
8. Bertrand, D.: Valeurs de fonctions thêta et hauteurs p-adiques, Séminaire de Théorie des Nombres, Paris 1980–81. Progr. Math. **22**, 1–12 (1982). Birkhäuser, Boston, Mass
9. Bianchi, F.: SageMath code. https://github.com/bianchifrancesca/QC_elliptic_imaginary_quadratic_rank_2
10. Bianchi, F.: Topics in the theory of p-adic heights on elliptic curves. Ph.D. thesis, University of Oxford (2019)
11. Bianchi, F.: Quadratic Chabauty for (bi)elliptic curves and Kim's conjecture. Algebra Number Theory (2019). to appear (arXiv:1904.04622)
12. Bosma, W., Cannon, J., Playoust, C.: The Magma algebra system. I. The user language. J. Symb. Comput. **24**(3–4), 235–265 (1997). https://doi.org/10.1006/jsco.1996.0125
13. Brown, J., Heras, D., James, K., Keaton, R., Qian, A.: Amicable pairs and aliquot cycles for elliptic curves over number fields. Rocky Mt. J. Math. **46**(6), 1853–1866 (2016). https://doi.org/10.1216/RMJ-2016-46-6-1853
14. Coleman, R.F., Gross, B.H.: p-adic heights on curves. In: Algebraic number theory, Advanced Studies in Pure Mathematics, vol. 17, pp. 73–81. Academic Press, Boston (1989)
15. Cremona, J.E., Prickett, M., Siksek, S.: Height difference bounds for elliptic curves over number fields. J. Number Theory **116**(1), 42–68 (2006). https://doi.org/10.1016/j.jnt.2005.03.001
16. Kim, M.: Massey products for elliptic curves of rank 1. J. Am. Math. Soc. **23**(3), 725–747 (2010)

17. Mazur, B., Tate, J.: The p-adic sigma function. Duke Math. J. **62**(3), 663–688 (1991). https://doi.org/10.1215/S0012-7094-91-06229-0

18. Schmitt, S., Zimmer, H.G.: Elliptic Curves. A Computational Approach. With an appendix by Attila Pethö, vol. 31 in de Gruyter Studies in Mathematics. Walter de Gruyter & Co. (2003). https://doi.org/10.1515/9783110198010

19. Siegel, C.: Über einige Anwendungen Diophantischer Approximationen. Abh. Preus. Acad. Wiss. **1** (1929). https://doi.org/10.1007/978-88-7642-520-2_2

20. Silverman, J.H., Stange, K.E.: Amicable pairs and aliquot cycles for elliptic curves. Exp. Math. **20**(3), 329–357 (2011). https://doi.org/10.1080/10586458.2011.565253

21. Silverman, J.: Advanced Topics in the Arithmetic of Elliptic Curves. Graduate Texts in Mathematics, vol. 151. Springer, New York (1994). https://doi.org/10.1007/978-1-4612-0851-8

22. Silverman, J.: The Arithmetic of Elliptic Curves. Graduate Texts in Mathematics, vol. 106, 2nd edn. Springer, Dordrecht (2009). https://doi.org/10.1007/978-0-387-09494-6

23. Smart, N.P., Stephens, N.M.: Integral points on elliptic curves over number fields. In: Mathematical Proceedings of the Cambridge Philosophical Society, vol. 122, no. 1, pp. 9–16 (1997). https://doi.org/10.1017/S0305004196001211

24. Smart, N.P.: The algorithmic Resolution of Diophantine Equations. London Mathematical Society Student Texts, vol. 41. Cambridge University Press, Cambridge (1998). https://doi.org/10.1017/CBO9781107359994

25. Stein, W., et al.: Sage Mathematics Software (Version 8.9). The Sage Development Team (2019). http://www.sagemath.org

A Numerical Approach for Computing Euler Characteristics of Affine Varieties

Xiaxin Li[1], Jose Israel Rodriguez[2(✉)] [iD], and Botong Wang[2]

[1] University of California, La Jolla, San Diego, CA 92093, USA
xil095@ucsd.edu
[2] University of Wisconsin-Madison, Madison, WI 53706, USA
{Jose,wang}@math.wisc.edu

Abstract. We develop a numerical nonlinear algebra approach for computing the Euler characteristic of an affine variety. Our approach is to relate Euler characteristics of a smooth affine variety with the number of critical points using Morse theory. In general, we stratify a variety into the union of smooth affine varieties to obtain results on singular varieties.

Keywords: Euler characteristic · Numerical algebraic geometry · Homotopy continuation

1 Introduction

The Euler characteristic is one of the most fundamental topological invariants. In the past decade, a series of work appeared which relate Euler characteristics of complex algebraic varieties with the complexity of algebraic optimization problems [1,5,15,20,21]. There are several existing approaches to compute the Euler characteristics of complex algebraic varieties [3,6,16,19], each having their own benefits. Our new approach has the following advantages.

1. Our methods directly compute the Euler characteristic of an affine variety without involving any compactification. This is useful because the closure of a smooth affine variety can have singularities along infinity.
2. We stratify and compute the Euler characteristics of smooth affine varieties. In theory, any d-dimensional affine variety can be stratified into the union of at most $d + 1$ smooth affine varieties. In contrast to the inclusion-exclusion principle, our method does not involve too many varieties.
3. We can tailor the stratification to reduce the degree of each stratum.

A standard method to compute Euler characteristics of complex algebraic varieties is to reduce to the projective hypersurface case. The drawback of this method is that to compute the Euler characteristic of a projective variety, the number of involved hypersurfaces grows exponentially in the codimension.

In contrast, we compute the Euler characteristic of a smooth equidimensional affine variety X by counting the critical points of $\dim(X) + 1$ algebraic functions.

BW is supported by NSF Grant DMS-1701305 and the Alfred P. Sloan foundation.

A. M. Bigatti et al. (Eds.): ICMS 2020, LNCS 12097, pp. 51–60, 2020.
https://doi.org/10.1007/978-3-030-52200-1_5

Given a singular complex affine variety, we stratify it into smooth affine varieties to reduce to the smooth case. In theory, we can always stratify a d-dimensional affine variety into $d+1$ smooth (possibly not connected) equidimensional affine varieties of dimension $d, d-1, \ldots, 1, 0$. So we need to compute the number of critical points of at most $(d+1)(d+2)/2$ algebraic functions. Our algorithms also have the practical feature of minimizing the degree of the algebraic functions at the expense of increasing the number of functions to consider.

This work is organized follows. In Sect. 2, we recall a theorem to determine the Euler characteristic of a smooth equidimensional variety with a general hyperplane removed by counting critical points of a function. In Sect. 3, we provide some key definitions from numerical algebraic geometry. In Sects. 4–5 we present algorithms for computing Euler characteristics.

2 Euler Characteristics and Critical Points

Let X be a topological space that is homotopy equivalent to a finite CW-complex. The Euler characteristic of X, denoted by $\chi(X)$, is the alternating sum of the Betti numbers of X [9, Page 146]. We are only interested in the situation where X is a (complex) affine algebraic variety [8, Corollary 6.10]. In this case, the Euler characteristic of X is an alternating sum of cardinalities of several sets of critical points as shown in Theorem 1.

Given a singular affine algebraic variety, it admits a stratification into locally closed smooth subvarieties. We can always refine the stratification into locally closed smooth affine subvarieties. Since the Euler characteristic is additive for a stratification of locally closed subvarieties, it is enough to compute the Euler characteristic of smooth affine varieties. In this paper, we discuss the following two objectives.

1. Compute the Euler characteristics of smooth affine varieties.
2. Find algorithms to stratify singular affine varieties into locally closed smooth subvarieties.

Let X be a smooth subvariety of $\mathbb{C}^n$. Let $f : \mathbb{C}^n \to \mathbb{C}$ be a regular function. If X is defined by polynomials $g_1, \ldots, g_l \in \mathbb{C}[x_1, \ldots, x_n]$, then the critical points of $f|_X$ for a polynomial $f \in \mathbb{C}[x_1, \ldots, x_n]$ are the points $P \in X$ such that the vector $(\frac{df}{dx_1}, \ldots, \frac{df}{dx_n})|_P$ is in the linear span of $\{(\frac{dg_i}{dx_1}, \ldots, \frac{dg_i}{dx_n})|_P : i = 1, \ldots, l\}$.

This theorem relates the number of critical points to the Euler characteristic.

Theorem 1 ([20]). *Let ℓ denote a general affine linear function $\ell : \mathbb{C}^n \to \mathbb{C}$ and let X denote a smooth equidimensional affine subvariety of $\mathbb{C}^n$. Then*

$$(-1)^{\dim(X)}\chi(X \setminus V(\ell)) = \#\{\text{critical points of } \ell|_X\}.$$

As a corollary we are able to determine the Euler characteristic of X itself.

Corollary 1. *Let X be a smooth subvariety of $\mathbb{C}^n$. For $i = 1, \ldots, \dim(X)$, let h_i denote a general affine linear function $\mathbb{C}^n \to \mathbb{C}$. Then we have the equality*

$$\chi(X) = p + \sum_{i=1}^{\dim(X)} (-1)^{\dim(X)-i+1}\eta_i$$

where η_i is the number of critical points of $h_i|_{X \cap V(h_1, \ldots, h_{i-1})}$ and p is the cardinality of $X \cap V(h_1, \ldots, h_{\dim(X)})$.

Proof. The additive property of Euler characteristic implies the equality

$$\chi(X) = \sum_{i=1}^{\dim(X)} \chi\big(X \cap V(h_1, \ldots h_{i-1}) \setminus V(h_i)\big) + \chi\big(X \cap V(h_1, \ldots, h_{\dim(X)})\big).$$

It follows from Bertini's theorem that each intersection $X \cap V(h_1, \ldots h_{i-1})$ is smooth and $X \cap V(h_1, \ldots, h_{\dim(X)})$ is a set of p points. By Theorem 1, we have $\eta_i = (-1)^{\dim(X)-i+1}\chi(X \cap V(h_1, \ldots h_{i-1})\setminus V(h_i))$. $\qquad\square$

3   Numerical Algebraic Geometry Basics

In this section we recall a witness set [4,23], which is a fundamental concept in numerical algebraic geometry. A witness set is used to analyze algebraic varieties and is manipulated using homotopy continuation [2], as seen in Sects. 3.2–3.3.

3.1   Witness Sets and Numerical Irreducible Decomposition

Let X be an equidimensional subvariety of affine space $\mathbb{C}^n$. As a consequence of Bertini's Theorem, there are two invariants, dimension and degree, of X that can be understood by intersecting X with a general linear space. The dimension $\dim(X)$ of a subvariety X of $\mathbb{C}^n$ is the codimension of a general affine linear space $\mathcal{L} \subseteq \mathbb{C}^n$ such that $X \cap \mathcal{L}$ is finite and nonempty. The degree $\deg(X)$ of X is the number of points in $X \cap \mathcal{L}$.

Definition 1 (Witness set). *Suppose X is an equidimensional subvariety of $\mathbb{C}^n$. A witness set for X is a triple (F, L, W), where F is a finite set of polynomials with each irreducible component of X being an irreducible component of $V(F)$, L is a set of $\dim(X)$ general[1] affine linear functions, and W is the set of points $X \cap V(L)$.*

In numerical algebraic geometry, W is called a *witness point set* for X. Since L consists of general affine linear functions, the affine linear space $V(L)$ is general and the cardinality of the set W is $\deg(X)$. Throughout, we assume the ideal generated by F in Definition 1 defines a reduced scheme by using *deflation* [13,18].

Given a (not necessarily equidimensional) subvariety X of $\mathbb{C}^n$, we denote by X_i the union of i-dimensional irreducible components of X. We call a set of witness sets of X_i for $i = 0, 1, \ldots, k$ a *numerical equidimensional decomposition of X*. This decomposition can be refined to a *numerical irreducible decomposition of X* by providing a witness set for each irreducible component of X (see [4]).

---

[1] By "general" here, we mean the intersection $X \cap V(L)$ is transverse and has cardinality $\deg(X)$.

Example 1 (Embedding). Suppose X is an equidimensional subvariety of $\mathbb{C}^n$ and h is a linear function $\mathbb{C}^n \to \mathbb{C}$. Let $\widehat{X}$ be the image of X under the closed embedding $\mathbb{C}^n \to \mathbb{C}^{n+1}$ given by $x \mapsto (x, h(x))$. Given a witness set (F, L, W) for X, we construct a witness set for $\widehat{X}$ as $(F \cup \{h - x_{n+1}\},\, L,\, \{(x, h(x)) : x \in W)\})$.

Example 2. We can also easily construct a witness set for the Cartesian product of two varieties. Suppose X_i is a subvariety of $\mathbb{C}^{n_i}$ for $i = 1, 2$. If (F_i, L_i, W_i) is a witness set for X_i, then $(F_1 \cup F_2, L_1 \cup L_2, W_1 \times W_2)$ is a witness set for $X_1 \times X_2 \subset \mathbb{C}^{n_1} \times \mathbb{C}^{n_2}$.

3.2   Witness Collections of Subvarieties in $\mathbb{C}^n \times \mathbb{C}^n$

A witness collection is a generalization of a witness set and is used to study varieties that are defined by polynomials with a natural multi-variable group structure. For a complete description of witness collections see [10,11,17]. For our purposes, it suffices to study the following special case.

A *witness collection* for a d-dimensional irreducible subvariety Z of $\mathbb{C}^n \times \mathbb{C}^n$ is the following collection of triples, which we call *multi-affine witness sets*:

$$\left(F, L^i \cup M^{d-i}, Z \cap (\mathcal{L}^i \times \mathcal{M}^{d-i})\right) \text{ for } i = 0, 1, \ldots, d,$$

where F is a set of polynomials such that $V(F)$ contains Z as an irreducible component; $\mathcal{L}^i$ and $\mathcal{M}^i$ are general codimension i affine linear spaces in $\mathbb{C}^n$ defined the sets of general linear functions L^i and M^i respectively.

Witness collections are used to understand the intersection of Z with a Cartesian product of linear spaces. Let $\mathcal{A}^i$ and $\mathcal{B}^i$ denote general codimension i affine linear spaces in $\mathbb{C}^n$. With homotopy continuation, we determine the isolated points in the intersection $Z \cap (\mathcal{A}^i \times \mathcal{B}^{d-i})$. These points are contained in the set of endpoints of the homotopy $H^i : \mathbb{C}^n \times \mathbb{C}^n \times \mathbb{C} \to \mathbb{C}^{N+d}$ with

$$(x, y, t) \mapsto \left(F(x,y), tL^i(x) + (1-t)A^i(x), tM^{d-i}(y) + (1-t)B^{d-i}(y)\right) \quad (1)$$

where $L^i, A^i, M^{d-i}, B^{d-i}$ are sets of affine linear functions defining $\mathcal{L}^i, \mathcal{A}^i, \mathcal{M}^{d-i}$ and $\mathcal{B}^{d-i}$ respectively. For more details see [10, Remark 1.3]. To conveniently denote linear functions, for $a, x \in \mathbb{C}^n$, we take $a \circ x$ to be the usual inner product.

Example 3 (Conormal variety). Let $X = V(f_1, \ldots, f_k)$ be a smooth equidimensional variety in $\mathbb{C}^n$ with $(f_1, \ldots, f_k)$ generating a radical ideal. The *(affine) conormal variety* of X is a subvariety $\mathcal{C}(X)$ in $\mathbb{C}^n \times \mathbb{C}^n$ with an ideal

$$\langle f_1, \ldots, f_k \rangle + \langle (1 + \mathrm{codim}(X))\text{-minors of } \mathrm{Jac}_x(x \circ y, f_1, \ldots, f_k) \rangle \subset \mathbb{C}[x, y],$$

where $\mathrm{Jac}_x(x \circ y, f_1, \ldots, f_k)$ is a $(k+1) \times n$ matrix of partial derivatives with respect to x. The dimension of $\mathcal{C}(X)$ is n. For a projective formulation, see [22].

A witness collection for the conormal variety of X is given by

$$\left(F, L^i \cup M^{n-i}, W_i\right) \text{ for } i = 0, \ldots, n, \quad (2)$$

where $V(F)$ contains $\mathcal{C}(X)$ as an irreducible component and

$$W_i := \mathcal{C}(X) \cap (\mathcal{L}^i \times \mathcal{M}^{n-i}).$$

Each multi-affine witness set $(F, L^i \cup M^{n-i}, W_i)$ has information about the variety X. For example, the dimension of X is the maximal i such that $W_i \neq \emptyset$ and the degree of X is the cardinality of $W_{\dim(X)}$. Moreover, for $i = 0$, the linear space $V(M^n) \subset \mathbb{C}^n$ contains a unique point, say c. The set of points (x, y) in W_0 such that $x \circ c \neq 0$ is the set of critical points of the general linear function $x \circ c$ on $X \setminus V(x \circ c)$. The cardinality of the set $W_0 \setminus V(x \circ c)$ is the Euler characteristic of $X \setminus V(x \circ c)$ up to a sign. In general, the number of points in W_i is an upperbound for the η_i appearing in Corollary 1.

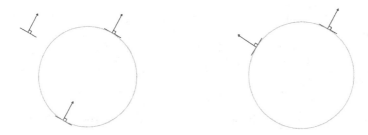

Fig. 1. We illustrate W_0 (left) and W_1 (right) for the circle $X = V(x_1^2 + x_2^2 - 1)$ in $\mathbb{C}^2$.

3.3 Regeneration and Removing a Hypersurface

Given a witness collection for an irreducible variety X and a polynomial g, regeneration determines a witness set for $X \cap V(g)$. One of two situations can occur. First, if X is contained in $V(g)$, then $X = X \cap V(g)$ and we are done. Second, if X is not contained in $V(g)$, then for $i = 1, \ldots, \deg(g)$ compute a witness set $(F \cup \{\ell_i\}, L, W_i)$ for $X \cap V(\ell_i)$ where $\ell_i : \mathbb{C}^n \to \mathbb{C}$ is a general affine linear function. This is easy to do using standard homotopy continuation methods when given the witness set $(F, L \cup \{\ell\}, W)$ for X. This produces a witness set $\left(F \cup \{\ell_1 \cdots \ell_{\deg(g)}\}, L, \cup_{i=1}^{\deg(g)} W_i\right)$ for $X \cap V(\ell_1 \cdots \ell_{\deg(g)})$. Finally, the homotopy $H(x, t) = (F(x), t\ell_1(x) \cdots \ell_{\deg(g)}(x) + (1 - t)g(x), L(x))$ provides a witness set for $X \cap V(g)$ when $t = 0$. This procedure for computing a witness set for $X \cap V(g)$ from a witness set for X is called *regeneration* [11,12,14].

Example 4. For $X \subset \mathbb{C}^n$, consider the embedding $\widehat{X} \subset \mathbb{C}^{n+1}$ as in Example 1. Fix a polynomial $g \in \mathbb{C}[x_1, \ldots, x_n]$. The affine variety $\widehat{X} \cap V(gx_{n+1} - 1)$ is isomorphic to $X \setminus V(g)$. A useful application of regeneration computes a witness set for $\widehat{X} \cap V(gx_{n+1} - 1)$. From a witness set for $\widehat{X} \subset \mathbb{C}^{n+1}$ and a polynomial $gx_{n+1} - 1$, regeneration produces a witness set for $\widehat{X} \cap V(gx_{n+1} - 1)$.

4   Euler Characteristics of Smooth Varieties

Theorem 1 leads to an algorithm that outputs the the Euler characteristic of an algebraic variety by computing critical points. The proof of correctness of the following algorithm is easily derived from Corollary 1.

Algorithm 1. Smooth X

1: **Input**
2:      X A smooth equidimensional affine variety in $\mathbb{C}^n$
3: **Output**
4:      $\chi(X)$ Euler characteristic of the variety X
5: **procedure**
6:      **for** $i = 1, \ldots, \dim X$ **do**
7:          $h_i \leftarrow$ general affine linear functions $\mathbb{C}^n \to \mathbb{C}$
8:          $\eta_i \leftarrow$ the number of critical points of h_i on $X \cap V(h_1, \ldots, h_{i-1}) \setminus V(h_i)$
9:      **end for**
10:      $p \leftarrow$ number of points in $X \cap V(h_1, \ldots, h_{\dim(X)})$
11:      $\chi(X) \leftarrow \sum_{i=1}^{\dim(X)} (-1)^{\dim(X)-i+1} \eta_i + p$
12:      **return** $\chi(X)$

We can compute these critical points from a witness set as follows.

Algorithm 2. Numerical Smooth X

1: **Input**
2:      W_X A witness set for a smooth equidimensional affine variety $X \subset \mathbb{C}^n$
3: **Output**
4:      $\chi(X)$ Euler characteristic of the variety X
5: **procedure**
6:      $W_{X \times \mathbb{C}^n} \leftarrow$ Witness set for $X \times \mathbb{C}^n$, computed by using W_X and methods described in example 2
7:      $W_{\mathcal{C}(X)} \leftarrow$ Witness collection for the conormal variety $\mathcal{C}(X) \subset \mathbb{C}^n \times \mathbb{C}^n$, by regeneration $\triangleright$ Recall a witness collection consists of multi-affine witness sets $(F, L^i \cup M^{n-i}, W_i)$ as described in equation (2).
8:      **for** $i = 1, \ldots, \dim X$ **do**
9:          $p_1, \ldots, p_i \leftarrow$ general points in $\mathbb{C}^n$
10:          $q_1, \ldots, q_{n-i} \leftarrow$ basis for the perpendicular complement of the linear span of $p_1, \ldots, p_i$
11:          $a, b \leftarrow$ general points in $\mathbb{C}^n$
12:          $A^i \leftarrow \{p_1 \circ (x - a), \ldots, p_i \circ (x - a)\}$ $\triangleright$ set of i affine linear functions
13:          $B^{n-i} \leftarrow \{q_1 \circ (y - b), \ldots, q_{n-i} \circ (y - b)\}$ $\triangleright$ set of $n - i$ affine linear functions
14:          $S_i \leftarrow$ the endpoints of the homotopy (1) with L^i, M^{n-i} and A^i, B^{n-i} as above and start points W_i.
15:          $\zeta_i \leftarrow$ the number of nonsingular isolated points (x, y) in S_i such that $x \circ b \neq 0$
16:      **end for**
17:      $\chi(X) \leftarrow \sum_{i=0}^{\dim(X)} (-1)^{\dim(X)-i} \zeta_i$
18:      **return** $\chi(X)$

Proof. (**Correctness sketch**). Example 3 explains how $\zeta_{\dim(X)} = p$ and $\zeta_0 = \eta_{\dim(X)}$. With some substitutions and algebra one shows $\zeta_i = \eta_{\dim(X)-i}$ for $i = 1, \ldots, \dim(X)$, and then the result follows. $\square$

Example 5. Consider the smooth curve $X = V(x_1^2 + x_2^2 - 1) \subset \mathbb{C}^2$ and its conormal variety $\mathcal{C}(X) = V(x_1^2 + x_2^2 - 1, x_2 y_1 - x_1 y_2) \subset \mathbb{C}^2 \times \mathbb{C}^2$. The algorithms find $(p, \eta_1, \eta_2) = (\zeta_2, \zeta_1, \zeta_0) = (2, 2, 0)$ and both output $\chi(X) = 0$. The points corresponding to p and η are illustrated in Fig. 1 by plotting points on X with the respective normal vectors. The homotopy used to determine ζ_i is given by $H^i : \mathbb{C}^2 \times \mathbb{C}^2 \times \mathbb{C} \to \mathbb{C}^4$. Concretely, for H^1, with general affine linear functions $\ell, m : \mathbb{C}^2 \to \mathbb{C}$, we have $H^1(x, y, t)$ is

$$\left(x_1^2 + x_2^2 - 1, x_2 y_1 - x_1 y_2, (1-t)\ell(x) + t p_1 \circ (x - a), (1 - t)m(y) + t q_1 \circ (y - b)\right).$$

5   Euler Characteristics of Singular Varieties

In this section, an excision-restriction method to compute $\chi(X)$ is presented.

Algorithm 3. Excision-restriction method

1: **Input**
2:   X An affine variety in $\mathbb{C}^n$
3: **Output**
4:   $\chi(X)$ Euler characteristic of the variety X
5: **procedure**
6:    **for** $i = 1, \ldots, \dim X$ **do**
7:        $X_i \leftarrow i$-dimensional irreducible components of X
8:    **end for**
9:    $S \leftarrow |X_0|$, the cardinality of X_0
10:   $Y \leftarrow \bigcup_{i=1}^{\dim(X)} X_i$
11:   **if** Y is smooth **then**
12:       **for** $i = 1, \ldots, \dim X$ **do**
13:           **if** $X_i \neq \emptyset$ **then**
14:               $S \leftarrow S + \chi(X_i)$ (Algorithm 1)
15:           **end if**
16:       **end for**
17:   **else**
18:       $C \leftarrow$ a union of irreducible components of $\mathrm{Sing}(Y) \cup \bigcup_{i=1}^{\dim(X)-1} X_i$
19:       Choose a polynomial g such that $X_{\dim(X)} \subsetneq V(g)$ and $V(g)$ contains C.
20:       $S_1 \leftarrow \chi(Y \cap V(g))$ $\triangleright$ Algorithm 3, recursion
21:       $S_2 \leftarrow \chi(Y \backslash V(g))$ $\triangleright$ Algorithm 3, recursion $\triangleright$ Here, $Y \backslash V(g)$ is considered as
      a subvariety of $\mathbb{C}^{n+1}$ with coordinate functions $x_1, \ldots, x_n, 1/g$ as in Example 4.
22:   **end if**
23:   $\chi(X) \leftarrow S + S_1 + S_2$
24:   **return** $\chi(X)$

The freedom of choosing g in step 19 is a feature of this algorithm. One way to find such a g is by taking a generic linear combination of minimal generators of the ideal of $\mathrm{Sing}(Y) \cup \bigcup_{i=1}^{\dim(X)-1} X_i$. An alternative numerical approach is to sample a set of points from an irreducible component of $\mathrm{Sing}(Y) \cup \bigcup_{i=1}^{\dim(X)-1} X_i$. to use numerical implicitization, which has been implemented in [7]. To minimize the number of recursions, choose a g vanishing on $\mathrm{Sing}(Y) \cup \bigcup_{i=1}^{\dim(X)-1} X_i$. Alternatively, we could choose g so that the degree is small as possible. This heuristic works well for a numerical approach.

Next, we tailor the previous algorithm for regeneration, which was described in Sect. 3.3.

Algorithm 4. Numerical Excision-Restriction

1: **Input**
2:    $\cup_i (F_i, L_i, W_i)$A numerical equidimensional decomposition for an affine variety $X \subset \mathbb{C}^n$ where (F_i, L_i, W_i) is a witness set for X_i, the union of i-dimensional irreducible components of X

3: **Output**
4:    $\chi(X)$ Euler characteristic of the variety X

5: **procedure**
6:    $S \leftarrow |W_0|$
7:    $Y \leftarrow \bigcup_{i=1}^{\dim(X)} X_i$
8:    **if** Y is smooth **then**
9:        **for** $i = 1, \ldots, \dim X$ **do**
10:           $S \leftarrow S + \chi(X_i)$ ▷ Input (F_i, L_i, W_i) to Algorithm 2.
11:       **end for**
12:   **else**
13:       $C \leftarrow$ a union of irreducible components of $\mathrm{Sing}(Y) \cup \bigcup_{i=1}^{\dim(X)-1} X_i$
14:       $g \leftarrow$ a polynomial such that $X_{\dim(X)} \subsetneq V(g)$ and $V(g)$ contains C
15:       $\mathcal{D}_1 \leftarrow$ a numerical decomposition for $Y \cap V(g)$, using methods in Section 3.3
16:       $S_1 \leftarrow \chi(Y \cap V(g))$ ▷ Input $\mathcal{D}_1$ to Algorithm 4, recursion
17:       $\mathcal{D}_2 \leftarrow$ a numerical decomposition for $Y \backslash V(g)$, using methods in Example 4
18:       $S_2 \leftarrow \chi(Y \backslash V(g))$ ▷ Input $\mathcal{D}_2$ to Algorithm 4, recursion
19:   **end if**
20:   $\chi(X) \leftarrow S + S_1 + S_2$
21:   **return** $\chi(X)$

Example 6. Consider the Whitney umbrella $X = V(x^2 y - z^2) \subset \mathbb{C}^3$. The variety X is two dimensional with a singular locus given by the y-axis. The output of the algorithm to compute $\chi(X)$ is summarized by the following equations,

$$\chi(X) = \chi(X \cap V(g)) + \chi(X \setminus V(g))$$
$$= \chi(X \cap V(g) \cap V(g')) + \chi(X \cap V(g) \setminus V(g')) + \chi(X \setminus V(g))$$
$$= 1 + 0 + 0,$$

where we have made the following choices: $g = 2x + 5z$ and $g' = y - x - 4/25$ such that $V(g) \supset \mathrm{Sing}(X)$ and $V(g') \supset \mathrm{Sing}(X \cap V(g))$.

We input a numerical equidimensional decomposition for X. Since X is itself irreducible, we have $Y = X$ and $S = |W_0| = 0$. Using regeneration we compute a numerical decomposition $\mathcal{D}_1$ for $X \cap V(g)$, which is a union of two lines that intersect at the point $(0, 4/25, 0)$.

In step 16, we apply Algorithm 4 to $\mathcal{D}_1$, and in this recursive step we let g' play the role of g. The variety $X \cap V(g) \cap V(g')$ is the point $(0, 4/25, 0)$ so we have $\chi(X \cap V(g) \cap V(g')) = 1$. On the other hand, $X \cap V(g) \setminus V(g')$ is smooth, so by using Algorithm 2 we get $\chi(X \cap V(g) \setminus V(g')) = 0$.

Lastly, in step 17, we find a numerical decomposition $\mathcal{D}_2$ for $X \setminus V(g)$ to do step 18. Since $X \setminus V(g)$ is smooth, we use Algorithm 2 to find $\chi(X \setminus V(g)) = 0$.

In this paper we focussed on algorithm development, and we are working towards implementing Algorithm 4. In addition, we are also working to implement Algorithm 3 using Grobner basis to compare with the numerical version and previous techniques.

Acknowledgements. We thank Martin Helmer for very helpful correspondences. We also thank the referees for their helpful comments and suggestions.

References

1. Adamer, M.F., Helmer, M.: Complexity of model testing for dynamical systems with toric steady states. Adv. Appl. Math. **110**, 42–75 (2019)
2. Allgower, E.L., Georg, K.: Introduction to Numerical Continuation Methods. Classics in Applied Mathematics, vol. 25. Society for Industrial and Applied Mathematics (SIAM), Philadelphia (2003)
3. Aluffi, P.: The Chern-Schwartz-MacPherson class of an embeddable scheme. In: Forum Mathematics Sigma, vol. 7, pp. e30 (2019)
4. Bates, D.J., Hauenstein, J.D., Sommese, A.J., Wampler, C.W.: Numerically Solving Polynomial Systems with Bertini. Software, Environments, and Tools, vol. 25. Society for Industrial and Applied Mathematics (SIAM), Philadelphia (2013)
5. Catanese, F., Hoşten, S., Khetan, A., Sturmfels, B.: The maximum likelihood degree. Am. J. Math. **128**(3), 671–697 (2006)
6. Chan, C.-Y.J.: A correspondence between Hilbert polynomials and Chern polynomials over projective spaces. Illinois J. Math. **48**(2), 451–462 (2004)
7. Chen, J., Kileel, J.: Numerical implicitization. J. Softw. Algebra Geom. **9**(1), 55–63 (2019)
8. Dimca, A.: Singularities and Topology of Hypersurfaces. Universitext. Springer, New York (1992). https://doi.org/10.1007/978-1-4612-4404-2
9. Hatcher, A.: Algebraic Topology. Cambridge University Press, Cambridge (2002)
10. Hauenstein, J., Leykin, A., Rodriguez, J.I., Sottile, F.: A numerical toolkit for multiprojective varieties (2019)
11. Hauenstein, J., Rodriguez, J.I.: Multiprojective witness sets and a trace test (2019)
12. Hauenstein, J.D., Sommese, A.J., Wampler, C.W.: Regenerative cascade homotopies for solving polynomial systems. Appl. Math. Comput. **218**(4), 1240–1246 (2011)
13. Hauenstein, J.D., Wampler, C.W.: Isosingular sets and deflation. Found. Comput. Math. **13**(3), 371–403 (2013)

14. Hauenstein, J.D., Wampler, C.W.: Unification and extension of intersection algorithms in numerical algebraic geometry. Appl. Math. Comput. **293**, 226–243 (2017)
15. Huh, J.: The maximum likelihood degree of a very affine variety. Compos. Math. **149**(8), 1245–1266 (2013)
16. Jost, C.: Computing characteristic classes and the topological Euler characteristic of complex projective schemes. J. Softw. Algebra Geom. **7**, 31–39 (2015)
17. Leykin, A., Rodriguez, J.I., Sottile, F.: Trace test. Arnold Math. J. **4**(1), 113–125 (2018)
18. Leykin, A., Verschelde, J., Zhao, A.: Higher-order deflation for polynomial systems with isolated singular solutions. In: Dickenstein, A., Schreyer, F.O., Sommese, A.J. (eds.) Algorithms in Algebraic Geometry. The IMA Volumes in Mathematics and its Applications, vol. 146, pp. 79–97. Springer, New York (2008). https://doi.org/10.1007/978-0-387-75155-9_5
19. Marco-Buzunáriz, M.A.: A polynomial generalization of the Euler characteristic for algebraic sets. J. Singul. **4**, 114–130 (2012). With an appendix by J. V. Rennemo
20. Maxim, L.G., Rodriguez, J.I., Wang, B.: Euclidean distance degree of the multiview variety. SIAM J. Appl. Algebra Geom. **4**(1), 28–48 (2020)
21. Rodriguez, J.I., Wang, B.: The maximum likelihood degree of mixtures of independence models. SIAM J. Appl. Algebra Geom. **1**(1), 484–506 (2017)
22. Rostalski, P., Sturmfels, B.: Dualities. Semidefinite Optimization and Convex Algebraic Geometry. MOS-SIAM Series on Optimization, vol. 13, pp. 203–249. SIAM, Philadelphia (2013)
23. Sommese, A.J., Wampler, C.W.I.: The Numerical Solution of Systems of Polynomials Arising in Engineering and Science. World Scientific Publishing Co. Pte. Ltd., Hackensack (2005)

Evaluating and Differentiating a Polynomial Using a Pseudo-witness Set

Jonathan D. Hauenstein and Margaret H. Regan$^{(\boxtimes)}$

Department of Applied and Computational Mathematics and Statistics,
University of Notre Dame, Notre Dame, USA
{hauenstein,mregan9}@nd.edu
http://www.nd.edu/~jhauenst, http://www.nd.edu/~mregan9

Abstract. Polynomials which arise via elimination can be difficult to compute explicitly. By using a pseudo-witness set, we develop an algorithm to explicitly compute the restriction of a polynomial to a given line. The resulting polynomial can then be used to evaluate the original polynomial and directional derivatives along the line at any point on the given line. Several examples are used to demonstrate this new algorithm including examples of computing the critical points of the discriminant locus for parameterized polynomial systems.

Keywords: Numerical algebraic geometry · Pseudo-witness set · Implicit polynomial · Directional derivatives · Critical points

1 Introduction

Parameterized polynomial systems arise in various applications in science and engineering, such as in computer vision [15,17,22], kinematics [14,23], and chemistry [1,19]. Often in these applications, real solutions are desired. The complement of the discriminant locus associated with the parameterized polynomial system consists of cells where the number of real solutions is constant. Elimination methods (e.g., see [8, Chap. 3]) theoretically provide an approach to explicitly compute a defining equation for the discriminant locus. If the discriminant locus is a curve or surface, there are several numerical methods that can be used to plot it, e.g., [6,7,18]. When the explicit expression is difficult to compute, this paper develops a numerical algebraic geometric approach based on pseudo-witness sets [13] for both evaluating implicitly defined polynomials and directional derivatives. In particular, the approach yields an explicit univariate polynomial equal to the defining equation restricted to a line which can then be evaluated or differentiated as needed. When the parameterized system and line have rational coefficients, the resulting univariate polynomial also has rational coefficients which can be computed exactly from the numerical data [2].

One application of this new approach is to compute the critical points of the discriminant polynomial which are outside of the discriminant locus without explicitly computing the discriminant. This set of critical points contains at least

© Springer Nature Switzerland AG 2020
A. M. Bigatti et al. (Eds.): ICMS 2020, LNCS 12097, pp. 61–69, 2020.
https://doi.org/10.1007/978-3-030-52200-1_6

one point in each compact cell in the complement of the discriminant locus [10] which can be useful for determining the possible number of real solutions as well as the real monodromy structure [11].

The remainder of the paper is as follows. Section 2 describes the approach based on using pseudo-witness sets. Section 3 presents an algorithm for performing the computations with some illustrative examples. Section 4 provides two examples of computing critical points.

2   Implicit Representation of a Polynomial

In numerical algebraic geometry, e.g., see [4,21], a witness point set for a hypersurface $\mathcal{H} \subset \mathbb{C}^n$ consists of the intersection points of $\mathcal{H}$ with a line $\mathcal{L} \subset \mathbb{C}^n$. Suppose that $f(x)$ is a given polynomial and $\mathcal{H}$ is the hypersurface defined by the vanishing of f. Then, the witness point set for $\mathcal{H}$ corresponds with the roots of the univariate polynomial obtained by restricting f to the line $\mathcal{L}$. Since every univariate polynomial is defined up to scale by its roots, one can recover $f|_{\mathcal{L}}$ by computing its roots along with knowing $f|_{\mathcal{L}}(T)$ for some value T which is not a root of $f|_{\mathcal{L}}$. The following is an illustration of this basic setup.

Example 1. Consider the polynomial $f(x, y) = y - x^2$ with corresponding hypersurface $\mathcal{H} \subset \mathbb{C}^2$ and the line $\mathcal{L} \subset \mathbb{C}^2$ defined parametrically by:

$$x(t) = t \qquad y(t) = 2t + 1.$$

Therefore, one can explicitly compute

$$f|_{\mathcal{L}}(t) = f(x(t), y(t)) = -t^2 + 2t + 1 = -\left(t - 1 + \sqrt{2}\right)\left(t - 1 - \sqrt{2}\right). \tag{1}$$

For $t_1 = 1 - \sqrt{2}$ and $t_2 = 1 + \sqrt{2}$, one has

$$\mathcal{H} \cap \mathcal{L} = \{(t_1, 2t_1 + 1), (t_2, 2t_2 + 1)\}. \tag{2}$$

Hence, $f|_{\mathcal{L}}(t) = s(t - t_1)(t - t_2)$ for some constant s which can be computed from, say, requiring $f|_{\mathcal{L}}(T) = 1$ where $T = 2$, i.e., $s = -1$. Therefore, one has recovered $f|_{\mathcal{L}}(t)$ in (1) from $\mathcal{H} \cap \mathcal{L}$ with $f|_{\mathcal{L}}(T) = 1$ as illustrated in Fig. 1(a).

The remainder of this section extends this idea using pseudo-witness sets when f is a polynomial over $\mathbb{C}$ that is not known explicitly, but the corresponding hypersurface $\mathcal{H}$ arises as the closure of a projection of an algebraic set. For simplicity of presentation, assume that $F : \mathbb{C}^N \to \mathbb{C}^r$ is a polynomial system and that V is a pure d-dimensional subset of $\mathcal{V}(F) = \{x \in \mathbb{C}^N \mid F(x) = 0\}$. Let $\pi(x_1, \ldots, x_N) = (x_1, \ldots, x_n)$ such that $\mathcal{H} = \overline{\pi(V)} \subset \mathbb{C}^n$. Note that one has $n - 1 \leq d \leq N - 1$. A pseudo-witness set [13] for $\mathcal{H}$, say $\{F, \pi, \mathcal{M}, W\}$, is a numerical algebraic geometric data structure that permits computations on $\mathcal{H}$ without knowing the defining polynomial f for $\mathcal{H}$. The last two items are a linear space $\mathcal{M} \subset \mathbb{C}^N$ and a finite set $W = V \cap \mathcal{M}$. In particular, $\mathcal{M} = \mathcal{L} \times \mathcal{L}'$ where

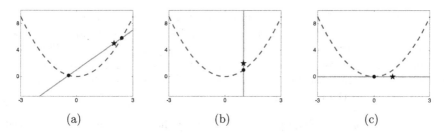

Fig. 1. A visual representation of the pseudo-witness set for $\mathcal{H}$ defined by $y - x^2$ with a linear slice, $\mathcal{L}$, that is (a) generic, (b) special with one root of multiplicity one, and (c) tangent. The black dots represent the roots $t_1, \ldots, t_k$ and the black stars represent T selected for scale.

$\mathcal{L}' \subset \mathbb{C}^{N-n}$ is a codimension $d - (n - 1)$ general linear space so that $\mathcal{M}$ has codimension d. Hence, $\pi(W) = \mathcal{H} \cap \mathcal{L}$ is a witness point set for $\mathcal{H}$ with respect to $\mathcal{L}$. With this setup, the local multiplicity of each point in $\mathcal{H} \cap \mathcal{L}$ can be easily computed via [5, Prop. 6] (see also [9, pg. 158]). Thus, parameterizing $\mathcal{L}$ by t and denoting $t_1, \ldots, t_k$ as the corresponding points in $\mathcal{H} \cap \mathcal{L}$ with multiplicity $m_1, \ldots, m_k$, respectively, yields

$$f|_{\mathcal{L}}(t) = \frac{f|_{\mathcal{L}}(T)}{(T - t_1)^{m_1} \cdots (T - t_k)^{m_k}} \cdot (t - t_1)^{m_1} \cdots (t - t_k)^{m_k} \tag{3}$$

as shown in the following.

Theorem 1. *The univariate polynomial describing f along the line $\mathcal{L}$ is correctly described by* (3).

Proof. The assumption on T is that $f|_{\mathcal{L}}(T) \neq 0$, i.e., $\mathcal{L} \not\subset \mathcal{H}$. Hence, $f|_{\mathcal{L}}$ is a nonzero polynomial which has finitely many roots, namely $t_1, \ldots, t_k$ with multiplicity $m_1, \ldots, m_k$, respectively. Thus, $m_i \geq 1$ with $\deg(f|_{\mathcal{L}}) = m_1 + \cdots + m_k$. Since the roots define the univariate polynomial up to scale, the leading coefficient is used to achieve the desired value at T and thus everywhere along $\mathcal{L}$.

The following illustrates a pseudo-witness set and Theorem 1.

Example 2. Consider the hypersurface $\mathcal{H} \subset \mathbb{C}^2$ from Example 1 under the assumption that we are given $\mathcal{H} = \overline{\pi(V)}$ where $\pi(x, y, s) = (x, y)$ and $V = \mathcal{V}(F)$ with

$$F(x, y, s) = \begin{bmatrix} x - s^2 \\ y - s^4 \end{bmatrix}.$$

Since $n = 2$ and $d = \dim V = 1$, we have $\mathcal{M} = \mathcal{L} \times \mathbb{C}$ with

$$W = V \cap \mathcal{M} = \{(t_1, 2t_1 + 1, \pm\sqrt{t_1}), (t_2, 2t_2 + 1, \pm\sqrt{t_2})\}$$

where t_1 and t_2 are as in Example 1 with $m_1 = m_2 = 1$. Hence, $\pi(W) = \mathcal{H} \cap \mathcal{L}$ as in (2). Therefore, with $T = 2$ and $f|_{\mathcal{L}}(T) = 1$, (3) simplifies to $f|_{\mathcal{L}}(t)$ in (1).

The only assumption on the line $\mathcal{L}$ is that $\mathcal{L} \not\subset \mathcal{H}$ so that one can find T such that $f|_{\mathcal{L}}(T) \neq 0$. Of course, one can check if $\mathcal{L} \subset \mathcal{H}$ by a pseudo-witness set membership test [12] in which case one would simply have $f_{\mathcal{L}}(t) \equiv 0$. Thus, $\mathcal{L}$ is not necessarily assumed to intersect $\mathcal{H}$ transversely, so the number of roots and multiplicities can vary for different choices of $\mathcal{L}$. Nonetheless, Theorem 1 applies as is illustrated in the following two examples.

Example 3. Reconsider Example 2 with $\mathcal{L}$ being the vertical line parametrized by

$$x(t) = 1 \qquad y(t) = t$$

as shown in Fig. 1(b). One has $\mathcal{M} = \mathcal{L} \times \mathbb{C}$ and $W = V \cap \mathcal{M} = \{(1, 1, \pm 1)\}$ with $t_1 = 1$ and $m_1 = 1$. For scale, consider $T = 2$ with $f|_{\mathcal{L}}(T) = 1$. Thus, (3) yields

$$f|_{\mathcal{L}}(t) = t - 1.$$

Example 4. Reconsider Example 2 with $\mathcal{L}$ being the horizontal line parametrized by

$$x(t) = t \qquad y(t) = 0$$

as shown in Fig. 1(c). One has $\mathcal{M} = \mathcal{L} \times \mathbb{C}$ and $W = V \cap \mathcal{M} = \{(0, 0, 0)\}$ with $t_1 = 0$ and $m_1 = 2$. For scale, consider $T = 1$ with $f|_{\mathcal{L}}(T) = -1$. Thus, (3) yields

$$f|_{\mathcal{L}}(t) = -t^2.$$

Clearly, once the univariate polynomial $f|_{\mathcal{L}}(t)$ in (3) is computed explicitly, one can easily determine the value of f at any point along $\mathcal{L}$ via evaluation. Moreover, if $\mathcal{L}$ is parameterized by $v \cdot t + u$, then $f|_{\mathcal{L}}^{(k)}(t)$ is equal to the kth directional derivative of f with respect to v at $v \cdot t + u$, denoted $D_v^{(k)} f(v \cdot t + u)$.

Example 5. For $\mathcal{L}$ in Example 3 and Example 4, one has $v = (0, 1)$ and $v = (1, 0)$, respectively. Hence, the corresponding directional derivatives are simply partial derivatives of $f(x, y) = y - x^2$ with respect to y and x, respectively. From Example 3, one obtains $\frac{\partial f}{\partial y}(1, t) = 1$ while Example 4 yields $\frac{\partial f}{\partial x}(t, 0) = -2t$ and $\frac{\partial^2 f}{\partial x^2}(t, 0) = -2$.

3   Algorithm

Theorem 1 immediately justifies Algorithm 1 for explicitly computing a polynomial restricted to a line. The following two examples exemplify this algorithm applied to the discriminant locus.

Example 6. Consider the discriminant locus $\mathcal{H} \subset \mathbb{C}^2$ for $g(x; b, c) = x^2 + bx + c$. Hence, $\mathcal{H} = \overline{\pi(V)}$ where $\pi(b, c, x) = (b, c)$ and $V = \mathcal{V}(F)$ with

$$F(b, c, x) = \begin{bmatrix} x^2 + bx + c \\ 2x + b \end{bmatrix}.$$

Algorithm 1. Computing a polynomial restricted to a line

Input: A line $\mathcal{L} \subset \mathbb{C}^n$ parameterized by t, a pseudo-witness set $\{F, \pi, \mathcal{M}, W\}$ for a hypersurface $\mathcal{H}$ defined by f such that $\mathcal{M} = \mathcal{L} \times \mathcal{L}'$, and T along with $f|_{\mathcal{L}}(T) \neq 0$.
Output: The univarite polynomial $f|_{\mathcal{L}}(t)$ corresponding to f restricted to $\mathcal{L}$.

1: Use the pseudo-witness set to extract the roots $t_1, \ldots, t_k$ of f along $\mathcal{L}$ and the corresponding multiplicities $m_1, \ldots, m_k$.
2: Compute the scale factor $s := \dfrac{f|_{\mathcal{L}}(T)}{(T - t_1)^{m_1} \cdots (T - t_k)^{m_k}}$.
3: Construct the univariate polynomial $f|_{\mathcal{L}}(t) := s \cdot (t - t_1)^{m_1} \cdots (t - t_k)^{m_k}$.
4: (Optional) If $\mathcal{L}$ and F are defined with rational coefficients and T and $f|_{\mathcal{L}}(T)$ are rational, expand $f|_{\mathcal{L}}(t)$ and use exactness recovery [2] to compute the exact rational coefficients of $f|_{\mathcal{L}}(t)$.
5: **Return** $f|_{\mathcal{L}}(t)$.

For the line $\mathcal{L} \subset \mathbb{C}^2$ parameterized by

$$b(t) = t \qquad c(t) = t/3$$

with $\mathcal{M} = \mathcal{L} \times \mathbb{C}$, one has $W = V \cap \mathcal{M} = \{(0,0,0), (4/3, 4/9, -2/3)\}$. The other input for Algorithm 1 is, say, $T = 3$ with $f|_{\mathcal{L}}(T) = 5$ to set the scale. This setup is illustrated in Fig. 2(a).

The pseudo-witness set yields $t_1 = 0$ and $t_2 = 4/3$ with $m_1 = m_2 = 1$. The corresponding scale factor is

$$s = \frac{5}{(3 - 0)(3 - 4/3)} = 1$$

so that Algorithm 1 returns $f|_{\mathcal{L}}(t) = t(t - 4/3) = t^2 - 4t/3$.

Of course, one can easily compute that the discriminant of g satisfying $f(b(T), c(T)) = 5$ is $f(b, c) = b^2 - 4c$ with $f|_{\mathcal{L}}(t) = f(b(t), c(t)) = t^2 - 4t/3$.

Example 7. Consider the discriminant locus $\mathcal{H} \subset \mathbb{C}^2$ for $g(x) = x^3 + bx + c$. Hence, $\mathcal{H} = \overline{\pi(V)}$ where $\pi(b, c, x) = (b, c)$ and $V = V(F)$ with

$$F(b, c, x) = \begin{bmatrix} x^3 + bx + c \\ 3x^2 + b \end{bmatrix}.$$

For the line $\mathcal{L} \subset \mathbb{C}^2$ parameterized by

$$b(t) = t \qquad c(t) = t + 3$$

with $\mathcal{M} = \mathcal{L} \times \mathbb{C}$, one has, rounded to 4 decimal places with $i = \sqrt{-1}$,

$$W = V \cap \mathcal{M} = \left\{ \begin{array}{c} (-1.9511, 1.0489, 0.8064), \\ (-2.3995 \pm 5.0378i, 0.6005 \pm 5.0378i, -1.1532 \pm 0.7281i) \end{array} \right\}.$$

The other input for Algorithm 1 is, say, $T = -3$ with $f|_{\mathcal{L}}(T) = -108$ for scale. This setup is illustrated in Fig. 2(b).

The pseudo-witness set yields $t_1 = -1.9511$, $t_2 = -2.3995 + 5.0378i$, and $t_3 = -2.3995 - 5.0378i$ with $m_1 = m_2 = m_3 = 1$. The corresponding scale factor is $s = 4$ so that $f|_{\mathcal{L}}(t) = 4t^3 + 27t^2 + 162t + 243$.

As in Example 6, one can easily compute that the discriminant of g satisfying $f(b(T), c(T)) = -108$ is $f(b, c) = 4b^3 + 27c^2$ with $f(b(t), c(t)) = f|_{\mathcal{L}}(t)$ as above.

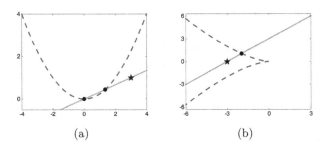

(a)   (b)

Fig. 2. Pseudo-witness set for the discriminant locus of (a) the quadratic $x^2 + bx + c$ and (b) the cubic $x^3 + bx + c$.

4   Computing Critical Points

When the line $\mathcal{L}$ is fixed, Algorithm 1 computes the restriction of a polynomial f to $\mathcal{L}$. The following presents two examples of combining this idea with homotopy continuation to compute critical points of f, namely $\nabla f = 0$. The set of real solutions to $\nabla f = 0$ with $f \neq 0$ contains at least one point in each compact cell of $\mathbb{R}^n \cap \{f \neq 0\}$ [10]. The website dx.doi.org/10.7274/r0-0mc0-gt33 contains the necessary files to perform these computations using Bertini [3].

4.1   Lemniscate

This first example demonstrates the approach given $f(x, y) = x^4 - x^2 + y^2$ which defines a lemniscate, but utilizes a pseudo-witness set for the computation. The aim is to compute all real solutions of $\nabla f = 0$ and $f \neq 0$. For genericity, replace $\nabla f = 0$ with the equivalent condition that the directional derivatives of f in both the $\alpha = (\alpha_1, \alpha_2)$ and $\beta = (\beta_1, \beta_2)$ directions, namely $D_\alpha f$ and $D_\beta f$, vanish for general α and β. We used $\alpha_1 = 1, \alpha_2 = 5 + 3\sqrt{-1}, \beta_1 = 4 + \sqrt{-1}$, and $\beta_2 = 1$ in our computation.

Since one is setting directional derivatives equal to zero, the scale factor is irrelevant and can be simply set to 1. We first compute a witness set for each of the cubic curves defined by $D_\alpha f = 0$ and $D_\beta f = 0$ where each of them are expressed in terms of univariate roots following Sect. 2. Then, we simply intersect these two cubic curves using a diagonal homotopy [20] that tracks $3^2 = 9$ paths. There are 3 finite endpoints corresponding with the 3 solutions of $\nabla f = 0$, all of which are real and shown in Fig. 3(a). Two of these have $f \neq 0$ with one in each of the two compact cells of $\mathbb{R}^2 \cap \{f \neq 0\}$.

4.2   Kuramoto Model

The Kuramoto model [16] is a mathematical model of an oscillating system to describe synchronization. After a standard conversion to polynomials, the discriminant locus for the steady states of the 3-oscillator Kuramoto model is modeled by $\mathcal{H} = \overline{\pi(V)}$ where $\pi(\omega_1, \omega_2, c_1, c_2, s_1, s_2) = (\omega_1, \omega_2)$ and $V = \mathcal{V}(F)$ with

$$F(\omega_1, \omega_2, c_1, c_2, s_1, s_2) = \begin{bmatrix} (s_1 c_2 - c_1 s_2) + (s_1 c_3 - c_1 s_3) - 3\omega_1 \\ (s_2 c_1 - c_2 s_1) + (s_2 c_3 - c_2 s_3) - 3\omega_2 \\ s_1^2 + c_1^2 - 1 \\ s_2^2 + c_2^2 - 1 \\ c_1^2 c_2 + c_1 c_2^2 + c_1 c_2 + s_1 s_2 c_1 + s_1 s_2 c_2 \end{bmatrix}. \qquad (4)$$

Letting f be a defining polynomial for $\mathcal{H}$, the aim is to compute the real solutions of $\nabla f = 0$ with $f \neq 0$ using a pseudo-witness set for $\mathcal{H}$. As in Sect. 4.1, we replace $\nabla f = 0$ with the equivalent condition that two general directional derivatives vanish. In this case, the vanishing of a general directional derivative of f yields a degree 11 curve, so a diagonal homotopy [20] to intersect the vanishing of two directional derivatives tracks $11^2 = 121$ paths. This yields 103 finite solutions consisting of 37 that satisfy $f \neq 0$ which can be verified using a membership test via a pseudo-witness set for $\mathcal{H}$ [12]. Sorting through these 37 yields 19 real critical points with $f \neq 0$. Figure 3(b) plots the real part of $\mathcal{H}$ along with these 19 real critical points on a contour plot of f showing that at least one is contained in each compact cell of $\mathbb{R}^2 \cap \{f \neq 0\}$.

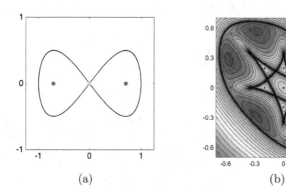

(a)   (b)

Fig. 3. (a) The lemniscate with 2 critical points satisfying $f \neq 0$ (red) and the other satisfying $f = 0$ (green), and (b) the discriminant locus (black) for the 3-oscillator Kuramoto model with contour plot and 19 real critical points (red) in the complement. (Color figure online)

Acknowledgments. All authors acknowledge support from NSF CCF-1812746. Additional support for was provided by ONR N00014-16-1-2722 (Hauenstein) and Schmitt Leadership Fellowship in Science and Engineering (Regan).

References

1. Al-Khateeb, A.N.: One-dimensional slow invariant manifolds for spatially homogeneous reactive systems. J. Chem. Phys. **131** (2009)
2. Bates, D.J., Hauenstein, J.D., McCoy, T.M., Peterson, C., Sommese, A.J.: Recovering exact results from inexact numerical data in algebraic geometry. Exp. Math. **22**(1), 38–50 (2013)
3. Bates, D.J., Hauenstein, J.D., Sommese, A.J., Wampler, C.W.: Bertini: software for numerical algebraic geometry. bertini.nd.edu
4. Bates, D.J., Hauenstein, J.D., Sommese, A.J., Wampler, C.W.: Numerically solving polynomial systems with Bertini. SIAM (2013)
5. Bernardi, A., Daleo, N.S., Hauenstein, J.D., Mourrain, B.: Tensor decomposition and homotopy continuation. Differ. Geom. Appl. **55**, 78–105 (2017)
6. Besana, G.M., Di Rocco, S., Hauenstein, J.D., Sommese, A.J., Wampler, C.W.: Cell decomposition of almost smooth real algebraic surfaces. Numer. Algor. **63**, 645–678 (2013)
7. Chen, C., Wu, W.: A numerical method for computing border curves of biparametric real polynomial systems and applications. In: Gerdt, V.P., Koepf, W., Seiler, W.M., Vorozhtsov, E.V. (eds.) CASC 2016. LNCS, vol. 9890, pp. 156–171. Springer, Cham (2016). https://doi.org/10.1007/978-3-319-45641-6_11
8. Cox, D.A., Little, J., O'Shea, D.: Ideals, Varieties, and Algorithms, 4th edn. Springer, Cham (2015). https://doi.org/10.1007/978-3-319-16721-3
9. Fischer, G.: Complex Analytic Geometry. Lecture Notes in Mathematics, vol. 538. Springer, Heidelberg (1976). https://doi.org/10.1007/BFb0080338
10. Harris, K., Hauenstein, J.D., Szanto, A.: Smooth points on semi-algebraic sets. arXiv:2002.04707 (2020)
11. Hauenstein, J.D., Regan, M.H.: Real monodromy action. Appl. Math. Comput. **373**, 124983 (2020)
12. Hauenstein, J.D., Sommese, A.J.: Membership tests for images of algebraic sets by linear projections. Appl. Math. Comput. **219**(12), 6809–6818 (2013)
13. Hauenstein, J.D., Sommese, A.J.: Witness sets of projections. Appl. Math. Comput. **217**(7), 3349–3354 (2010)
14. Hauenstein, J.D., Wampler, C.W., Pfurner, M.: Synthesis of three-revolute spatial chains for body guidance. Mech. Mach. Theory **110**, 61–72 (2017)
15. Irschara, A., Zach, C., Frahm, J.M., Bischof, H.: From structure-from-motion point clouds to fast location recognition. In: CVPR, pp. 2599–2606. IEEE (2009)
16. Kuramoto, Y.: Self-entrainment of a population of coupled non-linear oscillators. In: Araki, H. (ed.) International Symposium on Mathematical Problems in Theoretical Physics. Lecture Notes in Physics, vol. 39, pp. 420–422. Springer, Heidelberg (1975). https://doi.org/10.1007/BFb0013365
17. Leibe, B., Cornelis, N., Cornelis, K., Van Gool, L.: Dynamic 3D scene analysis from a moving vehicle. In: CVPR, pp. 1–8. IEEE (2007)
18. Lu, Y., Bates, D.J., Sommese, A.J., Wampler, C.W.: Finding all real points of a complex curve. Contemp. Math. **448**, 183–205 (2007)
19. Morgan, A.P.: Solving Polynomial Systems Using Continuation for Engineering and Scientific Problems. Prentice-Hall, Englewood Cliffs (1987)
20. Sommese, A.J., Verschelde, J., Wampler, C.W.: Solving polynomial systems equation by equation. In: Dickenstein, A., Schreyer, F.O., Sommese, A.J. (eds.) Algorithms in Algebraic Geometry. The IMA Volumes in Mathematics and its Applications, vol. 146, pp. 133–152. Springer, New York (2008). https://doi.org/10.1007/978-0-387-75155-9_8

21. Sommese, A.J., Wampler, C.W.: The Numerical Solution of Systems of Polynomials Arising in Engineering and Science. World Scientific, Hackensack (2005)
22. Snavely, N., Seitz, S.M., Szeliski, R.: Photo tourism: exploring photo collections in 3D. In: ACM SIGGRAPH, pp. 835–846 (2006)
23. Wampler, C.W., Sommese, A.J.: Numerical algebraic geometry and algebraic kinematics. Acta Numerica **20**, 469–567 (2011)

Computational Algebraic Analysis

Algorithms for Pfaffian Systems and Cohomology Intersection Numbers of Hypergeometric Integrals

Saiei-Jaeyeong Matsubara-Heo$^{(\boxtimes)}$ and Nobuki Takayama

Department of Mathematics, Graduate School of Science,
Kobe University, Kobe, Japan
{saiei,takayama}@math.kobe-u.ac.jp

Abstract. In the theory of special functions, a particular kind of multidimensional integral appears frequently. It is called the Euler integral. In order to understand the topological nature of the integral, twisted de Rham cohomology theory plays an important role. We propose an algorithm of computing an invariant *cohomology intersection number* of a given basis of the twisted cohomology group. We also develop an algorithm of computing the Pfaffian system that a given basis satisfies. These algorithms are based on the fact that the Euler integral satisfies GKZ system and utilizes algorithms to find rational function solutions of differential equations. We provide software to perform this algorithm.

Keywords: Cohomology intersection numbers · GKZ hypergeometric systems · Gröbner basis

1 Introduction

In the study of hypergeometric functions in several variables, one often considers the integral of the following form:

$$\langle \omega \rangle = \int_{\Gamma} h_1(x)^{-\gamma_1} \cdots h_k(x)^{-\gamma_k} x^c \omega, \tag{1}$$

where $h_l(x;z) = h_{l,z^{(l)}}(x) = \sum_{j=1}^{N_l} z_j^{(l)} x^{\mathbf{a}^{(l)}(j)}$ ($l = 1,\ldots,k$) are Laurent polynomials in torus variables $x = (x_1,\ldots,x_n)$, $\mathbf{a}^{(l)}(j) \in \mathbb{Z}^n$, $\gamma_l \in \mathbb{C}$ and $c = {}^t(c_1,\ldots,c_n) \in \mathbb{C}^n$ are parameters, $x^c = x_1^{c_1}\ldots x_n^{c_n}$, Γ is a suitable integration cycle, and ω is an algebraic n-form on $V_z = \{x \in \mathbb{C}^n \mid x_1\ldots x_n h_1(x)\ldots h_k(x) \neq 0\}$. As a function of the independent variable $z = (z_j^{(l)})_{j,l}$, the integral (1) defines a hypergeometric function. We call the integral (1) the Euler integral.

Supported by JSPS KAKENHI Grant Number 19K14554 (the first author) and JST CREST Grant Number JP19209317 (the first and the second authors).

A. M. Bigatti et al. (Eds.): ICMS 2020, LNCS 12097, pp. 73–84, 2020.
https://doi.org/10.1007/978-3-030-52200-1_7

We can naturally define the twisted de Rham cohomology group associated to the Euler integral (1). We set $N = N_1 + \cdots + N_k$, $\mathbb{G}_m^n = \mathrm{Specm}\ \mathbb{C}[x_1^{\pm}, \ldots, x_n^{\pm}]$, and $\mathbb{A}^N = \mathrm{Specm}\ \mathbb{C}[z_j^{(l)}]$. For any $z \in \mathbb{A}^N$, we can define an integrable connection $\nabla_x = d_x - \sum_{l=1}^k \gamma_l \frac{d_x h_l}{h_l} \wedge + \sum_{i=1}^n c_i \frac{dx_i}{x_i} \wedge : \mathcal{O}_{V_z} \to \Omega_{V_z}^1$. The algebraic de Rham cohomology group $\mathrm{H}_{\mathrm{dR}}^* (V_z; (\mathcal{O}_{V_z}, \nabla_x))$ is defined as the hypercohomology group

$$\mathrm{H}_{\mathrm{dR}}^* (V_z; (\mathcal{O}_{V_z}, \nabla_x)) = \mathbb{H}^* \left(V_z; (0 \to \mathcal{O}_{V_z} \xrightarrow{\nabla_x} \Omega_{V_z}^1 \xrightarrow{\nabla_x} \cdots \xrightarrow{\nabla_x} \Omega_{V_z}^n \to 0) \right). \quad (2)$$

Under a genericity assumption on the parameters γ_l and c, we have the vanishing result $\mathrm{H}_{\mathrm{dR}}^m (V_z; (\mathcal{O}_{V_z}, \nabla_x)) = 0$ $(m \neq n)$. Moreover, we can define a perfect pairing $\langle \bullet, \bullet \rangle_{ch} : \mathrm{H}_{\mathrm{dR}}^n (V_z; (\mathcal{O}_{V_z}, \nabla_x)) \times \mathrm{H}_{\mathrm{dR}}^n (V_z; (\mathcal{O}_{V_z}^\vee, \nabla_x^\vee)) \to \mathbb{C}$ which is called the cohomology intersection form. Here, $(\mathcal{O}_{V_z}^\vee, \nabla_x^\vee)$ is the dual connection of $(\mathcal{O}_{V_z}, \nabla_x)$.

The study of intersection numbers of twisted cohomology groups and twisted period relations for hypergeometric functions started with the celebrated work by K. Cho and K. Matsumoto [6]. They clarified that the cohomology intersection number appears naturally as a part of the quadratic relation, a class of functional relations of hypergeometric functions. They also developed a systematic method of computing the cohomology intersection number for 1-dimensional integrals. Since this work, several methods have been proposed to evaluate intersection numbers of twisted cohomology groups, see, e.g., [2,3,10,11,14,17,19] and references therein. All methods utilize comparison theorems of twisted cohomology groups and residue calculus.

When z belongs to a certain non-empty Zariski open subset of $\mathbb{A}^N$ (the non-singular locus), we proposed a new method in the paper [16] to obtain cohomology intersection numbers by constructing a rational function solution of a system of linear partial differential equations. One weak point of the method was that it was not algorithmic to construct the Pfaffian system (the explicit form of the integrable connection) for a given basis of the twisted cohomology group. We will give a new algorithm to construct the Pfaffian system for a given basis in this paper (Algorithm 1). To our knowledge, algorithms to find the Pfaffian system (or equation) with respect to a given basis of twisted cohomology group do not appear in the literature except the twisted logarithmic cohomology case[1]. Our algorithm works for a more general class of twisted cohomology groups. Moreover, it is more efficient by utilizing Saito's b-function [23] expressed in terms of facets of a polytope. The Sect. 2 is a brief overview of the paper [16]. The Sect. 3 is the main part and in the Sects. 4 and 5, we will give demonstrations of our implementation. As to the construction of rational function solutions, we utilize the algorithm and the implementation by M. Barkatou, T. Cluzeau, C. El Bacha, J.-A. Weil [5] (see also [4,18] and their references).

---

[1] K. Nishitani, master thesis 2011 (in Japanese), Kobe University.

2   General Results

2.1   The Cohomology Intersection Form

We denote by $\mathrm{H}^n_{dR,c}\left(V^{an}_z;(\mathcal{O}_{V^{an}_z},\nabla^{an}_x)\right)$ the analytic de Rham cohomology group with compact support. By Poincaré-Verdier duality, the bilinear pairing

$$
\mathrm{H}^n_{dR,c}\left(V^{an}_z;(\mathcal{O}_{V^{an}_z},\nabla^{an}_x)\right) \times \mathrm{H}^n_{dR}\left(V^{an}_z;(\mathcal{O}^\vee_{V^{an}_z},\nabla^{an\vee}_x)\right) \rightarrow \quad \mathbb{C}
$$
$$
\cup \qquad\qquad\qquad\qquad\qquad\qquad \cup \qquad\qquad (3)
$$
$$
(\phi,\psi) \qquad\qquad\qquad\qquad\qquad \mapsto \int_{V^{an}_z} \phi \wedge \psi
$$

is perfect. We say that the regularization condition is satisfied if the canonical morphism $\mathrm{H}^n_{dR,c}\left(V^{an}_z;(\mathcal{O}_{V^{an}_z},\nabla^{an}_x)\right) \rightarrow \mathrm{H}^n_{dR}\left(V^{an}_z;(\mathcal{O}_{V^{an}_z},\nabla^{an}_x)\right)$ is an isomorphism. In the following, we always assume that the regularization condition is satisfied. A criterion for this assumption is explained in Sect. 2.3. Since $(\mathcal{O}_{V_z},\nabla_x)$ is a regular connection, the canonical morphism $\mathrm{H}^n_{dR}\left(V_z;(\mathcal{O}_{V_z},\nabla_x)\right)$ $\rightarrow \mathrm{H}^n_{dR}\left(V^{an}_z;(\mathcal{O}_{V^{an}_z},\nabla^{an}_x)\right)$ is always an isomorphism by Deligne-Grothendieck comparison theorem ([7, Corollaire 6.3]). Therefore, we have a canonical isomorphism $\mathrm{reg}: \mathrm{H}^n_{dR}\left(V_z;(\mathcal{O}_{V_z},\nabla_x)\right) \rightarrow \mathrm{H}^n_{dR,c}\left(V^{an}_z;(\mathcal{O}_{V^{an}_z},\nabla^{an}_x)\right)$. Note that the Poincaré dual of the isomorphism reg is called a regularization map in the theory of special functions ([2, § 3.2]). Finally, we define the cohomology intersection form $\langle\bullet,\bullet\rangle_{ch}$ between algebraic de Rham cohomology groups by the formula

$$
\langle\bullet,\bullet\rangle_{ch}: \mathrm{H}^n_{dR}\left(V_z;(\mathcal{O}_{V_z},\nabla_x)\right) \times \mathrm{H}^n_{dR}\left(V_z;(\mathcal{O}^\vee_{V_z},\nabla^\vee_x)\right) \rightarrow \quad \mathbb{C}
$$
$$
\cup \qquad\qquad\qquad\qquad\qquad\qquad \cup \qquad\qquad (4)
$$
$$
(\phi,\psi) \qquad\qquad\qquad\qquad\qquad \mapsto \int_{V^{an}_z} \mathrm{reg}(\phi) \wedge \psi.
$$

The value above is called *the cohomology intersection number* of ϕ and ψ.

2.2   The Secondary Equation

Now, we treat z as a variable. Let $\pi : X = (\mathbb{G}_m)^n_x \times \mathbb{A}^N_z \setminus \bigcup^k_{l=1}\{(x,z) \mid h_{l,z^{(l)}}(x) = 0\} \rightarrow \mathbb{A}^N_z = Y$ be the natural projection where subscripts stand for coordinates. We define an $\mathcal{O}_Y$-module $\mathcal{H}^n_{dR}$ by the hypercohomology group

$$
\mathcal{H}^n_{dR} = \mathbb{H}^n\left(X;\left(0 \rightarrow \Omega^0_{X/Y} \xrightarrow{\nabla_x} \Omega^1_{X/Y} \xrightarrow{\nabla_x} \cdots \xrightarrow{\nabla_x} \Omega^n_{X/Y} \rightarrow 0\right)\right). \qquad (5)
$$

Here, $\Omega^\bullet_{X/Y}$ denotes the sheaf of relative differential forms $\oplus_{|I|=\bullet}\mathcal{O}_X dx^I$ with respect to the morphism π. Since Y is affine, $\mathcal{H}^n_{dR}$ is also identified with the sheaf $R^n\pi_*(\Omega^\bullet_{X/Y},\nabla_x)$. For any $z \in Y$, there is a natural evaluation morphism $\mathrm{ev}_z : \mathcal{H}^n_{dR} \rightarrow \mathrm{H}^n_{dR}\left(V_z;(\mathcal{O}_{V_z},\nabla_x)\right)$. We define the dual object $\mathcal{H}^{n\vee}_{dR}$ by replacing ∇_x by $\nabla^\vee_x$ in the construction above. By the general theory of relative de Rham cohomology, there exists a non-empty Zariski open subset U of Y such that $\mathcal{H}^n_{dR}\restriction_U \simeq \mathcal{O}^{\oplus r}_U$. Therefore, for any global sections ϕ of $\mathcal{H}^n_{dR}\restriction_U$ and ψ of $\mathcal{H}^{n\vee}_{dR}\restriction_U$, we can define the cohomology intersection number $\langle\phi,\psi\rangle_{ch}$ as a function of $z \in U$

by the formula $U \ni z \mapsto \langle \mathrm{ev}_z(\phi), \mathrm{ev}_z(\psi) \rangle_{ch} \in \mathbb{C}$. This actually defines a $\mathcal{O}_U$-bilinear map $\langle \bullet, \bullet \rangle_{ch} : \mathcal{H}_{dR}^n \restriction_U \times \mathcal{H}_{dR}^{n\vee} \restriction_U \to \mathcal{O}_U$.

We can equip $\mathcal{H}_{dR}^n$ with a structure of a $\mathcal{D}_Y$-module. For this purpose, we only need to define a connection $\nabla^{GM} : \mathcal{H}_{dR}^n \to \Omega_Y^1(\mathcal{H}_{dR}^n) := \Omega_Y^n \otimes \mathcal{H}_{dR}^n$. For any section $\phi \in \mathcal{H}_{dR}^n$, we define

$$\nabla^{GM} \phi = d_z \phi - \sum_{j,l} \gamma_l \frac{x^{\mathbf{a}^{(l)}(j)}}{h_{l,z^{(l)}}(x)} dz_j^{(l)} \wedge \phi. \tag{6}$$

Here, the superscript GM stands for "Gauß-Manin". The dual connection $\nabla^{\vee GM} : \mathcal{H}_{dR}^{n\vee} \to \Omega_Y^1(\mathcal{H}_{dR}^{n\vee})$ is defined by replacing γ_l by $-\gamma_l$ in (6).

The $\mathcal{D}_Y$-module structures of $\mathcal{H}_{dR}^n$ and $\mathcal{H}_{dR}^{n\vee}$ are compatible with the cohomology intersection form. Namely, for any local sections ϕ of $\mathcal{H}_{dR}^n \restriction_U$ and ψ of $\mathcal{H}_{dR}^{n\vee} \restriction_U$, we have

$$d_z \langle \phi, \psi \rangle_{ch} = \langle \nabla^{GM} \phi, \psi \rangle_{ch} + \langle \phi, \nabla^{\vee GM} \psi \rangle_{ch}. \tag{7}$$

We call (7) the secondary equation. Let us rewrite it in terms of local frames. Let $\{\phi_i\}_{i=1}^r$ (resp. $\{\psi_i\}_{i=1}^r$) be a free basis of $\mathcal{H}_{dR}^n \restriction_U$ (resp. $\mathcal{H}_{dR}^{n\vee} \restriction_U$). We set $I = I_{ch} = (\langle \phi_i, \psi_j \rangle_{ch})_{i,j}$ and call it the cohomology intersection matrix. On the other hand, there is a $r \times r$ matrix Ω (resp. $\Omega^\vee$) with values in 1-forms on U such that the connection ∇^{GM} (resp. $\nabla^{\vee GM}$) is trivialized as $d_z + \Omega\wedge$ (resp. $d_z + \Omega^\vee\wedge$). Then, the secondary equation is equivalent to the system

$$d_z I = {}^t\Omega I + I\Omega^\vee. \tag{8}$$

We also call (8) the secondary equation. The theorem which our algorithm is based on is the following

Theorem 1 [16]. *Under the regularization condition, all the entries of the cohomology intersection matrix I_{ch} are rational functions. Moreover, any rational function solution I of the secondary equation (8) is, up to a scalar multiplication, equal to I_{ch}.*

2.3 GKZ System Behind

In [16], it is discussed that Theorem 1 is true for more general direct image $\mathcal{D}$-modules. However, by employing the combinatorial structure behind our integrable connection $\mathcal{H}_{dR}^n \restriction_U$, we can show that the cohomology intersection number in question has a rational expression with respect to z and δ.

Let us recall the definition of GKZ system [8]. For a given $d \times N$ ($d < N$) integer matrix $A = (\mathbf{a}(1)|\cdots|\mathbf{a}(N))$ and a parameter vector $\delta \in \mathbb{C}^d$, GKZ system $M_A(\delta)$ is defined as a system of partial differential equations on $\mathbb{C}^N$ given by

$$M_A(\delta) : \begin{cases} E_i \cdot f(z) = 0 & (i = 1, \dots, d) & (a) \\ \Box_u \cdot f(z) = 0 & (u \in \mathrm{Ker}(A \times : \mathbb{Z}^{N \times 1} \to \mathbb{Z}^{d \times 1})), & (b) \end{cases} \tag{9}$$

where E_i and $\Box_u$ for $u = {}^t(u_1, \ldots, u_N)$ are differential operators defined by

$$E_i = \sum_{j=1}^{N} a_{ij} z_j \frac{\partial}{\partial z_j} + \delta_i, \quad \Box_u = \prod_{u_j > 0} \left(\frac{\partial}{\partial z_j} \right)^{u_j} - \prod_{u_j < 0} \left(\frac{\partial}{\partial z_j} \right)^{-u_j}. \tag{10}$$

For convenience, we assume an additional condition $\mathbb{Z}A \overset{def}{=} \mathbb{Z}\mathbf{a}(1) + \cdots + \mathbb{Z}\mathbf{a}(N) = \mathbb{Z}^d$. In our setting, we put $A_l = (\mathbf{a}^{(l)}(1)| \ldots |\mathbf{a}^{(l)}(N_l))$, $d = n + k$, $N = N_1 + \cdots + N_k$. We define an $(n + k) \times N$ matrix A by

$$A = \begin{pmatrix} 1 \cdots 1 & 0 \cdots 0 & \cdots & 0 \cdots 0 \\ \hline 0 \cdots 0 & 1 \cdots 1 & \cdots & 0 \cdots 0 \\ \hline \vdots & \vdots & \ddots & \vdots \\ \hline 0 \cdots 0 & 0 \cdots 0 & \cdots & 1 \cdots 1 \\ \hline A_1 & A_2 & \cdots & A_k \end{pmatrix}. \tag{11}$$

We put $\delta = \begin{pmatrix} \gamma \\ c \end{pmatrix}$. By abuse of notation, we also denote by $M_A(\delta)$ the quotient $\mathcal{D}_Y$-module $\mathcal{D}_Y / J$ where J is the left ideal of $\mathcal{D}_Y$ generated by the operators (10). It is known that GKZ system $M_A(\delta)$ is holonomic ([1]). We say that the parameter δ is non-resonant if it does not belong to any $\mathbb{C}\Gamma + \mathbb{Z}^d$ where Γ is any facet of the cone $\sum_{j=1}^{N} \mathbb{R}_{\geq 0} \mathbf{a}(j)$. $\mathcal{H}_{dR}^n$ (resp. $\mathcal{H}_{dR}^{n\vee}$) is isomorphic to GKZ system $M_A(\delta)$ (resp. $M_A(-\delta)$) and the regularization condition is true when the parameter vector δ is non-resonant and $\gamma_l \notin \mathbb{Z}$ (see [9, 2.9] and [15, Theorem 2.12]). We set $\frac{dx}{x} = \frac{dx_1}{x_1} \wedge \cdots \wedge \frac{dx_n}{x_n}$. The isomorphism $M_A(\delta) \simeq \mathcal{H}_{dR}^n$ is given by the correspondence $[1] \mapsto [\frac{dx}{x}]$. Thus, any section ϕ of $\mathcal{H}_{dR}^n$ can be written as $\phi = P \cdot [\frac{dx}{x}]$ for some linear differential operator $P \in \mathcal{D}_Y$. We define the field $\mathbb{Q}(\delta)$ as the field extension $\mathbb{Q}(\gamma_1, \ldots, \gamma_k, c_1, \ldots, c_n)$ of $\mathbb{Q}$.

Theorem 2 [16]. *Suppose that A as in (11) admits a unimodular regular triangulation T and δ is non-resonant and $\gamma_l \notin \mathbb{Z}$. Then, for any $P_1, P_2 \in \mathbb{Q}(\delta)\langle z, \partial_z \rangle$, the cohomology intersection number $\frac{\langle P_1 \cdot \frac{dx}{x}, P_2 \cdot \frac{dx}{x} \rangle_{ch}}{(2\pi\sqrt{-1})^n}$ belongs to the field $\mathbb{Q}(\delta)(z)$.*

3 An Algorithm of Finding the Pfaffian System for a Given Basis

In this section, we set $\beta := -\delta$. With this notation, we put $H_A(\beta) := M_A(\delta)$. This is because we use some results from [12] and [23] where the hypergeometric ideal is denoted by $H_A(\beta)$ while our main references [15, 16] denote it by $M_A(\delta)$.

Let ω_q be the differential form

$$\prod_{l=1}^{k} h_l^{-q_l'} x^{q''} \frac{dx}{x}, \quad q = (q', q'') \in \mathbb{Z}^k \times \mathbb{Z}^n. \tag{12}$$

It is known that there exists a basis of the twisted cohomology group of which elements are of the form ω_q when δ is generic (see, e.g., [13, Th 2]). Let $\{\omega_q \mid q \in$

Q} be a basis of the twisted cohomology group. We will give an algorithm to find the Pfaffian system $\frac{\partial}{\partial z_i}\omega = P_i\omega$ with respect to this basis $\omega = (\omega_q \mid q \in Q)^T$. Note that algorithms to translate a given holonomic ideal to a Pfaffian system are well known (see, e.g., [12, Chap 6]). However, as long as we know, algorithms to find the Pfaffian system with respect to a given basis of twisted cohomology group do not appear in the literature. Note that the pairing of the twisted homology and cohomology groups is perfect under our assumption. Then, the Pfaffian equation of the fundamental solution matrix of solutions of the GKZ system can be regarded as a relation of the twisted cycles.

Put $\partial_i = \frac{\partial}{\partial z_i}$. In this subsection, we use $\bullet$ to denote the action to avoid a confusion with the multiplication. The function $\langle \omega_q \rangle$ is a solution of the hypergeometric system $H_A(\beta - q)$. The main point of our method is of use of the following contiguity relation

$$\frac{1}{\mathbf{a}_i' \cdot (\beta - q)} \partial_i \bullet \langle \omega_q \rangle = \langle \omega_{q'} \rangle, \quad q' = q + \mathbf{a}_i \tag{13}$$

where $\mathbf{a}_i = \mathbf{a}(i)$ is the i-th column vector of A and $\mathbf{a}_i'$ is the column vector that the first k elements are equal to those of $\mathbf{a}_i$ and the last n elements are 0. For example, $\mathbf{a}_1' = (1,0,\ldots,0)$, $\mathbf{a}_2' = (1,0,\ldots,0)$, $\ldots$, $\mathbf{a}_{N_1+1}' = (0,1,0,\ldots,0)^T$, $\ldots$, $\mathbf{a}_N' = (0,\ldots,0,1)^T$. The relation (13) can be proved by differentiating $\langle \omega_q \rangle = \int_\Gamma h_1^{-\gamma_1-q_1'} \cdots h_k^{-\gamma_k-q_k'} x^{c+q''} \frac{dx}{x}$, with respect to z_i where we have $\beta - q = (-\gamma_1 - q_1', \ldots, -\gamma_k - q_k', -c_1 - q_1'', \ldots, -c_n - q_n'')^T$.

In [23, Algorithm 3.2], an algorithm to obtain an operator C_i satisfying

$$C_i \partial_i - b_i(\beta) = 0 \mod H_A(\beta) \tag{14}$$

is given. The polynomial b_i is a b-function in the direction i [23, Th 3.2]. Note that the algorithm outputs the operator C_i in $\mathbb{C}\langle z_1, \ldots, z_N, \partial_1, \ldots, \partial_N \rangle$, which does not depend on the parameter β. Since $\langle \omega_q \rangle$ is a solution of $H_A(\beta - q)$, we have the following inverse contiguity relation

$$\frac{\mathbf{a}_i' \cdot (\beta - q'')}{b_i(\beta - q'')} C_i \bullet \langle \omega_q \rangle = \langle \omega_{q''} \rangle, \quad q'' = q - \mathbf{a}_i. \tag{15}$$

Example 1. (Gauss hypergeometric function $_2F_1$.) Put

$$A = \left(\begin{array}{cc|cc} 1 & 1 & 0 & 0 \\ 0 & 0 & 1 & 1 \\ 0 & 1 & 0 & 1 \end{array} \right) \tag{16}$$

Then, $h_1 = z_1 + z_2 x$, $h_2 = z_3 + z_4 x$. We have

$$\langle \omega_{(1,0,0)} \rangle = \int_\Gamma h_1^{-\gamma_1} h_2^{-\gamma_2} x^c \frac{1}{h_1} \frac{dx}{x}. \tag{17}$$

We can show that $\{\omega_{(1,0,0)}, \omega_{(1,0,0)} - \omega_{(0,1,0)}\}$ is a basis of the twisted cohomolgy group. This A is normal and the b-function $b_4(s) \in \mathbb{Q}[s_1, s_2, s_3]$ for the direction

z_4 is $b_4(s) = s_2 s_3$. Then, $C_4 = z_2 z_3 \partial_1 + (\theta_2 + \theta_3 + \theta_4)z_4$ where $\theta_i = z_i \partial_i$ by reducing $(\theta_3 + \theta_4)(\theta_2 + \theta_4)$ by the toric ideal $I_A = \langle \partial_2 \partial_3 - \partial_1 \partial_4 \rangle$ (see Algorithm 3.2 of [23]).

Our algorithm to find a Pfaffian system with respect to a given basis of the twisted cohomology group is as follows.

Algorithm 1. Input: $\{\omega_q \,|\, q \in Q\}$, a basis of the twisted cohomology group. A direction (index) i.
Output: P_i, the coefficient matrix of the Pfaffian system $\partial_i - P_i$.

1. Compute a Gröbner basis G of $H_A(\beta)$ in the ring of differential operators with rational function coefficients. Let S be a column vector of the standard monomials with respect to G.
2. Put

$$
F(Q) = (F(q) \,|\, q \in Q)^T, \quad F(q) = \prod_{r_i < 0} C_i^{-r_i} \prod_{r_i > 0} \partial_i^{r_i} \frac{1}{BB'}, \quad q = \sum_{i=1}^{N} r_i a_i
$$
(18)

It is a vector with entries in the ring of differential operators and the order of the product is $i = N, N-1, \ldots, 3, 2, 1$. In other words, we apply operators from ∂_1. The polynomial B is derived from the coefficient of the contiguity relation (15) and is equal to

$$
B = \prod_{j=1, r_j < 0}^{N} \frac{b_j(\beta_j' + a_j)}{a_j' \cdot (\beta_j' + a_j)} \frac{b_j(\beta_j' + 2a_j)}{a_j' \cdot (\beta_j' + 2a_j)} \cdots \frac{b_j(\beta_j' + (-r_j)a_j)}{a_j' \cdot (\beta_j' + (-r_j)a_j)}, \quad (19)
$$

$$
\beta_j' = \beta - \sum_{r_l > 0} r_l a_l + \sum_{l=1, r_l < 0}^{j-1} (-r_l)a_l.
$$
(20)

The polynomial B' comes from the denominator of the contiguity relation (13) and is equal to

$$
B' = \prod_{j=1, r_j > 0}^{N} (a_j' \cdot (\beta_j')) (a_j' \cdot (\beta_j' - a_j)) \cdots (a_j' \cdot (\beta_j' - (r_j - 1)a_j)), \quad (21)
$$

$$
\beta_j' = \beta - \sum_{r_l > 0, l < j} r_l a_l.
$$
(22)

3. Compute the normal form of the vectors $\partial_i F(Q)$ and $F(Q)$. Write the normal forms of them as $P'S$ and $P''S$ respectively where P' and P'' are matrices with rational function entries.
4. Output $P_i = P'(P'')^{-1}$.

The matrix P'' is invertible if and only if the given set of differential forms $\{\omega_q\}$ is a basis of the twisted cohomology group.

We show the correctness of the algorithm. Take an element $q \in Q$. We express $\langle \omega_q \rangle$ in terms of $\langle \omega_0 \rangle$, which is a solution of $H_A(\beta)$, by the contiguity relations (13) and (15). Note that the contiguity relations for functions $\langle \omega_q \rangle$ give the contiguity relations for cohomology classes $[\omega_q]$ by virtue of the perfectness of the pairing between the twisted homology and the twisted cohomology groups. The point of the correctness is the following identity

$$F(q) \bullet \omega_0 = \omega_q. \tag{23}$$

Let us illustrate how to prove (23) by examples. We assume that $q = 2\mathbf{a_1} + \mathbf{a_2}$ and $N_1 \geq 2$. Then ω_q can be obtained by applying (13) with $i = 1$ for two times and that with $i = 2$. We have

$$\omega_{\mathbf{a_1}} = \frac{1}{\mathbf{a_1'} \cdot \beta} \partial_1 \bullet \omega_0, \tag{24}$$

$$\omega_{2\mathbf{a_1}} = \frac{1}{\mathbf{a_1'} \cdot (\beta - \mathbf{a_1})} \partial_1 \bullet \omega_{\mathbf{a_1}}, \tag{25}$$

$$\omega_{2\mathbf{a_1}+\mathbf{a_2}} = \frac{1}{\mathbf{a_2'} \cdot (\beta - 2\mathbf{a_1})} \partial_2 \bullet \omega_{2\mathbf{a_1}}. \tag{26}$$

Thus, we obtain the numbers (21) and then (23). Let us consider the case that $q = -2\mathbf{a_1} - \mathbf{a_2}$ and $N_1 \geq 2$. Then ω_q can be obtained by applying (15) with $i = 1$ for two times and that with $i = 2$. Since $\langle \omega_{-\mathbf{a_1}} \rangle$ is a solution of $H_A(\beta + \mathbf{a_1})$, we have

$$[c_1 \partial_1 - b_1(\beta + \mathbf{a_1})] \bullet \omega_{-\mathbf{a_1}} = 0 \tag{27}$$

from [23]. Then, we have

$$\omega_{-\mathbf{a_1}} = \frac{\mathbf{a_1'} \cdot (\beta + \mathbf{a_1})}{b_1(\beta + \mathbf{a_1})} c_1 \bullet \omega_0, \tag{28}$$

$$\omega_{-2\mathbf{a_1}} = \frac{\mathbf{a_1'} \cdot (\beta + 2\mathbf{a_1})}{b_1(\beta + 2\mathbf{a_1})} c_1 \bullet \omega_{-\mathbf{a_1}}, \tag{29}$$

$$\omega_{-2\mathbf{a_1}-\mathbf{a_2}} = \frac{\mathbf{a_2'} \cdot (\beta + 2\mathbf{a_1} + \mathbf{a_2})}{b_2(\beta + 2\mathbf{a_1} + \mathbf{a_2})} c_2 \bullet \omega_{-2\mathbf{a_1}}. \tag{30}$$

Thus, we obtain the numbers (19) and then (23). The general case can be shown by repeating these procedures. When the normal form $F(q)$ with respect to the Gröbner basis G is $\sum p_i'' s_i$ where $S = (s_i)$ and p_i'' is a rational function in z and β, we have $\omega_q = F(q) \bullet \omega_0 = \sum p_i'' s_i \bullet \omega_0$. The correctness of the last two steps follows from this fact.

Example 2. This is a continuation of Example 1. We have $(1,0,0)^T = \mathbf{a_1}$ and $(0,1,0)^T = \mathbf{a_3}$. Then, the basis of the twisted cohomology group $F(Q)$ is expressed as $F(Q) = (\partial_1/\beta_1, \partial_1/\beta_1 - \partial_3/\beta_2)^T$ and $\partial_4 F(Q) = (\partial_4\partial_1/\beta_1, \partial_4\partial_1/\beta_1 - \partial_4\partial_3/\beta_2)^T$. We can obtain a Gröbner basis whose set of the standard monomials is $\{\partial_4, 1\}$ by the graded reverse lexicographic order such that $\partial_i > \partial_{i+1}$. We multiply $\beta_1\beta_2$ to $F(Q)$ and $\partial_4 F(Q)$ in order to avoid rational polynomial arithmetic. Then, the normal form, for example, of $\beta_2\partial_1$ is

$\frac{1}{z_1 z_4 - z_2 z_3}\left((\beta_1(\beta_1 + \beta_2)z_4)\partial_4 - \beta_2^2\beta_3\right)$. By computing the other normal forms, we obtain the matrix

$$P_4 = \begin{pmatrix} \frac{-\beta_2(z_3 - z_1)}{z_1 z_4 - z_2 z_3} & \frac{\beta_2 z_3}{z_1 z_4 - z_2 z_3} \\ \frac{-((\beta_2 z_3 + (-\beta_2 + \beta_3)z_1)z_4 + (\beta_1 - \beta_3)z_2 z_3 - \beta_1 z_1 z_2)}{z_4(z_1 z_4 - z_2 z_3)} & \frac{(\beta_2 z_3 + \beta_3 z_1)z_4 + (\beta_1 - \beta_3)z_2 z_3}{z_4(z_1 z_4 - z_2 z_3)} \end{pmatrix}. \tag{31}$$

4   Implementation and Examples

We implemented our algorithms on the computer algebra system Risa/Asir [21] with a Polymake interface. Polymake (see, e.g., [20,22]) is a system for polyhedral geometry and it is used for an efficient computation of contiguity relations ([23, Algorithm 3.2]). Here is an input[2] to find the coefficient matrix P_4 for Example 1 with respect to the variable z_4 when $z_1 = z_2 = z_3 = 1$ (note that in our implementation x is used instead of z).

```
P4=pfaff_eq(A=[[1,1,0,0],[0,0,1,1],[0,1,0,1]],
        Beta=[-gamma1,-gamma2,-c],
        Ap = [[1,1,0,0],[0,0,1,1],[0,0,0,0]],
        Rvec = [[1,0,0,0],[0,0,1,0]],DirX=[dx4] //Rvec is the set of r's in Algorithm 1.
    | xrule=[[x1,1],[x2,1],[x3,1]],
        cg=matrix_list_to_matrix([[1,0], [1,-1]]));//get Pfaffian sys for cg*(the basis omega_q)
```

It outputs the following coefficient matrix

$$P_4 = \begin{pmatrix} 0 & \frac{-\gamma_2}{x_4 - 1} \\ \frac{c}{x_4} & \frac{(-c - \gamma_2)x_4 + c - \gamma_1}{(x_4 - 1)x_4} \end{pmatrix} \tag{32}$$

Example 3. ($_3F_2$, see, e.g., [24, p.224], [19].) Let $A = \begin{pmatrix} 1 & 1 & 0 & 0 & 0 & 0 \\ 0 & 0 & 1 & 1 & 0 & 0 \\ 0 & 0 & 0 & 0 & 1 & 1 \\ 1 & 0 & 0 & 1 & 0 & 0 \\ 0 & 0 & 1 & 0 & 0 & 1 \end{pmatrix}$. The integrals are

$$\int_\Gamma (z_1 x_1 + z_2)^{-\gamma_1}(z_3 x_2 + z_4 x_1)^{-\gamma_2}(z_5 + z_6 x_2)^{-\gamma_3}x_1^{c_1}x_2^{c_2}\omega_i \tag{33}$$

where

$$\omega_1 = \frac{dx_1 dx_2}{(z_1 x_1 + z_2)x_1 x_2}, \omega_2 = \frac{dx_1 dx_2}{(z_5 + z_6 x_2)x_1 x_2}, \omega_3 = \frac{dx_1 dx_2}{(z_3 x_2 + z_4 x_1)x_1 x_2} \tag{34}$$

When $z_2 = -1, z_3 = z_4 = z_5 = z_6 = 1$, the coefficient matrix for z_1 for the basis $(\langle\omega_1\rangle, \langle\omega_2\rangle, \langle\omega_3\rangle)^T$ is

$$P_1 = \begin{pmatrix} \frac{\beta_4 z_1 + \beta_2 + \beta_3 + \beta_4 + \beta_5}{z_1(z_1 - 1)} & \frac{\beta_3(\beta_4 - \beta_1 - \beta_2)}{\beta_1 z_1(z_1 - 1)} & \frac{(-\beta_4 + 1)\beta_2(-\beta_2 + \beta_4 + \beta_5 + 1)}{\beta_4 \beta_1 z_1(z_1 - 1)} \\ \frac{(\beta_2 + \beta_3 - \beta_5)\beta_1}{\beta_3(z_1 - 1)} & \frac{\beta_1 z_1 + \beta_2 - \beta_4}{z_1(z_1 - 1)} & \frac{(-\beta_4 + 1)\beta_2(-\beta_2 + \beta_4 + \beta_5 + 1)}{\beta_4 \beta_3 z_1(z_1 - 1)} \\ \frac{\beta_4(\beta_2 + \beta_3 - \beta_5)\beta_1}{(-\beta_4 + 1)\beta_2(z_1 - 1)} & \frac{\beta_4 \beta_3(\beta_1 + \beta_2 - \beta_4)}{(-\beta_4 + 1)\beta_2(z_1 - 1)} & \frac{(-\beta_2 + \beta_4 + \beta_5 + 1)}{z_1 - 1} \end{pmatrix} \tag{35}$$

The result can be obtained in a few seconds.

---

[2] The Algorithm 1 is implemented in `saito-b.rr` distributed at [25].

5   An Algorithm of Finding the Cohomology Intersection Matrix

Theorem 3 [16]. *Given a matrix $A = (a_{ij})$ as in (11) admitting a unimodular regular triangulation T. When parameters are non-resonant, $\gamma_l \notin \mathbb{Z}$ and moreover the set of series solutions by T is linearly independent, the intersection matrix of the twisted cohomology group of the GKZ system associated to the matrix A can be algorithmically determined.*

We denote by Ω_i the coefficient matrix of Ω with respect to the 1-form dz_i. The algorithm we propose is summarized as follows.

Algorithm 2. (A modified version of the algorithm in [16]).

    Input: Free bases $\{\phi_j\}_j \subset \mathcal{H}^n_{dR} \lceil_U$, $\{\psi_j\}_j \subset \mathcal{H}^{n\vee}_{dR} \lceil_U$ which are expressed as (12).

    Output: The secondary equation (8) and the cohomology intersection matrix $I_{ch} = (\langle \phi_i, \psi_j \rangle_{ch})_{i,j}$.

1. *Obtain a Pfaffian system with respect to the given bases $\{\phi_j\}_j$ and $\{\psi_j\}_j$, i.e., obtain matrices $\Omega_i = (\omega_{ijk})$ and $\Omega_i^\vee = (\omega_{ijk}^\vee)$ so that the equalities*

$$\partial_i \phi_j = \sum_k \omega_{ikj} \phi_k, \quad \partial_i \psi_j = \sum_k \omega_{ikj}^\vee \psi_k \qquad (36)$$

    hold by Algorithm 1.

2. *Find a non-zero rational function solution I of the secondary equation*

$$\partial_i I - {}^t\Omega_i I - I\Omega_i^\vee = 0, \quad i = 1, \ldots, N. \qquad (37)$$

    To be more precise, see, e.g., [4, 5, 18] and references therein.

3. *Determine the scalar multiple of I by [15, Theorem 8.1].*

Example 4. This is a continuation of Example 3. We want to evaluate the cohomology intersection matrix $I_{ch} = (\langle \omega_i, \omega_j \rangle_{ch})_{i,j=1}^3$. By solving the secondary equation (for example, using [5]), we can verify that $(1,1)$, $(1,2)$, $(2,1)$, $(2,2)$ entries of I_{ch} are all independent of z_1. Therefore, we can obtain the exact values of these entries by taking a unimodular regular triangulation $T = \{23456, 12456, 12346\}$ and substituting $z_1 = 0$ in [15, Theorem 8.1]. Thus, we get a correct normalization of I_{ch} and the matrix $\frac{I_{ch}}{(2\pi\sqrt{-1})^2}$ is given by

$$\begin{bmatrix} r_{11} & \frac{\beta_4+\beta_5}{\beta_5\beta_4(\beta_2-\beta_4-\beta_5)} \\ \frac{\beta_4+\beta_5}{\beta_5\beta_4(\beta_2-\beta_4-\beta_5)} & r_{22} \\ \frac{\beta_4(\beta_1+\beta_2-\beta_4-\beta_5)z_1-\beta_5\beta_3}{\beta_5(\beta_4-1)(\beta_2-\beta_4-\beta_5)(\beta_2-\beta_4-\beta_5-1)} & \frac{-\beta_4\beta_1 z_1+\beta_5(\beta_2+\beta_3-\beta_4-\beta_5)}{\beta_5(\beta_4-1)(\beta_2-\beta_4-\beta_5)(\beta_2-\beta_4-\beta_5-1)} \\ & \begin{array}{c} \frac{\beta_4(\beta_1+\beta_2-\beta_4-\beta_4)z_1-\beta_5\beta_3}{\beta_5(\beta_4+1)(\beta_2-\beta_4-\beta_5)(\beta_2-\beta_4-\beta_5+1)} \\ \frac{-(\beta_4\beta_1 z_1-\beta_5\beta_2-\beta_5\beta_3+\beta_5\beta_4+\beta_5^2)}{\beta_5(\beta_4+1)(\beta_2-\beta_4-\beta_5)(\beta_2-\beta_4-\beta_5+1)} \\ r_{33} \end{array} \end{bmatrix} \qquad (38)$$

where

$$r_{11} = -\frac{(\beta_4\beta_2 + (\beta_4 + \beta_5)\beta_3)\beta_1 + \beta_4\beta_2^2 + (\beta_4\beta_3 - \beta_4^2 - \beta_5\beta_4)\beta_2 + (-\beta_4^2 - \beta_5\beta_4)\beta_3}{\beta_5\beta_4\beta_1(\beta_2 - \beta_4 - \beta_5)(\beta_2 + \beta_3 - \beta_5)} \quad (39)$$

$$r_{22} = -\frac{(\beta_5\beta_2 + (\beta_4 + \beta_5)\beta_3 - \beta_5\beta_4 - \beta_5^2)\beta_1 + \beta_5\beta_2^2 + (\beta_5\beta_3 - \beta_5\beta_4 - \beta_5^2)\beta_2}{\beta_5\beta_4\beta_3(\beta_2 - \beta_4 - \beta_5)(\beta_1 + \beta_2 - \beta_4)} \quad (40)$$

$$r_{33} = -\frac{\beta_4\{a_0 z_1^2 - 2\beta_1\beta_3\beta_4\beta_5 z_1 + a_2\}}{\beta_5\beta_2(\beta_4 - 1)(\beta_4 + 1)(\beta_2 - \beta_4 - \beta_5)(\beta_2 - \beta_4 - \beta_5 - 1)(\beta_2 - \beta_4 - \beta_5 + 1)} \quad (41)$$

$$a_0 = (\beta_1\beta_2 - \beta_1\beta_5 + \beta_2^2 - \beta_2\beta_4 - 2\beta_2\beta_5 + \beta_4\beta_5 + \beta_5^2)\beta_1\beta_4 \quad (42)$$

$$a_2 = (\beta_2^2 + \beta_2\beta_3 - 2\beta_2\beta_4 - \beta_2\beta_5 - \beta_3\beta_4 + \beta_4^2 + \beta_4\beta_5)\beta_3\beta_5 \quad (43)$$

Example 5. Let $A = \begin{pmatrix} 1\,1\,1\,1\,1 \\ 0\,1\,0\,2\,0 \\ 0\,0\,1\,0\,2 \end{pmatrix}$. The integrals are

$$\int_\Gamma h_1^{-\gamma_1} x_1^{c_1} x_2^{c_2} \omega_i, \quad h_1 = z_1 + z_2 x_1 + z_3 x_2 + x_4 x_1^2 + z_5 x_2^2 \quad (44)$$

where

$$\omega_1 = \frac{dx_1 dx_2}{x_1 x_2}, \omega_2 = x_1 \omega_1 = \frac{dx_1 dx_2}{x_2}, \omega_3 = x_2^2 \omega_1 = \frac{x_2 dx_1 dx_2}{x_1}, \omega_4 = x_1 x_2 \omega_1 = dx_1 dx_2. \quad (45)$$

Note that this A is not normal. When $z_1 = z_4 = z_5 = 1$, we have obtained the coefficient matrices P_2 and P_3 in about 9 h 45 min on a machine with Intel(R) Xeon(R) CPU E5-4650 2.70 GHz and 256 GB memory. The $(1,1)$ element of P_2 is

$$\frac{((b_2 z_2^2 + b_{123})z_3^2 + b_2 z_2^4 + b_{132} z_2^2 - 32b_1 + 16b_2 + 16b_3 - 16)}{z_2(z_2 - 2)(z_2 + 2)(z_3^2 + z_2^2 - 4)} \quad (46)$$

where $b_1 = -\gamma_1$, $b_2 = -c_1$, $b_3 = -c_2$ and $b_{ijk} = 8b_i - 4b_j - 8b_k + 4$. A complete data of P_2 and P_3 is at [25]. The intersection matrix can be obtained by [5] in a few seconds when we specialize b_i's to rational numbers. See [25] as to Maple inputs for it.

References

1. Adolphson, A.: Hypergeometric functions and rings generated by monomials. Duke Math. J. **73**, 269–290 (1994)
2. Aomoto, K., Kita, M.: Theory of Hypergeometric Functions. Springer, Tokyo (1994). https://doi.org/10.1007/978-4-431-53938-4. (in Japanese)
3. Aomoto, K., Kita, M.: Theory of Hypergeometric Functions. Springer, Tokyo (2011). (English translation of [2])
4. Barkatou, M.: On rational solutions of systems of linear differential equations. J. Symb. Comput. **28**, 547–567 (1999)

5. Barkatou, M., Cluzeau, T., El Bacha, C., Weil, J.-A.: IntegrableConnections – a maple package for computing closed form solutions of integrable connections (2012). https://www.unilim.fr/pages_perso/thomas.cluzeau/Packages/IntegrableConnections/PDS.html

6. Cho, K., Matsumoto, K.: Intersection theory for twisted cohomologies and twisted Riemann's period relations I. Nagoya Math. J. **139**, 67–86 (1995)

7. Deligne, P.: Equations Différentielles à Points Singuliers Réguliers. LNM, vol. 163. Springer, Heidelberg (1970). https://doi.org/10.1007/BFb0061194

8. Gel'fand, I.M., Kapranov, M.M., Zelevinsky, A.V.: Hypergeometric functions and toric varieties. Funktsional. Anal. i Prilozhen. **23**(2), 12–26 (1989). Translation in Functional Analysis and Applications **23**(2), 94–106 (1989)

9. Gel'fand, I.M., Kapranov, M.M., Zelevinsky, A.V.: Generalized Euler integrals and A-hypergeometric functions. Adv. Math. **84**, 255–271 (1990)

10. Goto, Y., Kaneko, J., Matsumoto, K.: Pfaffian of Appell's hypergeometric system F_4 in terms of the intersection form of twisted cohomology groups. Publ. Res. Inst. Math. Sci. **52**, 223–247 (2016)

11. Goto, Y., Matsumoto, K.: Pfaffian equations and contiguity relations of the hypergeometric function of type $(k + 1, k + n + 2)$ and their applications. Funkcialaj Ekvacioj **61**, 315–347 (2018)

12. Hibi, T. (ed.): Gröbner Bases. Statistics and Software Systems. Springer, Tokyo (2013). https://doi.org/10.1007/978-4-431-54574-3. https://www.springer.com/gp/book/9784431545736

13. Hibi, T., Nishiyama, K., Takayama, N.: Pfaffian systems of A-hypergeometric equations I: bases of twisted cohomology groups. Adv. Math. **306**, 303–327 (2017)

14. Kita, M., Yoshida, M.: Intersection theory for twisted cycles. Mathematische Nachrichten **166**, 287–304 (1994)

15. Matsubara-Heo, S.-J.: Euler and Laplace integral representations of GKZ hypergeometric functions (2019). arXiv:1904.00565

16. Matsubara-Heo, S.-J., Takayama, N.: An algorithm of computing cohomology intersection number of hypergeometric integrals (2019). arXiv:1904.01253

17. Matsumoto, K.: Intersection numbers for logarithmic k-forms. Osaka J. Math. **35**, 873–893 (1998)

18. Oaku, T., Takayama, N., Tsai, H.: Polynomial and rational solutions of holonomic systems. J. Pure Appl. Algebra **164**, 199–220 (2001)

19. Ohara, K., Sugiki, Y., Takayama, N.: Quadratic relations for generalized hypergeometric functions $_pF_{p-1}$. Funkcialaj Ekvacioj **46**, 213–251 (2003)

20. Polymake – polymake is open source software for research in polyhedral geometry. https://polymake.org

21. Risa/Asir, a computer algebra system. http://www.math.kobe-u.ac.jp/Asir, http://www.openxm.org

22. Assarf, B.: Computing convex hulls and counting integer points with polymake. Math. Program. Comput. **9**, 1–38 (2017)

23. Saito, M., Sturmfels, B., Takayama, N.: Hypergeometric polynomials and integer programming. Compositio Mathematica **115**, 185–204 (1999)

24. Saito, M., Sturmfels, B., Takayama, N.: Gröbner Deformations of Hypergeometric Differential Equations. Springer, Heidelberg (2000). https://doi.org/10.1007/978-3-662-04112-3

25. Rational function solutions and intersection numbers (software appendix of this paper). http://www.math.kobe-u.ac.jp/OpenXM/Math/intersection2

Software for Number Theory and Arithmetic Geometry

Computations with Algebraic Surfaces

Andreas-Stephan Elsenhans[1(✉)] and Jörg Jahnel[2]

[1] Mathematisches Institut, Universität Würzburg, Emil-Fischer-Straße 30,
97074 Würzburg, Germany
stephan.elsenhans@mathematik.uni-wuerzburg.de
[2] Department Mathematik, Universität Siegen, Walter-Flex-Straße 3,
57068 Siegen, Germany
jahnel@mathematik.uni-siegen.de,
https://www.mathematik.uni-wuerzburg.de/computeralgebra/team/
elsenhans-stephan-prof-dr/, https://www.uni-math.gwdg.de/jahnel/

Abstract. Computations with algebraic number fields and algebraic curves have been carried out for a long time. They resulted in many interesting examples and the formation of various conjectures.

The aim of this talk is to report on some computations with algebraic surfaces that are currently possible.

Keywords: Algebraic surfaces · Computer algebra · Point counting

1 Introduction

Algebraic geometry is the study of the sets of solutions of systems of algebraic equations. In dimension zero, these sets consist of a finite number of points. From an arithmetic perspective, the points are usually not defined over the base field. Thus, a detailed inspection requires to work with algebraic number fields. Many algorithms for them are described in [9] and [10].

In dimension 1, the solution sets are algebraic curves. Projective curves are classified by the degree, abstract irreducible curves by the genus. A smooth plane curve of degree 1 or 2 has genus 0. Thus, from a geometric perspective, the curve is isomorphic to the projective line. However, the isomorphism is only defined over the base field if the curve has a rational point. The answer to this question can be found using the famous theorem of Hasse-Minkowski.

Smooth curves of degree 3 are of genus one. They have been studied by many authors from various perspectives. Most notable are the investigations towards the Birch and Swinnerton-Dyer conjecture [1, 2].

Increasing the dimension once more leads us to algebraic surfaces. Here, we have the Enriques-Kodaira classification, which is based on the Kodaira dimension. Prominent surfaces of Kodaira dimensions $-\infty$ are the projective plane, quadratic and cubic surfaces. Ruled surfaces, i.e. surfaces birationally equivalent to $\mathbf{P}^1 \times C$, for a curve C of arbitrary genus, are in this class too.

© Springer Nature Switzerland AG 2020
A. M. Bigatti et al. (Eds.): ICMS 2020, LNCS 12097, pp. 87–93, 2020.
https://doi.org/10.1007/978-3-030-52200-1_8

The most important family of surfaces of Kodaira dimension 0 are K3 surfaces. This family includes all smooth quartic surfaces. Further, abelian surfaces have Kodaira dimension 0, as well.

Finally, there are surfaces of Kodaira-dimensions 1 and 2. They will not be considered in this talk.

The algorithms presented in this article have been implemented by the authors over the past 10 years as a part of their research. The given examples are based on magma [4], version 2.25.

2   Computation with Cubic Surfaces

2.1   Definition

A cubic surface is a smooth algebraic surface in $\mathbf{P}^3$ given as the zero set of a homogeneous cubic form in four variables.

2.2   Properties of Cubic Surfaces

1. It is well known that every smooth cubic surface contains exactly 27 lines. As the lines generate the Picard group of the surface, many other properties of the surface relate to them [24].
2. The moduli stack of cubic surfaces if of dimension 4.

For a modern presentation of the geometry of cubic surfaces we refer the interested read to [12, Chap. 9].

2.3   Computational Questions

1. Given two cubic surfaces, can we test for isomorphy?
2. Given a cubic surface over a finite field, can we count the number of points on the surface efficiently?
3. Given a cubic surface over the rationals, what is known about the number of rational points on the surface? Is there a computational approach to this?

2.4   Invariants and Isomorphy Testing

As proven by Clebsch, the ring of invariants of even weight of cubic surfaces is generated by five invariants of degrees 8, 16, 24, 32, and 40 [8]. Further, there is an invariant of odd weight an degree 100. These invariants can be computed in magma:

```
r4<x,y,z,w>:= PolynomialRing(Rationals(),4);
f:= x^3 + y^3 + z^3 + w^3 + (3*x+3*z+4*w)^3;
time ClebschSalmonInvariants(f);
[ -2579, -46656, 0, 0, 0 ]
-708235798046072773554016875
```

```
Time: 0.010
> time Factorization(708235798046072773554016875);
[ <3, 27>, <5, 4>, <13, 4>, <2281, 2> ]
Time: 0.000
> time SkewInvariant100(f);
0
Time: 0.010
```

The last return value of `ClebschSalmonInvariants` is the discriminant of the surface. Thus, this surface has bad reduction only at 3, 5, 13 and 2281. Further, the degree 100 invariant vanishes. This shows that the surface has at least one non-trivial automorphism. Multiplication of y by a 3rd root of unity and interchanging x- and z-coordinate are automorphisms of the example.

A detailed description of the algorithm is given in [19]. As an isomorphism of smooth cubic surfaces is always given by a projective linear map, they are isomorphic if and only if the invariants coincide.

2.5   Counting Points over Finite Fields

The number of points on a variety over a finite field relates to the Galois module structure on its etale cohomology [25]. In the case of a cubic surface, the cohomology is generated by the lines on the surface. Using Gröbner bases, one can explicitly determine the lines on a cubic surface and compute the Galois module structure. In magma, this is available as follows:

```
r4<x,y,z,w>:= PolynomialRing(Rationals(),4);
f:= x^3 + y^3 + z^3 + w^3 + (x+2*x+3*z+4*w)^3;
p:= NextPrime(17^17);
time NumberOfPointsOnCubicSurface(PolynomialRing(GF(p),4)!f);
68432645088577503417220551894681908835253 9
Time: 0.170
```

The second return value is the action of the Frobenius on the lines encoded by its Swinnerton-Dyer number. A detailed description of the lines on a cubic surface and potential Frobenius actions are given in [24].

2.6   Rational Points on Cubic Surfaces

As soon as a smooth cubic surface over $\mathbb{Q}$ has one rational point, one can construct infinitely many other rational points. Further, there are numerous conjectures and questions towards the set of rational points.

If we fix a search bound B, then we can ask for the number of points,

$$n(B) := \#\{(x : y : z : w) \in \mathbf{P}^3(\mathbb{Q}) \mid x, y, z, w \in \mathbb{Z}, |x|, |y|, |z|, |w| < B,$$
$$f(x, y, z, w) = 0\},$$

on the surface, given by $f = 0$. For cubic surfaces, this question is covered by the Manin conjecture [20]. More precisely, when counting only the points that are not contained in any of the lines on the surface

$$n'(B) := \#\{(x : y : z : w) \in \mathbf{P}^3(\mathbb{Q}) \mid x, y, z, w \in \mathbb{Z}, |x|, |y|, |z|, |w| < B,$$
$$f(x, y, z, w) = 0, (x : y : z : w) \text{ not on a line of } f = 0\},$$

the conjecture predicts the existence of a constant C such that

$$n' \sim C \cdot B \log^{r-1}(B).$$

Here, r is the rank of the arithmetic Picard group. A conjecture for the value of C is presented in [26]. Today, we have a lot of numerical and theoretical evidence for this conjecture. Some numerical examples of smooth cubics are given in [15]. Examples such that a more complex set than just a fixed finite collection of lines on the variety needs to be excluded from the count are given in [13] and [14].

Finally, the conjecture is proven for some singular surfaces [5]. The interested reader my also consult [6] for a general introduction to the Manin conjecture.

3   Computations with K3 Surfaces

3.1   Definition

A K3 surface is a smooth algebraic surface which is simply connected and has trivial canonical class.

3.2   Examples

As the definition of K3 surfaces is abstract, they arise in various forms.

1. Let $f_6(x, y, z) = 0$ be a smooth plane curve of degree 6. Then the double cover of $\mathbf{P}^2$, given by
$$w^2 = f_6(x, y, z),$$
   is a K3 surface of degree 2 in $\mathbf{P}(1, 1, 1, 3)$.
2. A smooth quartic surface in $\mathbf{P}^3$ is a K3 surface.
3. A smooth complete intersection of a quadric and a cubic in $\mathbf{P}^4$ is a K3 surface of degree 6.
4. A smooth complete intersection of three quadrics in $\mathbf{P}^5$ is a K3 surface of degree 8.

If a surface of the shape above has only ADE-singularites, then the minimal resolution of singularities is still a K3 surface.

3.3   Questions Towards K3 Surfaces

1. Can we test isomorphy of K3 surfaces?
2. What is known about its cohomology?
3. Can we count point over finite fields on K3 surfaces efficiently?
4. What is known about the rational points on a K3 surface defined over $\mathbb{Q}$?

3.4   Invariants and Isomorphy

For none of the above models of K3 surfaces, a complete system of invariants is known. In contrast to cubic surfaces, an isomorphism of K3 surfaces may not be given by a projective linear map. Thus, the isomorphy test is harder in this instance as different embeddings have to be taken into account.

3.5   Cohomology of K3 Surfaces

The cohomology $H^2(V, \mathbb{Z})$ of a K3 surface V over $\mathbb{Q}$ is isomorphic to $\mathbb{Z}^{22}$. All algebraic cycles defined over $\overline{\mathbb{Q}}$ generate a sublattice called the geometric Picard group. Its rank $r \in \{1, \ldots, 20\}$ is called the geometric Picard rank.

3.6   Counting Points over Finite Fields

Counting points over finite fields can always be done naively by enumeration. However, there are more efficient methods available. Most notable are the p-adic methods developed by Kedlaya, Harvey, and others [11, 21, 22]. The following is available in magma:

```
r3<x,y,z>:= PolynomialRing(Rationals(),3);
f:= x^6+y^6+z^6+(x+2*y+3*z)^6;
time WeilPolynomialOfDegree2K3Surface(f,31);
Time: 72.700
t^22 + 58*t^21 + 372*t^20 - 55738*t^19 - 1549132*t^18
    - 12929294*t^17 - 572583020*t^16 - 15975066258*t^15
    + 495227053998*t^14 + 10234692449292*t^13 - 608111309695433*t^12
    + 584394968617311113*t^10 - 9451953405462597132*t^9
    - 4395158333354010766638*t^8 + 136249908339743337765778*t^7
    + 4693052398368937185990020*t^6 + 101839237044605936935987341*t^5
    + 1172606072256462645291511372*t^4
    + 4054510995994461223527187397 8*t^3
    - 260047946639644754336571329652*t^2
    - 389638506715067723580962708928 58*t
    - 6455906981951380730367330401 38561
```

This gives us the characteristic polynomial of the Frobenius on the etale cohomology and the number of points over $\mathbb{F}_{31^d}$ is encoded in this. Some details on this function are given in [18].

3.7   Computing Algebraic Cycles

As explained above, the algebraic cycles on a K3 surface form a lattice of rank $r = 1, \ldots, 20$. Thus, a first step to determine its rank is a computation of lower and upper bounds. Lower bounds can be generated by enumerating cycles. Upper bounds can be derived by point counting [23]. Applying WeilPolynomialToRankBound to the above example gives us the bound $r \leq 10$.

Combining this with a modulo 71 computation and `ArtinTateFormula`, one can sharpen this bound to $r \le 9$. These functions were implemented as part of the research described in [16].

As worked out by Charles [7], primes resulting in sharp upper bounds have positive density, as long as the surface does not have real multiplication. The first explicit family V_t of K3 surfaces such that the approach fails was given in [17]:

$$V_t \colon w^2 = q_1 q_2 q_3$$

with

$$q_1 := \left(\frac{1}{8}t^2 - \frac{1}{2}t + \frac{1}{4}\right) y^2 + (t^2 - 2t + 2)yz + (t^2 - 4t + 2)z^2,$$

$$q_2 := \left(\frac{1}{8}t^2 + \frac{1}{2}t + \frac{1}{4}\right) x^2 + (t^2 + 2t + 2)xz + (t^2 + 4t + 2)z^2,$$

$$q_3 := 2x^2 + (t^2 + 2)xy + t^2 y^2.$$

3.8 Rational Points

For the structure of the set of rational points on K3 surfaces, only conjectures are known. Most notable is a conjecture of Bogomolov [3]: *Every rational point on a K3 surfaces lies on some rational curve on the surface.*

Up to the authors knowledge, there are no computational investigations on this conjecture.

References

1. Birch, B.J., Swinnerton-Dyer, H.P.F.: Notes on elliptic curves. I. J. Reine Angew. Math. **212**, 7–25 (1963)
2. Birch, B.J., Swinnerton-Dyer, H.P.F.: Notes on elliptic curves. II. J. Reine Angew. Math. **218**, 79–108 (1965)
3. Bogomolov, F., Tschinkel, Y.: Rational curves and points on K3 surfaces. Am. J. Math. **127**(4), 825–835 (2005)
4. Bosma, W., Cannon, J., Playoust, C.: The Magma algebra system. I. The user language. J. Symb. Comput. **24**, 235–265 (1997)
5. de la Bretèche, R., Browning, T.D., Derenthal, U.: On Manin's conjecture for a certain singular cubic surface. Annales Scientifiques de l'École Normale Supérieure **40**(1), 1–50 (2007)
6. Browning, T.D.: An overview of Manin's conjecture for del Pezzo surfaces. In: Analytic Number Theory, Clay Mathematics Proceedings, vol. 7, pp. 39–55. American Mathematical Society, Providence (2007)
7. Charles, F.: On the Picard number of K3 surfaces over number fields. Algebra Number Theory **8**(1), 1–17 (2014)
8. Clebsch, A.: Ueber eine Transformation der homogenen Functionen dritter Ordnung mit vier Veränderlichen. J. für die Reine und Angew. Math. **58**, 109–126 (1861)

9. Cohen, H.: A Course in Computational Algebraic Number Theory. Graduate Texts in Mathematics, vol. 138. Springer, Heidelberg (1993). https://doi.org/10.1007/978-3-662-02945-9

10. Cohen, H.: Advanced Topics in Computational Number Theory. Graduate Texts in Mathematics, vol. 193. Springer, New York (2000). https://doi.org/10.1007/978-1-4419-8489-0

11. Costa, E.: Effective computations of Hasse-Weil zeta functions. Ph.D. thesis (2015)

12. Dolgachev, I.V.: Classical Algebraic Geometry: A Modern View. Cambridge University Press, Cambridge (2012)

13. Elsenhans, A.-S.: Rational points on some Fano quadratic bundles. Exp. Math. **20**(4), 373–379 (2011)

14. Elsenhans, A.-S., Jahnel, J.: The asymptotics of points of bounded height on diagonal cubic and quartic threefolds. In: Hess, F., Pauli, S., Pohst, M. (eds.) ANTS 2006. LNCS, vol. 4076, pp. 317–332. Springer, Heidelberg (2006). https://doi.org/10.1007/11792086_23

15. Elsenhans, A.-S., Jahnel, J.: Experiments with general cubic surfaces. In: Tschinkel, Y., Zarhin, Y. (eds.) Algebra, Arithmetic, and Geometry: Volume I: In Honor of Yu. I. Manin. Progress in Mathematics, vol. 269, pp. 637–653. Birkhäuser, Boston (2009). https://doi.org/10.1007/978-0-8176-4745-2_14

16. Elsenhans, A.-S., Jahnel, J.: On Weil polynomials of *K*3 surfaces. In: Hanrot, G., Morain, F., Thomé, E. (eds.) ANTS 2010. LNCS, vol. 6197, pp. 126–141. Springer, Heidelberg (2010). https://doi.org/10.1007/978-3-642-14518-6_13

17. Elsenhans, A.-S., Jahnel, J.: Examples of K3 surfaces with real multiplication. In: Proceedings of the ANTS XI Conference (Gyeongju 2014) (2014). LMS Journal of Computation and Mathematics 17, 14–35

18. Elsenhans, A.-S., Jahnel, J.: Point counting on K3 surfaces and an application concerning real and complex multiplication. In: Proceedings of the ANTS XII Conference (Kaiserslautern 2016) (2016). LMS Journal of Computation and Mathematics 19, 12–28

19. Elsenhans, A.-S., Jahnel, J.: Computing invariants of cubic surfaces. In: Le Matematiche (to appear)

20. Franke, J., Manin, Y.I., Tschinkel, Y.: Rational points of bounded height on Fano varieties. Invent. Math. **95**(2), 421–435 (1989). https://doi.org/10.1007/BF01393904

21. Harvey, D.: Computing zeta functions of arithmetic schemes. Proc. Lond. Math. Soc. **111**(6), 1379–1401 (2015)

22. Kedlaya, K.S.: Computing zeta functions via *p*-adic cohomology. In: Buell, D. (ed.) ANTS 2004. LNCS, vol. 3076, pp. 1–17. Springer, Heidelberg (2004). https://doi.org/10.1007/978-3-540-24847-7_1

23. van Luijk, R.: K3 surfaces with Picard number one and infinitely many rational points. Algebra Number Theory **1**(1), 1–15 (2007)

24. Manin, Y.I.: Cubic Forms. North Holland, Amsterdam (1986)

25. Milne, J.S.: Etale Cohomology. Princeton University Press, Princeton (1980)

26. Peyre, E.: Hauteurs et mesures de Tamagawa sur les variétés de Fano. Duke Math. J. **79**(1), 101–218 (1995)

Evaluating Fractional Derivatives
of the Riemann Zeta Function

Ricky E. Farr, Sebastian Pauli[(✉)][(iD)], and Filip Saidak

Department of Mathematics and Statistics, University of North Carolina Greensboro,
Greensboro, NC 27402, USA
{s_pauli,f_saidak}@uncg.edu

Abstract. We present a method for evaluating the reverse Grünwald-Letnikov fractional derivatives of the Riemann Zeta function $\zeta(s)$ and use it to explore the location of zeros of integral and fractional derivatives on the left half plane.

1 Introduction

The Riemann zeta function $\zeta(s)$ and its derivatives $\zeta^{(k)}(s)$ are defined by

$$\zeta(s) = \sum_{n=1}^{\infty} \frac{1}{n^s} \quad \text{and} \quad \zeta^{(k)}(s) = (-1)^k \sum_{n=2}^{\infty} \frac{(\log n)^k}{n^s} \tag{1}$$

everywhere in the half-plane $\Re(s) > 1$. By a process of analytic continuation these functions can be extended to meromorphic functions with a single pole at $s = 1$. Moreover, $\zeta(s)$ has the Laurent series expansion:

$$\zeta(s) = \frac{1}{s-1} + \sum_{n=0}^{\infty} \frac{(-1)^n \gamma_n}{n!} (s-1)^n, \tag{2}$$

where γ_0 is the Euler constant and for $n \geq 1$ γ_n are the Stieltjes constants.

Unlike $\zeta(s)$ itself, the functions $\zeta^{(k)}(s)$ have neither Euler products nor functional equations. Thus their nontrivial zeros do not lie on a line, but appear to be distributed seemingly at random with most zeros located to the right of the critical line $\sigma = \frac{1}{2}$. Speiser [16] was the first to show, in 1934, that the Riemann Hypothesis is equivalent to the fact that $\zeta'(s)$ has no zeros with $0 < \sigma < \frac{1}{2}$. Spira [17] noticed that the zeros of $\zeta'(s)$ and $\zeta''(s)$ seem to come in pairs, where a zero of $\zeta''(s)$ is located to the right of a zero of $\zeta'(s)$. More recently, with the help of extensive computations, Skorokhodov [15] observed this behavior for higher derivatives as well.

Our results from [2] support a straightforward one-to-one correspondence between the zeros of $\zeta^{(k)}(s)$ and $\zeta^{(m)}(s)$ for large k and m on the right half plane. Furthermore in [3] we have observed an interesting behavior of the zeros of $\zeta^{(k)}(s)$ on the left half plane, namely they seem to lie on curves which are extensions of chains of zeros of $\zeta^{(k)}(s)$ that were observed on the right half plane.

A. M. Bigatti et al. (Eds.): ICMS 2020, LNCS 12097, pp. 94–101, 2020.
https://doi.org/10.1007/978-3-030-52200-1_9

Also some of the zeros of $\zeta^{(k)}(s)$ on the negative real axis appeared to be part the chains.

We are investigating this correspondence between the zeros of different derivatives by considering curves of zeros of fractional derivatives $\zeta^{(k)}(s)$ that connect the zeros of integral derivatives. We have found that among the multitude of existing definitions of fractional derivatives, the reverse Grünwald-Letnikov fractional derivative works best for situations dealing with $\zeta(s)$.

In [4] we have applied it in a proof of a conjecture by Kreminski [10] and in [5] we have been able to apply some of the properties of the fractional Stieltjes constants to prove that the zero free region of $\zeta(s)$ of radius one about $s = 1$ generalizes to fractional derivatives.

In [14] we present generalizations of the zero free regions of integral derivatives of $\zeta(s)$ on the right half plane from [2] to fractional derivatives. This yields the existence of curves of zeros of fractional derivatives on the right half plane. Here we conduct numerical investigations of the zeros of fractional derivatives where we concentrate our attention to the left half plane.

2 Grünwald-Letnikov Fractional Derivatives of $\zeta(s)$

The fractional derivative introduced by Grünwald [7] in 1867 was simplified both in approach and notation, by Letnikov in 1869 [11, 12]. For $N \in \mathbb{N}$ and $h > 0$, let

$$\Delta_h^N f(z) = (-1)^N \sum_{k=0}^{N} (-1)^k \binom{N}{k} f(z + kh)$$

be the N-th finite difference of f. Then for all $n \in \mathbb{N}$ we have:

$$f^{(n)}(z) = \lim_{h \to 0} \frac{\Delta_h^n f(z)}{h^n}$$

This can be naturally extended to the fractional case with the generalization of $\Delta_h^N f(z)$

$$\Delta_h^\alpha f(z) = (-1)^\alpha \sum_{k=0}^{\infty} (-1)^k \binom{\alpha}{k} f(z + kh),$$

where $\alpha \in \mathbb{C}$ and $\binom{\alpha}{k} = \frac{\Gamma(\alpha+1)}{\Gamma(k+1)\Gamma(\alpha-k+1)}$. The reverse α^{th} Grünwald-Letnikov derivative of a function $f(z)$ is now defined as:

$$D_z^\alpha [f(z)] = \lim_{h \to 0^+} \frac{\Delta_h^\alpha f(z)}{h^\alpha} = \lim_{h \to 0^+} \frac{(-1)^\alpha \sum_{k=0}^{\infty} (-1)^k \binom{\alpha}{k} f(z + kh)}{h^\alpha}, \qquad (3)$$

whenever the limit exists.

Defined this way, the fractional derivatives $D_s^\alpha [f(s)]$ coincides with the integral derivatives for all $\alpha \in \mathbb{N}$. Furthermore, they satisfy $D_s^0[f(s)] = f(s)$ and

$D_s^\alpha \left[D_s^\beta \left[f(s) \right] \right] = D_s^{\alpha+\beta} \left[f(s) \right]$, for all $\alpha, \beta \in \mathbb{C}$. For $c \in \mathbb{C}$ we have that $D_s^\alpha[c] = 0$ and for $m \neq 0$ we have $D_s^\alpha \left[e^{ms} \right] = m^\alpha e^{ms}$. So for $s \in \mathbb{C}$ with $\Re(s) > 1$ and $\alpha > 0$ we have as the generalization of (1) that

$$\zeta^{(\alpha)}(s) = D_s^\alpha \left[\zeta(s) \right] = (-1)^\alpha \sum_{n=1}^\infty \frac{\log^\alpha(n+1)}{n^s}. \tag{4}$$

We have already used the generalization of (2) to the fractional domain in our proof [4] of a conjecture of Kreminski [10]. For $1 \neq s \in \mathbb{C}$ and $\alpha > 0$ we have

$$\zeta^{(\alpha)}(s) = D_s^\alpha \left[\zeta(s) \right] = (-1)^{-\alpha} \frac{\Gamma(\alpha+1)}{(s-1)^{\alpha+1}} + \sum_{n=0}^\infty \frac{(-1)^n \gamma_{\alpha+n}}{n!} (s-1)^n,$$

where the γ_α are the fractional Stieltjes constants. Because of the branch cut of the complex logarithm there is a discontinuity along $(-\infty, 1]$ for $\alpha \notin \mathbb{N}$. On $\mathbb{C} \setminus (-\infty, 1]$ the fractional derivative is analytic. As a direct consequence we obtain the following useful property:

Proposition 1. *Let α be a positive real number.*

1. *If $\sigma \in (1, \infty)$ and $\alpha \notin \mathbb{N}$ then $D_\sigma^\alpha \left[\zeta(\sigma) \right]$ is non-real.*
2. *For $s \in \mathbb{C} \setminus (-\infty, 1]$ we have $D_\sigma^\alpha \left[\zeta(\bar{s}) \right] = (-1)^{2\alpha} \overline{D_\sigma^\alpha \left[\zeta(s) \right]}$.*

While this establishes symmetry for the location of the zeros $D_\sigma^\alpha \left[\zeta(s) \right]$ in $\mathbb{C}$, with respect to the real axis, the symmetry is not perfect. It only refers to the location, and not the actual mirroring of properties, or the dynamics surrounding the zeros. Nevertheless it asserts that chains of zeros can be observed on the upper as well as lower half plane.

3   Evaluating $D_s^\alpha \left[\zeta(s) \right]$

One of the most effective ways for evaluating (4) and its analytic continuations to the regions where $\sigma < 1$ is Euler-Maclaurin summation. We use the following form of the summation formula:

$$\sum_{k=m}^N g(k) = \int_m^N g(x)dx + \sum_{k=1}^v \frac{(-1)^k B_k}{k!} g^{(k-1)}(x) \Big|_{x=m}^N + (-1)^{v+1} \int_m^N P_v(x) g^{(v)}(x)dx,$$

where $g(x) \in C^v[m, n]$, $v \in \mathbb{N}$, B_k denotes the k-th Bernoulli number, and $P_k(x) = \frac{B_k(x - \lfloor x \rfloor)}{k!}$ is the kth periodic Bernoulli polynomial. If $g(x)$ decreases rapidly enough for $N \to \infty$, then

$$\sum_{k=2}^\infty g(k) = \sum_{k=2}^{m-1} g(k) + \int_m^\infty g(x)dx + \sum_{k=1}^v \frac{(-1)^k B_k}{k!} g^{(k-1)}(x) \Big|_{x=m}^\infty \tag{5}$$
$$+ (-1)^{v+1} \int_m^\infty P_v(x) g^{(v)}(x)dx$$

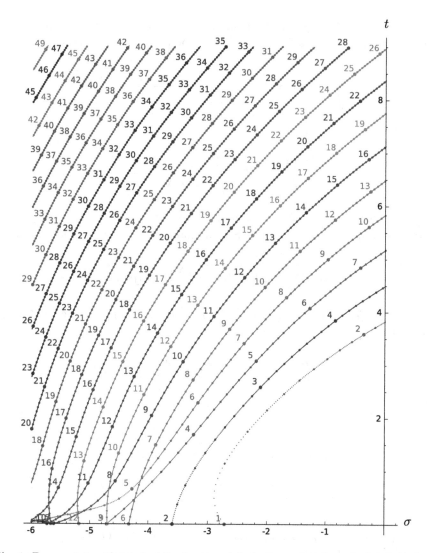

Fig. 1. Zeros $\sigma + it$ with $t \geq 0$ of the fractional derivatives of $\zeta(s)$ on the left half plane. For $k \in \mathbb{N}$ zeros of $\zeta^{(k)}(s)$ are labeled with k. Not all zeros on the real axis are shown. The values for α are $1/100$ apart. For details about $\sigma = -6$ see Fig. 2.

We now use this to approximate $\zeta^{(\alpha)}(s) = D_s^\alpha \left[\zeta(s) \right] = (-1)^\alpha \sum_{k=2}^{\infty} \frac{\log^\alpha k}{k^s}$ where $s \in \mathbb{C}$ with $\Re(s) > 1$. Let $g(x) = \frac{\log^{(\alpha)}(x)}{x^s}$. Then $\sum_{k=2}^{\infty} g(k)$ converges for $\Re(s) > 1$. We assume that v is even. We evaluate the first summand of (5) as is, namely as

$$G_s^\alpha(m) = \sum_{k=2}^{m-1} g(k) = \sum_{k=2}^{m-1} \frac{\log^\alpha k}{k^s}$$

The second term of the right hand side of (5) can be written in terms of the Upper Incomplete Gamma function $\Gamma(\alpha, s)$ (compare [6, p. 346] and [1, 6.5.3]):

$$I_s^\alpha(m) = \int\limits_m^\infty g(x)dx = \int\limits_m^\infty \frac{\log^\alpha x}{x^s}dx = \frac{\Gamma(\alpha + 1, (s-1)\log(m))}{(s-1)^{\alpha+1}}$$

For the third term we get:

$$B_s^\alpha(m, v) = \frac{1}{2}\frac{\log^\alpha(m)}{m^s} - \sum_{j=1}^{v/2}\frac{B_{2j}}{(2j)!}\left(\frac{\log^\alpha(x)}{x^s}\right)^{(2j-1)}$$

Now we determine a bound for the fourth term of (5). We denote the falling factorial by $(\alpha)_i = \frac{\Gamma(\alpha+1)}{\Gamma(\alpha-i+1)}$ and the Stirling numbers of the first kind by $s(j, i)$. Let

$$S(k, i, s) = \sum_{j=0}^{k-i}(-1)^{k-i+j}(-1)^k\binom{k}{j}(-\alpha)_j s(k-j, i) \qquad (6)$$

be the the non-central Stirling numbers. The derivatives of g can be written as [8, Theorem 1]:

$$g^{(k)}(x) = \left(\frac{\log^\alpha x}{x^s}\right)^{(k)} = \sum_{i=0}^{k}S(k, i, s)(\alpha)_i\frac{\log^{\alpha-i}(x)}{x^{s+k}}$$

Writing $s = \sigma + it$ and

$$E_s^\alpha(m, v) = \frac{1}{v!}\int\limits_m^\infty P_v(x)g^{(v)}(x)dx$$

we obtain

$$|E_s^\alpha(m, v)| = \left|\frac{1}{v!}\int_m^\infty P_v(x)g^{(v)}(x)dx\right| \leq \frac{|B_v|}{v!}\int_m^\infty |g^{(v)}(x)|dx$$

$$\leq \frac{|B_v|}{v!}\sum_{j=0}^{v}\int_m^\infty\left|S(v, j, s)(\alpha)_j\frac{\log^{\alpha-j}(x)}{x^{s+v}}\right|dx$$

$$\leq \frac{|B_v|}{v!}\left(\sum_{j=0}^{v}|S(v, j, s)(\alpha)_j|\right)\left(\int_m^\infty\frac{\log^k(x)}{x^{\sigma+v}}dx\right)$$

$$= \frac{|B_v|}{v!}\left(\sum_{j=0}^{v}|S(v, j, s)(\alpha)_j|\right)\frac{\Gamma(\alpha+1, (\sigma+v-1)\log(m))}{(\sigma+v-1)^{\alpha+1}}$$

The error term $E_s^\alpha(m, v)$ converges for $\sigma + v > 1$ and $m > 2$.

For all $s \in \mathbb{C} \setminus (\infty, 1]$ we can choose $m \in \mathbb{N}$ and $v \in \mathbb{N}$ such that $|E_s^\alpha(m, v)|$ becomes arbitrarily small. We can thus approximate $D_s^\alpha [\zeta(s)]$ as

$$D_s^\alpha [\zeta(s)] \approx (-1)^\alpha (G_s^\alpha(m) + I_s^\alpha(m) + B_s^\alpha(m, v))$$

where the error is $|E_s^\alpha(m, v)|$.

We have implemented the method described above in the computer algebra system SageMath [18] using the library mpmath [9]. A considerable increase in speed was obtained by caching the values of the non-central Stirling numbers, which we evaluate by their recurrence relation. Figures 1 and 2 were generated with our implementation.

4   Exploring the Left Half Plane

With our implementation of the approximation to $\zeta^{(\alpha)}(s)$, see Sect. 3, we have investigated the distribution of the zeros on the left half plane. We observe, see Fig. 1, that the zeros on the left half plane given in [3] appear to be connected in a similar manner as on the right half plane.

Furthermore they connect to zeros of integral derivatives on the negative real axis. Note that there is a discontinuity of $\zeta^{(\alpha)}(s)$ for $\alpha \notin \mathbb{N} \cup \{0\}$ on the real axis $\sigma < 1$. We find different patterns how zeros of integral derivatives are connected, see Fig. 2. Some of the curves start and stop at zeros of integral derivatives on the left real axis such as shown in the first plot in Fig. 2. Further to the left we find curves touching (or crossing) the real line at integral derivative and jumping between those points, as shown in the second, third, and fourth plot in Fig. 2.

Levinson and Montgomery [13] have shown that $\zeta^{(k)}(s)$ for $k \in \mathbb{N}$ has only finitely many non-real zeros on the left half plane. Taking derivatives of the Laurent series expansion (2) of $\zeta(s)$ one immediately sees that the order of the pole of the k-th derivative of $\zeta(s)$ is $k + 1$. Thus the argument of $\zeta^{(k)}(\gamma(t))$ on a curve $\gamma : [0, 2\pi) \to \mathbb{C}$ around $s = 1$ whose interior does not contain any zeros cycles through all of $[0, 2\pi)$ exactly $k + 1$ times. Each of these cycles "spawns" at most 2 zeros of $\zeta^{(k)}(s)$. If those zeros were evenly distributed, there would be at most $\frac{k+1}{2}$ such zeros in the upper left half plane. Experiments suggest that this is indeed an upper bound for the count of such zeros (see Table 1) and that these are the only non-real zeros on the upper left half plane (see Fig. 1). This leads us to conjecture:

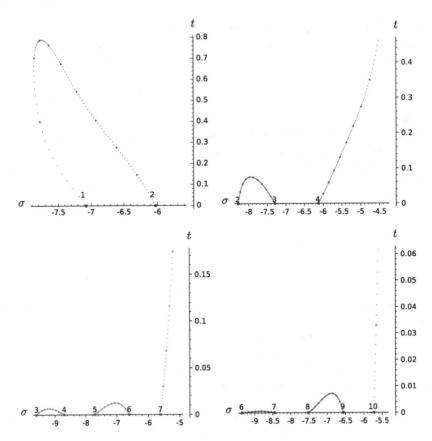

Fig. 2. Selected zeros of the fractional derivatives of $\zeta(s)$ on the upper left half plane. For $k \in \mathbb{N}$ zeros of $\zeta^{(k)}(s)$ are labeled with k. The values for α are $1/100$ apart.

Conjecture 1. Let $k \in \mathbb{N}$. The number of pairs of non-real zeros of $\zeta^{(k)}(s)$ with $\sigma \leq 0$ is at most $\frac{k+1}{2}$.

Fig. 2 shows that this is not the case for fractional derivatives.

Table 1. The number N of pairs of non-real zeros of $\zeta^{(k)}(s)$ for $\Re(s) < 1$.[†] Levinson and Montgomery [13, Theorem 9], [‡] Yıldırım [19, Theorems 2 and 3]. The values for $k > 3$ are experimental.

k	0	1	2	3	4	5	6	7	8	9	10	11	12	13	14	15	16
$\lfloor \frac{k+1}{2} \rfloor$	0	1	1	2	2	3	3	4	4	5	5	6	6	7	7	8	8
N	0	0[†]	1[‡]	1[‡]	2	3	3	3	3	4	4	4	4	5	5	5	6

References

1. Abramowitz, M., Stegun, I.A.: Handbook of mathematical functions with formulas, graphs, and mathematical tables, National Bureau of Standards Applied Mathematics Series, vol. 55. U.S. Government Printing Office, Washington, D.C, For sale by the Superintendent of Documents (1964)
2. Binder, T., Pauli, S., Saidak, F.: Zeros of high derivatives of the Riemann zeta function. Rocky Mt. J. Math. **45**(3), 903–926 (2015). https://doi.org/10.1216/RMJ-2015-45-3-903
3. Farr, R., Pauli, S.: More zeros of the derivatives of the riemann zeta function on the left half plane. In: Rychtář, J., Gupta, S., Shivaji, R., Chhetri, M. (eds.) Topics from the 8th Annual UNCG Regional Mathematics and Statistics Conference. SPMS, vol. 64, pp. 93–104. Springer, New York (2013). https://doi.org/10.1007/978-1-4614-9332-7_10
4. Farr, R.E., Pauli, S., Saidak, F.: On fractional Stieltjes constants. Indag. Math. (N.S.) **29**(5), 1425–1431 (2018). https://doi.org/10.1016/j.indag.2018.07.005
5. Farr, R.E., Pauli, S., Saidak, F.: A zero free region for the fractional derivatives of the Riemann zeta function. NZJM **50**, 1–9 (2018). http://nzjm.math.auckland.ac.nz/index.php/Azero-freeregionforthefractionalderivativesoftheRiemannzetafunction
6. Gradshteyn, I.S.: Table of Integrals, Series, and Products. Academic Press, Cambridge (2007)
7. Grünwald, A.K.: Über begrenzte Derivation und deren Anwendung. Z. Angew. Math. Phys **12**, 441–480 (1867)
8. Janjic, M.: On non-central Stirling numbers of the first kind (2009). arXiv preprint arXiv:0901.2655
9. Johansson, F., et al.: mpmath: a Python library for arbitrary-precision floating-point arithmetic (version 0.18) (2013). http://mpmath.org/
10. Kreminski, R.: Newton-Cotes integration for approximating Stieltjes (generalized Euler) constants. Math. Comput. **72**(243), 1379–1397 (2003). https://doi.org/10.1090/S0025-5718-02-01483-7. (electronic)
11. Letnikov, A.V.: Historical development of the theory of differentiation of fractional order. Mat. Sbornik **3**, 85–119 (1868)
12. Letnikov, A.V.: Theory of differentiation of fractional order. Mat. Sbornik **3**, 1–68 (1868)
13. Levinson, N., Montgomery, H.L.: Zeros of the derivatives of the Riemann zeta-function. Acta Math. **133**, 49–65 (1974). https://doi.org/10.1007/BF02392141
14. Pauli, S., Saidak, F.: Zeros of fractional derivatives of the Riemann zeta function. preprint (2020)
15. Skorokhodov, S.L.: Padé approximants and numerical analysis of the Riemann zeta function. Zh. Vychisl. Mat. Mat. Fiz. **43**(9), 1330–1352 (2003)
16. Speiser, A.: Geometrisches zur Riemannschen Zetafunktion. Math. Ann. **110**, 514–521 (1934)
17. Spira, R.: Zero-free regions of $\zeta^{(k)}(s)$. J. Lond. Math. Soc. **40**, 677–682 (1965). https://doi.org/10.1112/jlms/s1-40.1.677
18. The Sage Developers: SageMath, the Sage Mathematics Software System (2019). https://www.sagemath.org/
19. Yıldırım, C.Y.: Zeros of $\zeta''(s)$ & $\zeta'''(s)$ in $\sigma\frac{1}{2}$. Turkish J. Math. **24**(1), 89–108 (2000)

Groups and Group Actions

Towards Efficient Normalizers
of Primitive Groups

Sergio Siccha[(✉)] [iD]

Department of Mathematics, University of Kaiserslautern, Kaiserslautern, Germany
`siccha@mathematik.uni-kl.de`

Abstract. We present the ideas behind an algorithm to compute normalizers of primitive groups with non-regular socle in polynomial time. We highlight a concept we developed called permutation morphisms and present timings for a partial implementation of our algorithm. This article is a collection of results from the author's PhD thesis.

Keywords: Normalizers · Primitive groups · Permutation group algorithms

1 Introduction

One of the tools to study the internal structure of groups is the normalizer. For two groups G and H, which are contained in a common overgroup K, we call the *normalizer of G in H*, denoted $N_H(G)$, the subgroup of H consisting of those elements that leave G invariant under conjugation.

We only consider finite sets, finite groups, and permutation groups acting on finite sets. We assume permutation groups to always be given by generating sets and say that a problem for permutation groups can be solved in *polynomial time*, if there exists an algorithm which, given permutation groups of degree n, solves it in time bounded polynomially in n and in the sizes of the given generating sets. While many problems for permutation groups can be solved efficiently both in theory and in practice, no polynomial time algorithm to compute normalizers of permutation groups is known.

A transitive permutation group G acting on a set Ω is called *primitive* if there exists no non-trivial G-invariant partition of Ω. Primitive groups have a rich and well-understood structure. Hence many algorithms use the natural recursion from general permutation groups to transitive and in turn to primitive ones. For two permutation groups $G, H \leq \mathrm{Sym}\,\Omega$ computing the normalizer of G in H in general is done by searching for the normalizer of G in the symmetric group $\mathrm{Sym}\,\Omega$ and simultaneously computing the intersection with H. We focus on computing the normalizer of a primitive group $G \leq \mathrm{Sym}\,\Omega$ in $\mathrm{Sym}\,\Omega$. Being able to compute normalizers for primitive groups efficiently may lead to improved algorithms for more general situations.

Our results build substantially on the O'Nan-Scott classification of primitive groups, see [17], and on the classification of finite simple groups (CFSG).

© Springer Nature Switzerland AG 2020
A. M. Bigatti et al. (Eds.): ICMS 2020, LNCS 12097, pp. 105–114, 2020.
https://doi.org/10.1007/978-3-030-52200-1_10

Recall that the *socle* of a group G, denoted $\operatorname{Soc} G$, is the subgroup generated by all minimal normal subgroups of G. Our theoretical main result is the following theorem.

Theorem 1 ([23, **Theorem 9.1**]). *Let a primitive group $G \leq \operatorname{Sym} \Omega$ with non-regular socle[1] be given. Then we can compute $N_{\operatorname{Sym} \Omega}(G)$ in polynomial time.*

As is often the case in computational group theory, ideas from theoretical algorithms can be employed in practical algorithms and vice versa. While the algorithms in [23] are primarily theoretical ones, we also provide probabilistic nearly-linear time versions where possible. The author is developing the GAP package `NormalizersOfPrimitiveGroups`, hosted at https://github.com/ssiccha/NormalizersOfPrimitiveGroups[2], with the aim to implement practical versions of the algorithms developed in [23]. Until now, algorithms concerning permutation morphisms and primitive groups of type PA are implemented. First experiments indicate that already for moderate degrees these outperform the GAP built-in algorithm `Normalizer` by several orders of magnitude, see Table 1.

Since no polynomial time solutions are known for the normalizer problem, the generic practical algorithms resolve to backtracking over the involved groups in one way or another. The fundamental framework of modern backtrack algorithms for permutation groups is Leon's partition backtrack algorithm [16], which generalizes previous backtrack approaches [5,6,12,24] and generalizes ideas of *nauty* [19] to the permutation group setting. Partition backtrack is implemented in GAP [9] and Magma [4]. Recently, the partition backtrack approach was generalized to a "graph backtrack" framework [14].

Theißen developed a normalizer algorithm which uses orbital graphs to prune the backtrack search [25]. Chang is currently developing specialized algorithms for highly intransitive permutation groups, her PhD thesis should appear shortly. It is to expect that the work in [14] can also be extended to normalizer problems. Hulpke also implemented normalizer algorithms in [13] using group automorphisms and the GAP function `NormalizerViaRadical` based on [10].

In Sect. 2 we outline the strategy behind our algorithms. In Sect. 3 we recall the O'Nan-Scott Theorem. We present our new concept of permutation morphisms in Sect. 4. In Sect. 5 we sketch how we use our results to obtain Theorem 1. In Sect. 6 we discuss our implementation.

2   Strategy

We describe the strategy of the theoretical algorithm behind Theorem 1. Comments regarding the implementation of its building blocks are given at the end of each following section.

In this section let $G \leq \operatorname{Sym} \Omega$ be a primitive group with non-regular socle H. The normalizer of H in $\operatorname{Sym} \Omega$ plays a central role in our algorithm, in this

---

[1] This excludes groups of affine and of twisted wreath type.

[2] May move to https://github.com/gap-packages/NormalizersOfPrimitiveGroups.

section we denote it by M. Observe that to compute $N_{\mathrm{Sym}\,\Omega}(G)$ it suffices to compute $N_M(G)$ since the former is contained in M.

The socle H is isomorphic to T^ℓ for some finite non-abelian simple group T and some positive integer ℓ. The group G is isomorphic to a subgroup of the wreath product $\mathrm{Aut}(T) \wr S_\ell$, see Sect. 3 for a definition of wreath products. By the O'Nan-Scott Theorem the respective isomorphism extends to an embedding[3] of the normalizer M into $\mathrm{Aut}(T) \wr S_\ell$. Furthermore ℓ is of the order $O(\log|\Omega|)$. Hence the index of G in M, and thus also the search-space of the normalizer computation $N_M(G)$, is tiny in comparison to the index of G in $\mathrm{Sym}\,\Omega$.

Our approach can be divided into two phases. First we compute M, this is by far the most labor intensive part. To this end we compute a sufficiently well-behaved conjugate of G, such that we can exhibit the wreath structure mentioned above. In [23] we make this more precise and define a *weak canonical form* for primitive groups. Using that conjugate and the O'Nan-Scott Theorem we can write down generators for M. In the second phase, we compute a reduction homomorphism $\rho : M \to S_k$ with $k \leq 6\log|\Omega|$. After this logarithmic reduction, we use Daniel Wiebking's simply exponential time algorithm [26,27], which is based on the canonization framework [22], to compute $N_{S_k}(\rho(G))$. Note that the running time of a simply exponential time algorithm called on a problem of size $\log n$ is $2^{O(\log n)}$ and thus is bounded by $2^{c\log n} = n^c$ for some constant $c > 0$. Then we use Babai's famous quasipolynomial time algorithm for graph-isomorphism [1,2] to compute the group intersection $N_{\rho(M)}(\rho(G)) = \rho(M) \cap N_{S_k}(\rho(G))$. Notice that since we perform these algorithms on at most $6\log n$ points they run in time polynomial in n. The homomorphism ρ is constructed in such a way, that computing the preimage of the above normalizer $N_{\rho(M)}(\rho(G))$ yields $N_M(G)$. Recall that $N_M(G)$ is equal to $N_{\mathrm{Sym}\,\Omega}(G)$.

In our implementation we do not use the algorithms by Wiebking and Babai since these are purely theoretical. Instead we use the partition backtrack implemented in GAP.

3   The O'Nan-Scott Theorem

The goal of this and the next section is to illustrate how we use the O'Nan-Scott Theorem to prove the following theorem. In this article we limit ourselves to groups of type PA, which we define shortly.

Theorem 2 ([23, **Theorem 8.1**]). *Let a primitive group $G \leq \mathrm{Sym}\,\Omega$ with non-abelian socle be given. Then we can compute $N_{\mathrm{Sym}\,\Omega}(\mathrm{Soc}\,G)$ in polynomial time.*

Proof. For groups of type PA this will follow from Corollary 5 and Lemma 6.

The O'Nan-Scott Theorem classifies how the socles of primitive groups can act, classifies the normalizers of the socles, and determines criteria to decide which subgroups of these normalizers act primitively. We follow the division of

---

[3] For twisted wreath type the situation is slightly more complicated.

primitive groups into eight O'Nan-Scott types as it was suggested by László G. Kovács and first defined by Cheryl Praeger in [21]. In this section we define the types AS and PA and recall some of their basic properties. In particular we describe the normalizer of the socle for groups of type PA and how to construct the normalizer of the socle, if the group is given in a sufficiently well-behaved form.

The version of the O'Nan-Scott Theorem we use, for a proof see [17], is:

Theorem 3. *Let G be a primitive group on a set Ω. Then G is a group of type HA, AS, PA, HS, HC, SD, CD, or TW.*

The abbreviation AS stands for **A**lmost **S**imple. A group is called *almost simple* if it contains a non-abelian simple group and can be embedded into the automorphism group of said simple group. A primitive group G is of *AS type* if its socle is a non-regular non-abelian simple group.

The abbreviation PA stands for **P**roduct **A**ction. The groups of AS type form the building blocks for the groups of PA type. To define this type, we shortly recall the notion of wreath products and their product action.

The *wreath product* of two permutation groups $H \le \mathrm{Sym}\,\Delta$ and $K \le S_d$ is denoted by $H \wr K$ and defined as the semidirect product $H^d \rtimes K$ where K acts per conjugation on H^d by permuting its components. We identify H^d and K with the corresponding subgroups of $H \wr K$ and call them the *base group* and the *top group*, respectively.

For two permutation groups $H \le \mathrm{Sym}\,\Delta$ and $K \le S_d$ the *product action* of the wreath product $H \wr K$ on the set of tuples Δ^d is given by letting the base group act component-wise on Δ^d and letting the top group act by permuting the components of Δ^d.

Definition 4. *Let $G \le \mathrm{Sym}\,\Omega$ be a primitive group. We say that G is of* type PA *if there exist an $\ell \ge 2$ and a primitive group $H \le \mathrm{Sym}\,\Delta$ of type AS such that G is permutation isomorphic to a group $\widehat{G} \le \mathrm{Sym}\,\Delta^\ell$ with*

$$(\mathrm{Soc}\,H)^\ell \le \widehat{G} \le H \wr S_\ell$$

in product action on Δ^ℓ.

The product action wreath products $A_5 \wr \langle (1,2,3) \rangle$ and $A_5 \wr \langle (1,2) \rangle$ are examples for primitive groups of type PA.

Let $\widehat{G} \le \mathrm{Sym}(\Delta^\ell)$ and $H \le \mathrm{Sym}\,\Delta$ be as in Definition 4. We sketch how to construct the normalizer of the socle of $\widehat{G}$. Let $T := \mathrm{Soc}\,H \le \mathrm{Sym}\,\Delta$. Since $\widehat{G}$ is given acting in product action we can read off H and thus compute T. By [8, Lemma 4.5A] we know that the normalizer of $\mathrm{Soc}\,\widehat{G}$ in $\mathrm{Sym}\,\Delta^\ell$ is $N_{\mathrm{Sym}\,\Delta}(T) \wr S_\ell$. By recent work of Luks and Miyazaki we can compute the normalizer of T, in polynomial time [18, Corollary 3.24]. More precisely this approach yields the following corollary:

Corollary 5. *Let $G \le \mathrm{Sym}(\Delta^\ell)$ be a primitive group of type PA with socle T^ℓ in component-wise action on Δ^ℓ. Then $N_{\mathrm{Sym}(\Delta^\ell)}(T^\ell)$ can be computed in polynomial time.*

In the practical implementation we use the GAP built-in algorithm to compute the normalizer of T in Sym Δ. Our long-term goal is to use the constructive recognition provided by the `recog` package [20]. Computing the normalizer of T in Sym Δ is then only a matter of iterating through representatives for the outer automorphisms of T.

4   Permutation Morphisms

In general a group of PA type might be given on an arbitrary set and needs only be permutation isomorphic to a group in product action. In this section we discuss how to construct such a permutation isomorphism:

Lemma 6. *Let $G \leq$ Sym Ω be a primitive group of type PA. Then we can compute a non-abelian simple group $T \leq$ Sym Δ, a positive integer ℓ, and a permutation isomorphism from G to a permutation group $\widehat{G} \leq$ Sym(Δ^ℓ) such that the socle of $\widehat{G}$ is T^ℓ in component-wise action on Δ^ℓ.*

To this end we present the notion of *permutation morphisms* developed in [23]. They arise from permutation isomorphisms by simply dropping the condition that the domain map and the group homomorphism be bijections. We illustrate how to use them to prove Lemma 6.

4.1   Basic Definitions

For two maps $f : A \to B$ and $g : C \to D$ we denote by $f \times g$ the product map $A \times C \to B \times D$, $(a,c) \mapsto (f(a), g(c))$. For a right-action $\rho : \Omega \times G \to \Omega$ of a group G and $g \in G$, $\omega \in \Omega$ we also denote $\rho(\omega, g)$ by ω^g.

Definition 7. *Let G and H be permutation groups on sets Ω and Δ, respectively, let $f : \Omega \to \Delta$ be a map, and let $\varphi : G \to H$ be a group homomorphism. Furthermore let ρ and τ be the natural actions of G and H on Ω and Δ, respectively. We call the pair (f, φ) a permutation morphism from G to H if the following diagram commutes:*

$$
\begin{array}{ccc}
\Omega \times G & \xrightarrow{\ \rho\ } & \Omega \\
{\scriptstyle f \times \varphi}\big\downarrow & & \big\downarrow{\scriptstyle f} \\
\Delta \times H & \xrightarrow{\ \tau\ } & \Delta ,
\end{array}
$$

that is if $f(\omega^g) = f(\omega)^{\varphi(g)}$ holds for all $\omega \in \Omega$, $g \in G$. We call φ the group homomorphism of (f, φ) and f the domain map of (f, φ).

It is immediate from the definition, that the component-wise composition of two permutation morphisms again yields a permutation morphism. In particular we define the *category of permutation groups* as the category with all permutation groups as objects, all permutation morphisms as morphisms, and the component-wise composition as the composition of permutation morphisms. We rely on this categorical perspective in many of our proofs.

We denote a permutation morphism F from a permutation group G to a permutation group H by $F : G \to H$. When encountering this notation keep in mind that F itself is not a map but a pair of a domain map and a group homomorphism. We use capital letters for permutation morphisms.

It turns out that a permutation morphism F is a mono-, epi-, or isomorphism in the categorical sense if and only if both its domain map and group homomorphism are injective, surjective, or bijective, respectively.

For a permutation group $G \leq \operatorname{Sym} \Omega$ we call a map $f : \Omega \to \Delta$ *compatible with G* if there exists a group homomorphism φ such that $F = (f, \varphi)$ is a permutation morphism. We say that a partition Σ of Ω is *G-invariant* if for all $A \in \Sigma$ and $g \in G$ we have $A^g \in \Sigma$.

Lemma 8 ([23, **Lemma 4.2.10**]).   *Let $G \leq \operatorname{Sym} \Omega$ be a permutation group and $f : \Omega \to \Delta$ a map. Then f is compatible with G if and only if the partition of Ω into the non-empty fibers $\{f^{-1}(\{\delta\}) \mid \delta \in f(\Omega)\}$ is G-invariant.*

If G is transitive, then the G-invariant partitions of Ω are precisely the block systems of G. Hence for a given blocksystem we can define a compatible map f by sending each point to the block it is contained in.

Let $G \leq \operatorname{Sym} \Omega$ be a permutation group and $f : \Omega \to \Delta$ a surjective map compatible with G. Then there exist a unique group $H \leq \operatorname{Sym} \Delta$ and a unique group homomorphism $\varphi : G \to H$ such that $F := (f, \varphi)$ is a permutation epimorphism, see [23, Corollary 4.2.7]. We call F the *permutation epimorphism* and φ the *group epimorphism of G induced by f*.

Example 9. Let $\Omega = \{1, \ldots, 4\}$, $a := (1,2)(3,4)$, $b := (1,3)(2,4)$, and $V := \langle a, b \rangle$. Further consider the set $\Omega_1 := \{1, 2\}$, the map $p_1 : \Omega \to \Omega_1$, $1, 3 \mapsto 1$, $2, 4 \mapsto 2$, and the following geometric arrangement of the points $1, \ldots, 4$:

$$1. \quad 2.$$
$$3. \quad 4.$$

Observe that a acts on Ω by permuting the points horizontally, while b acts on Ω by permuting the points vertically. The map p_1 projects Ω vertically or "to the top". Notice how the fibers of p_1 correspond to a block-system of V. We determine the group epimorphism π_1 of V induced by p_1. By definition $\pi_1(a)$ is the permutation which makes the following square commute:

$$
\begin{array}{ccc}
\Omega & \xrightarrow{\ a\ } & \Omega \\
\downarrow{\scriptstyle p_1} & & \downarrow{\scriptstyle p_1} \\
\Omega_1 & \xrightarrow{\ \pi_1(a)\ } & \Omega_1
\end{array}
$$

Take $1 \in \Omega_1$. We have $p_1^{-1}(\{1\}) = \{1,3\}$, $a(\{1,3\}) = \{2,4\}$, and $p_1(\{2,4\}) = \{2\}$. Hence $\pi_1(a) = (1,2)$. Correspondingly we get $\pi_1(b) = \mathrm{id}_{\Omega_1}$.

4.2   Products of Permutation Morphisms

For two permutation groups $H \leq \mathrm{Sym}\,\Delta$ and $K \leq \mathrm{Sym}\,\Gamma$ we define the *product of the permutation groups H and K* as the permutation group given by $H \times K$ in component-wise action on $\Delta \times \Gamma$. Correspondingly, for an additional permutation group $G \leq \mathrm{Sym}\,\Omega$ and two permutation morphisms (f, φ) and (g, ψ) from G to H and K, respectively, we define the *product permutation morphism* $G \to H \times K$ as $(f \times g, \varphi \times \psi)$. Iteratively, we define the product of several permutation groups or permutation morphisms.

To prove Lemma 6 it suffices to be able to compute the following: given the socle $H \leq \mathrm{Sym}\,\Omega$ of a PA type group compute a non-abelian simple group $T \leq \mathrm{Sym}\,\Delta$ and permutation epimorphisms, think projections, $P_1, \ldots, P_\ell : H \to T$ such that the product morphism $P : H \to T^\ell$ is an isomorphism. Since every surjective map compatible with H induces a unique permutation epimorphism, it in turn suffices to compute suitable maps $p_i : \Omega \to \Delta$.

Example 10. Consider the situation in Example 9. Let $P_1 := (p_1, \pi_1)$ and $\Omega_2 := \{1,3\}$. Then the map $p_2 : \Omega \to \Omega_2$, $1,2 \mapsto 1$, $3,4 \mapsto 3$ is compatible with V and induces the permutation epimorphism $P_2 : V \to \langle (1,3) \rangle$. The product maps $p_1 \times p_2 : \Omega \to \Omega_1 \times \Omega_2$ and $P_1 \times P_2 : V \to \langle (1,2) \rangle \times \langle (1,3) \rangle$ are isomorphisms of sets and permutation groups, respectively.

We illustrate how to construct one of the needed projections for PA type groups.

Example 11. Let $\Delta = \{1, \ldots, 5\}$ and $H := A_5 \times A_5 \leq \mathrm{Sym}(\Delta^2)$. We denote by $\mathbf{1}_\Delta$ the trivial permutation group on Δ. The subgroup $H_1 := A_5 \times \mathbf{1}_\Delta \leq \mathrm{Sym}(\Delta^2)$ is normal in H. Let us denote sets of the form $\{(\delta, x_2) \mid \delta \in \Delta\}$ by $\{(*, x_2)\}$. Then partitioning Δ^2 into orbits under H_1 yields the block system

$$\Sigma = \{\{(*, \delta_2)\} \mid \delta_2 \in \Delta\}.$$

Note how mapping each $x \in \Delta^2$ to the block of Σ it is contained in is equivalent to mapping each x to x_2. Thus we have essentially constructed the map $p_2 : \Delta^2 \to \Delta$, $x \mapsto x_2$. Observe that we only used the group theoretic property that H_1 is a maximal normal subgroup of H and thus in particular did not use the actual product structure of Δ^2.

Analogously we can construct the map $p_1 : \Delta^2 \to \Delta$, $x \mapsto x_1$. For $i = 1, 2$ let $P_i : H \to A_5$ be the permutation epimorphisms of H induced by p_1 and p_2, respectively. Since $p_1 \times p_2$ is an isomorphism, $P_1 \times P_2$ must be a monomorphism. By order arguments $P_1 \times P_2$ is thus an isomorphism.

In general the above construction does not yield permutation epimorphisms with identical images. We can alleviate this by computing elements of the given group which conjugate the minimal normal subgroups of its socle to each other. For the general construction see the *(homogenized) product decomposition by minimal normal subgroups* in [23, Definitions 5.1.3 and 5.1.5]. Lemma 6 then follows from [23, Corollary 5.19].

5   Reduction Homomorphism

Recall from Sect. 2 that a key ingredient of our second phase is a group homomorphism which reduces the original problem on n points to a problem on less or equal than $6 \log n$ points. We illustrate shortly how to construct this homomorphism, for the details refer to [23, Theorem 9.1.6].

Let $G \leq \text{Sym}\, \Omega$ be a primitive group with non-regular socle and $T \leq \text{Sym}\, \Delta$ be a *socle-component of* G, confer [23, Chapters 5 and 7] for a definition. Then T is a non-abelian simple group, there exists a positive integer ℓ such that Soc G is isomorphic to T^{ℓ}, and by [23, Lemma 2.6.1] we have $|\Omega| = |\Delta|^{s}$ for some $s \in \{\ell/2, \ldots, \ell\}$. Denote by R the permutation group induced by the right-regular action of Out T on itself. We show that we can evaluate the following two group homomorphisms: first an embedding $N_{\text{Sym}\, \Omega}(\text{Soc}\, G) \to \text{Aut}\, T \wr S_{\ell}$ and second an epimorphism $\text{Aut}\, T \wr S_{\ell} \to R \wr S_{\ell}$, where $R \wr S_{\ell}$ is the imprimitive wreath product and thus acts on $|R| \cdot \ell$ points.

We sketch the proof that $|R| \cdot \ell \leq 6 \log n$. Let $m := |\Delta|$ and $r := |R|$. Note that for ℓ we have $\ell \leq 2s = 2 \log_{m} n$. Since R is regular, we have $r = |\text{Out}\, T|$. Since T is a socle-component of G, we have $|\text{Out}\, T| \leq 3 \log m$ by [11, Lemma 7.7]. In total we have $r \cdot \ell \leq 3 \log m \cdot 2 \log_{m} n = 6 \log n$.

In our implementation we use a modified version of this reduction. For groups of type PA we can directly compute an isomorphism from the product action wreath product into the corresponding imprimitive wreath product.

6   Implementation

A version of our normalizer algorithm for groups of type PA is implemented in the GAP package `NormalizersOfPrimitiveGroups`.

Table 1 shows a comparison of runtimes of our algorithm and the GAP function `Normalizer`. At the time of writing, there are two big bottlenecks in the implementation. First, the GAP built-in algorithm to compute the socle of a group is unnecessarily slow. State-of-the-art algorithms as in [7] are not yet implemented. Secondly, computing a permutation which transforms a given product decomposition into a so-called natural product decomposition currently also is slow. The latter may be alleviated by implementing the corresponding routines in for example C [15] or Julia [3]. Note that the actual normalizer computation inside the normalizer of the socle appears to be no bottleneck: in the example with socle type $(A_5)^7$ it took only 40 ms!

Table 1. Table with runtime comparison.

Socle type	Degree	Our algorithm	GAP built-in alg.
$(A_5)^2$	25	24 ms	200 ms
$(A_5)^3$	125	50 ms	1500 ms
$(A_5)^4$	625	300 ms	29400 ms
$(A_5)^7$	78125	67248 ms	–
$PSL(2,5)^2$	36	40 ms	300 ms
$PSL(2,5)^3$	216	90 ms	1900 ms
$PSL(2,5)^4$	1296	400 ms	64000 ms
$(A_7)^2$	49	38 ms	900 ms
$(A_7)^3$	343	200 ms	16800 ms
$(A_7)^4$	2401	1400 ms	839000 ms

Acknowledgments. Substantial parts of the work presented in this article were written while the author was supported by the German Research Foundation (DFG) research training group "Experimental and constructive algebra" (GRK 1632) and employed by the Lehr- und Forschungsgebiet Algebra, RWTH Aachen University and the Department of Mathematics, University of Siegen.

References

1. Babai, L.: Graph isomorphism in quasipolynomial time. arXiv e-prints arXiv:1512.03547, December 2015
2. Babai, L.: Graph isomorphism in quasipolynomial time [extended abstract]. In: Proceedings of the Forty-Eighth Annual ACM Symposium on Theory of Computing, STOC 2016, pp. 684–697. ACM (2016). https://doi.org/10.1145/2897518.2897542
3. Bezanson, J., Edelman, A., Karpinski, S., Shah, V.B.: Julia: a fresh approach to numerical computing. SIAM Rev. **59**(1), 65–98 (2017). https://doi.org/10.1137/141000671
4. Bosma, W., Cannon, J., Playoust, C.: The magma algebra system. I. The user language. J. Symbolic. Comput. **24**(3–4), 235–265 (1997). https://doi.org/10.1006/jsco.1996.0125. Computational algebra and number theory (London, 1993)
5. Butler, G.: Computing in permutation and matrix groups. ii. Backtrack algorithm. Math. Comput. **39**, 671–680 (1982). https://doi.org/10.1090/S0025-5718-1982-0669659-5
6. Butler, G.: Computing normalizers in permutation groups. J. Algorithms **4**(2), 163–175 (1983). https://doi.org/10.1016/0196-6774(83)90043-3
7. Cannon, J., Holt, D.: Computing chief series, composition series and socles in large permutation groups. J. Symbolic Comput. **24**(3), 285–301 (1997). https://doi.org/10.1006/jsco.1997.0127
8. Dixon, J.D., Mortimer, B.: Permutation Groups, vol. 163. Springer, New York (1996). https://doi.org/10.1007/978-1-4612-0731-3
9. The GAP-Group: GAP - Groups, Algorithms, and Programming, Version 4.11.0 (2020). https://www.gap-system.org

10. Glasby, S.P., Slattery, M.C.: Computing intersections and normalizersin soluble groups. J. Symbolic Comput. **9**(5), 637–651 (1990). https://doi.org/10.1016/S0747-7171(08)80079-X

11. Guralnick, R.M., Maróti, A., Pyber, L.: Normalizers of primitive permutation groups. Adv. Math. **310**, 1017–1063 (2017). https://doi.org/10.1016/j.aim.2017.02.012

12. Holt, D.: The computation of normalizers in permutation groups. J. Symbolic Comput. **12**(4), 499–516 (1991). https://doi.org/10.1016/S0747-7171(08)80100-9

13. Hulpke, A.: Normalizer calculation using automorphisms. In: Computational Group Theory and the Theory of Groups. AMS special session On Computational Group Theory, Davidson, USA, pp. 105–114 (2007). https://doi.org/10.1090/conm/470

14. Jefferson, C., Pfeiffer, M., Waldecker, R., Wilson, W.A.: Permutation group algorithms based on directed graphs. arXiv preprint arXiv:1911.04783 (2019)

15. Kernighan, B.W., Ritchie, D.M.: The C Programming Language, 2nd edn. Prentice Hall, Englewood Cliffs (1988)

16. Leon, J.S.: Permutation group algorithms based on partitions i theory and algorithms. J. Symbolic Comput. **12**, 533–583 (1991)

17. Liebeck, M.W., Praeger, C.E., Saxl, J.: On the o'nan-scott theorem for finite primitive permutation groups. J. Aust. Math. Soc. Ser. A. Pure Math. Stat. **44**(3), 389–396 (1988). https://doi.org/10.1017/S144678870003216X

18. Luks, E., Miyazaki, T.: Polynomial-time normalizers. Discrete Math. Theor. Comput. Sci. **13**(4), 61–96 (2011)

19. McKay, B.D., Piperno, A.: Practical graph isomorphism, ii. J. Symbolic Comput. **60**, 94–112 (2014). https://doi.org/10.1016/j.jsc.2013.09.003

20. Neunhöffer, M., et al.: Recog, a collection of group recognition methods, Version 1.3.2, April 2018. https://gap-packages.github.io/recog. GAP package

21. Praeger, C.E.: The inclusion problem for finite primitive permutation groups. Proc. Lond. Math. Soc. **3**(1), 68–88 (1990). https://doi.org/10.1112/plms/s3-60.1.68

22. Schweitzer, P., Wiebking, D.: A unifying method for the design of algorithms canonizing combinatorial objects. arXiv e-prints arXiv:1806.07466, June 2018

23. Siccha, S.: Normalizers of primitive groups with non-regular socle in polynomial time. Ph.D. thesis, RWTH Aachen University. to appear

24. Sims, C.C.: Determining the conjugacy classes of permutation groups. Comput. Algebra Number Theo. **4**, 191–195 (1971)

25. Theißen, H.: Eine methode zur normalisatorberechnung in permutationsgruppen mit anwendungen in der konstruktion primitiver gruppen. Ph.D. thesis, RWTH Aachen University (1997)

26. Wiebking, D.: Normalizers and permutational isomorphisms in simply-exponential time. arXiv e-prints arXiv:1904.10454, April 2019

27. Wiebking, D.: Normalizers and permutational isomorphisms in simply-exponential time. In: Proceedings of the Thirty-First Annual ACM-SIAM Symposium on Discrete Algorithms, SODA 2020, 5–8 January 2020, Salt Lake City, Utah, USA (2020)

Homomorphic Encryption and Some Black Box Attacks

Alexandre Borovik[1] and Şükrü Yalçınkaya[2($\boxtimes$)]

[1] Department of Mathematics, University of Manchester, Manchester M13 9PL, UK
alexandre@borovik.net
[2] Department of Mathematics, Istanbul University, Istanbul, Turkey
sukru.yalcinkaya@istanbul.edu.tr
http://www.borovik.net

Abstract. This paper is a compressed summary of some principal definitions and concepts in the approach to the black box algebra being developed by the authors [6–8]. We suggest that black box algebra could be useful in cryptanalysis of homomorphic encryption schemes [11], and that homomorphic encryption is an area of research where cryptography and black box algebra may benefit from exchange of ideas.

Keywords: Homomorphic encryption · Black box groups · Probabilistic methods

1 Homomorphic Encryption

"Cloud computing" appears to be a hot topic in information technology; in a nutshell, this is the ability of small and computationally weak devices to delegate hard resource-intensive computations to third party (and therefore untrusted) computers. To ensure the privacy of the data, the untrusted computer should receive data in an encrypted form but still being able to process it. It means that encryption should preserve algebraic structural properties of the data. This is one of the reasons for popularity of the idea of *homomorphic encryption* [1,2,10, 11,13,14,18,19,21–23] which we describe here with some simplifications aimed at clarifying connections with black box algebra (as defined in Sect. 2.1).

1.1 Homomorphic Encryption: Basic Definitions

Let A and X denote the sets of plaintexts and ciphertexts, respectively, and assume that we have some (say, binary) operators $\square_A$ on A needed for processing data and corresponding operators $\square_{\mathsf{X}}$ on X. An encryption function E is *homomorphic* if

$$E(a_1 \square_A a_2) = E(a_1) \square_{\mathsf{X}} E(a_2)$$

for all plaintexts a_1, a_2 and all operators on A.

Suppose that Alice is the owner of data represented by plaintexts in A which she would like to process using operators $\square_A$ but has insufficient computational

© Springer Nature Switzerland AG 2020
A. M. Bigatti et al. (Eds.): ICMS 2020, LNCS 12097, pp. 115–124, 2020.
https://doi.org/10.1007/978-3-030-52200-1_11

resources, while Bob has computational facilities for processing ciphertexts using operators $\square_X$. Alice may wish to enter into a contract with Bob; in a realistic scenario, Alice is one of the many customers of the encrypted data processing service run by Bob, and all customers use the same ambient structure A upto isomorphism and formats of data and operators which are for that reason are likely to be known to Bob. What is not known to Bob is the specific password protected encryption used by Alice. This is what is known in cryptology as Kerckhoff's Principle: *obscurity is no security*, the security of encryption should not rely on details of the protocol being held secret; see [11] for historic details.

Alice encrypts plaintexts a_1 and a_2 and sends ciphertexts $E(a_1)$ and $E(a_2)$ to Bob, who computes

$$x = E(a_1) \, \square_X \, E(a_2)$$

without having access to the content of plaintexts a_1 and a_2, then return the output x to Alice who decrypts it using the decryption function E^{-1}:

$$E^{-1}(x) = a_1 \, \square_A \, a_2$$

In this set-up, we say that the homomorphic encryption scheme is *based* on the algebraic structure A or the homomorphism E is a *homomorphic encryption of* the algebraic structure A.

To simplify exposition, we assume that the encryption function E is deterministic, that is, E establishes a one-to-one correspondence between A and X. Of course, this is a strong assumption in the cryptographic context; it is largely unnecessary for our analysis, but, for the purposes of this paper, allows us to avoid technical details and makes it easier to explain links with the black box algebra.

1.2 Back to Algebra

In algebraic terms, A and X as introduced above are algebraic structures with operations on them which we refer to as *algebraic operations* and $E : A \longrightarrow X$ is a homomorphism. In this paper we assume that the algebraic structure A is finite as a set. This is not really essential for our analysis, many observations are relevant for the infinite case as well, but handling probability distributions (that is, random elements) on infinite sets is beyond the scope of the present paper.

We discuss a class of potential attacks on homomorphic encryption of A. Our discussion is based on a simple but fundamental fact of algebra that a map $E : A \longrightarrow X$ of algebraic structures of the same type is a homomorphism if and only if its graph

$$\Gamma(E) = \{(a, E(a)) \mid a \in A\}$$

is a substructure of $A \times X$, that is, closed under all algebraic operations on $A \times X$. Obviously, $\Gamma(E)$ is isomorphic to A and we shall note the following observation:

if an algebraic structure A has a rich internal configuration (has many substructures with complex interactions between them),

the graph $\Gamma(E)$ of a homomorphic encryption $E : A \longrightarrow X$ also has a rich (admittedly hidden) internal configuration, and this could make it vulnerable to an attack from Bob.

We suggest that

before attempting to develop a homomorphic encryption scheme based on a particular algebraic structure A, the latter needs to be examined by black box theory methods – as examples in this paper show, it could happen that *all* homomorphic encryption schemes on A are insecure.

2   Black Box Algebra

2.1   Axiomatic Description of Black Box Algebraic Structures

A *black box algebraic structure* X is a black box (device, algorithm, or oracle) which produces and operates with 0–1 strings of uniform length $l(X)$ encrypting (not necessarily in a unique way) elements of some fixed algebraic structure A: if x is one of these strings then it corresponds to a unique (but unknown to us) element $\pi(x) \in A$. Here, π is the decrypting map, not necessarily known to us in advance. We call the strings produced or computed by X *cryptoelements*.

Our axioms for black boxes are the same as in [6–8], but stated in a more formal language.

BB1 On request, X produces a 'random' cryptoelement x as a string of fixed length $l(X)$, which depends on X, which encrypts an element $\pi(x)$ of some fixed explicitly given algebraic structure A; this is done in time polynomial in $l(X)$. When this procedure is repeated, the elements $\pi(x_1), \pi(x_2), \ldots$ are independent and uniformly distributed in A.

To avoid messy notation, we assume that operations on A are unary or binary; a general case can be treated in exactly the same way.

BB2 On request, X performs algebraic operations on the encrypted strings which correspond to operations in A in a way which makes the map π (unknown to us!) a homomorphism: for every binary (unary case is similar) operation $\boxdot$ and strings x and y produced or computed by X,

$$\pi(x \boxdot y) = \pi(x) \boxdot \pi(y).$$

It should be noted that we do not assume the existence of an algorithm which allows us to decide whether a specific string can be potentially produced by X; requests for operations on strings can be made only in relation to cryptoelements previously output by X. Also, we do not make any assumptions on probabilistic distribution of cryptoelements.

BB3 On request, X determines, in time polynomial in $l(X)$, whether two cryptoelements x and y encrypt the same element in A, that is, check whether $\pi(x) = \pi(y)$.

We say in this situation that a black box X *encrypts* the algebraic structure A and we denote this as $X \vDash A$.

Clearly, in black box problems, the decrypting map π is not given in advance. However, it is useful to think about any algebraic structure (say, a finite field) implemented on a computer as a trivial black box, with π being the identity map, and with random elements produced with the help of a random number generator. In this situation, obviously, the axioms BB1–BB3 hold.

In our algorithms, we have to build new black boxes from existing ones and work with several black box structures at once: this is why we have to keep track of the length $l(X)$ on which a specific black box X operates. For example, it turns out in [8] that it is useful to consider an automorphism of A as a graph in $A \times A$. This produces an another algebraic structure isomorphic to A which can be seen as being encrypted by a black box Z producing, and operating on, certain pairs of strings from X, see [8] for more examples. In this case, clearly, $l(Z) = 2l(X)$.

2.2 Morphisms

Given two black boxes X and Y encrypting algebraic structures A and B, respectively, we say that a map ϕ which assigns strings produced by X to strings produced by Y is a *morphism* of black boxes, if

- the map ϕ is computable in time polynomial in $l(X)$ and $l(Y)$, and
- there is a homomorphism $\phi : A \to B$ such that the following diagram is commutative:

$$
\begin{array}{ccc}
X & \xrightarrow{\ \phi\ } & Y \\
\vdots{\scriptstyle\pi_X} & & \vdots{\scriptstyle\pi_Y} \\
A & \xrightarrow{\ \phi\ } & B
\end{array}
$$

where π_X and π_Y are the canonical projections of X and Y onto A and B, respectively.

We say in this situation that a morphism ϕ *encrypts* the homomorphism ϕ and call ϕ bijective, injective, etc., if ϕ has these properties.

2.3 Construction and Interpretation

Construction of a new black box Y in a given black box $X \vDash A$ can be formally described as follows.

Strings of Y are concatenated n-tuples of strings $(x_1, \ldots, x_n)$ from X produced by a polynomial time algorithm which uses operations on X; new operations on

Y are also polynomial time algorithms running on X, as well as the algorithm for checking the new identity relation $=_Y$ on Y.

If this is done in a consistent way and axioms BB1–BB3 hold in Y, then Y encrypts an algebraic structure B which can be obtained from the structure A by a similar construction, with algorithms replaced by description of their outputs by formulae of first order language in the signature of A. At this point we are entering the domain of model theory, and full discussion of this connection can be found in our forthcoming paper [9]. Here we notice only that in model theory B is said to be *interpreted* in A, and if A is in its turn interpreted in B then A and B are called *bi-intrepretable*. A recent result on bi-interpretability between Chevalley groups and rings, relevant to our project is [20].

2.4   A Few Historic Remarks

Black box algebraic structures had been introduced by Babai and Szemerédi [4] in the special case of groups as an idealized setting for randomized algorithms for solving permutation and matrix group problems in computational group theory. Our Axioms BB1–BB3 are a slight modification – and generalization to arbitrary algebraic structures – of their original axioms.

So far, it appears that only finite groups, fields, rings, and, very recently, projective planes (in our paper [8]) got a black box treatment. In the case of finite fields, the concept of a black box field can be traced back to Lenstra Jr [16] and Boneh and Lipton [5], and in the case of rings – to Arvind [3].

A higher level of abstraction introduced in our papers produces new tools allowing us to solve problems which previously were deemed to be intractable. For example, recently, a fundamental problem of constructing a unipotent element in black box groups encrypting PSL_2 was solved in odd characteristics via constructing a black box projective plane and its underlying black box field [8]. There is an analogous recognition algorithm for the black box groups encrypting PSL_2 in even characteristic [15].

2.5   Recognition of Black Box Fields

A *black box* (finite) *field* K is a black box operating on 0-1 strings of uniform length which encrypts some finite field $\mathbb{F}$. The oracle can compute $x + y$, xy, and x^{-1} (the latter for $x \neq 0$) and decide whether $x = y$ for any strings $x, y \in K$. Notice in this definition that the characteristic of the field is not known. Such a definition is needed in our paper [8] to produce black box group algorithms which does not use characteristic of the underlying field. If the characteristic p of K is known then we say that K is a *black box field of known characteristic* p. We refer the reader to [5,17] for more details on black box fields of known characteristic and their applications to cryptography.

The following theorem is a reformulation of the fundamental results in [17].

Theorem 1. *Let* $K \vDash \mathbb{F}_{p^n}$ *be a black box field of known characteristic* p *and* K_0 *the prime subfield of* K. *Then the problem of finding two way morphisms between* K *and* $\mathbb{F}_{p^n}$ *can be reduced to the same problem for* K_0 *and* $\mathbb{F}_p$. *In particular,*

– a morphism $K_0 \longrightarrow \mathbb{F}_p$ can be extended in time polynomial in the input length $l(K)$ to a morphism $K \longrightarrow \mathbb{F}_{p^n}$;
– there is a morphism $\mathbb{F}_{p^n} \longrightarrow K$ computable in time polynomial in $l(K)$.

Here and in the rest of the paper, "efficient" means "computable in time polynomial in the input length".

In our terminology (Sect. 2.6), Theorem 1 provides a *structural proxy* for black box fields of known characteristic. Indeed, if K is a black box field of known characteristic p, then we can construct an isomorphism $\mathbb{F}_p = \mathbb{Z}/p\mathbb{Z} \longrightarrow K_0$ by the map

$$m \mapsto 1 + 1 + \cdots + 1 \quad (m \text{ times})$$

where 1 is the unit in K_0; it is computable in linear in $\log p$ time by double-and-add method. We say that p is *small* if it is computationally feasible to make a lookup table for the inverse $K_0 \longrightarrow \mathbb{F}_p$ of this map. Construction of a morphism $\mathbb{F}_p \longleftarrow K_0$ remains an open problem. However, we can observe that

Corollary 1. *Let $K \vDash \mathbb{F}_{p^n}$, where p is a known small prime number. Then there exist two way morphisms between K and $\mathbb{F}_{p^n}$.*

2.6 Construction of a Structural Proxy

Most groups of Lie type (we exclude ${}^2B_2$, ${}^2F_4$ and ${}^2G_2$ to avoid technical details) can be seen as functors $G : \mathcal{F} \longrightarrow \mathcal{G}$ from the category of fields $\mathcal{F}$ with an automorphism of order $\leqslant 2$ to the category of groups $\mathcal{G}$. There are also other algebraic structures which can be defined in a similar way as functors from $\mathcal{F}$, for example projective planes or simple Lie algebras (viewed as rings). The following problem is natural and, as our results show, useful in this context.

Construction of a structural proxy: Suppose that we are given a black box structure $X \vDash A(\mathbb{F})$. Construct, in time polynomial in $l(X)$,
- a black box field $K \vDash \mathbb{F}$, and
- two way bijective morphisms $A(K) \longleftrightarrow X$.

If we construct a black box field K by using X as a computational engine, then we can construct the natural representation $A(K)$ of the structure A over the black box field K. By Theorem 1, we can construct a polynomial time isomorphism $\mathbb{F}_q \longrightarrow K$ which further provides an isomorphism $A(\mathbb{F}_q) \longrightarrow A(K)$ completing *a structure recovery* of X.

Structural proxies and structure recovery play a crucial role for algorithms developed in Theorem 3. We summarize relevant results about constructing structural proxies of black box algebraic structures from our papers [6, 8].

Theorem 2. *We can construct structural proxies for the following black box structures.*

(a) $P \vDash \mathbb{P}^2(\mathbb{F})$, *a projective plane with a polarity encrypting a projective plane $\mathbb{P}^2(\mathbb{F})$ over a finite field $\mathbb{F}$ of odd characteristic.*

(b) $X \vDash SO_3(\mathbb{F}), (P)SL_2(\mathbb{F})$ *over a finite field* $\mathbb{F}$ *of unknown odd characteristic, under the assumption that we know a global exponent* E *of* X, *that is,* E *such that* $x^E = 1$ *for all* $x \in X$ *and* $\log E$ *is polynomially bounded in terms of* $l(X)$.

(c) $R \vDash M_{2 \times 2}(\mathbb{F}_q)$, *a black box ring encrypting the ring of* 2×2 *matrices over the known finite field* $\mathbb{F}_q$ *of odd characteristic.*

2.7 Black Boxes Associated with Homomorphic Encryption

As explained in Subsection 1.1, we assume that the algebraic structure A of plaintexts is represented in some standard form known to Bob. In agreement with the standard language of algebra – and with our terminology in [8] – we shall use the words *plain element* or just *element* in place of 'plaintext' and *cryptoelement* in place of 'ciphertext'.

Let A be a set of plain elements, X a set of cryptoelements, and E be the encryption function, that is, an isomorphism $E : A \longrightarrow X$.

Supply of random cryptoelements from X postulated in Axiom BB1 can be achieved by sampling a big dataset of cryptoelements provided by Alice, or computed on request from Alice. The computer system controlled by Bob performs algebraic operations referred to in Axiom BB2.

Axiom BB3 is redundant under the assumption that $E : A \longrightarrow X$ is a bijection but it gives us more freedom to construct new black boxes, for example, homomorphic images of X. Axiom BB3 could also be useful for handling another quite possible scenario: For Alice, the cost of computing homomorphisms E and E^{-1} could be higher than the price charged by Bob for processing cryptoelements. In that case, it could be cheaper to transfer initial data to Bob (in encrypted form) and ask Bob to run a computer programme which uses the black box but does not send intermediate values back to Alice, returns only the final result; checking equality of cryptoelements becomes unavoidable.

3 A Black Box Attack on Homomorphic Encryption

We assume that Bob can accumulate a big dataset of cryptoelements sent from/to Alice, or intermediate results from running Alice's programme, and that he can feed, without Alice's knowledge, cryptoelements into a computer system (the *black box*) which performs operations on them, and retain the outputs for peruse – again without Alice's knowledge. Bob's aim is to compute the decryption function E^{-1} efficiently, that is, in time polynomial in terms of the lengths of plain elements and cryptoelements involved.

3.1 Bob's Attack

As we discussed in Sect. 1.1, we can assume that Bob knows the algebraic structure A. Bob's aim is to find an efficient algorithm which maps cryptoelements from X to elements in A and vice versa while preserving the algebraic operations on X and A. This means solving the *constructive recognition problem* for X, that is, finding bijective morphisms

$$\alpha : \mathsf{X} \longrightarrow A \text{ and } \beta : A \longrightarrow \mathsf{X}$$

such that $\alpha \circ \beta$ is the identity map on A.

Assume that Bob solved the constructive recognition problem and can efficiently compute α and β.

Alice's encryption function is a map $E : A \longrightarrow \mathsf{X}$; the composition $\delta = \alpha \circ E$ is an automorphism of A. Therefore Bob reads not Alice's plaintexts $a \in A$, but their images $\delta(a) = \alpha(E(a))$ under an automorphism δ of A still unknown to him. This means that

solving the constructive recognition problem for X reduces the problem of inverting the encryption homomorphism $E : A \longrightarrow \mathsf{X}$ to a much simpler problem of inverting the automorphism

$$\delta : A \longrightarrow A.$$

We are again in the situation of homomorphic encryption, but this time the sets of plaintexts and ciphertexts are the same. One would expect that this encryption is easier to break. For example, if Bob can guess the plaintexts of a few cryptoelements, and if the automorphism group Aut A of A is well understood, computation of δ and δ^{-1} could be a more accessible problem than the constructive recognition for X. For example, automorphism groups of finite fields are very small, and in that case δ^{-1} can be found by direct inspection.

As soon as δ^{-1} is known, Bob knows $E^{-1} = \delta^{-1} \circ \alpha$ and can decrypt everything. Moreover, since $E = \beta \circ \delta$, the map E is also known and allows Bob to return to Alice cryptoelements which encrypt plaintexts of Bob's choice.

We suggest that this approach to analysis of homomorphic encryption is useful because it opens up connections to black box algebra. Indeed the theory of black box structures is reasonably well developed for groups and fields, and its methods could provide insight into assessment of security of other algebraic structures if any are proposed for use in homomorphic encryption.

4   Application of Theorem 2 to Homomorphic Encryption

The procedures described in Theorem 3 below are reformulations of the principal results of our Theorem 2 in a homomorphic encryption setup. They demonstrate the depth of structural analysis involved and suggest that a similarly deep but revealing structural theory can be developed for other algebraic structures if they are sufficiently rich ('rich' here can mean, for example, 'bi-interpretable with a finite field'). Also, it is worth noting that the procedures do not use any assumptions about the encryption homomorphism E, the analysis is purely algebraic.

Theorem 3. *Assume that Alice and Bob run a homomorphic encryption protocol over the group $A = \mathrm{SL}_2(\mathbb{F}_q)$, q odd, with Bob doing computations with cryptoelements using a black box $X \vDash A$. Assume that Bob knows A, including the representation of the field $\mathbb{F}_q$ used by Alice. Then, by Theorem 2, Bob can construct a structural proxy $\mathrm{SL}_2(\mathsf{K})$ for X. Moreover:*

(a) *If, in addition, Bob has two way bijective morphisms between a black box field K and an explicitly given field $\mathbb{F}_q$ (see Corollary 1), he gets two way bijective morphisms $X \longleftrightarrow \mathrm{SL}_2(\mathbb{F}_q)$.*

(b) *Under assumptions of (a), Bob gets an image of Alice's data transformed by an automorphism $\delta : \mathrm{SL}_2(\mathbb{F}_q) \longrightarrow \mathrm{SL}_2(\mathbb{F}_q)$ since Alice's group A is an explicitly given $\mathrm{SL}_2(\mathbb{F}_q)$.*

(c) *Automorphisms of the group $\mathrm{SL}_2(\mathbb{F}_q)$ are well known: every automorphism is a product of an inner automorphism and a field automorphism induced by an automorphism of the field $\mathbb{F}_q$. Therefore if Bob can run a few instances of known plaintexts attacks against Alice, he can compute the automorphism δ and after that read plaintexts of all Alice's cryptoelements.*

(d) *Moreover, under assumptions of (a) and (c), Bob can compute the inverse of δ and pass to Alice, as answers to Alice's requests, values of his choice.*

Items (c) and (d) in Theorem 3 look as serious vulnerabilities of homomorphic encryptions of the groups $\mathrm{SL}_2(\mathbb{F}_q)$. We conclude that homomorphic encryption of groups $\mathrm{SL}_2(\mathbb{F}_q)$ is no more secure than homomorphic encryption of the field $\mathbb{F}_q$. As a consequence of Theorem 1, homomorphic encryption of $\mathrm{SL}_2(\mathbb{F}_q), q = p^k$, does not survive a known plaintext attack when the prime $p > 2$ is small.

We think that this is a manifestation of a more general issue: for small odd primes p, there are no secure homomorphic encryption schemes based on sufficiently rich (say, bi-interpretable with finite fields) algebraic structures functorially defined over finite fields of characteristic p.

Acknowledgement. The authors worked on this paper during their visits to the Nesin Mathematics Village, Turkey. We thank Jeff Burdges, Adrien Deloro, Alexander Konovalov, and Chris Stephenson for fruitful advice, and the referees for their most perceptive comments.

References

1. Acar, A., Aksu, H., Uluagac, A.S., Conti, M.: A survey on homomorphic encryption schemes: theory and implementation. ACM Comput. Surv. **51**(4), 79 (2018)
2. Aguilar-Melchor, C., Fau, S., Fontaine, C., Gogniat, G., Sirdey, R.: Recent advances in homomorphic encryption: a possible future for signal processing in the encrypted domain. IEEE Sig. Process. Mag. **30**(2), 108–117 (2013)
3. Arvind, V., Das, B., Mukhopadhyay, P.: The complexity of black-box ring problems. In: Chen, D.Z., Lee, D.T. (eds.) COCOON 2006. LNCS, vol. 4112, pp. 126–135. Springer, Heidelberg (2006). https://doi.org/10.1007/11809678_15
4. Babai, L., Szemerédi, E.: On the complexity of matrix group problems. In: Proceedings of 25th IEEE Symposium Foundations Computer Science, pp. 229–240 (1984)

5. Boneh, D., Lipton, R.J.: Algorithms for black-box fields and their application to cryptography. In: Koblitz, N. (ed.) CRYPTO 1996. LNCS, vol. 1109, pp. 283–297. Springer, Heidelberg (1996). https://doi.org/10.1007/3-540-68697-5_22

6. Borovik, A., Yalçınkaya, Ş.: Natural representations of black box groups $SL_2(\mathbb{F}_q)$. http://arxiv.org/abs/2001.10292

7. Borovik, A., Yalçınkaya, Ş.: New approaches in black box group theory. In: Hong, H., Yap, C. (eds.) ICMS 2014. LNCS, vol. 8592, pp. 53–58. Springer, Heidelberg (2014). https://doi.org/10.1007/978-3-662-44199-2_10

8. Borovik, A., Yalçınkaya, Ş.: Adjoint representations of black box groups $PSL_2(\mathbb{F}_q)$. J. Algebra **506**, 540–591 (2018)

9. Borovik, A., Yalçınkaya, Ş.: Black box algebra: model-theoretic connections (in preparation)

10. Dyer, J., Dyer, M., Xu, J.: Practical homomorphic encryption over the integers. http://arxiv.org/abs/1702.07588

11. Fontaine, C., Galand, F.: A survey of homomorphic encryption for nonspecialists. J. Inform. Secur. **1**, 41–50 (2009)

12. The GAP Group: GAP - Groups, Algorithms, and Programming, Version 4.11.0 (2020). http://www.gap-system.org

13. Gentry, C.: Fully homomorphic encryption using ideal lattices. In: Proceedings of the Forty-First Annual ACM Symposium on Theory of Computing, New York, NY, USA, pp. 169–178. STOC 2009. ACM (2009)

14. Gentry, C., Halevi, S.: Implementing Gentry's fully-homomorphic encryption scheme. In: Paterson, K.G. (ed.) EUROCRYPT 2011. LNCS, vol. 6632, pp. 129–148. Springer, Heidelberg (2011). https://doi.org/10.1007/978-3-642-20465-4_9

15. Kantor, W.M., Kassabov, M.: Black box groups isomorphic to $PGL(2, 2^e)$. J. Algebra **421**, 16–26 (2015)

16. Lenstra Jr., H.W.: Finding isomorphisms between finite fields. Math. Comput. **56**(193), 329–347 (1991)

17. Maurer, U., Raub, D.: Black-box extension fields and the inexistence of field-homomorphic one-way permutations. In: Kurosawa, K. (ed.) ASIACRYPT 2007. LNCS, vol. 4833, pp. 427–443. Springer, Heidelberg (2007). https://doi.org/10.1007/978-3-540-76900-2_26

18. Prasanna, B.T., Akki, C.B.: A comparative study of homomorphic and searchable encryption schemes for cloud computing. http://arxiv.org/abs/1505.03263

19. Rass, S.: Blind turing-machines: arbitrary private computations from group homomorphic encryption (2013). http://arxiv.org/abs/1312.3146

20. Segal, D., Tent, K.: Defining R and $G(R)$. http://arxiv.org/abs/2004.13407

21. Sen, J.: Homomorphic encryption: theory & applications. http://arxiv.org/abs/1305.5886

22. Sharma, I.: Fully homomorphic encryption scheme with symmetric keys. http://arxiv.org/abs/1310.2452

23. Tebaa, M., Hajji, S.E.: Secure cloud computing through homomorphic encryption. http://arxiv.org/abs/1409.0829

Nilpotent Quotients of Associative Z-Algebras and Augmentation Quotients of Baumslag-Solitar Groups

Tobias Moede[(✉)] [iD]

Technische Universität Braunschweig, 38106 Braunschweig, Germany
t.moede@tu-bs.de
https://www.tu-braunschweig.de/iaa/personal/moede

Abstract. We describe the functionality of the package *zalgs* for the computer algebra system GAP. The package contains an implementation of the nilpotent quotient algorithm for finitely presented associative Z-algebras described in [3]. As an application of this algorithm we calculate augmentation quotients, i.e. successive quotients of powers of the augmentation ideal $I(G)$ of the integral group ring $\mathbb{Z}G$, where G is a finitely presented group. We apply these methods to obtain conjectures for augmentation quotients of the Baumslag-Solitar groups $BS(m,n)$ with $|m-n|$ equal to $0, 1$ or a prime p.

Keywords: Associative algebras · Augmentation quotients ·
Computer algebra · Group theory · Nilpotent quotient algorithm

1 Introduction

An associative Z-algebra A is called *nilpotent of class $c \in \mathbb{N}$* if its series of power ideals has the form $A = A^1 > A^2 > \ldots > A^c > A^{c+1} = \{0\}$. The power ideal A^i, $i \in \mathbb{N}$ is the ideal in A generated by all products of length i. In [3] we introduced so called *nilpotent presentations* to describe such algebras in a way that exhibits their nilpotent structure. We also introduced a nilpotent quotient algorithm, which computes a nilpotent presentation for the class-c quotient A/A^{c+1} for a given finitely presented associative Z-algebra A and a non-negative integer c. An implementation of this algorithm is available in the package *zalgs* [5] for the computer algebra system GAP [4].

The purpose of this paper is to describe the functionality of the *zalgs* package and to exhibit applications of the nilpotent quotient algorithm. In particular, we apply the algorithm in the calculation of *augmentation quotients*, i.e. the quotients $Q_k(G) = I^k(G)/I^{k+1}(G)$, where G is a finitely presented group and $I(G)$ denotes the augmentation ideal of the integral group ring $\mathbb{Z}G$. The augmentation ideal $I(G)$ is defined as the kernel of the augmentation map

$$\varepsilon \colon \mathbb{Z}G \to \mathbb{Z}, \quad \sum_i a_i g_i \mapsto \sum_i a_i,$$

© Springer Nature Switzerland AG 2020
A. M. Bigatti et al. (Eds.): ICMS 2020, LNCS 12097, pp. 125–130, 2020.
https://doi.org/10.1007/978-3-030-52200-1_12

where $a_i \in \mathbb{Z}$ and $g_i \in G$. The augmentation quotients are interesting objects studied in the integral representation theory of groups. We present conjectures on the augmentation quotients of certain Baumslag-Solitar groups, which are based on computer experiments using the *zalgs* package.

2 Nilpotent Presentations and Nilpotent Quotient Systems

As all algebras considered in this paper will be associative $\mathbb{Z}$-algebras, we will simply refer to them as algebras. For completeness we recall several important definitions from [3].

Definition 1. *Let A be a finitely generated algebra of class c and $s \in \mathbb{N}$. We call $(b_1, \ldots, b_s)$ a weighted generating sequence for A with powers $(e_1, \ldots, e_s)$ and weights $(w_1, \ldots, w_s)$ if*

(a) $A = \langle b_1, \ldots, b_s \rangle$, *i.e. A is the $\mathbb{Z}$-span of $b_1, \ldots, b_s$.*
(b) $b_i b_j \in \langle b_{\max\{i,j\}}, \ldots, b_s \rangle$ *for $1 \leq i, j \leq s$.*
(c) e_i *is minimal in $\mathbb{N}$ with respect to the property that $e_i b_i \in \langle b_{i+1}, \ldots, b_s \rangle$, or $e_i = 0$, if such an $e_i \in \mathbb{N}$ does not exist.*
(d) *The elements $b_i + A^{k+1}$ with $1 \leq i \leq s$ such that $w_i = k$ generate A^k / A^{k+1} for $1 \leq k \leq c$.*

Definition 2. *A consistent weighted nilpotent presentation for a finitely generated nilpotent algebra A is given by a weighted generating sequence $(b_1, \ldots, b_s)$ with powers $(e_1, \ldots, e_s)$, weights $(w_1, \ldots, w_s)$ and relations of the following form:*

(a) $e_i b_i = x_{i,i+1} b_{i+1} + \ldots + x_{i,s} b_s$ *for all $1 \leq i \leq s$ where $e_i > 0$.*
(b) $b_i b_j = y_{i,j,l+1} b_{l+1} + \ldots + y_{i,j,s} b_s$ *for $1 \leq i, j \leq s$ and $l = \max\{i, j\}$.*
(c) *The $x_{i,k}$ and $y_{i,j,k}$ are integers with $0 \leq x_{i,k}, y_{i,j,k} < e_k$ if $e_k > 0$.*

We note that every finitely generated nilpotent algebra has a consistent weighted nilpotent presentation, see [3, Theorem 7]. In an algebra A given by a consistent weighted nilpotent presentation, we can determine a normal form for each $a \in A$, i.e. there are uniquely determined $z_i \in \mathbb{Z}$ with

$$a = z_1 b_1 + \ldots + z_s b_s$$

and $0 \leq z_i < e_i$ if $e_i > 0$.

Definition 3. *Let $A = \langle x_1, \ldots, x_n \mid R_1, \ldots, R_t \rangle$ be a finitely presented algebra, $c \in \mathbb{N}$ and let $\varphi: A \to A/A^{c+1}$ be the natural homomorphism. A nilpotent quotient system describes φ using the following data:*

(a) *A consistent weighted nilpotent presentation for A/A^{c+1} with generators $(b_1, \ldots, b_s)$, powers $(e_1, \ldots, e_s)$, weights $(w_1, \ldots, w_s)$, multiplication relations for $b_i b_j$ and power relations $e_i b_i$ if $e_i > 0$.*

(b) Images $\varphi(x_i)$ for $1 \leq i \leq n$ given in normal form.

(c) Definitions $(d_1, \ldots, d_s)$, with d_i being an integer or a pair of integers, s.t.
- If d_i is an integer, then $w_i = 1$ and $b_i = \varphi(x_{d_i})$.
- If $d_i = (k, j)$, then $b_i = b_k b_j$, where $w_k = 1$ and $w_j = w_i - i$.

The description of φ using this data is very useful for computational purposes and usually the output of our calculations will be in the form of nilpotent quotient systems. The following example shall illustrate the definition of nilpotent quotient systems.

Example 1. Consider the finitely presented algebra given by

$$A = \langle x_1, x_2 \mid 2x_1, x_2^2, x_1^2 - x_1 x_2 \rangle.$$

Then a nilpotent quotient system for $\varphi \colon A \to A/A^2$ consists of:

- generators (b_1, b_2) with powers $(2, 0)$ and weights $(1, 1)$,
- the power relation $2b_1 = 0$ and the multiplication relations $b_i b_j = 0$ for $1 \leq i, j \leq 2$,
- images $\varphi(x_1) = b_1$ and $\varphi(x_2) = b_2$, and
- definitions $(1, 2)$.

A nilpotent quotient system for $\varphi \colon A \to A/A^3$ consists of:

- generators (b_1, b_2, b_3, b_4) with powers $(2, 0, 2, 2)$ and weights $(1, 1, 2, 2)$,
- the power relations $2b_1 = 2b_3 = 2b_4 = 0$ and the multiplication relations $b_1 b_1 = b_3$, $b_1 b_2 = b_3$, $b_2 b_1 = b_4$ and $b_i b_j = 0$ for all other $1 \leq i, j \leq 4$,
- images $\varphi(x_1) = b_1$ and $\varphi(x_2) = b_2$, and
- definitions $(1, 2, (1, 2), (2, 1))$.

3   Functionality

The central functionality provided by the *zalgs* package is the function

▷ `NilpotentQuotientFpZAlgebra(A, c)`,

which takes as input a finitely presented associative $\mathbb{Z}$-algebra A and a non-negative integer c. The output is a nilpotent quotient system for $\varphi \colon A \to A/A^{c+1}$.

Example 2. The following is an example calculation of a nilpotent quotient system in GAP for the class-2 quotient of the associative $\mathbb{Z}$-algebra considered in Example 1 above, i.e.

$$A = \langle x_1, x_2 \mid 2x_1, x_2^2, x_1^2 - x_1 x_2 \rangle.$$

To carry out the computation, we start by defining A as the quotient of the free associative $\mathbb{Z}$-algebra on two generators by the given relations.

```
gap> F := FreeAssociativeAlgebra(Integers, 2);;
gap> x1 := F.1;; x2 := F.2;;
gap> A := F / [2*x1, x2^2, x1^2-x1*x2];;
```

We then call `NilpotentQuotientFpZAlgebra(A, 2)` to compute the class-2 quotient. The output contains lists for the definitions `dfs`, the powers `pows`, the weights `wgs` and an integer `dim` indicating the dimension of the quotient. The entries for the images `imgs`, power relations `ptab` and multiplication relations `mtab` are to be interpreted as coefficients of normal forms. For computational purposes there is an additional entry `rels` in the output.

```
gap> NilpotentQuotientFpZAlgebra(A, 2);
rec( dfs := [1,2,[1,2],[2,1]],
     dim := 4,
     imgs := [ [1,0,0,0], [0,1,0,0] ],
     mtab := [ [[0,0,1,0], [0,0,1,0], [0,0,0,0], [0,0,0,0]],
               [[0,0,0,1], [0,0,0,0], [0,0,0,0], [0,0,0,0]],
               [[0,0,0,0], [0,0,0,0], [0,0,0,0], [0,0,0,0]],
               [[0,0,0,0], [0,0,0,0], [0,0,0,0], [0,0,0,0]] ],
     pows := [2,0,2,2],
     ptab := [ [0,0,0,0], [0,0,0,0], [0,0,0,0], [0,0,0,0] ],
     rels := [ <- omitted -> ],
     wgs := [1,1,2,2] )
```

In [3, Section 5], we describe how to obtain a presentation P for an algebra, such that $I(G)/I^{c+1}(G)$ is isomorphic to the class-c quotient of P. The nilpotent quotient algorithm can now be applied to determine this nilpotent quotient. The following functions are available to calculate the class-c quotient of the augmentation ideal of integral group rings.

▷ `AugmentationQuotientFpGroup(G, c)`,
▷ `AugmentationQuotientPcpGroup(G, c)`,

which take as input a finitely presented group or a polycyclically presented group G, respectively, and a non-negative integer c. The output in both cases is a nilpotent quotient system for $I(G)/I^{c+1}(G)$. Note that the augmentation quotients $Q_n(G)$ for $n \leq c$ can be read off from this.

4   Augmentation Quotients of Baumslag-Solitar Groups

In [3, Section 5], we describe how to obtain, for a given finitely presented group G, a presentation for an algebra A, such that $I(G)/I^{c+1}(G)$ is isomorphic to the class-c quotient of A. We apply these methods to compute augmentation quotients of the *Baumslag-Solitar groups* $BS(m,n)$, which for $m, n \in \mathbb{Z} \setminus \{0\}$ are given by the presentations

$$BS(m,n) = \langle a, b \mid ba^m b^{-1} = a^n \rangle.$$

These one-relator groups form an interesting set of groups with applications in combinatorial and geometric group theory, e.g. the group $BS(1,1)$ is the free abelian group on two generators and $BS(1,-1)$ arises as the fundamental group of the Klein bottle. The Baumslag-Solitar groups were introduced in [2] as examples of non-Hopfian groups and the isomorphism problem for these groups has been considered in [6].

We carried out computer experiments to gain some insight into the structure of the augmentation quotients $Q_k(BS(m, n))$ for $|m|, |n| \leq 10$ and small values of k. Our computations suggest the following conjectures for certain special cases:

Conjecture 1. Let p be a prime.

(a) If $|m - n| = 0$ and $|m| = |n| = p$, then

$$Q_k(BS(m, n)) = \begin{cases} \mathbb{Z}^{k+1} \oplus \mathbb{Z}_p^{A(k,1)}, & \text{if } k \leq p + 2, \\ \mathbb{Z}^{k+1} \oplus \mathbb{Z}_p^{A(k,1)-C(k)}, & \text{if } k > p + 2, \end{cases}$$

where

$$A(u, v) = \sum_{\ell=0}^{v} (-1)^\ell \binom{u+1}{\ell}(v + 1 - \ell)^u$$

are the Eulerian numbers and the values $C(k)$ are given by the recursion

$$C(1) = 1, \quad C(k) = C(k-1) + B(k+8) \text{ for all } k \geq 2,$$

where $B(u)$ is the number of 8-element subsets of $\{1, \dots, u\}$ whose elements sum to a triangular number, i.e. a number of the form $T_w = \binom{w+1}{2}$, $w \in \mathbb{N}$.

(b) If $|m - n| = 1$, then for all $k \in \mathbb{N}$:

$$Q_k(BS(m, n)) \cong \mathbb{Z}.$$

(c) If $|m - n| = p$, then for all $k \in \mathbb{N}$:

$$Q_k(BS(m, n)) \cong \mathbb{Z} \oplus \mathbb{Z}_p^{T_k},$$

where $T_k = \binom{k+1}{2}$ is the k-th triangular number.

The behaviour appears to be more complicated if $|m - n|$ contains several (not necessarily distinct) prime factors, as is illustrated in the following example.

Example 3. Let G be the Baumslag-Solitar group $BS(5, -1)$. Then the first few augmentation quotients are as follows:

$$Q_1(G) \cong \mathbb{Z} \oplus \mathbb{Z}_6$$
$$Q_2(G) \cong \mathbb{Z} \oplus \mathbb{Z}_6^3$$
$$Q_3(G) \cong \mathbb{Z} \oplus \mathbb{Z}_6^6$$
$$Q_4(G) \cong \mathbb{Z} \oplus \mathbb{Z}_2 \oplus \mathbb{Z}_6^8 \oplus \mathbb{Z}_{18}$$
$$Q_5(G) \cong \mathbb{Z} \oplus \mathbb{Z}_2^3 \oplus \mathbb{Z}_6^{10} \oplus \mathbb{Z}_{18} \oplus \mathbb{Z}_{54}$$
$$Q_6(G) \cong \mathbb{Z} \oplus \mathbb{Z}_2^5 \oplus \mathbb{Z}_6^{13} \oplus \mathbb{Z}_{18} \oplus \mathbb{Z}_{54}^2$$
$$Q_7(G) \cong \mathbb{Z} \oplus \mathbb{Z}_2^8 \oplus \mathbb{Z}_6^{16} \oplus \mathbb{Z}_{18} \oplus \mathbb{Z}_{54}^2 \oplus \mathbb{Z}_{162}$$
$$Q_8(G) \cong \mathbb{Z} \oplus \mathbb{Z}_2^{11} \oplus \mathbb{Z}_6^{20} \oplus \mathbb{Z}_{18} \oplus \mathbb{Z}_{54}^2 \oplus \mathbb{Z}_{162}^2$$
$$Q_9(G) \cong \mathbb{Z} \oplus \mathbb{Z}_2^{15} \oplus \mathbb{Z}_6^{23} \oplus \mathbb{Z}_{18}^2 \oplus \mathbb{Z}_{54}^2 \oplus \mathbb{Z}_{162}^3$$
$$Q_{10}(G) \cong \mathbb{Z} \oplus \mathbb{Z}_2^{19} \oplus \mathbb{Z}_6^{27} \oplus \mathbb{Z}_{18}^3 \oplus \mathbb{Z}_{54}^2 \oplus \mathbb{Z}_{162}^4$$

5 Further Aims

Bachmann and Grünenfelder [1] showed that for finite groups G the sequence $Q_n(G)$ for $n \in \mathbb{N}$ is virtually periodic, i.e. there exist $N \in \mathbb{N}$ and $k \in \mathbb{N}$ such that $Q_n(G) \cong Q_{n+k}(G)$ for all $n \geq N$. It will be interesting to extend our methods to allow the determination of these parameters, which in theory allows to determine all augmentation quotients for a given finite group G.

Furthermore, we plan to extend our algorithms to compute nilpotent presentations for the largest associative $\mathbb{Z}$-algebra on d generators so that every element a of the algebra satisfies $a^n = 0$, i.e. compute $\mathbb{Z}$-algebra analogues of Burnside groups and Kurosh algebras.

References

1. Bachmann, F., Grünenfelder, L.: The periodicity in the graded ring associated with an integral group ring. J. Pure Appl. Algebra **5**, 253–264 (1974)
2. Baumslag, G., Solitar, D.: Some two-generator one-relator non-hopfian groups. Bull. Am. Math. Soc. **68**, 199–201 (1962)
3. Eick, B., Moede, T.: A nilpotent quotient algorithm for finitely presented associative $\mathbb{Z}$-algebras and its application to integral groups rings. Accepted for Mathematics of Computation (2020)
4. The GAP Group: GAP - Groups, Algorithms, and Programming, Version 4.11.0 (2020). https://www.gap-system.org
5. Moede, T.: zalgs - a GAP package for the computation of nilpotent quotients of finitely presented $\mathbb{Z}$-algebras (2020). https://www.tu-braunschweig.de/iaa/personal/moede
6. Moldavanskii, D.: Isomorphism of the Baumslag-Solitar groups. Ukr. Math. J. **43**, 1569–1571 (1991)

The GAP Package LiePRing

Bettina Eick[1]([⊠])[iD] and Michael Vaughan-Lee[2]

[1] TU Braunschweig, Braunschweig, Germany
beick@tu-bs.de
[2] Christ Church, Oxford, England
michael.vaughan-lee@chch.ox.ac.uk
http://www.iaa.tu-bs.de/beick/, http://users.ox.ac.uk/~vlee

Abstract. A symbolic Lie p-ring defines a family of Lie rings with p^n elements for infinitely many different primes p and a fixed positive integer n. Symbolic Lie p-rings are used to describe the classification of isomorphism types of nilpotent Lie rings of order p^n for all primes p and all $n \leq 7$. This classification is available as the LiePRing package of the computer algebra system GAP. We give a brief description of this package, including an approach towards computing the automorphism group of a symbolic Lie p-ring.

Keywords: Lie ring · Automorphism group · Finite p-group

1 Introduction

A Lie ring is an additive abelian group with a multiplication, denoted by $[.,.]$, that is bilinear, alternating and satisfies the Jacobi identity. A Lie p-ring is a nilpotent Lie ring with p^n elements for some prime power p^n. Such a Lie p-ring of order p^n can be described by a presentation $P(A)$ on n generators $b_1, \ldots, b_n$ with coefficients $A = (a_{ijk}, a_{ik} \mid 1 \leq i < j < k \leq n)$, so that a_{ijk} and a_{ik} are integers in the range $\{0, \ldots, p-1\}$ and the following relations hold:

$$[b_j, b_i] = \sum_{k=j+1}^{n} a_{ijk} b_k \quad \text{for} \quad 1 \leq i < j \leq n, \quad \text{and}$$

$$p b_i = \sum_{k=i+1}^{n} a_{ik} b_k \quad \text{for} \quad 1 \leq i \leq n.$$

We generalize this type of presentation so that it defines a family of Lie p-rings for various different primes. For this purpose let p be an indeterminate, let $R = \mathbb{Z}[w, x_1, \ldots, x_m]$ be a polynomial ring in $m + 1$ commuting variables and let a_{ijk} and a_{ik} in R. In some (rare) cases it is convenient to allow some of the coefficients a_{ijk} and a_{ik} to be rational functions over R; note that we use this only for coefficients a_{ijk} or a_{ik} if $p b_k = 0$ so that b_k is an element of order p.

If a fixed prime P and integers $X_1, \ldots, X_m$ are given, then we *specify* the a polynomial $a \in R$ at these values by choosing W to be the smallest primitive

© Springer Nature Switzerland AG 2020
A. M. Bigatti et al. (Eds.): ICMS 2020, LNCS 12097, pp. 131–140, 2020.
https://doi.org/10.1007/978-3-030-52200-1_13

root mod P and evaluating $\bar{a} = a(W, X_1, \ldots, X_m)$ in $\mathbb{Z}$. We *specify* a rational function a/b with $a, b \in R$ by specifying the polynomials a and b to $\bar{a}$ and $\bar{b}$ in $\mathbb{Z}$, and then we determine $\bar{a}\,\bar{c}$ where $\bar{c} \in \{1, \ldots P-1\}$ satisfies $\bar{c}\bar{b} \equiv 1 \bmod P$. Note that only choices of $W, X_1, \ldots, X_m$ with $P \nmid \bar{b}$ are valid.

Let $\mathbb{P}$ be an infinite set of primes, let $m \in \mathbb{N}_0$ and for $P \in \mathbb{P}$ let

$$\Sigma_P \subseteq \{(X_1, \ldots, X_m) \in \mathbb{Z}^m \mid 0 \leq X_i < P\}.$$

Then the presentation $P(A)$ defines a *symbolic* Lie p-ring with respect to $\mathbb{P}$ and Σ_P if for each $P \in \mathbb{P}$ and each $(X_1, \ldots, X_m) \in \Sigma_P$ the presentation $P(A)$ specified at these points is a finite Lie p-ring of order P^n.

A symbolic Lie p-ring describes a family of finite Lie p-rings: for each $P \in \mathbb{P}$ this contains $|\Sigma_P| \leq P^m$ members. Symbolic Lie p-rings are used to describe the complete classification up to isomorphism of all Lie p-rings of order dividing p^7 for $p > 3$ as obtained by Newman, O'Brien and Vaughan-Lee [6,7]. This is available in computational form in the LiePRing package [4] of the computer algebra system GAP [9]. The following exhibits an example.

Example 1. We consider the symbolic Lie p-ring $\mathcal{L}$ with generators $b_1, \ldots, b_7$ and the (non-trivial) relations

$$\begin{aligned}
[b_2, b_1] &= b_4, & pb_1 &= b_4 + b_6 + x_2 b_7, \\
[b_3, b_1] &= b_5, & pb_3 &= x_1 b_6. \\
[b_3, b_2] &= b_6, \\
[b_5, b_1] &= b_6, \\
[b_5, b_3] &= b_7,
\end{aligned}$$

Let $\mathbb{P}$ be the set of all primes and let

$$\Sigma_P = \{(X_1, X_2) \mid 0 < X_1 < P, 0 \leq X_2 < P\}.$$

Then $\mathcal{L}$ defines a family of $P(P-1)$ Lie p-rings of order P^7 for each $P \in \mathbb{P}$.

The LiePRing package allows symbolic computations with symbolic Lie p-rings $\mathcal{L}$. "Symbolic computations" means that it computes with $\mathcal{L}$ as if computing with all Lie p-rings L in the family defined by $\mathcal{L}$ simultaneously. For example, it allows us

- to compute series of ideals such as the lower central series of L,
- to describe the automorphism group of L, and
- to determine the Schur multiplier of L, see [3].

Let P be a prime and let $n \in \mathbb{N}$ with $n \leq P$. The Lazard correspondence [5] associates to each Lie p-ring L of order P^n a group $G(L)$ of order P^n. This correspondence translates Lie ring isomorphisms to group isomorphisms and vice versa. Cicalo, de Graaf and Vaughan-Lee [2] determined an effective version of the Lazard correspondence and implemented this in the LieRing package [1] of GAP.

The following sections give a brief overview of some of the algorithms in the LiePRing package and they exhibit how the Lazard correspondence can be evaluated in GAP in this setting.

2   Elementary Computations

In this section we investigate computations with elements, subrings and ideals. Throughout, let $\mathcal{L}$ be a symbolic Lie p-ring with respect to Σ_P, let L be a finite Lie p-ring in the family defined by $\mathcal{L}$ and let P be the prime of L. We write $P(A)$ for the defining presentation in the finite and in the symbolic case. Thus depending on the context A is an integer matrix or a matrix over the ring $Quot(R)$ of rational functions over the polynomial ring R.

2.1   Ring Invariants

The definition of Σ_P can often be used for computations with $\mathcal{L}$. For example, if $\Sigma_P = \{(x_1, x_2, x_3) \in \mathbb{Z}_P^3 \mid x_1 \neq 1, x_3 = \pm 1\}$, then $(x_1 - 1)$ specifies to an invertible element in L and $(x_3 - 1)(x_3 + 1) = (x_3^2 - 1)$ specifies to 0. Hence we can treat $(x_1 - 1)$ as a unit and $(x_3^2 - 1)$ as zero. The following example illustrates this for $\Sigma_P = \{(x, y) \mid x \neq 0, y \in \{1, w\}\}$.

```
gap> L := LiePRingsByLibrary(7)[3195];
<LiePRing of dimension 7 over prime p with parameters [ x, y ]>
gap> ViewPCPresentation(L);
p*12 = x*17, p*13 = 15 + y*17, p*14 = 16,
[12,11] = 15, [13,11] = 16, [14,11] = 17
gap> RingInvariants(L);
rec( units := [ x, y ], zeros := [ w*y-y^2-w+y ] )
```

2.2   The Word Problem

Consider the case of a finite Lie p-ring L and let a be an arbitrary word in the generators of $P(A)$. Then the relations in $P(A)$ readily allow us to rewrite a to a *unique* equivalent *normal form*

$$c_1 b_1 + \ldots + c_n b_n \quad \text{with} \quad c_i \in \{0, \ldots, P-1\} \text{ for } 1 \leq i \leq n.$$

Now consider the case of a symbolic Lie p-ring $\mathcal{L}$ and let a be a word in the generators of $P(A)$. Then the relations and the zeros of $\mathcal{L}$ allow us to translate this to an equivalent *reduced form*; that is, a linear combination of the form

$$c_1 b_1 + \ldots + c_n b_n \quad \text{with} \quad c_i \in R \quad \text{for } 1 \leq i \leq n,$$

where $c_1, \ldots, c_n \in R$ are reduced modulo the polynomials in *zeros*; that is, the polynomial division algorithm dividing c_i by the polynomials in zeros yields only trivial quotients. If $c_1 = \ldots = c_k = 0$ and $c_{k+1} \neq 0$, then $k+1$ is the *depth* of this reduced form and c_{k+1} is its *leading coefficient*. We say that $(c_1, \ldots, c_n)$ *represents* the element a.

Example 2. We continue Example 1.

(1) Consider the element $a = pb_1 - [b_2, b_1] - [b_3, b_2] - [[b_3, b_1], b_3]$. Using the relations of $\mathcal{L}$ this reduces to $a = b_4 + b_6 + x_2 b_7 - b_4 - b_6 - [b_5, b_3] = x_2 b_7 - b_7 = (x_2 - 1)b_7$. Note that a can be zero and non-zero in the Lie p-rings in the family defined by $\mathcal{L}$, depending on the choice of x_2.

(2) Consider the element $a = pb_3$. Then $a = x_1 b_6$ and hence, since $x_1 \neq 0$ in $\mathcal{L}$, it follows that a is a non-zero element in each Lie p-ring in the family defined by $\mathcal{L}$.

2.3   Subrings, Ideals and Series

Let $\mathcal{L}$ be a symbolic Lie p-ring, let $w_1, \ldots, w_k$ be words in the generators $b_1, \ldots, b_n$ of $P(A)$ and let U be the subring of $\mathcal{L}$ generated by these words. Our aim is to determine an *echelon generating set* for U; that is, a generating set $v_1, \ldots, v_l$ so that each v_i is a reduced form in the generators with leading coefficient 1, the depths satisfy $d(v_1) < \ldots < d(v_l)$ and each element in U is a linear combination in $v_1, \ldots, v_l$ with coefficients in $Quot(R)$. This may require the distinction of finitely many cases, as the following example indicates.

Example 3. We continue Example 1.

(1) Let $U = \langle [b_3, b_1], pb_3 \rangle$. As $[b_3, b_1] = b_5$ and $pb_3 = x_1 b_6$ with $x_1 \neq 0$, it follows that $U = \langle b_5, b_6 \rangle$ in each Lie ring in the family defined by $\mathcal{L}$.

(2) Let $U = \langle pb_1 - b_4 - b_6, [b_3, b_2] \rangle$. Then using the relations of $\mathcal{L}$ it follows that $U = \langle x_2 b_7, b_6 \rangle$. Hence $U = \langle b_6, b_7 \rangle$ if $x_2 \neq 0$ and $U = \langle b_6 \rangle$ otherwise. Thus a case distinction is necessary to determine an echelon generating set for U.

Ideals are subrings that are closed under multiplication and hence they can also be described via echelon generating sets (subject to a case distinction). In turn, this then allows us to determine series such as the lower central series and the derived series of $\mathcal{L}$. The following example illustrates the handling of case distinctions in GAP.

```
gap> L := LiePRingsByLibrary(6)[267];
<LiePRing of dimension 6 over prime p with parameters [x,y,z,t]>
gap> ViewPCPresentation(L);
p*11 = t*15 + x*16, p*12 = y*15 + z*16,
[12,11] = 14, [13,11] = 16, [14,11] = 15,
[13,12] = w*15, [14,12] = 16
gap> RingInvariants(L);
rec( units := [ -x*y+z*t ], zeros := [ ] )
gap> S := LiePRecSubring(L, [p*b[1]]);
[<LiePRing of dimension 1 over prime p with parameters [x,y,z,t]>,
 <LiePRing of dimension 1 over prime p with parameters [x,y,z,t]>]
```

Here the LiePRing package returns two new symbolic Lie p-rings $S[1]$ and $S[2]$. These have different ring invariants and different bases:

```
gap> RingInvariants(S[1]);
rec( units := [ y, x ], zeros := [ t ] )
gap> BasisOfLiePRing(S[1]);
[ 16 ]
gap> RingInvariants(S[2]);
rec( units := [ -x*y+z*t, t ], zeros := [ ] )
gap> BasisOfLiePRing(S[2]);
[ 15 + x/t*16 ]
```

In particular, in $S[2]$ the polynomial t is a unit and the rational function x/t turns up as coefficient for the basis element l_6.

3   Automorphism Groups

Given a symbolic Lie p-ring $\mathcal{L}$, we show how to determine a generic description for $Aut(L)$ for each finite Lie p-ring L in the family defined by $\mathcal{L}$. The following gives a first illustration.

Example 4. We continue Example 1.

We note that $\mathcal{L}$ is generated by b_1, b_2, b_3. This allows us to describe each automorphism of $\mathcal{L}$ via its images of b_1, b_2, b_3 and the same holds for each finite Lie p-ring in the family defined by $\mathcal{L}$. Write g_r for the image of b_r. Then $g_r = g_{r1}b_1 + \ldots + g_{r7}b_7$ for certain integers g_{rs}. We say that the automorphism is represented by the 3×7 matrix (g_{rs}). Note that different matrices may represent the same automorphism for a finite Lie p-ring L; for example, if P is the prime of L, then b_7 has order P and g_{37} and $g_{37} + P$ give the same automorphism. We expand on this below.

Our algorithm determines that each automorphism of $\mathcal{L}$ corresponds to a matrix of the form

$$\begin{pmatrix} g_{11} & g_{12} & 0 & g_{14} & g_{15} & g_{16} & g_{17} \\ 0 & 1 & 0 & g_{24} & 0 & g_{26} & g_{27} \\ 0 & g_{32} & g_{11} & g_{34} & g_{35} & g_{36} & g_{37} \end{pmatrix}$$

with $g_{11} = \pm 1$ and g_{rs} arbitrary otherwise. If P is prime and L is a finite Lie p-ring over P, then we can choose $g_{rs} \in \{0, \ldots, P-1\}$ for $(r, s) \neq (1, 1)$ and thus $Aut(L)$ has order $2P^{13}$.

Given a finite Lie p-ring L with prime P, we define its radical $R(L)$ as the ideal of L generated by $\{[b_j, b_i], Pb_k \mid 1 \leq i < j \leq n, 1 \leq k \leq n\}$. The additive group of $L/R(L)$ is an elementary abelian group of order P^d, say, and the Lie ring multiplication of $L/R(L)$ is trivial. Burnside's Basis theorem (for example, see [8, page 140]) for finite p-groups translates readily to the following.

Lemma 1. *Let L be a finite Lie p-ring and let $\varphi : L \to L/R(L)$ the natural ring homomorphism.*

(a) $R(L)$ is the intersection of all maximal Lie subrings of L.

(b) *Each minimal generating set of L has d elements and maps under φ onto a minimal generating set of $L/R(L)$.*

(c) *Each list of preimages under φ of a minimal generating set of $L/R(L)$ is a minimal generating set of L.*

Next, let $P(A)$ be the presentation for the finite Lie p-ring L with generators $b_1, \ldots, b_n$ so that $R(L) = \langle b_{d+1}, \ldots, b_n \rangle$. Then $b_1, \ldots, b_d$ is a minimal generating set of L. Thus each automorphism α of L is defined by its images on $b_1, \ldots, b_d$. These have the general form

$$\alpha(b_r) = g_{r1}b_1 + \ldots + g_{rn}b_n \quad \text{for} \quad 1 \leq r \leq d,$$

with integer coefficients g_{rs}. For $k > d$ we note that $b_k \in R(L)$. This allows us to write b_k as a word in the ideal generators $[b_j, b_i]$ and Pb_i of $R(L)$ and that, in turn, allows us to determine the image $\alpha(b_k)$ in the form

$$\alpha(b_k) = w_{k1}b_1 + \ldots + w_{kn}b_n,$$

where w_{kj} is a word in $\{g_{rs}\}$.

Theorem 1. *The matrix $(g_{rs})_{1 \leq r \leq d, 1 \leq s \leq n}$ defines an automorphism α of L if and only if*

(a) *$det(G) \not\equiv 0 \bmod P$, where $G = (g_{rs})_{1 \leq r, s \leq d}$, and*

(b) *the images $\alpha(b_1), \ldots, \alpha(b_n)$ satisfy the relations of L.*

Proof. First recall that a map $b_i \mapsto v_i$ for $1 \leq i \leq n$ with $v_1, \ldots, v_n \in L$ extends to a Lie ring homomorphism $L \to L$ if and only if $v_1, \ldots, v_n$ satisfy the defining relations of L. This is von Dyck's theorem (for example, see [8, page 51]) in the case of finitely presented groups and it translates readily to other algebraic objects such as Lie rings.

$\Rightarrow$: Suppose that the coefficients g_{rs} define an automorphism α. Then α induces an automorphism $\beta : L/R(L) \to L/R(L)$. As $L/R(L) \cong \mathbb{Z}_P^d$ with trivial multiplication, it follows that $Aut(L/R(L)) \cong GL(d, \mathbb{Z}_P)$. Hence $det(G) \not\equiv 0 \bmod P$ so (a) follows. (b) follows from von Dyck's theorem.

$\Leftarrow$: Suppose that (a) and (b) hold. As (b) holds, von Dyck's theorem asserts that α is a Lie ring homomorphism. As $P \nmid det(G)$. it follows that the images of $b_1, \ldots, b_d$ generate L as Lie ring. Hence α is surjective. Since L is finite, it follows that α is also injective and hence an automorphism.

This allows us to determine a generic description for $Aut(L)$. Suppose that we have an automorphism given by indeterminates $\{g_{rs} \mid 1 \leq r \leq d, 1 \leq s \leq n\}$ and write $g_i = g_{i1}b_1 + \ldots + g_{in}b_n$ for $1 \leq i \leq d$. For $k > d$ write b_k as a word w_k in the generators $b_1, \ldots, b_d$ and use this to determine $g_k = w_k(g_1, \ldots, g_d)$. Evaluate the defining relations $R_1, \ldots, R_m$ of L in $g_1, \ldots, g_n$. For each relation R_i this leads to an expression

$$\overline{R}_i = R_i(g_1, \ldots, g_n) = w_{i\,d+1}b_{d+1} + \ldots + w_{in}b_n,$$

with w_{ij} a polynomial in the indeterminates $\{g_{rs} \mid 1 \leq r \leq d, 1 \leq s \leq n\}$.

Lemma 2. *Let P be a prime and k minimal with $P^k b_i = 0$ for $d < i \leq n$. If $w_{ij} \equiv 0 \bmod P^k$ for all i, j and if $\det(G) \not\equiv 0 \bmod P$, then the matrix $(g_{rs})_{1 \leq r \leq d, 1 \leq s \leq n}$ defines an automorphism.*

Proof. The generators that appear in the relations $\overline{R}_i = 0$ all lie in the radical, and so $w_{ij} \equiv 0 \bmod P^k$ ensures that $w_{ij} b_j = 0$ for all i, j. Hence the conditions of Theorem 1 are satisfied and the matrix $(g_{rs})_{1 \leq r \leq d, 1 \leq s \leq n}$ defines an automorphism.

The integer P^k in Lemma 2 is called the characteristic of $R(L)$. If $k = 1$, then the conditions in Lemma 2 clearly determine all automorphisms of L. If $k > 1$, then the conditions in Lemma 2 may miss some automorphisms and there are examples where

$$\overline{R}_i = w_{i\,d+1} b_{d+1} + \ldots + w_{in} b_n = 0,$$

but some of the summands $w_{ij} b_j$ are non-zero. So it seems possible that restricting our search to integer matrices (g_{rs}) which satisfy the equations $w_{ij} = 0 \bmod P^k$ could miss some automorphisms in some cases. In practice, we have not found a case where this happens.

Example 5. We continue Example 1 for a specific prime P.

Since the radical has characteristic P our method shows that the matrix

$$\begin{pmatrix} g_{11} & 0 & 0 & 0 & 0 & 0 & 0 \\ 0 & 1 & 0 & 0 & 0 & 0 & 0 \\ 0 & 0 & g_{11} & 0 & 0 & 0 & 0 \end{pmatrix}$$

gives an automorphism if and only if $g_{11}^2 = 1 \bmod P$. Let $P = 5$. Then

$$B = \begin{pmatrix} 4 & 0 & 0 & 0 & 0 & 0 & 0 \\ 0 & 1 & 0 & 0 & 0 & 0 & 0 \\ 0 & 0 & 4 & 0 & 0 & 0 & 0 \end{pmatrix}$$

gives an automorphism. There was no need in this case to look for solutions to $g_{11}^2 = 1 \bmod P^2$, but it is easy to "lift" B to a matrix $C = (h_{rs})$ which gives the same automorphism as B, but where $h_{11}^2 = 1 \bmod 25$. The first row of the matrix B represents the element $4b_1$. Now $5b_1 = b_4 + b_6 + x_2 b_7$ and so the vector $(-1, 0, 0, 1, 0, 1, x_2)$ also represents $4b_1$. Similarly the vector $(0, 0, -1, 0, 0, x_1, 0)$ represents the same element of L as the third row of B. So

$$C = \begin{pmatrix} -1 & 0 & 0 & 1 & 0 & 1 & x_2 \\ 0 & 1 & 0 & 0 & 0 & 0 & 0 \\ 0 & 0 & -1 & 0 & 0 & x_1 & 0 \end{pmatrix}$$

gives the same automorphism as B, but the $(1,1)$ entry in C satisfies the equation $x^2 = 1 \bmod 25$. Note that B gives an automorphism, but does not have the form specified in Example 4, whereas C gives the same automorphism as B, but does have the form specified.

More generally, in every case of Lie p-rings from our database that we have examined, we can show that if B is an integer matrix which gives an automorphism of L for some prime P, and if k is any positive integer, then B can be "lifted" to an integer matrix $C = (h_{rs})$ which gives the same automorphism as B but where the entries h_{rs} satisfy all the equations $w_{ij} = 0 \bmod P^k$. So in every case that we have examined our method finds the full automorphism group.

We do not have a proof that our method always finds the full automorphism group. But there are several general criteria (such as the radical having characteristic P) which imply that our method does not miss any automorphisms. So in most cases our program is able to issue a "certificate of correctness". In some cases it may be necessary to examine the output from our program to prove that it has found the full automorphism group.

Example 6. We consider the symbolic Lie p-ring $\mathcal{L}$ on 7 generators with the non-trivial relations

$$
\begin{aligned}
[b_2, b_1] &= b_3, & pb_1 &= b_5 + xb_7, \\
[b_3, b_1] &= b_4, & pb_2 &= w^2 b_6 + yb_7, \\
[b_3, b_2] &= b_5, & pb_3 &= w^2 b_7. \\
[b_4, b_1] &= b_6, \\
[b_5, b_2] &= -w^2 b_7, \\
[b_6, b_1] &= b_7,
\end{aligned}
$$

Then $R(\mathcal{L}) = \langle b_3, \ldots, b_7 \rangle$ and each Lie p-ring L in the family of $\mathcal{L}$ is generated by $\{b_1, b_2\}$. We define

$$g_1 = g_{11}b_1 + \ldots + g_{17}b_7 \quad \text{and} \quad g_2 = g_{21}b_1 + \ldots + g_{27}b_7.$$

Next, we write $b_3, \ldots, b_7$ as words in $\{b_1, b_2\}$. It can be read off from the defining relations that $b_3 = [b_2, b_1], b_4 = [b_3, b_1], b_5 = [b_3, b_2], b_6 = [b_4, b_1], b_7 = [b_6, b_1]$. Using this, we expand the mapping defined by $\{g_{rs}\}$ to the remaining generators $b_3, \ldots, b_7$. For example, for b_3 this yields

$$
\begin{aligned}
g_3 &= [g_2, g_1] \\
&= (g_{11}g_{22} - g_{12}g_{21})b_3 + (g_{11}g_{23} - g_{13}g_{21})b_4 + (g_{12}g_{23} - g_{13}g_{22})b_5 \\
&\quad + (g_{11}g_{24} - g_{14}g_{21})b_6 + (-g_{12}g_{25}w^2 + g_{15}g_{22}w^2 + g_{11}g_{26} - g_{16}g_{21})b_7.
\end{aligned}
$$

We now evaluate the defining relations of $\mathcal{L}$ in $g_1, \ldots, g_n$. For example $pb_1 = b_5 + xb_7$ evaluates to

$$
\begin{aligned}
pg_1 - g_5 - xg_7 = 0b_1 + 0b_2 + 0b_3 \\
+ (-g_{11}g_{21}g_{22} + g_{12}g_{21}^2)b_4 \\
+ (-g_{11}g_{22}^2 + g_{12}g_{21}g_{22} + g_{11})b_5 \\
+ (-g_{11}g_{21}g_{23} + g_{12}w^2 + g_{13}g_{21}^2)b_6 \\
+ (-g_{11}^4 g_{22}x + g_{11}^3 g_{12}g_{21}x + g_{12}g_{22}g_{23}w^2 - g_{13}g_{22}^2 w^2 - g_{11}g_{21}g_{24} \\
+ g_{13}w^2 + g_{14}g_{21}^2 + g_{11}x + g_{12}y)b_7
\end{aligned}
$$

Note that the coefficient of b_3 in this relation is zero. More generally, if R_i is any of the relations then

$$\overline{R}_i = w_{i4}b_4 + \ldots + w_{i7}b_7 = 0,$$

and b_4, b_5, b_6, b_7 all have order p. So we obtain an automorphism at the prime P if and only if $w_{ij} = 0 \bmod P$ ($j = 4, 5, 6, 7$) for all relations R_i.

Now let L be a finite Lie p-ring in the family defined by $\mathcal{L}$ and let P be its prime. If the integer coefficients g_{rs} define an automorphism of L, then $det(G)$ is coprime to P. Hence, examining the coefficient of b_4 in the relation above we see that

$$-g_{11}g_{21}g_{22} + g_{12}g_{21}^2 = -g_{21}det(G) \equiv 0 \bmod P$$

is equivalent to $g_{21} \equiv 0 \bmod P$. In turn, this can now be used to simplify the remaining coefficients. Using $g_{21} \equiv 0 \bmod P$ now yields

$$-g_{11}g_{22}^2 + g_{12}g_{21}g_{22} + g_{11} = -g_{11}g_{22}^2 + g_{11} = -g_{11}(g_{22}^2 - 1)$$

As $det(G) \equiv g_{11}g_{22} \bmod P$ via $g_{21} \equiv 0 \bmod P$, it follows that g_{11} is coprime to P and $g_{22}^2 = 1 \bmod P$. We now iterate this approach. Introducing another indeterminate D with $Ddet(G) \equiv 1 \bmod P$ we finally obtain that

$$g_{21}, g_{12}, x(g_{22} - 1), g_{22}^2 - 1, y(g_{11} - 1), y(D - g_{22}), Dg_{22} - g_{11}^2,$$
$$Dg_{11} - g_{22}, D^2 - g_{11}, x(g_{11}^2 - D), g_{11}^2g_{22} - D, g_{11}^3 - 1$$

evaluate to 0 modulo P. We use this to eliminate indeterminates in the descriptions of g_1, g_2; for example, we can replace g_{21} by 0. We obtain

$$g_1 = (D^2 \quad 0 \quad g_{13} \ g_{14} \ g_{15} \ g_{16} \ g_{17})$$
$$g_2 = (0 \quad D^3 \quad g_{23} \ g_{24} \ g_{25} \ g_{26} \ g_{27})'$$

subject to the additional condition that the polynomials

$$(D - 1)xy, (D^2 - 1)y, (D^3 - 1)x, D^6 - 1$$

must evaluate to 0 mod P. This is the resulting description of the automorphism groups of the Lie p-rings L in the family defined by $\mathcal{L}$. It implies that $|Aut(L)| = kP^{10}$, where $k \in \{1, 2, 3, 6\}$. The precise value of k depends on the two parameters x, y. When $P \equiv 1 \bmod 3$, if $x = y = 0$ then $k = 6$; if $x = 0$ and $y \neq 0$ then $k = 2$; if $x \neq 0$ and $y = 0$ then $k = 3$; finally if $x, y \neq 0$ then $k = 1$. When $P \equiv 2 \bmod 3$ then $k = 1$ or 2.

4   The Lazard Correspondence

The final example of this abstract illustrates how the Lazard correspondent $G(L)$ to a finite Lie p-ring L can be determined using the LieRing package [1].

```
gap> L := LiePRingsByLibrary(7)[300];
<LiePRing of dimension 7 over prime p with parameters [ x ]>
gap> NumberOfLiePRingsInFamily(L);
p
gap> LiePRingsInFamily(L, 7);
[ <LiePRing of dimension 7 over prime 7>,
...
gap> List(last, x -> PGroupByLiePRing(x));
[ <pc group of size 823543 with 7 generators>,
...
gap> List(last, x -> Size(AutomorphismGroup(x)));
[80707214,80707214,80707214,80707214,80707214,80707214,80707214]
gap> a := AutGroupDescription(L);
rec( auto := [ [ A11, A12, A13, A14, A15, A16, A17 ],
               [  0,   1, A23,   0, A25, A26, A27 ] ],
     eqns := [ A11^2-1, A12*w*x-A11*A26 ] )
gap> 2*7^9;
80707214
```

References

1. Cicalò, S., de Graaf, W.A.: Liering, a GAP 4 package, see [9] (2010)
2. Cicalò, S., de Graaf, W.A., Vaughan-Lee, M.R.: An effective version of the lazard correspondence. J. Algebra **352**, 430–450 (2012)
3. Eick, B., Jalaleean, T.: Computing the Schur multiplier of a symbolic lie ring (2020, submitted)
4. Eick, B., Vaughan-Lee, M.: LiePRing version 2.5, : a GAP 4 package, see [9]. Version **2**, 5 (2020). http://www.iaa.tu-bs.de/beick/soft/liepring
5. Lazard, M.: Sur les groupes nilpotents et les anneaux de Lie. Ann. Sci. Ecole Norm. Sup. **3**(71), 101–190 (1954)
6. Newman, M.F., O'Brien, E.A., Vaughan-Lee, M.R.: Groups and nilpotent Lie rings whose order is the sixth power of a prime. J. Alg. **278**, 383–401 (2003)
7. O'Brien, E.A., Vaughan-Lee, M.R.: The groups with order p^7 for odd prime p. J. Algebra **292**(1), 243–258 (2005)
8. Robinson, D.J.S.: A Course in the Theory of Groups. Graduate Texts in Math, vol. 80. Springer, New York (1982)
9. The GAP Group: GAP - Groups, Algorithms and Programming, Version 4.10 (2019). http://www.gap-system.org

The Classification Problem in Geometry

Classifying Simplicial Dissections of Convex Polyhedra with Symmetry

Anton Betten$^{(\boxtimes)}$ ⓘ and Tarun Mukthineni ⓘ

Department of Mathematics, Colorado State University,
Fort Collins, CO 80523-1874, USA
{betten,tarun}@math.colostate.edu

Abstract. A convex polyhedron is the convex hull of a finite set of points in $\mathbb{R}^3$. A triangulation of a convex polyhedron is a decomposition into a finite number of 3-simplices such that any two intersect in a common face or are disjoint. A simplicial dissection is a decomposition into a finite number of 3-simplices such that no two share an interior point. We present an algorithm to classify the simplicial dissections of a regular polyhedron under the symmetry group of the prolyhedron.

Keywords: Dissection · Triangulation · Polyhedra · Geometry · Classification · Computational group theory

1 Introduction

A convex polyhedron is the convex hull of a finite set of points in $\mathbb{R}^3$. A triangulation of a convex polyhedron is a decomposition into a finite number of 3-simplices such that any two intersect in a common face or are disjoint. A simplicial dissection is a decomposition into a finite number of 3-simplices such that no two share an interior point. A simplicial dissection is a triangulation but not conversely. The problem is that the intersection of two simplices in a dissection may not be face.

Standard implementations for enumerating triangulations include TOPCOM and mptopcom [10] (neither one can enumerate dissections, though). A parallel algorithm to classify regular triangulations with applications in tropical geometry is described in [6]. Regarding the enumeration of all triangulations, see [4]. For minimal dissections, see [1].

The goal of this paper is to present an efficient algorithm to classify the simplicial dissections of a regular polyhedron under the symmetry group (or automorphism group) G of the polyhedron $\mathfrak{P}$. Two dissections of $\mathfrak{P}$ are equivalent is there is a symmetry $g \in G$ which maps one to the other. The classification algorithm utilizes the concept of a partially ordered set under a group action, using the theory developed by Plesken [9] as a framework. The partially ordered set is the search space, which is to be partitioned into orbits. The ranking of the poset introduces level sets, and the orbits partition these level sets. The efficiency

© Springer Nature Switzerland AG 2020
A. M. Bigatti et al. (Eds.): ICMS 2020, LNCS 12097, pp. 143–152, 2020.
https://doi.org/10.1007/978-3-030-52200-1_14

of the orbit algorithm is based on an effective use of isomorph rejection. This is the problem of deciding when two objects belong to the same G-orbit. Isomorph rejection is necessary to avoid duplicates, and it helps reduce the number of objects in the search space that have to be examined. The ultimate goal of the classification algorithm is to establish the poset of orbits of G. Isomorphism testing is expensive, and the algorithm that we propose avoids backtracking at the cost of memory. Such trade-off between time complexity and space complexity is common in algorithm design, and it has proved to be useful for other classification problems before. The first author has previously used this technique to classify objects like cubic surfaces, packings in projective space and other objects.

In this note, we will develop an efficient algorithm to classify the simplicial dissections of a polyhedra. As an application, we compute and classify the simplicial dissections of the cube. We use the binary representation of the integers from 0 to 7 to denote the vertices of the cube (cf. Figure 1), with two vertices adjacent if their Hamming distance is one.

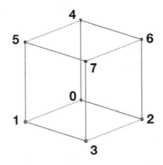

Fig. 1. The cube with labels

The Hamming distance is the number of components which differ in the binary expansion. The automorphism group of the cube has order 48 and is generated by the three permutations

$$(0,1,3,2)(4,5,7,6), \ (0,1,5,4)(2,3,7,6), \ (0,1)(2,3)(4,5)(6,7).$$

The tetrahedra are encoded using the lexicographic rank of their vertex set among the set of 4-subsets of $\{0,\dots,7\}$:

$0 = \{0,1,2,3\}$	$6 = \{0,1,3,5\}$	$12 = \{0,1,5,6\}$	$18 = \{0,2,3,7\}$
$1 = \{0,1,2,4\}$	$7 = \{0,1,3,6\}$	$13 = \{0,1,5,7\}$	$19 = \{0,2,4,5\}$
$2 = \{0,1,2,5\}$	$8 = \{0,1,3,7\}$	$14 = \{0,1,6,7\}$	$20 = \{0,2,4,6\}$
$3 = \{0,1,2,6\}$	$9 = \{0,1,4,5\}$	$15 = \{0,2,3,4\}$	$21 = \{0,2,4,7\}$
$4 = \{0,1,2,7\}$	$10 = \{0,1,4,6\}$	$16 = \{0,2,3,5\}$	$22 = \{0,2,5,6\}$
$5 = \{0,1,3,4\}$	$11 = \{0,1,4,7\}$	$17 = \{0,2,3,6\}$	$23 = \{0,2,5,7\}$

$24 = \{0,2,6,7\}$ $36 = \{1,2,3,5\}$ $48 = \{1,3,5,6\}$ $60 = \{2,3,6,7\}$
$25 = \{0,3,4,5\}$ $37 = \{1,2,3,6\}$ $49 = \{1,3,5,7\}$ $61 = \{2,4,5,6\}$
$26 = \{0,3,4,6\}$ $38 = \{1,2,3,7\}$ $50 = \{1,3,6,7\}$ $62 = \{2,4,5,7\}$
$27 = \{0,3,4,7\}$ $39 = \{1,2,4,5\}$ $51 = \{1,4,5,6\}$ $63 = \{2,4,6,7\}$
$28 = \{0,3,5,6\}$ $40 = \{1,2,4,6\}$ $52 = \{1,4,5,7\}$ $64 = \{2,5,6,7\}$
$29 = \{0,3,5,7\}$ $41 = \{1,2,4,7\}$ $53 = \{1,4,6,7\}$ $65 = \{3,4,5,6\}$
$30 = \{0,3,6,7\}$ $42 = \{1,2,5,6\}$ $54 = \{1,5,6,7\}$ $66 = \{3,4,5,7\}$
$31 = \{0,4,5,6\}$ $43 = \{1,2,5,7\}$ $55 = \{2,3,4,5\}$ $67 = \{3,4,6,7\}$
$32 = \{0,4,5,7\}$ $44 = \{1,2,6,7\}$ $56 = \{2,3,4,6\}$ $68 = \{3,5,6,7\}$
$33 = \{0,4,6,7\}$ $45 = \{1,3,4,5\}$ $57 = \{2,3,4,7\}$ $69 = \{4,5,6,7\}$
$34 = \{0,5,6,7\}$ $46 = \{1,3,4,6\}$ $58 = \{2,3,5,6\}$
$35 = \{1,2,3,4\}$ $47 = \{1,3,4,7\}$ $59 = \{2,3,5,7\}$

Theorem 1. *The number of equivalence classes of simplicial dissections of the cube under its automorphism group of order 48 is exactly 10. Six of these are triangulations as described in [5]. A system of representatives is given in Table 1, together with the order of the automorphism group. A more detailed drawing of the representatives is shown in Table 2.*

Table 1. The simplicial dissections of the cube

i	R_i	$|\mathrm{Aut}(R_i)|$
1	{1,38,41,52,63}	24
2	{1,35,45,56,65,68}	4
3	{1,35,45,56,66,67}	2
4	{1,35,45,57,63,66}	2
5	{1,35,45,58,61,68}	2
6	{1,35,45,59,61,64}	1
7	{1,35,45,59,62,63}	2
8	{1,35,47,52,57,63}	6
9	{2,19,36,56,66,67}	4
10	{2,19,36,59,61,64}	12

A tretrahedron is spatial if it has positive volume. Out of the list of 70 tetrahedra, 12 are flat. The remaining 58 are spatial and can be used for triangulating the cube. Following Takeuchi and Imai [12], triangulations can be identified using large cliques in a certain graph Γ, which we call the disjointness graph. This terminology is somewhat abusive, since the graph measures if the interior point sets of the tetrahedra are disjoint: Boundary points may or may not intersect. The vertices of Γ are the spatial tetrahedra. Two vertices are adjacent if the associated tetrahedra are non-overlapping, i.e. they do not share an interior point. The adjacency matrix of Γ is shown in Table 2.

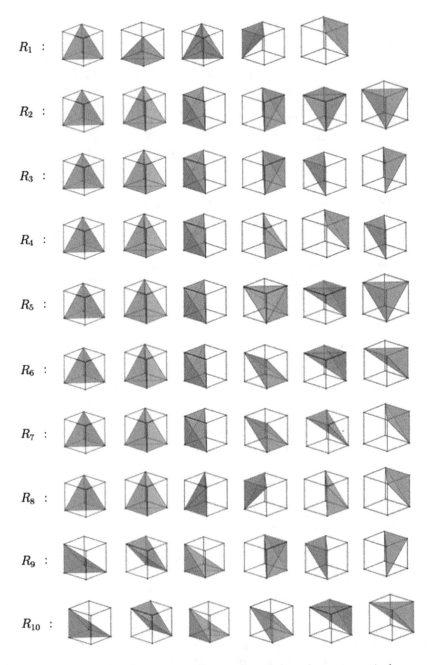

Fig. 2. The simplicial dissections $R_1, \ldots, R_{10}$ of the cube up to equivalence

Table 2. The adjacency matrix of the disjointness graph

2   The Types

As pointed out by De Loera et al. [5], there are four types of tetrahedra under the action of the group. The four types are listed in Table 3.

The table lists the volume of each tetrahedron, based on a cube of side length one. The type vector of a triangulation is the vector (a, b, c, d) where a is the number of Cores, b is the number of Corners, c is the number of Staircases, and d is the number of Slanted pieces. The Corner, Staircase and Slanted pieces each have volume $\frac{1}{6}$, whereas the Core piece has volume $\frac{1}{3}$. From this it follows that a triangulation or dissection either has 5 tetrahedra and includes a Core piece, or it has 6 tetrahedra, none of which are Core. This means that the type vector satisfies

$$2a + b + c + d = 6, \quad a \leq 1.$$

Table 3. The types of tetrahedra

Name	Representative	Numeric	#	Ago	Volume
Corner	$= \{0, 1, 2, 4\}$	1	8	6	$\frac{1}{6}$
Staircase	$= \{0, 1, 2, 5\}$	2	24	2	$\frac{1}{6}$
Slanted	$= \{1, 3, 4, 7\}$	47	24	2	$\frac{1}{6}$
Core	$= \{1, 2, 4, 7\}$	41	2	24	$\frac{1}{3}$

Table 4. The dissections and triangulations of the cube with tetrahedra sorted by type

i	Core	Corner	Staircase	Slanted	Type	$|\mathrm{Aut}(R_i)|$	DL
1	41	1, 38, 52, 63			(1, 4, 0, 0)	24	1
2		1, 68	45, 56	35, 65	(0, 2, 2, 2)	4	4
3		1	45, 56, 66, 67	35	(0, 1, 4, 1)	2	5
4		1, 63	45, 66	35, 57	(0, 2, 2, 2)	2	3
5		1, 68	45, 61	35, 58	(0, 2, 2, 2)	2	dissection
6		1	45, 59, 61, 64	35	(0, 1, 4, 1)	1	dissection
7		1, 63	45, 59	35, 62	(0, 2, 2, 2)	2	dissection
8		1, 52, 63		35, 47, 57	(0, 3, 0, 3)	6	2
9			2, 19, 36, 56, 66, 67		(0, 0, 6, 0)	4	dissection
10			2, 19, 36, 59, 61, 64		(0, 0, 6, 0)	12	6

In Table 4, the list of dissections from Theorem 1 is listed, with tetrahedra separated out by type. The type vector is listed in the column headed type. For triangulations, the De Loera number is in the column headed DL. This will be discussed in Sect. 4.

3   Poset Classification

Poset classification is a technique to classify combinatorial objects. Canonical augmentation due to McKay [7] is a very popular technique. McKay introduces the idea of a canonical predecessor to achieve the isomorph classification. McKay's work relies on the notion of a canonical form. His computer package Nauty [8] can compute canonical forms for graphs efficiently. This has led many authors to reduce the classification of different types of combinatorial structures to that of graphs. The original combinatorial objects are equivalent if and only if the associated graphs are isomorphic. By using Nauty to solve the isomorphism problem for the associated graphs, the combinatorial objects at hand are classified as well. For many combinatorial objects, this reduction is efficient and works very well. However, there are combinatorial objects for which this reduction is inefficient. Also, there is an interest in solving the isomorphism problem for the original combinatorial objects at hand directly, and avoiding the reduction to graphs altogether.

A second approach to classify combinatorial objects is losely based on ideas of Schmalz [11] for the enumeration of double cosets in groups. This has been adapted to the problem of classifying the orbits of groups on various posets. The critical operation in any poset orbit classification algorithm is the isomorphism testing. Using the ideas of Schmalz, backtracking can be avoided at the expense of higher memory complexity. The poset is examined breadth-first, using the rank of the combinatorial objects at hand. For most combinatorial objects, such rank functions are implicit. For instance, for orbits on sets, the size of the set is the rank of the set. In order to do isomorphism test in linear time, previously computed data in lower levels of the poset is utilized when constructing the next level in the poset. For a recent description of this technique, including some comparisons with canonical augmentation, see [3].

Let $(\mathcal{P}, \prec)$ be a partially ordered set with rank function. Assume that G is a group that acts on $\mathcal{P}$ (with the action written on the right). This means that for all $g \in G$ and all $a, b \in \mathcal{P}$ we have

$$a \prec b \iff ag \prec bg.$$

Let $\mathcal{P}_i$ be the set of objects at level i in $\mathcal{P}$. The poset of orbits for the action of G on $\mathcal{P}$ has as elements the orbits of G. Two orbits $\mathcal{O}_1$ and $\mathcal{O}_2$ are related with there exists $a \in \mathcal{O}_1$ and $b \in \mathcal{O}_2$ with $a \prec b$.

For computing dissections of a polyhedron $\mathfrak{P}$ with automorphism group G, let $\mathcal{P}$ be the set of partial dissections. A partial dissection is a set of spacial tetrahedra (simplices) which do not intersect in an interior point. The poset $\mathcal{P}$ is ordered with respect to inclusion. The group G of the polyhedron acts on this poset. The rank of a dissection is the number of simpices in it. The level set $\mathcal{P}_i$ contains all partial dissection size i. The dissections of the cube can be recognized using the volume function from Table 3. Dissections containing a Core tetrahedra have rank 5. All other dissections have rank 6. As all partial dissections correspond to cliques in the disjointness graph Γ of Table 2, the problem of finding dissections is reduced to that of finding suitable cliques in the graph Γ.

Let us present some results from the classification, computed using Orbiter [2]. The number of orbits of G on each of the levels $\mathcal{P}_i$ for $i = 0, \ldots, 6$ is shown in Table 5.

Table 5. The number of orbits on the poset by level

Level	#	Aut distribution
0	1	(48)
1	4	$(24, 6, 2^2)$
2	24	$(12, 6, 4^4, 2^9, 1^9)$
3	59	$(6^4, 4, 2^{15}, 1^{39})$
4	72	$(24, 8, 6, 4^6, 3, 2^{19}, 1^{43})$
5	32	$(24, 6, 2^{11}, 1^{19})$
6	9	$(12, 6, 4^2, 2^4, 1)$

The poset of orbits for the action of the group of the cube on the partial dissections is shown in Fig. 3.

The labeling of the representatives of the dissections is as in Table 1.

4   Comparison with the Types of de Loera et al.

De Loera, Rambau and Santos [5] list six types of triangulations of the cube. In Table 6, the De Loera triangulations are listed and identified with orbits in Table 4. An isomorphism from the representative picked by De Loera to the representative in Table 4 is given.

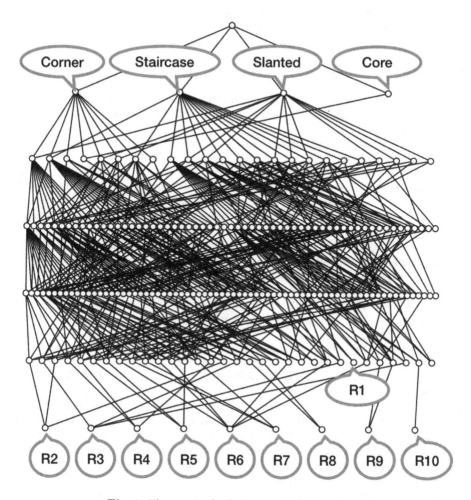

Fig. 3. The poset of orbits on partial dissections

Table 6. The triangulations listed by De Loera et al.

i	Representative	Type	Table 4	Isomorphism
1	{1, 38, 41, 52, 63}	(1, 4, 0, 0)	1	id
2	{4, 11, 21, 38, 52, 63}	(0, 3, 0, 3)	8	(0, 3, 5)(2, 7, 4)
3	{6, 17, 29, 30, 32, 33}	(0, 2, 2, 2)	4	(0, 4, 5, 1)(2, 6, 7, 3)
4	{6, 18, 21, 29, 32, 63}	(0, 2, 2, 2)	2	(0, 4, 5, 1)(2, 6, 7, 3)
5	{6, 18, 24, 29, 32, 33}	(0, 1, 4, 1)	3	(0, 4, 6, 7, 3, 1)(2, 5)
6	{8, 13, 18, 24, 32, 33}	(0, 0, 6, 0)	10	(0, 2, 3, 1)(4, 6, 7, 5)

Acknowledgement. The authors thank the three referees for helpful comments which have increased the clarity of the exposition.

References

1. Below, A., Brehm, U., De Loera, J.A., Richter-Gebert, J.: Minimal simplicial dissections and triangulations of convex 3-polytopes. Discrete Comput. Geom. **24**(1), 35–48 (2000)
2. Betten, A.: Orbiter - a program to classify discrete objects (2019). https://github.com/abetten/orbiter
3. Betten, A.: How fast can we compute orbits of groups? In: Davenport, J.H., Kauers, M., Labahn, G., Urban, J. (eds.) ICMS 2018. LNCS, vol. 10931, pp. 62–70. Springer, Cham (2018). https://doi.org/10.1007/978-3-319-96418-8_8
4. de Loera, J.A., Hoşten, S., Santos, F., Sturmfels, B.: The polytope of all triangulations of a point configuration. Doc. Math. **1**(04), 103–119 (1996)
5. De Loera, J., Rambau, J., Santos, F.: Triangulations. Structures for Algorithms and Applications. Algorithms and Computation in Mathematics, vol. 25. Springer, Heidelburg (2010). https://doi.org/10.1007/978-3-642-12971-1
6. Jordan, C., Joswig, M., Kastner, L.: Parallel enumeration of triangulations. Electron. J. Combin. **25**(3), 27 (2018). Paper 3.6
7. Brendan, D.: McKay. Isomorph-free exhaustive generation. J. Algorithms **26**(2), 306–324 (1998)
8. McKay, B.: Nauty User's Guide (Version 2.6). Australian National University (2016)
9. Plesken, W.: Counting with groups and rings. J. Reine Angew. Math. **334**, 40–68 (1982)
10. Rambau, J.: TOPCOM: triangulations of point configurations and oriented matroids. In: Arjeh, M.C., Gao, X.-S., Takayama, N. (eds.) Mathematical Software–ICMS 2002, pp. 330–340. World Scientific (2002)
11. Schmalz, B.: Verwendung von Untergruppenleitern zur Bestimmung von Doppelnebenklassen. Bayreuth. Math. Schr. **31**, 109–143 (1990)
12. Takeuchi, F., Imai, H.: Enumerating triangulations for products of two simplices and for arbitrary configurations of points. In: Jiang, T., Lee, D.T. (eds.) COCOON 1997. LNCS, vol. 1276, pp. 470–481. Springer, Heidelberg (1997). https://doi.org/10.1007/BFb0045114

Classification Results for Hyperovals of Generalized Quadrangles

Bart De Bruyn[✉][iD]

Ghent University, 9000 Gent, Belgium
Bart.DeBruyn@UGent.be

Abstract. A hyperoval of a point-line geometry is a nonempty set of points meeting each line in either 0 or 2 points. We discuss a combination of theoretical and practical techniques that are helpful for classifying hyperovals of generalized quadrangles. These techniques are based on the connection between hyperovals, even sets and pseudo-embeddings of point-line geometries.

Keywords: Generalized quadrangle · Hyperoval · Pseudo-embedding · Even set · Ideal

1 Introduction

A *(point-line) geometry* is a triple $\mathcal{S} = (\mathcal{P}, \mathcal{L}, I)$ consisting of a nonempty point set $\mathcal{P}$, a line set $\mathcal{L}$ and an incidence relation $I \subseteq \mathcal{P} \times \mathcal{L}$ between these sets. One of the most important classes of geometries are the so-called *(axiomatic) projective planes* [17]. A finite projective plane π contains $n^2 + n + 1$ points and $n^2 + n + 1$ lines for some $n \in \mathbb{N}$, called the *order* of π. The standard examples are the Desarguesian projective planes $\mathrm{PG}(2, q)$ with q some prime power. Axiomatic projective planes have been intensively investigated, in particular several construction and classification results have been obtained about them. Some of these results have been obtained by means of computer computations, like the classifications of all projective planes of order 8, 9 and 10 [15,18,19].

Besides classification results and constructions, also special sets of points in projective planes have been investigated. Certain of these sets have relationships with other mathematical areas, like coding theory, or certain geometries can be constructed from them, like partial geometries and generalized quadrangles. One of the substructures of finite projective planes that have been thoroughly investigated are the *hyperovals*. These are nonempty sets of points meeting each line in either 0 or 2 points, in which case it can be shown that the hyperoval has size $n + 2$ with n the (necessarily even) order of the plane. The classical examples of hyperovals here are those in $\mathrm{PG}(2, q)$, q even, by adding to an irreducible conic $\mathcal{C}$ its *nucleus*, that is the point that lies in all tangent lines of $\mathcal{C}$. The construction and classification problem of hyperovals in arbitrarily not necessarily Desarguesian projective planes has been intensively studied. Hyperovals

© Springer Nature Switzerland AG 2020
A. M. Bigatti et al. (Eds.): ICMS 2020, LNCS 12097, pp. 153–161, 2020.
https://doi.org/10.1007/978-3-030-52200-1_15

also play a crucial role in the nonexistence proof for the projective plane of order 10 [19]. Indeed this proof essentially relies on the fact that a plane of order 10 cannot have hyperovals [20].

The concept of a hyperoval, namely a nonempty set of points meeting each line in either 0 or 2 points, can be defined for general point-line geometries. Two families of point-line geometries that have attracted attention here are the *generalized quadrangles (GQ's)* [25] and the *polar spaces* [2]. The standard examples of polar spaces are related to symplectic polarities, quadrics and Hermitian varieties in projective spaces [16], but also every generalized quadrangle is an example of a polar space. A *generalized quadrangle of order* (s, t), or shortly a $GQ(s, t)$, is defined as a geometry that satisfies the following three properties:

1. Every two distinct points are incident with at most one line.
2. Every line is incident with exactly $s + 1$ points and every point is incident with precisely $t + 1$ lines.
3. For every non-incident point-line pair (x, L), there exists a unique point y on L collinear with x (i.e. y is in some line together with x).

Hyperovals of polar spaces, in particular of GQ's, are not only interesting point sets. They are also related to other combinatorial structures in finite geometry. Hyperovals (or *local subspaces*) of polar spaces were first considered in [1] because of their connection with so-called *locally polar spaces*. Hyperovals of GQ's have a number of additional applications. They naturally arise in the study of *extended generalized quadrangles* and play a fundamental role in their study, see [3, 21–23]. Lower and upper bounds for the size of a hyperoval H in a $GQ(s, t)$ were obtained in [3, Lemmas 3.9 and 3.11] and [14, Theorems 2.1 and 2.2]. The size $|H|$ is even and satisfies $\max(2(t + 1), (t - s + 2)(s + 1)) \leq |H| \leq 2(st + 1)$.

In recent years, many construction and classification results for hyperovals in GQ's have been obtained. These regard theoretical constructions of infinite families [4–9, 14, 24], or computer backtrack searches as in [22, 23]. We will emphasise here on a number of techniques that can help in studying and classifying hyperovals, both from a theoretical as a computational point of view. Hyperovals are special cases of *even sets*, these are sets of points that meet each line in an even number of points. The intention is to discuss some tools for classifying hyperovals inside the family of all even sets. The complements of the even sets were coined *pseudo-hyperplanes* in [11]. There exist close relationships between pseudo-hyperplanes and certain representations of the geometry in projective spaces, called *pseudo-embeddings*. Some of these relationships will be mentioned in Sect. 2. Via the connection with pseudo-embeddings, we show in Sect. 3 that the family of hyperovals is related to certain ideals in polynomial rings and that Gröbner bases can sometimes help in their study and/or classification.

2   Pseudo-embeddings, Pseudo-hyperplanes and Even Sets

Suppose $\mathcal{S} = (\mathcal{P}, \mathcal{L}, \mathrm{I})$ is a geometry for which the number of points on each line is finite and at least 3. A *pseudo-embedding* of $\mathcal{S}$ is a map ϵ from $\mathcal{P}$ to the point set of a projective space $\mathrm{PG}(V)$ defined over the field $\mathbb{F}_2$ of order 2 such that:

- the image of ϵ generates the whole projective space $\mathrm{PG}(V)$;
- ϵ maps every line $L \in \mathcal{L}$ to a *frame* of a subspace of $\mathrm{PG}(V)$, i.e. $\epsilon(L)$ is a set of the form $\{\langle \bar{v}_1 \rangle, \langle \bar{v}_2 \rangle, \ldots, \langle \bar{v}_k \rangle\}$, where k is the size of L, $\bar{v}_1, \bar{v}_2, \ldots, \bar{v}_{k-1}$ are $k-1$ linearly independent vectors of V and $\bar{v}_k = \bar{v}_1 + \bar{v}_2 + \ldots + \bar{v}_{k-1}$.

We denote such a pseudo-embedding also by $\epsilon : \mathcal{S} \to \mathrm{PG}(V)$. A pseudo-embedding thus maps the lines of a geometry $\mathcal{S}$ to frames of subspaces of a projective space $\mathrm{PG}(V)$. This is different from the notion of an (ordinary) embedding of $\mathcal{S}$ which maps the lines of $\mathcal{S}$ to lines of $\mathrm{PG}(V)$.

Two pseudo-embeddings $\epsilon_1 : \mathcal{S} \to \mathrm{PG}(V_1)$ and $\epsilon_2 : \mathcal{S} \to \mathrm{PG}(V_2)$ of the same point-line geometry $\mathcal{S}$ are called *isomorphic* if there exist a linear isomorphism θ between the vector spaces V_1 and V_2 such that $\epsilon_2 = \theta \circ \epsilon_1$.

If $\epsilon : \mathcal{S} \to \mathrm{PG}(V)$ is a pseudo-embedding, then projecting the image of ϵ from a (suitable) subspace on a complementary subspace can give rise to another pseudo-embedding ϵ', which is called a *projection* of ϵ. If ϵ_1 and ϵ_2 are two pseudo-embeddings of the same point-line geometry $\mathcal{S}$, then we write $\epsilon_1 \geq \epsilon_2$ if ϵ_2 is isomorphic to a projection of ϵ_1. If $\tilde{\epsilon}$ is a pseudo-embedding of $\mathcal{S}$ such that $\tilde{\epsilon} \geq \epsilon$ for any other pseudo-embedding ϵ of $\mathcal{S}$, then $\tilde{\epsilon}$ is called *universal*. If $\mathcal{S}$ has pseudo-embeddings, then it also has a universal pseudo-embedding which is moreover unique, up to isomorphism. The vector dimension of the universal pseudo-embedding is called the *pseudo-embedding rank*, and (in case $|\mathcal{P}| < \infty$) is equal to $|\mathcal{P}| - \dim(C)$, where C is the binary code of length $|\mathcal{P}|$ generated by the characteristic vectors of the lines of $\mathcal{S}$. Note that $\dim(C)$ equals the $\mathbb{F}_2$-rank of an incidence matrix of $\mathcal{S}$. We thus see that there exist connections between pseudo-embeddings and coding theory. There also exist connections between pseudo-embeddings and modular representation theory of groups.

Pseudo-hyperplanes and hence also even sets are closely related to pseudo-embeddings as the following theorem shows.

Theorem 1 ([11]). *If $\epsilon : \mathcal{S} \to \mathrm{PG}(V)$ is a pseudo-embedding, then for every hyperplane Π of $\mathrm{PG}(V)$, the set $\epsilon^{-1}(\epsilon(\mathcal{P}) \cap \Pi)$ is a pseudo-hyperplane of $\mathcal{S}$. Every pseudo-hyperplane of $\mathcal{S}$ arises in this way from the universal pseudo-embedding of $\mathcal{S}$.*

More background information about pseudo-embeddings, pseudo-hyperplanes and the above facts can be found in [10–13]. In [11] it was also shown that all GQ's have pseudo-embeddings and hence also universal pseudo-embeddings.

Hyperovals of GQ's can often be computationally classified without implementing a backtrack algorithm. One way to achieve this goal is to determine all (isomorphism classes of) even sets, and subsequently to verify which even sets are also hyperplanes. The number of even sets can be determined in advance: it equals 2^k, with k the pseudo-embedding rank. As soon as a computer model of the geometry has been implemented along with its automorphism group (e.g. with GAP [27]), it·is easy to generate even sets, the size of the orbit to which a given even set belongs can readily be computed, and it can easily be verified whether two hyperovals are isomorphic. Based on these three principles, it is often easy to compute all isomorphism classes of even sets. This has been illustrated in the papers [12,13]. We mention two reasons why it is so easy to generate even sets with a computer:

1. An even set can be found as a set whose characteristic vector is $\mathbb{F}_2$-orthogonal with all characteristic vectors of the lines.
2. The symmetric difference of any two even sets is again an even set.

The above method (as well as a backtrack search) has the disadvantage that it does not provide unified and explicit descriptions for the hyperovals. The method which we will discuss in the following section does have this potential. It is still based on the connection with even sets but it also takes into account a description of the universal pseudo-embedding.

3   Related Ideals in Polynomial Rings

The material discussed in this section is new with exception of Theorem 4, which is taken from [13, Corollary 1.3]. We continue with the notation in Sect. 2. We suppose that S has pseudo-embeddings and we denote by $\widetilde{\epsilon} : S \to \mathrm{PG}(\widetilde{V})$ the universal pseudo-embedding of S. If $k := \dim(\widetilde{V})$, then there exist k maps $f_i : \mathcal{P} \to \mathbb{F}_2$ ($i \in \{1, 2, \ldots, k\}$) such that $\widetilde{\epsilon}$ maps a point p of S to the point $(f_1(p), f_2(p), \ldots, f_k(p))$ of $\mathrm{PG}(\widetilde{V})$. Using these f_i's, Theorem 1 can now be rephrased as follows.

Theorem 2. *The even sets of S are precisely the subsets of $\mathcal{P}$ satisfying an equation of the form $\sum_{i=1}^{k} a_i f_i(p) = 1$ with $a_1, a_2, \ldots, a_k \in \mathbb{F}_2$.*

We denote by $E(\bar{a})$ the even set corresponding to a tuple $\bar{a} = (a_1, a_2, \ldots, a_k)$. Suppose $\alpha = \{p_1, p_2, \ldots, p_l\}$ is a line of S. The condition that the point p_i of α belongs to $E(\bar{a})$ implies by Theorem 2 that a certain linear combination $L_i(\bar{a})$ of the a_i's is equal to 1. If $E(\bar{a})$ is a hyperoval of S, then the number of i's for which $L_i(\bar{a})$ is equal to 1 is therefore either 0 to 2.

Theorem 3. *There exists a $g_\alpha(a_1, a_2, \ldots, a_k) \in \mathbb{F}_2[a_1, a_2, \ldots, a_k]$ such that the following two conditions are equivalent for any $\bar{a} = (a_1, a_2, \ldots, a_k) \in \mathbb{F}_2^k$:*

- *the number of i's for which $L_i(\bar{a})$ is equal to 1 is either 0 to 2;*
- *$g_\alpha(a_1, a_2, \ldots, a_k) = 0$.*

Proof. We define $h(a_1, a_2, \ldots, a_k) := (L_1(\bar{a}) + 1)(L_2(\bar{a}) + 1) \cdots (L_l(\bar{a}) + 1) + 1$ and $h_{uv}(a_1, a_2, \ldots, a_k) := 1 + L_u(\bar{a}) \cdot L_v(\bar{a}) \cdot \prod_{w \notin \{u,v\}} (L_w(\bar{a}) + 1)$ for all $u, v \in \{1, 2, \ldots, l\}$ with $u < v$. Then the following hold:

- $h(a_1, a_2, \ldots, a_k) = 0$ if and only if there are no i's for which $L_i(\bar{a}) = 1$;
- $h_{uv}(a_1, a_2, \ldots, a_k) = 0$ if and only if u, v are the only i's for which $L_i(\bar{a}) = 1$.

We can then put $g_\alpha(a_1, a_2, \ldots, a_k)$ equal to the product of h and all h_{uv}'s with $1 \leq u < v \leq l$.

There exists such a polynomial $g_\alpha(a_1, a_2, \ldots, a_k) \in \mathbb{F}_2[a_1, a_2, \ldots, a_k]$ for each line α of $\mathcal{S}$. Such a polynomial is not unique. If I is the ideal generated by the polynomials $a_i^2 + a_i$, $i \in \{1, 2, \ldots, k\}$, then any polynomial in $g_\alpha(a_1, a_2, \ldots, a_k) + I$ also satisfies the required property. By the above discussion, we know:

Corollary 1. *The even set $E(\bar{a})$ with $\bar{a} \in \mathbb{F}_2^k \setminus \{\bar{o}\}$ is a hyperoval if and only if $g_\alpha(a_1, a_2, \ldots, a_k) = 0$ for all $\alpha \in \mathcal{L}$.*

If we know all g_α's, we can directly determine all $\bar{a} \in \mathbb{F}_2^k$ for which $E(\bar{a})$ is a hyperoval. From a computational point of view, this can go faster (see example later) than verifying which of the sets $E(\bar{a})$ with $\bar{a} \in \mathbb{F}_2^k$ intersects each line of the geometry in either 0 or 2 points. In the latter approach we first need to determine the set $E(\bar{a})$ by solving the equation mentioned in Theorem 2 (with respect to p) before verifying that $E(\bar{a})$ intersects each of the lines in 0 or 2 points. The method of working with the polynomials g_α has two additional benefits.

1. If ϕ is an automorphism of $\mathcal{S}$, then the fact that $\tilde{e}$ is so-called homogeneous (see e.g. [12]) implies that there exists a linear automorphism ϕ' of $\mathbb{F}_2^k$ such that ϕ maps the even set $E(\bar{a})$ to the even set $E(\bar{a}^{\phi'})$. If α and β are lines of $\mathcal{S}$ such that $\alpha = \beta^\phi$, then we have $g_\beta(\bar{a}) = g_\alpha(\bar{a}^{\phi'})$. Information about automorphisms of $\mathcal{S}$ and their corresponding actions on $\mathbb{F}_2^k$ thus implies that certain of the g_α's can be derived from others. In particular, if we have such information for a set of automorphisms that generate a line-transitive automorphism group, then one of the g_α's determines all the others.
2. If we take the ideal $\mathcal{G}$ generated by I and all g_α's, then any polynomial in $\mathcal{G}$ determines a necessary condition for a set $E(\bar{a})$ to be a hyperoval. In particular, we can look for polynomials that have a simple form. Such polynomials can often be found with the aid of Gröbner bases (implemented in computer algebra systems), and can be useful for theoretical and computational purposes.

Both benefits are illustrated by the following example. Consider in the projective space PG(3, 4) the Hermitian variety $\mathcal{H}$ with equation $X_1 X_2^2 + X_2 X_1^2 + X_3 X_4^2 + X_4 X_3^2 = 0$. The points and lines contained in $\mathcal{H}$ then define a generalized quadrangle $H(3, 4)$ of order $(4, 2)$ [25]. The universal pseudo-embedding of $H(3, 4)$ was described in [13, Section 1] and has vector dimension 24. From this description, we easily deduce the following (see also [13, Corollary 1.3]).

Theorem 4. *The even sets of $H(3,4)$ are precisely the subsets of $\mathcal{H}$ satisfying an equation of the form*

$$\Sigma_1(a_iX_i^3) + a_5(\omega X_3X_4^2 + \omega^2 X_4X_3^2) + a_6\big((X_1^3 + X_2^3 + X_1^3X_2^3)(X_3^3 + X_4^3 + X_3^3X_4^3) + 1\big)$$
$$+ \Sigma_2(b'_{ij}X_iX_j^2 + (b'_{ij})^2X_jX_i^2) + \Sigma_3\left(b'_{ijk}X_iX_jX_k + (b'_{ijk})^2X_i^2X_j^2X_k^2\right) = 1,$$

with the a_i's belonging to $\mathbb{F}_2$ and the b'_{ij}'s and b'_{ijk}'s belonging to $\mathbb{F}_4$.

In Theorem 4, $\mathbb{F}_4 = \{0, 1, \omega, \omega^2\}$ is the finite field of order 4, Σ_1 denotes the summation over all $i \in \{1, 2, 3, 4\}$, Σ_2 denotes the summation over all $i, j \in \{1, 2, 3, 4\}$ with $i < j$ and $(i, j) \neq (3, 4)$, and Σ_3 denotes the summation over all $i, j, k \in \{1, 2, 3, 4\}$ with $i < j < k$. We can now put $b'_{ij} = b_{ij} + \omega c_{ij}$ and $b'_{ijk} = b_{ijk} + \omega c_{ijk}$, where all b_{ij}'s, c_{ij}'s, b_{ijk}'s and c_{ijk}'s belong to $\mathbb{F}_2$. Using the terminology of Theorem 2, the maps $f_i(p)$ with $i \in \{1, 2, \ldots, 24\}$ and $p = (X_1, X_2, X_3, X_4)$ can then be taken as follows:

$$f_1(p) = X_1^3, f_2(p) = X_2^3, f_3(p) = X_3^3, f_4(p) = X_4^3, f_5(p) = \omega X_3X_4^2 + \omega^2 X_3^2X_4,$$

$$f_6(p) = (X_1^3 + X_2^3 + X_1^3X_2^3)(X_3^3 + X_4^3 + X_3^3X_4^3) + 1, f_7(p) = X_1X_2^2 + X_2X_1^2,$$

$$f_8(p) = \omega X_1X_2^2 + \omega^2 X_2X_1^2, \ldots, f_{24}(p) = \omega X_2X_3X_4 + \omega^2 X_2^2X_3^2X_4^2.$$

We now determine one of the g_α's.

Theorem 5. *If α is the line of $H(3,4)$ with equation $X_2 = X_4 = 0$, then g_α is equal to $a_1 + a_3 + (b_{13} + c_{13} + b_{13}c_{13})(a_1 + a_3 + a_6 + a_1a_3 + a_1a_6 + a_3a_6)$.*

Proof. The even set determined by the tuple $(a_1, a_2, \ldots, c_{234}) \in \mathbb{F}_2^{24}$ intersects α in either 0 or 2 points if the equation $a_1X_1^3 + a_3X_3^3 + a_6(X_1^3X_3^3 + 1) + b'_{13}X_1X_3^2 + (b'_{13})^2X_3X_1^2 = 0$ has 0 or 2 solutions for $(X_1, X_3) \in \{(0, 1), (1, x) \mid x \in \mathbb{F}_4\}$. This means that precisely two of the equations $a_3 + a_6 = 1$, $a_1 + a_6 = 1$, $a_1 + a_3 + b'_{13} + (b'_{13})^2 = 1$, $a_1 + a_3 + b'_{13}\omega^2 + (b'_{13})^2\omega = 1$, $a_1 + a_3 + b'_{13}\omega + (b'_{13})^2\omega^2 = 1$ are satisfied. We denote these equations respectively by (1), (2), (3), (4) and (5).

Suppose $b'_{13} = 0$. If $a_1 + a_3 = 1$, then (3), (4) and (5) imply that at least three of the equations are satisfied which is impossible. So, $a_1 + a_3 = 0$, but then (3), (4) and (5) are never satisfied. As $a_1 + a_3 = 0$, either (1), (2) are satisfied or none of them is satisfied. So, if $b'_{13} = 0$, then necessarily $a_1 + a_3 = 0$.

Suppose $b'_{13} \neq 0$ and $a_1 + a_3 = 1$. Then precisely one of (1), (2) is satisfied. As precisely one of b'_{13}, $b'_{13}\omega^2$, $b'_{13}\omega$ belongs to $\mathbb{F}_2$, we also see that precisely one of (3), (4), (5) is satisfied. So, this case is always OK.

Suppose $b'_{13} \neq 0$ and $a_1 + a_3 = 0$. As precisely one of b'_{13}, $b'_{13}\omega^2$, $b'_{13}\omega$ belongs to $\mathbb{F}_2$, precisely two of the equations (3), (4), (5) are satisfied. So, none of (1), (2) can be satisfied. This implies that $a_3 + a_6 = 0$.

The overall condition is thus $((b'_{13})^3 + 1)(a_1 + a_3) + (b'_{13})^3((a_1 + a_3 + 1)(a_3 + a_6)) = 0$ which simplifies to $a_1 + a_3 + (b'_{13})^3(a_1 + a_3 + a_6 + a_1a_3 + a_1a_6 + a_3a_6) = a_1 + a_3 + (b_{13} + c_{13} + b_{13}c_{13})(a_1 + a_3 + a_6 + a_1a_3 + a_1a_6 + a_3a_6)$.

In Section 5 of [13], we described a list of 6 generators $\phi_1, \phi_2, \ldots, \phi_6$ for the (line-transitive) automorphism group of $H(3,4)$, along with their corresponding actions on the even sets $E(\bar{a})$, see [13, Tables 1 and 2]. From this information, the corresponding actions of $\phi_1', \phi_2', \ldots, \phi_6'$ on $\mathbb{F}_2^{24}$ (see above) can easily be derived:

- $\bar{a}^{\phi_1'} = (a_3, a_4, a_1, a_2, c_{12}, a_6, b_{12}, a_5, b_{13}+c_{13}, c_{13}, b_{23}+c_{23}, c_{23}, b_{14}+c_{14}, c_{14}, b_{24}+c_{24}, c_{24}, b_{134}, c_{134}, b_{234}, c_{234}, b_{123}, c_{123}, b_{124}, c_{124})$;
- $\bar{a}^{\phi_2'} = (a_1, a_2, a_3, a_4, a_5, a_6, b_{12}, c_{12}, c_{13}, b_{13}+c_{13}, c_{14}, b_{14}+c_{14}, c_{23}, b_{23}+c_{23}, c_{24}, b_{24}+c_{24}, b_{123}+c_{123}, b_{123}, b_{124}+c_{124}, b_{124}, c_{134}, b_{134}+c_{134}, c_{234}, b_{234}+c_{234})$;
- $\bar{a}^{\phi_3'} = (a_1+a_3+a_6+c_{13}, a_2, a_3, a_4+a_2+a_6+c_{24}, a_5+c_{23}+c_{234}, a_6, b_{12}+b_{123}+c_{123}+b_{234}, c_{12}+c_{123}+c_{23}, b_{13}+a_3+a_6, c_{13}, b_{14}+b_{12}+b_{23}+c_{23}+b_{123}+c_{123}+b_{124}+b_{134}+c_{134}+b_{234}, c_{14}+a_5+c_{12}+c_{23}+c_{123}+c_{124}+c_{134}+c_{234}, b_{23}, c_{23}, b_{24}+a_2+a_6, c_{24}, b_{123}, c_{123}, b_{124}+a_6+b_{234}, c_{124}+c_{234}, b_{134}+a_6+b_{123}, c_{134}+c_{123}, b_{234}, c_{234})$;
- $\bar{a}^{\phi_4'} = (a_1, a_2, a_3, a_4+a_3+a_5, a_5, a_6, b_{12}+a_3, c_{12}, b_{13}, c_{13}, b_{14}+b_{13}+b_{134}, c_{14}+c_{13}+c_{134}, b_{23}, c_{23}, b_{24}+b_{23}+b_{234}, c_{24}+c_{23}+c_{234}, b_{123}, c_{123}, b_{124}+b_{123}, c_{124}+c_{123}, b_{134}, c_{134}, b_{234}, c_{234})$;
- $\bar{a}^{\phi_5'} = (a_1, a_2, a_3+a_4+a_5, a_4, a_5, a_6, b_{12}+a_4, c_{12}, b_{13}+b_{14}+b_{134}, c_{13}+c_{14}+c_{134}, b_{14}, c_{14}, b_{23}+b_{24}+b_{234}, c_{23}+c_{24}+c_{234}, b_{24}, c_{24}, b_{123}+b_{124}, c_{123}+c_{124}, b_{124}, c_{124}, b_{134}, c_{134}, b_{234}, c_{234})$;
- $\bar{a}^{\phi_6'} = (a_1, a_2, a_3, a_4, a_5, a_6, b_{12}+c_{12}+a_5, c_{12}, b_{13}+c_{13}, c_{13}, b_{14}+c_{14}, c_{14}, b_{23}+c_{23}, c_{23}, b_{24}+c_{24}, c_{24}, b_{123}+c_{123}, c_{123}, b_{124}+c_{124}, c_{124}, b_{134}+c_{134}, c_{134}, b_{234}+c_{234}, c_{234})$.

Based on this information, we have computed with the aid of SageMath [26] all g_α's. The ideal $\mathcal{G}$ generated by I and the g_α's contains polynomials that have fewer terms than the g_α's themselves. These have been found by computing Gröbner bases of ideals generated by some of these g_α's. Specifically, $\mathcal{G}$ contains the eight polynomials that are obtained from $a_1 a_3 b_{13} + a_1 a_6 b_{13} + a_3 a_6 b_{13} + a_6 b_{13}$ and $a_1 a_3 c_{13} + a_1 a_6 c_{13} + a_3 a_6 c_{13} + a_6 c_{13}$ by applying one of the permutations $(), (12), (34), (12)(34)$ on the subindices. $\mathcal{G}$ also contains the eight polynomials that are obtained from $a_1 b_{13} c_{13} + a_6 b_{13} c_{13} + a_1 a_3 + a_1 a_6 + a_3 a_6 + a_1 b_{13} + a_6 b_{13} + a_1 c_{13} + a_6 c_{13} + a_1$ by applying one of the permutations $(), (12), (13), (34), (132), (143), (12)(34), (14)(23)$ on the subindices.

4 Summary

We have discussed here three methods by which hyperovals can be computed:

(1) via the connection with even sets discussed at the end of Sect. 2;
(2) by finding all $\bar{a} \in \mathbb{F}_2^k$ for which $E(\bar{a})$ is a hyperoval (via Theorem 2);
(3) by finding all $\bar{a} \in \mathbb{F}_2^k$ for which $g_\alpha(\bar{a}) = 0$ holds for all lines $\alpha \in \mathcal{L}$.

For the example of hyperovals of $H(3,4)$, our implementation of the methods (1) and (2) had similar performances ($\pm$ 1h40min, iMac, 2.7 GHz Intel Core i5-4570R processor). Methods (1) and (2) were already used in [13] to show that

$H(3,4)$ has 23 nonisomorphic hyperovals. The third method was almost three times faster. Note also that the three polynomials of $\mathcal{G}$ mentioned at the end of Sect. 3 give the conditions $(a_1 + a_6)(a_3 + a_6)b_{13} = (a_1 + a_6)(a_3 + a_6)c_{13} = (a_1 + a_6)(b_{13}c_{13} + b_{13} + c_{13} + a_1 + a_3) = 0$, and that the remaining polynomials give similar equations. This means that certain of the entries of $\bar{a}$ are 0 or can be expressed in terms of the others, a fact that would allow to speed up further the computations for the third method. Some of the code (in SageMath [26] and GAP [27]) used in our computations can be found on https://cage.ugent.be/geometry/preprints.php.

Our main intention here was to discuss theoretical and computational techniques that are useful for classifying hyperovals of generalized quadrangles. These techniques suffice so far for classifying all hyperovals of all finite generalized quadrangles of order (s,t) with $s \leq 4$. These GQ's comprise the 3×3, 4×4 and 5×5-grids as well as the GQ's $W(2)$, $Q(5,2)$, $W(3)$, $Q(4,3)$, $GQ(3,5)$, $Q(5,3)$, $H(3,4)$, $W(4)$, $GQ(4,6)$, $H(4,4)$ and $GQ(5,4)$ (see [25] for definitions). With exception of the GQ's $W(4)$, $GQ(4,6)$, $H(4,4)$ and $Q(5,4)$, these classifications have already appeared in the literature (below).

Our work on classifying hyperovals of generalized quadrangles is work in progress where on the one hand we try to obtain additional classification results (for larger GQ's) and on the other hand we try to obtain computer free uniform descriptions for all the hyperovals of a given GQ. As in [13], the latter problem can involve that algebraic descriptions of the universal pseudo-embeddings need to be found.

References

1. Buekenhout, F., Hubaut, X.: Locally polar spaces and related rank 3 groups. J. Algebra **45**, 391–434 (1977)
2. Buekenhout, F., Shult, E.: On the foundations of polar geometry. Geometriae Dedicata **3**, 155–170 (1974)
3. Cameron, P.J., Hughes, D.R., Pasini, A.: Extended generalized quadrangles. Geom. Dedicata **35**, 193–228 (1990)
4. Cossidente, A.: Hyperovals on $H(3,q^2)$. J. Combin. Theor. Ser. A **118**, 1190–1195 (2011)
5. Cossidente, A., King, O.H., Marino, G.: Hyperovals of $H(3,q^2)$ when q is even. J. Combin. Theor Ser. A **120**, 1131–1140 (2013)
6. Cossidente, A., King, O.H., Marino, G.: Hyperovals arising from a Singer group action on $H(3,q^2)$, q even. Adv. Geom. **16**, 481–486 (2016)
7. Cossidente, A., Marino, G.: Hyperovals of Hermitian polar spaces. Des. Codes Crypt. **64**, 309–314 (2012)
8. Cossidente, A., Pavese, F.: Hyperoval constructions on the Hermitian surface. Finite Fields Appl. **25**, 19–25 (2014)
9. Cossidente, A., Pavese, F.: New infinite families of hyperovals on $H(3,q^2)$, q odd. Des. Codes Crypt. **73**, 217–222 (2014)
10. De Bruyn, B.: The pseudo-hyperplanes and homogeneous pseudo-embeddings of $AG(n,4)$ and $PG(n,4)$. Des. Codes Crypt. **65**, 127–156 (2012)

11. De Bruyn, B.: Pseudo-embeddings and pseudo-hyperplanes. Adv. Geom. **13**, 71–95 (2013)
12. De Bruyn, B.: The pseudo-hyperplanes and homogeneous pseudo-embeddings of the generalized quadrangles of order $(3, t)$. Des. Codes Crypt. **68**, 259–284 (2013)
13. De Bruyn, B., Gao, M.: The homogeneous pseudo-embeddings and hyperovals of the generalized quadrangle $H(3, 4)$. Linear Algebra Appl. **593**, 90–115 (2020)
14. Del Fra, A., Ghinelli, D., Payne, S.E.: $(0, n)$-sets in a generalized quadrangle. In: Annals of Discrete Mathematics, vol. 52, pp. 139–157, Gaeta (1990), North-Holland (1992)
15. Hall Jr., M., Swift, J.D., Walker, R.J.: Uniqueness of the projective plane of order eight. Math. Tables Aids Comput. **10**, 186–194 (1956)
16. Hirschfeld, J.W.P., Thas, J.A.: General Galois Geometries. Springer Monographs in Mathematics. Springer, London (2016). https://doi.org/10.1007/978-1-4471-6790-7
17. Hughes, D.R., Piper, F.C.: Projective planes. In: Graduate Texts in Mathematics, vol. 6. Springer, New York (1973)
18. Lam, C.W.H., Kolesova, G., Thiel, L.: A computer search for finite projective planes of order 9. Discrete Math. **92**, 187–195 (1991)
19. Lam, C.W.H., Thiel, L., Swiercz, S.: The nonexistence of finite projective planes of order 10. Canad. J. Math. **41**, 1117–1123 (1989)
20. Lam, C.W.H., Thiel, L., Swiercz, S., McKay, J.: The nonexistence of ovals in a projective plane of order 10. Discrete Math. **45**, 319–321 (1983)
21. Makhnev, A.A.: Extensions of $GQ(4, 2)$, the description of hyperovals. Discrete Math. Appl. **7**, 419–435 (1997)
22. Pasechnik, D.V.: The triangular extensions of a generalized quadrangle of order $(3, 3)$. Bull. Belg. Math. Soc. Simon Stevin **2**, 509–518 (1995)
23. Pasechnik, D.V.: The extensions of the generalized quadrangle of order $(3, 9)$. Eur. J. Combin. **17**, 751–755 (1996)
24. Pavese, F.: Hyperovals on $H(3, q^2)$ left invariant by a group of order $6(q + 1)^3$. Discrete Math. **313**, 1543–1546 (2013)
25. Payne, S.E., Thas, J.A.: Finite generalized quadrangles. In: EMS Series of Lectures in Mathematics, 2nd edn. European Mathematical Society (2009)
26. Sage Mathematics Software (Version 6.3), The Sage Developers (2014). http://www.sagemath.org
27. The GAP Group, GAP - Groups, Algorithms, and Programming, Version 4.7.5 (2014). (http://www.gap-system.org)

Isomorphism and Invariants of Parallelisms of Projective Spaces

Svetlana Topalova[ID] and Stela Zhelezova[✉][ID]

Institute of Mathematics and Informatics,
Bulgarian Academy of Sciences, Sofia, Bulgaria
{svetlana,stela}@math.bas.bg

Abstract. We consider the computer-aided constructive classification of parallelisms with predefined automorphism groups in small finite projective spaces. The usage of a backtrack search algorithm makes it very important to filter away equivalent partial solutions as soon as possible and to use a fast method for checking for isomorphism of any two parallelisms. The rejection of most of the equivalent solutions can be done by a test which uses the normalizer of the predefined automorphism group. We consider the applicability and effectiveness of such a test, and present sensitive invariants of resolutions of Steiner 2-designs. They can be used to facilitate any type of test for isomorphism of parallelisms.

Keywords: Resolutions of combinatorial designs · Parallelisms of projective spaces · Classification · Isomorphism · Invariants

1 Introduction

1.1 The Isomorphism Problem and Invariants

The approaches to solving the isomorphism problem are of major importance for the success of computer-aided classifications up to isomorphism of various combinatorial structures, such as graphs, designs, design resolutions, codes, Hadamard matrices, etc. Since the number of isomorphic solutions might be extremely big, classification is usually impossible without an efficient method for their rejection. Many authors consider this problem (recently [1,8,9,17,19,21,24]) and software solving it is available, for instance [4,23,25,32]. Regardless of the difference in methods, they all make use of suitable invariants, namely functions which yield the same value for all members of an isomorphism class. An invariant is *complete* if its value is provably different for members of different isomorphism classes. The invariants that are usually applied are not complete. Structures with different

The research of the first author is partially supported by the National Scientific Program "Information and Communication Technologies for a Single Digital Market in Science, Education and Security (ICTinSES)", financed by the Ministry of Education and Science, and of the second author by the Bulgarian National Science Fund under Contract No KP-06-N32/2-2019.

A. M. Bigatti et al. (Eds.): ICMS 2020, LNCS 12097, pp. 162–172, 2020.
https://doi.org/10.1007/978-3-030-52200-1_16

invariants, however, can obviously not be isomorphic to each other, and this is very useful. The *sensitivity* of an invariant is measured by the ratio of the number of classes it distinguishes to the number of non-isomorphic objects under consideration [10]. A complete invariant has sensitivity 1. In a less formal manner, an invariant with relatively high sensitivity is called *sensitive*. Very often, however, sensitive invariants need considerable time to be calculated, and a considerable amount of memory to be stored.

The present paper describes invariants of design resolutions and parallelisms of projective spaces. These invariants are quite simple. They can be calculated and compared relatively fast and do not need much memory to be stored. We do not know previous papers presenting them. We show their effectiveness on some of our recent classifications of parallelisms with a predefined automorphism group. The invariants are used after the rejection of most of the isomorphic solutions by a normalizer-based minimality test (NM test). We present the main principles of such a test and point out the cases in which it may not establish that two parallelisms are isomorphic, and therefore the calculation of good invariants might be very helpful.

1.2   Design Resolutions and Parallelisms of Projective Spaces

The basic concepts and notations concerning designs and resolutions, and spreads and parallelisms in projective spaces, can be found, for instance, in [14, 16, 38].

A *t-spread* in the projective space $PG(n, q)$ is a set of distinct t-dimensional subspaces which partition the point set. A *t-parallelism* is a partition of the set of t-dimensional subspaces by t-spreads. Usually 1-spreads and 1-parallelisms are called line spreads (parallelisms) or just spreads (parallelisms). There can be line spreads and parallelisms if n is odd. Two parallelisms are *isomorphic* if there exists an automorphism of the projective space which maps each spread of the first parallelism to a spread of the second one.

Let v, k, and λ be positive integers, $1 < k < v/2$. Let $V = \{P_i\}_{i=1}^v$ be a finite set of *points*, and $\mathcal{B} = \{B_j\}_{j=1}^b$ a finite collection of k-element subsets of V, called *blocks*. $D = (V, \mathcal{B})$ is a *2-design* with parameters 2-(v,k,λ) if any 2-subset of V is contained in exactly λ blocks of $\mathcal{B}$. If $\lambda = 1$ the design is called a *Steiner 2-design*. A *parallel class* is a partition of the point set by blocks. A *resolution* of the design is a partition of the collection of blocks by parallel classes. Two resolutions are *isomorphic* if there exists an automorphism of the design which maps each parallel class of the first resolution to a parallel class of the second one.

Proposition 1. [37, 2.35-2.36] *The incidence of the points and t-dimensional subspaces of $PG(n, q)$ defines a 2-design. There is a one-to-one correspondence between the t-parallelisms of $PG(n, q)$ and the resolutions of this design.*

Example 1. **A parallelism of $PG(3, 2)$.** The projective space $PG(3, 2)$ has 15 points and 35 lines with 3 points each. A parallelisms has 7 spreads consisting of 5 disjoint lines. The point-line incidence defines a 2-(15, 3, 1) design (Fig. 1) whose points and blocks correspond respectively to the points and lines of the projective space. The parallelisms of $PG(3,2)$ correspond to resolutions of this design (Fig. 2).

P\B	1	2	3	4	5	6	7	8	9	10	11	12	13	14	15	16	17	18	19	20	21	22	23	24	25	26	27	28	29	30	31	32	33	34	35
1	1	1	1	1	1	1	1	0	0	0	0	0	0	0	0	0	0	0	0	0	0	0	0	0	0	0	0	0	0	0	0	0	0	0	0
2	1	0	0	0	0	0	0	1	1	1	1	1	1	0	0	0	0	0	0	0	0	0	0	0	0	0	0	0	0	0	0	0	0	0	0
3	1	0	0	0	0	0	0	0	0	0	0	0	0	1	1	1	1	1	1	0	0	0	0	0	0	0	0	0	0	0	0	0	0	0	0
4	0	1	0	0	0	0	0	1	0	0	0	0	0	1	0	0	0	0	0	1	1	1	1	0	0	0	0	0	0	0	0	0	0	0	0
5	0	1	0	0	0	0	0	0	1	0	0	0	0	0	1	0	0	0	0	0	0	0	0	1	1	1	1	0	0	0	0	0	0	0	0
6	0	0	1	0	0	0	0	1	0	0	0	0	0	0	1	0	0	0	0	0	0	0	0	0	0	0	0	1	1	1	1	0	0	0	0
7	0	0	1	0	0	0	0	0	1	0	0	0	0	1	0	0	0	0	0	0	0	0	0	0	0	0	0	0	0	0	0	1	1	1	1
8	0	0	0	1	0	0	0	0	0	1	0	0	0	0	0	1	0	0	0	1	0	0	0	1	0	0	0	1	0	0	0	1	0	0	0
9	0	0	0	1	0	0	0	0	0	0	1	0	0	0	0	0	1	0	0	0	1	0	0	0	1	0	0	0	1	0	0	0	1	0	0
10	0	0	0	0	1	0	0	0	0	1	0	0	0	0	0	0	1	0	0	0	0	1	0	0	0	1	0	0	0	1	0	0	0	1	0
11	0	0	0	0	1	0	0	0	0	0	1	0	0	0	0	1	0	0	0	0	0	0	1	0	0	0	1	0	0	0	1	0	0	0	1
12	0	0	0	0	0	1	0	0	0	0	0	1	0	0	0	0	0	1	0	1	0	0	0	0	1	0	0	0	0	1	0	0	0	0	1
13	0	0	0	0	0	1	0	0	0	0	0	0	1	0	0	0	0	0	1	0	1	0	0	1	0	0	0	0	0	0	1	0	0	1	0
14	0	0	0	0	0	0	1	0	0	0	0	1	0	0	0	0	0	0	1	0	0	1	0	0	0	0	1	1	0	0	0	0	1	0	0
15	0	0	0	0	0	0	1	0	0	0	0	0	1	0	0	0	0	1	0	0	0	0	1	0	0	1	0	0	1	0	0	1	0	0	0

Fig. 1. $PG(3,2)$ - the point-line incidence defines a 2-(15, 3, 1) design

C_1

P \ B	1	20	26	31	33
1	1	0	0	0	0
2	1	0	0	0	0
3	1	0	0	0	0
4	0	1	0	0	0
5	0	0	1	0	0
6	0	0	0	1	0
7	0	0	0	0	1
8	0	1	0	0	0
9	0	0	0	0	1
10	0	0	1	0	0
11	0	0	0	1	0
12	0	1	0	0	0
13	0	0	0	1	0
14	0	0	0	0	1
15	0	0	1	0	0

C_1

bl	points
B_1	1,2,3
B_{20}	4,8,12
B_{26}	5,10,15
B_{31}	6,11,13
B_{33}	7,9,14

C_2

bl	points
B_2	1,4,5
B_{10}	2,8,10
B_{19}	3,13,14
B_{29}	6,9,15
B_{35}	7,11,12

C_3

bl	points
B_3	1,6,7
B_{11}	2,9,11
B_{18}	3,12,15
B_{22}	4,10,14
B_{24}	5,8,13

C_4

bl	points
B_4	1,8,9
B_{12}	2,12,14
B_{15}	3,5,6
B_{23}	4,11,15
B_{34}	7,10,13

C_5

bl	points
B_5	1,10,11
B_{13}	2,13,15
B_{14}	3,4,7
B_{25}	5,9,12
B_{28}	6,8,14

C_6

bl	points
B_6	1,12,13
B_8	2,4,6
B_{17}	3,9,10
B_{27}	5,11,14
B_{32}	7,8,15

C_7

bl	points
B_7	1,14,15
B_9	2,5,7
B_{16}	3,8,11
B_{21}	4,9,13
B_{30}	6,10,12

Fig. 2. A parallelism of $PG(3,2)$ - a resolution of the 2-(15, 3, 1) point-line design. The 7 spreads correspond to the 7 parallel classes $C_1, C_2, \ldots, C_7$.

There are several theoretical constructions of infinite families of parallelisms that are presently known [2,7,11,15,28,42]. A full classification is available in $PG(3,2)$ and $PG(3,3)$ [3], but it is currently out of reach in the other projective spaces. There are computer aided classifications of parallelisms with predefined automorphism groups [5,6,13,29–31,34,36,39–41].

1.3   The Present Paper

Section 2 briefly describes the specifics of the classification problem for parallelisms and the approach that we have used in several recent works. One of the major difficulties in all these cases is the final test for isomorphism of the obtained parallelisms. Since parallelisms can be considered as resolutions of the point-line design of $PG(n, q)$, Sect. 3 is devoted to invariants of design resolutions. The invariants that we offer, are fast to calculate, do not need too much memory to store, and partition the parallelisms to numerous invariant classes. Section 4 is a comment on their further usability for the classification of parallelisms with bigger parameters, and on their applicability to other problems.

2   Computer-Aided Classification of Parallelisms

The software for construction of parallelisms is based on the exhaustive backtrack search techniques. A lexicographic order can be defined both on the parallelisms, and on the partial solutions. This allows the rejection of partial solutions which are not minimal with respect to the lexicographic order. Such a technique for classification of various combinatorial structures is known as *orderly generation* [12], [21, chapter 4], [33]. One way to implement the method is by applying a *minimality test* to some of the partial solutions and to all full solutions.

The automorphism group $G \cong P\Gamma L(n+1, q)$ of the projective space $PG(n, q)$, however, is very rich, and this makes the minimality test too slow. That is why only a *normalizer-based minimality test (NM test)* is usually applied when parallelisms invariant under some predefined automorphism group G_c are classified. The NM test checks if the normalizer $N(G_c) = \{g \in G \mid gG_cg^{-1} = G_c\}$ contains an element which maps the constructed (partial) parallelism to a lexicographically smaller (partial) solution. If so, the current (partial) solution is discarded. We briefly explain below how the normalizer can help us remove isomorphic solutions.

Proposition 2. *Let G_c be a Sylow subgroup of the automorphism group of $PG(n, q)$, and let each of the parallelisms P and P' be invariant under G_c. Then to establish an isomorphism of P and P' it is enough to check if there is an element of the normalizer $N_G(G_c)$ of G_c in G which maps P to P'.*

Proof. The statement was first proved in [31] for parallelisms of $PG(3, 5)$ with automorphisms of order 31. We present here the main ideas for the general case. Denote by G_P the full automorphism group of P. We have to check if there is some $\varphi \in G$ such that $P' = \varphi P$. Let $\alpha \in G_c$. Then $\varphi P = \alpha \varphi P$ and thus $P = \varphi^{-1}\alpha\varphi P$. Namely $\varphi^{-1}\alpha\varphi$ is also an automorphism of P. That is why P is invariant both under G_c and under $\varphi^{-1}G_c\varphi$. If $\varphi^{-1}G_c\varphi = G_c$, then φ is in the normalizer $N_G(G_c)$ of G_c in G. If φ is not in $N_G(G_c)$, then $\varphi^{-1}G_c\varphi$ is a conjugate subgroup of G_c in G, and $|G_P| > |G_c|$. If G_c and $\varphi^{-1}G_c\varphi$ are conjugate in G_P too (for instance, this always holds if G_c is a Sylow subgroup of G), then since φ is not an automorphism of P, there must exist an automorphism $\beta \in G_P$, such that $\varphi^{-1}G_c\varphi = \beta G_c\beta^{-1}$. Then $G_c = \varphi\beta G_c\beta^{-1}\varphi^{-1}$ and therefore $\varphi\beta \in N_G(G_c)$. Since $\varphi\beta P = \varphi P$, the statement follows.

If the predefined group is not a Sylow subgroup of G, the normalizer-based minimality test may not succeed in removing all isomorphic parallelisms. That is why a further test for isomorphism must be applied.

Our experience shows that a very small number of isomorphic parallelisms remain if an NM test has been applied to them. In [40] and [41] we classify parallelisms with predefined groups which are of prime order, but not Sylow, and observe that the NM test removes all isomorphic solutions. Constructing parallelisms of $PG(3, 4)$ invariant under cyclic groups of order 4 [6] we obtain 253344 parallelisms after the NM test, and 252738 after a full test for isomorphism, i.e. the latter removed 0.23% of the solutions obtained after the NM test.

If the number of parallelisms is big, we can first determine the order of their full automorphism groups. Whatever algorithm or software we use for that purpose, invariants of the points, lines and spreads of the parallelism might be very helpful, because only points (lines, spreads) with the same invariants can be mapped to one another. When we know the full automorphism groups, we can look for isomorphisms only among parallelisms with $|G_P| > |G_c|$. In the cases that we have considered, the percentage of these parallelisms is very small. Their number, however, might be quite big, and therefore it might be very slow to test for isomorphisms any two of them. If we can easily calculate sensitive invariants of the parallelisms, we can only check for isomorphisms among parallelisms with the same invariants.

Parallelisms can be considered as resolutions of the point-line design of the projective space. The next section presents the invariants we use, as invariants of resolutions of Steiner 2-designs, because we believe that they can also have various applications outside the parallelisms classification problem.

3   Invariants of Resolutions and Parallelisms

Resolutions of designs with small parameters have been classified in many papers (for instance, [20,22,26,27]). The invariants that are usually used, are the order of the automorphism group and some properties of the underlying design.

The resolution isomorphism problem can be transformed to graph isomorphism problems (for instance, Betten's approach in [5]). This makes it possible to use Nauty [25] or some other graph isomorphism software [4,18,23,32], and to use graph invariants to distinguish resolutions in a way similar to that for designs (for example, [22,24]). These invariants, however, do not take in consideration the fact that we deal with resolutions and are therefore quite complex.

Invariants of resolutions (not of their underlying design, or related graph) are used in the works of Morales and Velarde [26,27] and Kaski et al. [20] who construct matrices of the intersections between the parallel classes. The invariants and their usage are different from those that we describe. Invariants of resolutions of 2-$(v, 3, 1)$ designs (Kirkman triple systems) are applied by Stinson and

Vanstone in [35]. They are defined by a function which maps 3-subsets of points to 3-subsets of parallel classes and are different from the invariants we present.

Our first aim is to describe the relation of each block to each resolution class. Consider a resolution of a 2-(v, k, λ) design with b blocks and r parallel classes. Denote by $B_1, B_2, \ldots B_b$ the blocks of the design, and by $C_1, C_2, \ldots C_r$ the parallel classes of the resolution. Denote by N_j^d the number of blocks in the parallel class of block B_j, which are different from B_j and disjoint with all the blocks of class C_d that contain some of the points of B_j. The number N_j^d remains unchanged by a permutation of the points of the design, and a permutation of the blocks which maps parallel classes to parallel classes defines a permutation on the numbers N_j^d, where $d = 1, 2, \ldots r$ and $j = 1, 2, \ldots b$.

For each block B_j define a vector $\Omega_j = (b_0^j, b_1^j, \ldots b_{v/k-1}^j)$, where b_m^j is the number of parallel classes C_d for which $N_j^d = m$. A resolution isomorphism maps parallel classes to parallel classes. That is why the vector Ω_j will be the same for the image of B_j under a resolution isomorphism. So we will call these vectors resolution block invariant vectors, or just block invariants. To compare block invariants we define a lexicographic order such that $\Omega_{j'} < \Omega_{j''}$ if $b_k^{j'} < b_k^{j''}$ and $b_i^{j'} = b_i^{j''}$ for $i < k$. Let n_b be the number of the different block invariants and let us denote them by $\beta_1, \beta_2, \ldots \beta_{n_b}$, and the set they form, by β. The block invariants are relatively easy to calculate, because they are based on the relation of each block to the r parallel classes, while the most used block invariants of designs [22] present the relation of each block to all the (b-1)(b-2)/2 pairs of different other blocks. For the design considered in Example 2, for instance, $r = 21$ and $b = 357$.

Each point P_i is in r blocks. Their block invariants become the elements of a resolution point invariant vector $\Pi_i = (\beta_{i_1}, \beta_{i_2}, \ldots \beta_{i_r})$, such that $\beta_{i_j} \leq \beta_{i_{j+1}}$. We next find the number n_p of the different point invariants and denote them by $\pi_1, \pi_2, \ldots \pi_{n_p}$, and the set of these invariants by π. Each parallel class C_l contains v/k blocks. Their block invariants make up a resolution class invariant vector $\Gamma_l = (\beta_{l_1}, \beta_{l_2}, \ldots \beta_{l_{v/k}})$, such that $\beta_{l_j} \leq \beta_{l_{j+1}}$. We denote the different class invariants by $\gamma_1, \gamma_2, \ldots \gamma_{n_c}$ and the set they comprise by γ.

Example 2. **Parallelisms of** $PG(3, 4)$. They can be considered as resolutions of the 2-(85, 5, 1) point-line design. There are 21 parallel classes with 17 blocks each. Table 1 presents three of the parallel classes of a resolution with automorphism group order 960. The blocks are given by their points.

In this example $N_2^2 = 12$, because each of the blocks $B_6, B_7, \ldots, B_{17}$ of parallel class C_1 is disjoint with each of the blocks $B_{18}, B_{19}, \ldots, B_{22}$ (these blocks of C_2 contain points of B_2). In the same manner $N_2^3 = 6$ because blocks $B_1, B_3, B_6, B_7, B_8, B_9$ are pairwise disjoint with blocks $B_{40}, B_{44}, B_{46}, B_{49}, B_{50}$, and $N_2^1 = 16$ because $B_1, B_3, B_4, \ldots, B_{17}$ are disjoint with B_2. In the same way we find N_2^d for $d = 4, 5, \ldots, 21$ (for the classes that are not given in Table 1). In total, the value is 6 for 15 classes, 12 for five, and 16 for C_1, namely $\Omega_2 = (0, 0, 0, 0, 0, 0, 15, 0, 0, 0, 0, 0, 5, 0, 0, 0, 1)$. We proceed with finding Ω_j for each $j = 1, 3, 4, \ldots 357$, and establish that there are $n_b = 5$ different vectors

Table 1. Three parallel classes of a resolution of a 2-$(85, 5, 1)$ design.

C_1			C_2			C_3		
Block	Points	Inv	Block	Points	Inv	Block	Points	Inv
B_1	1,2,3,4,5	1	B_{18}	1,6,7,8,9	3	B_{35}	1,10,11,12,13	3
B_2	6,22,38,54,70	2	B_{19}	2,22,26,30,34	4	B_{36}	2,24,28,32,36	4
B_3	7,26,43,60,77	2	B_{20}	3,38,43,48,53	4	B_{37}	3,40,45,46,51	4
B_4	8,30,48,65,79	2	B_{21}	4,54,60,65,67	4	B_{38}	4,57,59,62,68	4
B_5	9,34,53,67,84	2	B_{22}	5,70,77,79,84	4	B_{39}	5,73,74,80,83	4
B_6	10,28,40,68,80	2	B_{23}	10,29,41,69,81	5	B_{40}	6,23,39,55,71	5
B_7	11,24,45,62,83	2	B_{24}	11,25,44,63,82	5	B_{41}	7,27,42,61,76	5
B_8	12,36,46,59,73	2	B_{25}	12,37,47,58,72	5	B_{42}	8,31,49,64,78	5
B_9	13,32,51,57,74	2	B_{26}	13,33,50,56,75	5	B_{43}	9,35,52,66,85	5
B_{10}	14,33,41,61,85	2	B_{27}	14,31,39,59,83	5	B_{44}	14,30,38,58,82	5
B_{11}	15,37,44,55,78	2	B_{28}	15,35,42,57,80	5	B_{45}	15,34,43,56,81	5
B_{12}	16,25,47,66,76	2	B_{29}	16,23,49,68,74	5	B_{46}	16,22,48,69,75	5
B_{13}	17,29,50,64,71	2	B_{30}	17,27,52,62,73	5	B_{47}	17,26,53,63,72	5
B_{14}	18,35,39,63,75	2	B_{31}	18,36,40,64,76	5	B_{48}	18,37,41,65,77	5
B_{15}	19,31,42,69,72	2	B_{32}	19,32,45,66,71	5	B_{49}	19,33,44,67,70	5
B_{16}	20,27,49,56,82	2	B_{33}	20,28,46,55,85	5	B_{50}	20,29,47,54,84	5
B_{17}	21,23,52,58,81	2	B_{34}	21,24,51,61,78	5	B_{51}	21,25,50,60,79	5

Table 2. The set β of the different block invariants, $n_b = 5$

#inv	b_0	b_1	b_2	b_3	b_4	b_5	b_6	b_7	b_8	b_9	b_{10}	b_{11}	b_{12}	b_{13}	b_{14}	b_{15}	b_{16}
β_1	0	0	0	0	0	0	0	0	0	0	0	0	20	0	0	0	1
β_2	0	0	0	0	0	0	15	0	0	0	0	0	5	0	0	0	1
β_3	0	0	0	0	19	0	0	0	0	0	0	0	1	0	0	0	1
β_4	0	0	8	3	4	4	0	0	0	0	0	0	1	0	0	0	1
β_5	4	0	2	4	9	0	1	0	0	0	0	0	0	0	0	0	1

$\beta_1, \beta_2, \ldots, \beta_5$. They are presented in Table 2. The number m in column inv of Table 1 means that the block invariant is β_m.

We next calculate the point invariants and establish that $n_p = 2$, namely $\Pi_i = \pi_1 = (1, 3, \ldots, 3, 4, 4, 4, 4)$ for $i \leq 5$ and $\Pi_i = \pi_2 = (2, 3, 3, 3, 3, 4, 5, \ldots, 5)$ for $i > 5$. This means that the first five points, for instance, are in one block with invariant β_1, 12 blocks with invariant β_3, and 4 blocks with invariant β_4. Finally we calculate the invariants of the classes and obtain that $\Gamma_1 = \gamma_1 = (1, 2, \ldots, 2)$ for the first parallel class and $\Gamma_l = \gamma_2 = (3, 3, 3, 3, 4, 5, \ldots, 5)$ for $l > 1$.

Interested readers can obtain the whole example using the invariant calculation C++ source available at http://www.moi.math.bas.bg/moiuser/~stela and the example files going with it.

The invariant sets $\beta = \{\beta_1, \beta_2, \ldots, \beta_{n_b}\}$, $\pi = \{\pi_1, \pi_2, \ldots, \pi_{n_p}\}$ and $\gamma = \{\gamma_1, \gamma_2, \ldots, \gamma_{n_c}\}$ make up an invariant of the resolution. We calculated the resolution invariants for some of the known nonisomorphic parallelisms of $PG(3,4)$ [5,6,39–41]. They partition the parallelisms to invariant classes that contain either one, or two parallelisms. The results are presented in Table 3, where $|G_P|$ is the order of the full automorphism group, I the number of invariant classes, N the number of isomorphism classes, and S the sensitivity of the invariant, $S = I/N$.

Table 3. PG(3,4)

$|G_P|$	4	5	6	7	10	12	15	20	24	30	48	60	96	960	All
I	251836	31648	4488	482	72	40	26	52	14	20	12	4	2	3	288699
N	251836	31830	4488	482	76	52	40	52	14	38	12	8	2	4	288934
S	1	0.9943	1	1	0.9474	0.7692	0.65	1	1	0.5263	1	0.5	1	0.75	0.9992

The most complex part of the invariant calculation is the determination of the block invariants $\Omega_1, \Omega_2, \ldots, \Omega_b$. It can be done, for instance, as shown in Example 3, where the main operations are repeated $\frac{b^3}{r}$ times, b is the number of blocks, and r the number of parallel classes. The complexity is $O(b^3)$, but the actual performance is faster because $\frac{b}{r} = \frac{v}{k}$ is much smaller than b.

Example 3. **Calculation of** $\Omega_1, \Omega_2, \ldots, \Omega_b$. All arrays in this code segment are of integer type, except covered (boolean); pclass[j] is the parallel class of block j, cpoints[j1][j2] is the number of common points of blocks $j1$ and $j2$, Omega[j] is Ω_j, and covered and inv are auxiliary arrays.

```
for(j=1; j<=b; j++) // b iterations
{
  for(i=0; i<=r; i++) inv[i]=0;
  for(i=0; i<b/r; i++) Omega[j][i]=0;
  for(jj=1; jj<=b; jj++) if(jj!=j && pclass[jj]==pclass[j])//b/r iterations
  {
    for(i=1; i<=r; i++) covered[i] = false;
    for(j1=1; j1<=b; j1++) if (!covered[pclass[j1]]) //at most b iterations
      if(cpoints[j1][j]>0 && cpoints[j1][jj]>0) covered[pclass[j1]]=true;
    for(i=1; i<=r; i++) if(!covered[i]) inv[i]++;
  }
  for(i=1; i<=r; i++) Omega[j][inv[i]]++;
}
```

4   Comments

- Our experience with classification of parallelisms with predefined automorphism groups, shows that the normalizer-based minimality test is a powerful fast way of filtering away most of the isomorphic solutions.
- The invariants presented in Sect. 3 are very useful for the classification of the parallelisms we applied them to, because they partition them to numerous small invariant classes. We believe that they will be helpful to future classifications of parallelisms with bigger parameters too.
- We suppose that these invariants will work well on the resolutions of any $2\text{-}(v, k, 1)$ design (Steiner 2-design). For resolutions of designs with $\lambda \neq 1$, however, modifications of the block invariants might be more suitable, such that the exact number of common points of two blocks is encountered (not only if these blocks are disjoint or not).

References

1. Al-ogaidi, A., Betten, A.: Large Arcs in Small Planes, accepted for publication in Congressus Numerantium (2020)
2. Baker, R.D.: Partitioning the planes of $AG_{2m}(2)$ into 2-designs. Discrete Math. **15**, 205–211 (1976)
3. Betten, A.: The packings of $PG(3,3)$. Des. Codes Cryptogr. **79**(3), 583–595 (2016)
4. Betten, A.: Orbiter - A program to classify discrete objects, 2016–2018. https://github.com/abetten/orbiter. Accessed 24 Jan 2019
5. Betten, A., Topalova, S., Zhelezova, S.: Parallelisms of $PG(3,4)$ invariant under a Baer involution. In: Proceedings of the 16th International Workshop on Algebraic and Combinatorial Coding Theory, Svetlogorsk, Russia, pp. 57–61 (2018). http://acct2018.skoltech.ru/
6. Betten, A., Topalova, S., Zhelezova, S.: Parallelisms of PG(3,4) invariant under cyclic groups of order 4. In: Ćirić, M., Droste, M., Pin, J.É. (eds.) CAI 2019. LNCS, vol. 11545, pp. 88–99. Springer, Cham (2019). https://doi.org/10.1007/978-3-030-21363-3_8
7. Beutelspacher, A.: On parallelisms in finite projective spaces. Geom. Dedicata. **3**(1), 35–40 (1974)
8. Bouyukliev, I.: About the code equivalence. In: Advances in Coding Theory and Cryptography. Series on Coding Theory and Cryptology, pp. 126–151 (2007)
9. Bouyukliev, I., Dzhumalieva-Stoeva, M.: Representing equivalence problems for combinatorial objects. Serdica J. Comput. **8**(4), 327–354 (2014)
10. Colbourn, M.: Algorithmic aspects of combinatorial designs: a survey. Ann. Discret. Math. **26**, 67–136 (1985)
11. Denniston, R.H.F.: Some packings of projective spaces. Atti Accad. Naz. Lincei Rend. Cl. Sci. Fis. Mat. Natur. **52**(8), 36–40 (1972)
12. Faradžev, I.A.: Constructive enumeration of combinatorial objects. In: Problèmes Combinatoires et Théorie des Graphes, pp. 131–135. CNRS, Paris, (Université d'Orsay, 9–13 July 1977) (1978)
13. Fuji-Hara, R.: Mutually 2-orthogonal resolutions of finite projective space. Ars Combin. **21**, 163–166 (1986)

14. Colbourn, C., Dinitz, J. (eds.) Handbook of Combinatorial Designs. Discrete mathematics and its applications, 2nd edn. CRC Press, Boca Raton (2007)
15. Johnson, N.L.: Some new classes of finite parallelisms. Note Mat. **20**(2), 77–88 (2000)
16. Johnson, N.L.: Combinatorics of Spreads and Parallelisms. Chapman & Hall Pure and Applied Mathematics. CRC Press, Boca Raton (2010)
17. Jungnickel, D., Tonchev, V.D.: New invariants for incidence structures. Des. Codes Cryptogr. **68**, 163–177 (2013)
18. Junttila, T., Kaski, P.: Bliss: a tool for computing automorphism groups and canonical labelings of graphs. http://www.tcs.hut.fi/Software/bliss/. Accessed 24 Jan 2019
19. Junttila, T., Kaski, P.: Engineering an efficient canonical labeling tool for large and sparse graphs. In: Proceedings of the Ninth Workshop on Algorithm Engineering and Experiments (ALENEX07), pp. 135–149. SIAM (2007)
20. Kaski, P., Morales, L.B., Östergård, P., Rosenblueth, D.A., Velarde, C.: Classification of resolvable 2-(14,7,12) and 3-(14,7,5) designs. J. Comb. Math. Comb. Comput. **47**, 65–74 (2003)
21. Kaski, P., Östergård, P.: Classification Algorithms for Codes and Designs. Springer, Berlin (2006)
22. Kramer, E.S., Magliveras, S.S., Mathon, R.: The Steiner systems $S(2,4,25)$ with nontrivial automorphism group. Discret. Math. **77**(1–3), 137–157 (1989)
23. Magma. The Computational Algebra Group within the School of Mathematics and Statistics of the University of Sydney (2004). https://www.magmasoft.com/en/company/about-magma/
24. McKay, B.D., Piperno, A.: Practical graph isomorphism, II. J. Symb. Comput. **60**, 94–112 (2014)
25. McKay, B.: Nauty User's Guide (Version 2.6). Australian National University (2016)
26. Morales, L.B., Velarde, C.: A complete classification of (12,4,3)-RBIBDs. J. Comb. Des. **9**(6), 385–400 (2001)
27. Morales, L.B., Velarde, C.: Enumeration of resolvable 2-(10,5,16) and 3-(10,5,6) designs. J. Comb. Des. **13**(2), 108–119 (2005)
28. Penttila, T., Williams, B.: Regular packings of $PG(3,q)$. Eur. J. Combin. **19**(6), 713–720 (1998)
29. Prince, A.R.: Parallelisms of $PG(3,3)$ invariant under a collineation of order 5. In: Johnson, N.L. (ed.) Mostly Finite Geometries. Lecture Notes in Pure and Applied Mathematics, vol. 190, pp. 383–390. Marcel Dekker, New York (1997)
30. Prince, A.R.: Uniform parallelisms of PG(3,3). In: Hirschfeld, J., Magliveras, S., Resmini, M. (eds.) Geometry, Combinatorial Designs and Related Structures, London Mathematical Society Lecture Note Series, vol. 245, pp. 193–200. Cambridge University Press, Cambridge (1997)
31. Prince, A.R.: The cyclic parallelisms of $PG(3,5)$. Eur. J. Combin. **19**(5), 613–616 (1998)
32. Q-extension, a software for classification of linear codes over small finite fields. http://www.moi.math.bas.bg/moiuser/~data/. Accessed 24 Jan 2020
33. Read, R.C.: Every one a winner; or, How to avoid isomorphism search when cataloguing combinatorial configurations. Ann. Discret. Math. **2**, 107–120 (1978)
34. Sarmiento, J.: Resolutions of $PG(5,2)$ with point-cyclic automorphism group. J. Combin. Des. **8**(1), 2–14 (2000)
35. Stinson, D.R., Vanstone, S.A.: Some non-isomorphic Kirkman triple systems of orders 39 and 51. Utilitas Math. **27**, 199–205 (1985)

36. Stinson, D.R., Vanstone, S.A.: Orthogonal packings in $PG(5,2)$. Aequationes Math. **31**(1), 159–168 (1986)
37. Storme, L.: Finite geometry. In: Colbourn, C., Dinitz, J. (eds.) Handbook of Combinatorial Designs. 2nd edn. Rosen, K. (eds.) Discrete mathematics and its applications, pp. 702–729. CRC Press, Boca Raton (2007)
38. Tonchev, V.D.: Combinatorial Configurations. Longman Scientific and Technical, New York (1988)
39. Topalova, S., Zhelezova, S.: On transitive parallelisms of $PG(3,4)$. Appl. Algebra Engrg. Comm. Comput. **24**(3–4), 159–164 (2013)
40. Topalova, S., Zhelezova, S.: On point-transitive and transitive deficiency one parallelisms of $PG(3,4)$. Des. Codes Cryptogr. **75**(1), 9–19 (2015)
41. Topalova, S., Zhelezova, S.: New parallelisms of $PG(3,4)$. Electron. Notes Discret. Math. **57**, 193–198 (2017)
42. Zaicev, G., Zinoviev, V., Semakov, N.: Interrelation of Preparata and Hamming codes and extension of Hamming codes to new double-error-correcting codes. In: Proceedings of the Second International Symposium on Information Theory, (Armenia, USSR, 1971), Budapest, Academiai Kiado, pp. 257–263 (1973)

Classification of Linear Codes
by Extending Their Residuals

Stefka Bouyuklieva[1]([⊠]) [iD] and Iliya Bouyukliev[2] [iD]

[1] St. Cyril and St. Methodius University of Veliko Tarnovo, Veliko Tarnovo, Bulgaria
stefka@ts.uni-vt.bg
[2] Institute of Mathematics and Informatics, Bulgarian Academy of Sciences,
P.O. Box 323, Veliko Tarnovo, Bulgaria
iliyab@math.bas.bg

Abstract. An approach for classification of linear codes with given parameters starting from their proper residual codes or subcodes is presented. The base of the algorithm is the concept of canonical augmentation which is important for parallel implementations. The algorithms are implemented in the programs LENGTHEXTENSION and DIMEXTENSION of the package QEXTNEWEDITION. As an application, the nonexistence of binary $[41, 14, 14]$ codes is proved.

Keywords: Linear code · Classification · Residual code

1 Introduction

The paper is a contribution to the problem of classifying linear codes with given parameters over finite fields with q elements. Many authors have considered this problem before [2,3,5,10], and it is known to be very hard. The structure of the codes for classification is very important in the generation process. We discuss an algorithm that solves the following problem: Find all inequivalent codes with given parameters if the set of all residual codes with respect to a codeword with a given weight is given. The extension of the generator matrix of a given residual code can be done row by row or column by column. We consider in more details the problem how to generate only inequivalent codes and obtain all of needed codes. To do this, we use the concept of canonical augmentation [10,12]. This concept is very important for parallel implementations. We also mention the dual problem namely the classification of linear codes by extending their proper subcodes.

The algorithms presented in this paper are implemented in the programs LENGTHEXTENSION and DIMEXTENSION of the package QEXTNEWEDITION.

Supported by Grant DN 02/2/13.12.2016 of the Bulgarian National Science Fund.
I. Bouyukliev—The research of the second author was supported, in part, by the Bulgarian Ministry of Education and Science by Grant No. DO1-221/03.12.2018 for NCHDC, a part of the Bulgarian National Roadmap on RIs.

© Springer Nature Switzerland AG 2020
A. M. Bigatti et al. (Eds.): ICMS 2020, LNCS 12097, pp. 173–180, 2020.
https://doi.org/10.1007/978-3-030-52200-1_17

Restrictions on the dual distance, minimum distance, etc. can be applied. The program will be available on the webpage

http://www.moi.math.bas.bg/moiuser/~data/Software/QextNewEdition

2   Preliminaries

Let q be a prime power and $\mathbb{F}_q$ the finite field with q elements, $\mathbb{F}_q^* = \mathbb{F}_q \setminus \{0\}$. A linear code of length n, dimension k, and minimum distance d over $\mathbb{F}_q$ is called an $[n, k, d]_q$ code. Two linear codes of the same length and dimension are equivalent if one can be obtained from the other by a sequence of the following transformations: (1) a permutation of the coordinate positions of all codewords; (2) a multiplication of a coordinate of all codewords with a nonzero element from $\mathbb{F}_q$; (3) a field automorphism. A sequence of the transformations given above that maps a code C to itself is called an automorphism of C. The set of all automorphisms of C forms a group, called the automophism group of the code and denoted by $\mathrm{Aut}(C)$. The action of $\mathrm{Aut}(C)$ on the code partitions the set of its codewords into orbits.

The defined equivalence relation in the set of all linear $[n, k, d]_q$ codes partitions this set into equivalence classes. We choose a canonical representative of each equivalence class. If C is a linear $[n, k, d]_q$ code, we call the canonical representative of its equivalence class the canonical form of C and denote it by $\rho(C)$. If two codes C_1 and C_2 are equivalent they have the same canonical form, or $\rho(C_1) = \rho(C_2)$.

Let C be an $[n, k, d]_q$ code and let c be a codeword of weight w. Then the residual code of C with respect to c, denoted $Res(C; c)$, is the code of length $n - w$ punctured on the set of coordinates on which c is nonzero. If only the weight w of c is of importance, we will denote it by $Res(C; w)$. The next result gives a lower bound for the minimum distance of residual codes.

Theorem 1. [8] *Let C be an $[n, k, d]$ code over $\mathbb{F}_q$ and let c be a codeword of weight $w < qd/(q - 1)$. Then $Res(C; c)$ is an $[n - w, k - 1, d']$ code, where $d' \geq d - w + \lceil w/q \rceil$.*

We need also the following theorem

Theorem 2. *Let C be an $[n, k, d]$ code over $\mathbb{F}_q$ and $x, y \in C$ be codewords of the same weight $w < qd/(q-1)$ such that $y = \phi(x)$ for an automorphism $\phi \in Aut(C)$. Then the residual codes $Res(C; x)$ and $Res(C; y)$ are equivalent.*

Proof. Let $\phi = \mathrm{diag}(\gamma_1, \ldots, \gamma_n)\pi$, where $\gamma_i \in \mathbb{F}_q^*$, $\pi \in S_n$. Then for any $v = (v_1, v_2, \ldots, v_n) \in C$ we have

$$\phi(v) = (\gamma_1 v_1, \ldots, \gamma_n v_n)\pi = (\gamma_{1\pi^{-1}} v_{1\pi^{-1}}, \ldots, \gamma_{n\pi^{-1}} v_{n\pi^{-1}}) \in C.$$

Without loss of generality we can take $x = (00 \cdots 0 \underbrace{11 \cdots 1}_{w})$. Then the support of $y = \phi(x)$ will be $\{(n - w + 1)\pi^{-1}, \ldots, n\pi^{-1}\}$. If v is a codeword in C

then $(v_1, \ldots, v_{n-w}) \in Res(C; x)$ and $(\gamma_{1\pi^{-1}} v_{1\pi^{-1}}, \ldots, \gamma_{(n-w)\pi^{-1}} v_{(n-w)\pi^{-1}}) \in Res(C; y)$. Hence the restriction of ϕ on the first $n - w$ coordinates maps $Res(C; x)$ to $Res(C; y)$.

To see the connection to the dual code, we use a theorem that gives the relation between a punctured of a code C and a shortened of its dual code $C^\perp$. A code C can be punctured on a coordinate set T of size t. We denote the resulting code by C^T. Consider the set $C(T)$ of codewords whose i-th coordinate is 0 if $i \in T$. $C(T)$ is a subcode of C. Shortening $C(T)$ on T gives a code of length $n - t$ called shortened code of C on T and denoted by C_T. If we take T to be the support of the codeword $c \in C$ of weight w, then C^T is the residual code of $Res(C; c)$ with respect to c.

Theorem 3 ([9, **Theorem 1.5.7**]). *Let C be an $[n, k, d]$ code and T be a set of t coordinates. Then:*

   (i) $(C^\perp)_T = (C^T)^\perp$ and $(C^\perp)^T = (C_T)^\perp$;
   (ii) if $t < d$, then C^T and $(C^\perp)_T$ have dimensions k and $n - t - k$, respectively;
   (iii) if $t = d$ and T is the set of coordinates where a minimum weight codeword is nonzero, then C^T and $(C^\perp)_T$ have dimensions $k - 1$ and $n - d - k + 1$, respectively.

As a corollary we obtain

Corollary 1. *Let C be an $[n, k, d]$ code over $\mathbb{F}_q$ with dual distance $d^\perp$ and let c be a codeword of weight $w < qd/(q-1)$. If T is the support of c then $Res(C; c) = C^T$ is a linear $[n - w, k - 1, d']$ code and $Res(C; c)^\perp = (C^\perp)_T$ is a linear $[n - w, n - w - k + 1, \geq d^\perp]$ code.*

Since $Res(C; c)^\perp$ is a shortened code of $C^\perp$, its minimum distance is at least $d^\perp$. Therefore we consider all $[n - w, k - 1, d' \geq d - w + \lceil w/q \rceil]_q$ codes with dual distance $\geq d^\perp$ as residual codes and then extend them to the linear $[n, k, d]_q$ codes with dual distance $\geq d^\perp$.

We developed a second algorithm which extends all possible $[n-w, k-w+1, \geq d]$ shortened codes to the $[n, k, d]$ codes provided that their dual codes contain codewords of weight w, $w < qd^\perp/(q-1)$. The theoretical base of this algorithm is the following corollary.

Corollary 2. *If C is a linear $[n, k, d]_q$ code whose dual code $C^\perp$ contains a codeword of weight w, $w < qd^\perp/(q-1)$, then C has a shortened code with parameters $[n - w, k - w + 1, \geq d]_q$ and dual distance $d' \geq d^\perp - w + \lceil w/q \rceil$.*

Proof. Let $x \in C^\perp$ be a vector of weight w. According to Theorem 2, its residual code $Res(C^\perp; x)$ has parameters $[n - w, n - k - 1, d']$ where $d' \geq d^\perp - w + \lceil w/q \rceil$. Then $Res(C^\perp; x)^\perp$ is a shortened code of C with parameters $[n-w, k-w+1, \geq d]$ (see Theorem 3 and Corollary 1).

Corollary 3. *Let C be a linear $[n, k, d]_q$ code with dual distance $d^\perp$. If no linear $[n - i, k - i + 1, \geq d]_q$ codes exist for $1 \leq i \leq w - 1$ then $d^\perp \geq w$.*

Proof. Suppose that $d^\perp = i < w$ and $x \in C^\perp$ is a vector of weight $d^\perp$. Then $Res(C^\perp; x)^\perp$ is a shortened code of C with parameters $[n - i, k - i + 1, \geq d]_q$ which is not possible. Hence $d^\perp \geq w$.

3   The Construction

We are looking for all inequivalent linear codes with length n, dimension k, minimum distance d and dual distance at least $d^\perp \geq 2$. We propose two algorithms depending on the input codes.

The input in the first algorithm is a set of all inequivalent linear $[n-w, k-1, \geq d']_q$ codes with dual distance $\geq d^\perp$ where $d' > d - w + \lceil w/q \rceil$. These codes are all possible residual codes of $[n, k, d]_q$ linear codes with dual distance at least $d^\perp$ with respect to a codeword of weight w.

Without loss of generality, we can consider the generator matrices in the form

$$\left(\begin{array}{c|c} 00\cdots0 & 11\cdots1 \\ \hline G_{res} & G_1 \end{array} \right)$$

where G_{res} is a $(k-1) \times (n-w)$ matrix that generates the residual code $Res(C; x)$, $x = (00\cdots0, 11\cdots1) \in C$, $wt(x) = w$. We construct the matrix G_1 row by row in the same way as it is in the program QEXT_L of the package Q-EXTENSION [3]. The main question is which of the constructed in this way codes to take in our set of representatives of the equivalence classes. To do this, we use canonical augmentation [10,12]. The presentation that follows differs from the original McKay's paper [12] but the idea is the same.

First, we find the canonical form and the automorphism group of the constructed $[n, k, d]$ code C. The orbits are ordered in the way described in [1] and this ordering depends on the canonical form $\rho(C)$ and the automorphism group $Aut(C)$. Then we check if the vector x is in the first orbit in the set of all codewords of weight w in C. If not, we reject it (it can be obtained by another residual code), if yes we say that this code passes the parent test. Finally, we check for equivalence the codes obtained from the same residual code that have passed the parent test. A pseudocode is presented in Algorithm 1.

Theorem 4. *The set M, obtained by Algorithm 1, consists of all inequivalent $[n, k, d]_q$ codes with dual distance $\geq d^\perp$ that have codewords of weight w.*

Proof. We have to prove that (1) any $[n, k, d]_q$ code with the needed dual distance is equivalent to a code in the set M, and (2) the codes in M are not equivalent.

(1) Let C be an $[n, k, d]_q$ code with dual distance $\geq d^\perp$. The set of all codewords of weight w is partitioned into orbits under the action of $Aut(C)$. These orbits are ordered depending on the canonical form $\rho(C)$ (see [1] for details). Take a codeword x in the first orbit and the residual code $Res(C; x)$. There is a code $B \cong Res(C; x)$ in the set R. If ϕ maps $Res(C; x)$ into B, we can extend the map ϕ to $\overline{\phi} : C \to C'$, $C' = \overline{\phi}(C)$. If $x' = \overline{\phi}(x)$, then

$B = Res(C', x')$ and the code C' passes the parent test (the codeword $x' \in C'$ belongs to the first orbit in the partition of the set of all codewords of weight w in C' since $\rho(C) = \rho(C')$). Hence there is a code that is equivalent to C, has a residual code in the set R and passes the parent test.

(2) If $C_1 \cong C_2$ are two codes with the needed parameters, $x_i \in C_i$, $i = 1, 2$ are vectors of weight w, and both codes pass the parent test, then their residuals $Res(C_1, x_1)$ and $Res(C_2, x_2)$ are also equivalent (see Theorem 2).

Algorithm 1: Extension of a residual code.

Input: The set R of all inequivalent linear $[n-w, k-1, \geq d']_q$ codes with dual distance at least $d^\perp$.

Output: A set M of all inequivalent linear $[n, k, d]_q$ codes with dual distance $\geq d^\perp$

begin $M = \emptyset$
| for all codes $B \in R$ do
| $M_B = \emptyset$;
| for all constructed codes C with a residual code B do:
| Obtain $\rho(C)$ and $Aut(C)$;
| if $x \in O_1$ then $M_B = M_B \cup C$
| end for;
| Remove equivalent codes from the set M_B
| $M = M \cup M_B$;
| end for;
end.

The second algorithm extends all $[n-w, k-w+1, \geq d]$ codes to the $[n, k, \geq d]$ codes with dual distance $d^\perp$ whose dual codes contain codewords of weight w. The generator matrices of the considered codes have the form

$$\begin{pmatrix} & & & 1 & \\ I_{w-1} & & \vdots & & A \\ & & & 1 & \\ \hline & O & & & G_0 \end{pmatrix}$$

where I_{w-1} is the identity matrix, O is the $(k-w+1) \times w$ zero matrix, A and G_0 are $(w-1) \times (n-w)$ and $(k-w+1) \times (n-w)$ matrices, respectively. We fill out the matrix A row by row in a similar way as it is done in [4]. The dual code $C^\perp$ has a generation matrix $\left(\begin{array}{c|c} 11 \cdots 1 & 00 \cdots 0 \\ \hline G_1 & G_2 \end{array} \right)$ where G_2 generates the residual code of $C^\perp$ with respect to the codewords $(11 \cdots 100 \cdots 0)$ of weight w and it is the dual code of C_0. To take only inequivalent codes, we apply Algorithm 1 to the dual codes.

4   Examples

We use the presented algorithms implemented in the programs LENGTHEXTEN-SION and DIMEXTENSION to obtain a systematic classification of linear codes with specific properties and parameters over fields with 2, 3 and 4 elements. Besides specifying the parameters such as length (n), dimension (k) and minimum distance (d), many other constraints can be considered. We give two examples, both over the filed $\mathbb{F}_2$, but the first one uses the program LENGTHEXTENSION and the second one DIMEXTENSION. All calculations have been done on $2 \times$ INTEL XEON E5-2620 V4, 32 thread computer.

Example 1. We construct all inequivalent $[45, 8, 20]_2$ codes from their residual $[25, 7, 10]_2$ codes with respect to a codeword of minimum weight 20. Since no $[44, 8, 20]_2$ code exists, the dual distance $d^{\perp}$ must be at least 2. Using the program GENERATION, we obtain 188572 inequivalent $[25, 7, 10]_2$ codes. Six of these codes have dual distance 1 (these codes have a zero coordinate) and therefore we cannot use them as residual codes. The other 188566 have dual distances 2 (30522 codes), 3 (158036 codes), and 4 (only 8 codes). Considering these codes as residual codes, the program LENGTHEXTENSION constructs 424208 inequivalent $[45, 8, 20]_2$ codes. The calculations took 459 min. All doubly-even $[45, 8, 20]_2$ codes are classified in [11] and their number is 424207. There is only one code (up to equivalence) with these parameters which is not doubly-even. This code has a generator matrix

$$\begin{pmatrix} 111111111111111111110000000000000000000000000 \\ 000000000001111111100001111111111111111111000000 \\ 000000011110011111101110000000001111110100000 \\ 000111100010100011100110000111110001110010000 \\ 011001100100101100100110011001110110010001000 \\ 100110101101000101100010101110011010010000100 \\ 101011010010110101000101100010110110100000010 \\ 101010101011010110001001010110101000110000001 \end{pmatrix}$$

and weight enumerator $W(y) = 1 + 99y^{20} + 90y^{22} + 15y^{24} + 45y^{28} + 6y^{30}$. Its automorphism group is isomorphic to $(C_{15} : C_4) \times S_3$, where $C_{15} : C_4$ is the semidirect product of the cyclic groups of orders 15 and 4, and S_3 is the symmetric group (calculated by GAP COMPUTER ALGEBRA SYSTEM [6]). The group acts transitively on the coordinates and has order 360. The code is not self-orthogonal.

The following proposition allows one to reduce the number of cases that need to be considered for an exhaustive search for a certain class of codes.

Proposition 1. *If binary linear* $[n, k, 2d]$ *codes exist then at least one of these codes is even.*

Proof. Let C be a binary linear $[n, k, 2d]$ code. Suppose that C contains codewords of odd weight. If C^* is the punctured code of C on the right-most coordinate then C^* is an $[n - 1, k, d^*]$ code where $d^* = 2d - 1$ or $2d$. Then we extend

C^* with one coordinate by adding an overall parity check. The resulting code $\widehat{C}^*$ is even and its parameters are $[n, k, 2d]$.

Proposition 2. *Binary linear* $[41, 14, 14]$ *codes do not exist.*

Proof. According to Proposition 1, it is enough to prove the nonexistence of even codes with these parameters. Feulner proved in [5] that binary $[35, 10, 13]$ code does not exist. We prove that binary $[36, 11, 13]$ and $[37, 12, 13]$ codes do not exist. The nonexistence of codes with these parameters proves that binary linear $[36, 10, 14]$, $[37, 11, 14]$ and $[38, 12, 14]$ codes do not exist. This gives us that no linear $[41 - i, 15 - i, 14]_2$ codes exist for $1 \leq i \leq 5$. According to Corollary 3, the dual distance of a binary $[41, 14, 14]$ must be at least 6. Since no $[41, 27, \geq 7]_2$ codes exist [7], $d^{\perp} = 6$. Therefore we are looking for binary even $[41, 14, 14]$ codes with dual distance 6 and we try to construct them by extending all possible even $[35, 9, 14]_2$ codes with dual distance ≥ 3. The program GENERATION shows that there are exactly 209 inequivalent even $[35, 9, 14]_2$ codes with needed dual distance. Then we try to extend them using the program DIMEXTENSION. The result is 'RES 0, Elapsed time: 432m' which means that these codes cannot be extended to $[41, 14, 14]$ codes and this result is obtained in 432 min. $\square$

Remark 1. The table of optimal codes [7] indicates that the existence of $[40, 13, 14]$ binary codes is also unknown. If a code with these parameters exists, its dual distance can be 5 or 6. If C is a $[40, 13, 14]$ binary even code with dual distance 5, it contains an even $[35, 9, 14]$ shortened code with dual distance ≥ 3. By the program DIMEXTENSION, we obtain that these codes cannot be extended to $[40, 13, 14]$ binary codes. This means that if a $[40, 13, 14]$ binary even code exists, its dual distance is 6. Then this code contains a shortened code with parameters $[34, 8, 14]$ and dual distance ≥ 3. There are 10 607 917 inequivalent $[34, 8, 14]$ codes with needed dual distance. We were not able to extend all these codes for a reasonable time and therefore we have no result for the codes with parameters $[40, 13, 14]$.

Acknowledgements. We are greatly indebted to the unknown referees for their useful suggestions.

References

1. Bouyukliev, I.: About the code equivalence. In: Shaska, T., Huffman, W., Joyner, D., Ustimenko, V. (eds.) Advances in Coding Theory and Cryptology, pp. 126–151 (2007)
2. Bouyukliev, I., Bouyuklieva, S., Kurz, S.: Computer classification of linear codes. arXiv:2002.07826 [cs.IT] (2020)
3. Bouyukliev, I., Simonis, J.: Some new results for optimal ternary linear codes. IEEE Trans. Inf. Theory **48**(4), 981–985 (2002)
4. Bouyuklieva, S., Bouyukliev, I.: Classification of the extremal formally self-dual even codes of length 30. Adv. Math. Commun. **4**(3), 433–439 (2010)
5. Feulner, T.: Classification and nonexistence results for linear codes with prescribed minimum distances. Des. Codes Cryptogr. **70**, 127–138 (2014)

6. The GAP Group: GAP - Groups, Algorithms, and Programming, Version 4.11.0 (2020). https://www.gap-system.org
7. Grassl, M.: Bounds on the minimum distance of linear codes and quantum codes. http://www.codetables.de. Accessed 10 Mar 2020
8. Hill, R., Newton, D.E.: Optimal ternary linear codes. Des. Codes Crypt. **2**, 137–157 (1992)
9. Huffman, W.C., Pless, V.: Fundamentals of Error-Correcting Codes. Cambridge University Press, Cambridge (2003)
10. Kaski, P., Östergård, P.R.: Classification Algorithms for Codes and Designs. Springer, Heidelberg (2006)
11. Kurz, S.: The $[46, 9, 20]_2$ code is unique. arXiv:1906.02621v2 (2020)
12. McKay, B.: Isomorph-free exhaustive generation. J. Algorithms **26**, 306–324 (1998)

The Program GENERATION in the Software Package QEXTNEWEDITION

Iliya Bouyukliev[(⊠)]

Institute of Mathematics and Informatics, Bulgarian Academy of Sciences,
P.O. Box 323, Veliko Tarnovo, Bulgaria
iliyab@math.bas.bg

Abstract. This paper is devoted to the program GENERATION which is a self-containing console application for classification of linear codes. It can be used for codes over fields with $q < 8$ elements and with wide-range parameters. The base of the implemented algorithm is the concept of canonical augmentation.

Keywords: Linear code · Classification · Software

1 Introduction

The classification problem of linear codes is important and difficult. Computer algorithms have been used to find the best linear codes for given length and dimension. There are many computational results for classification of linear codes over finite fields (see for example [3,12,13]), but there is not much related software available (for example MAGMA [4], GUAVA [1], ORBITER [2], Q-EXTENSION [5]). Our paper is a contribution to this research.

The system QEXTNEWEDITION is a software package consisting of several user interface programs for classification of linear codes over finite fields, along with the necessary supporting functions. Here we describe the program GENERATION for classification of linear codes over fields with $q < 8$ elements and with wide-range parameters. Despite its simple interface, it allows a lot of restrictions on the considered codes. It gives the possibility to classify not only codes with fixed parameters but also all codes with a given length n and dimensions from k_0 to k for given integers $1 \leq k_0 \leq k$. To use the program, a knowledge of a programming language is not needed. This program is supported by many different basic functions which implement complicated (in some cases) algorithms and has specific data organizations. The most important of these functions give the minimum distance, a list of codewords with weights smaller than a given integer (for large dimensions Brouwer-Zimmerman algorithm is applied), canonical form, automorphism group.

Supported by Grant DN 02/2/13.12.2016 of the Bulgarian National Science Fund.

A. M. Bigatti et al. (Eds.): ICMS 2020, LNCS 12097, pp. 181–189, 2020.
https://doi.org/10.1007/978-3-030-52200-1_18

This paper tries to give answers to the following questions:

- What type of algorithms are implemented and why they allow parallel implementation?
- What is the data organization?
- What is the difference with the previous version?
- How can the interface be used to enter parameters and what type of restrictions are possible and suitable for different cases?
- What can be expected from the program?

The remaining part of the paper is organized as follows. Section 2 contains the needed definitions. In Sects. 3 and 4 we describe the main algorithms in the program GENERATION and its basic functions and data organization, respectively. Sections 5 and 6 answer the questions how we can use GENERATION for code classification and what can be expected from the program. Finally, we draw a brief conclusion in Sect. 7.

2 Basic Definitions

In this section we present some definitions following [11]. Let $\mathbb{F}_q^n$ denote the vector space of n-tuples over the q-element field $\mathbb{F}_q$. A q-ary linear code C of length n and dimension k, or an $[n, k]_q$ code, is a k-dimensional subspace of $\mathbb{F}_q^n$. A $k \times n$ matrix G whose rows form a basis of C is called a generator matrix of C. The number of nonzero coordinates of a vector $\mathbf{x} \in \mathbb{F}_q^n$ is called its Hamming weight $\mathrm{wt}(\mathbf{x})$. The Hamming distance $d(\mathbf{x}, \mathbf{y})$ between two vectors $\mathbf{x}, \mathbf{y} \in \mathbb{F}_q^n$ is defined by $d(\mathbf{x}, \mathbf{y}) = \mathrm{wt}(\mathbf{x} - \mathbf{y})$. The minimum distance of a linear code C is

$$d(C) = \min\{d(\mathbf{x}, \mathbf{y}) \mid \mathbf{x}, \mathbf{y} \in C, \mathbf{x} \neq \mathbf{y}\} = \min\{\mathrm{wt}(\mathbf{c}) \mid \mathbf{c} \in C, \mathbf{c} \neq \mathbf{0}\}.$$

A q-ary linear code of length n, dimension k and minimum distance d is said to be an $[n, k, d]_q$ code. Let A_i denote the number of codewords in C of weight i. Then the $n + 1$-tuple $(A_0, \ldots, A_n)$ is called the *weight spectrum* of the code C.

An inner product $(\mathbf{x}, \mathbf{y})$ of vectors $\mathbf{x}, \mathbf{y} \in \mathbb{F}_q^n$ defines orthogonality: Two vectors are said to be orthogonal if their inner product is 0. The set of all vectors of $\mathbb{F}_q^n$ orthogonal to all codewords in C is called the orthogonal code $C^\perp$ to C:

$$C^\perp = \{\mathbf{x} \in \mathbb{F}_q^n \mid (\mathbf{x}, \mathbf{y}) = 0 \text{ for any } \mathbf{y} \in C\}.$$

It is well-known that the code $C^\perp$ is a linear $[n, n-k]_q$ code. If $C \subseteq C^\perp$, the code C is called *self-orthogonal*. Self-orthogonal codes with $n = 2k$ are of particular interest, then $C = C^\perp$ and these codes are called *self-dual*.

The program GENERATION has an option for classification of self-orthogonal codes over fields with 2, 3 and 4 elements. In the binary and ternary cases, we consider *Euclidean inner product* defined by $u \cdot v = u_1 v_1 + u_2 v_2 + \cdots + u_n v_n \in \mathbb{F}_q$ for $\mathbf{u} = (u_1, u_2, \ldots, u_n)$, and $\mathbf{v} = (v_1, v_2, \ldots, v_n)$. For $q = 4$ the considered inner product is the Hermitian inner product defined by $u \cdot v = u_1 v_1^2 + u_2 v_2^2 + \cdots + u_n v_n^2 \in \mathbb{F}_4$ where $\mathbf{u} = (u_1, u_2, \ldots, u_n), \mathbf{v} = (v_1, v_2, \ldots, v_n) \in \mathbb{F}_4^n$.

Two linear q-ary codes C_1 and C_2 are said to be *equivalent* if the codewords of C_2 can be obtained from the codewords of C_1 via a sequence of transformations of the following types:

1. permutation of coordinates;
2. multiplication of the elements in a given coordinate by a nonzero element of $\mathbb{F}_q$;
3. application of a field automorphism to the elements in all coordinates simultaneously.

This equivalence may not preserve self-orthogonality over fields with $q \geq 5$ elements, for that reason we exclude the classification of self-orthogonal codes over fields with 5 and 7 elements.

An *automorphism* of a linear code C is a sequence of such transformations that maps each codeword of C onto a codeword of C. The automorphisms of a code C form a group, called the automorphism group of the code and denoted by $\mathrm{Aut}(C)$.

Practically, we will identify a linear code with its generator matrix. We consider the code classification problem as follows. Given a set of parameters q, n, k, d find generator matrices of all inequivalent $[n, k, d]$ q-ary codes. In geometrical aspect, we can define an $[n, k, d]_q$ code C as a multiset of n points in $PG(k - 1, q)$ such that (a) each hyperplane of $PG(k - 1, q)$ meets C in at most $n - d$ points and (b) there is a hyperplane meeting C in exactly $n - d$ points. This definition is equivalent to the one given in [9].

Codes which are equivalent belong to the same equivalence class. Every code can serve as a representative for its equivalence class. We use the concept for a canonical representative, selected on the base of some specific conditions.

Let G be a group that acts on a set Ω. This action defines an equivalence relation in Ω as two elements $X, Y \in \Omega$ are equivalent, $X \cong Y$, if they belong to the same orbit. A canonical representative map for this action is a function $\rho : \Omega \to \Omega$ that satisfies the following two properties: (1) for all $X \in \Omega$ it holds that $\rho(X) \cong X$; (2) for all $X, Y \in \Omega$ it holds that $X \cong Y$ implies $\rho(X) = \rho(Y)$. We take Ω to be the set of all linear $[n, k]_q$ codes. For a code $C \in \Omega$, the code $\rho(C)$ is the canonical form of C with respect to ρ. Analogously, C is in canonical form if $\rho(C) = C$. The code $\rho(C)$ is the canonical representative of its equivalence class with respect to ρ. Let $\gamma_C : C \to \rho(C)$ maps the code C to its canonical form, or $\gamma_C(C) = \rho(C)$. According to the definition given above, γ_C induces a permutation of the coordinates which we denote by π_C. The permutation π_C defines an ordering of the coordinates and the orbits of C with respect to the action of $\mathrm{Aut}(C)$.

To find the canonical form and the automorphism group of C, we need a sufficiently large set $M(C)$ of codewords of the code C (we will call it *sufficient set*) with the following properties:

- $M(C)$ generates the code C;
- $M(C)$ is stable with respect to $\mathrm{Aut}(C)$;
- if $C' \cong C''$ and $\psi(C') = C''$ then $\psi(M(C')) \equiv M(C'')$, $\psi \in G$.

This set is not uniquely determined. Usually, we can accept as a sufficient set the set of all codewords with minimum weight. If the rank of this set is smaller than the dimension of the code, a larger set of codewords is used.

3   Main Algorithms in the Program GENERATION

In the program GENERATION any linear code is represented by its generator matrix. The program has two main parts. The first one implements a construction method for generator matrices. This method is based on row by row backtracking with $k \times k$ identity matrix as a fixed part. In the m-th step the considered matrices have the following form

$$G = (I_k \ A') = \left(\begin{array}{c|c|c} I_m & O & A_m \\ \hline O & I_{k-m} & X \end{array} \right)$$

where the columns of the matrix A_m are lexicographically ordered, and X is the unknown part of G. In that case any vector v_m of length $n - k$ which fits for the m-th row of A_m strictly depends on one of the vectors put on the previous rows. Consider, for example, the binary case. If $m = 1$ there are only two options for columns of matrix A_1, namely 0 and 1, four options for $m = 2$, namely $(00)^T$, $(01)^T$, $(10)^T$ and $(11)^T$, and so on. Let the matrix A_{m-2} already be constructed. We define a set T_{m-1} of all suitable vectors for the last row in the next matrix A_{m-1}. Taking $v_{m-1} \in T_{m-1}$, we obtain the matrix A_{m-1}. The vector v_{m-1} defines an ordered partition $\Pi_{v_{m-1}}$ of the set $S = \{k + 1, k + 2, \ldots, n\}$. The possibilities for the next m-th row correspond to the refinement partitions of $\Pi_{v_{m-1}}$ induced by the vectors in T_{m-1}.

Example 1. Let us try to construct all $[11, 3, 6]$ binary even codes taking their generator matrices in a systematic form: $G = (I_3|X)$. Any row in the unknown matrix X must have 5 or 7 nonzero coordinates. For the set T_1 we have $T_1 = \{(00011111), (01111111)\}$. The current possible matrices G are

$$\left(\begin{array}{c|c} 100 & 00011111 \\ 010 & X \\ 001 & \end{array} \right) , \quad \text{and} \quad \left(\begin{array}{c|c} 100 & 01111111 \\ 010 & X \\ 001 & \end{array} \right) .$$

Take $v_1 = (00011111)$. The vector v_1 induces the partition $\Pi_{v_1} = \{\{1, 2, 3\}, \{4, 5, 6, 7, 8\}\}$ and the set $T_2 = \{t_1 = (01100111), t_2 = (11100011), t_3 = (11101111)\}$. Let fix $v_2 = t_1$,

$$A_2 = \left(\begin{array}{c} 0 \ 00 \ 11 \ 111 \\ 0 \ 11 \ 00 \ 111 \end{array} \right) \quad \text{and} \quad G = \left(\begin{array}{c|c} 100 & 00011111 \\ 010 & 01100111 \\ 001 & X \end{array} \right) .$$

Then $\Pi_{v_1, v_2} = \{\{1\}, \{2, 3\}, \{4, 5\}, \{6, 7, 8\}\}$. Now we have to find the solutions for the last row. The first vector $t_1 \in T_2$ and Π_{v_1} give the information that we have to put two 1's in the first three coordinate positions, and three 1's in the last 5 positions. We obtain the following possibilities (taking in mind also the lexicographical ordering and the partition Π_{v_1, v_2}) (01100111), (01101011), (01111001), (10100111), (10101011), (10111001). Only the last two vectors give $[11, 3, 6]$ even codes (the other four codes have minimum distance ≤ 4). In the same way we consider the second and the third vectors in T_2. By exhausting all possibilities in this way, we get all inequivalent codes we are looking for.

As a result, we obtain only one (up to equivalence) binary even $[11, 3, 6]$ code. This example explains the construction part given above.

There are several advantages of this approach:

- the number of equivalent candidates in the search tree becomes smaller,
- the construction of generator matrices is very effective,
- it allows us to consider codes with relatively large length - more than a hundred in the binary case.

Moreover, this construction is also appropriate to the other part of the program that determines inequivalent objects. In fact, this has been a key idea in the package Q-EXTENSION (more detail for this approach and the implementation can be found in [8]). To the rest of construction part we add functions for minimum and dual distances, orthogonality check and restrictions on weights.

The second part of the program is related to the identification of non-equivalent objects in the whole generation process. The general method which we apply is known as *canonical augmentation* [13,14]. Description for this specific case is given in [6]. The basic idea is to accept only non-equivalent objects without an equivalence test (in some cases with a small number of tests) at every step of the generation process. Instead of an equivalence test, a canonical form of the objects and a canonical ordering of orbits are used. So for every vector v_m in the construction that fits as a m-th row (we call these vectors *possible solutions*), the algorithm decides acceptance (possible solution becomes *real*) or rejection. In this model, the different branches of the search tree are independent and therefore it is easy for parallel implementation.

The main algorithms are developed by the basic functions of the package. Some of them are presented in the next section.

4   Basic Functions and Data Organization

To present the basic functions used in the program GENERATION, we have to give some information for the whole package QEXTNEWEDITION. It contains several hierarchically ordered modules with functions written in C/C++. Each module depends on the previous one and makes it possible to realize the functions of the next one. The interface programs (like GENERATION) stays on the top of this hierarchy.

The first module deals with the safe allocation of dynamic memory for the whole package. The main structures of the package are matrices (two dimensional arrays) of different types. These structures are used to store generator matrices, check matrices, sets of generator matrices with non-intersecting information sets, sets of all or some of the codewords of considered linear codes, sufficient sets, their corresponding binary matrices, the canonical forms and so on. The concept of the package is to investigate linear codes one by one (in consecutive execution).

Therefore, it is convenient to use some global variables. The size of the dynamic variables for different types of data related to linear codes changes when the main function considers the next object. In the beginning, the first module allocates memory for the first object. If this memory is not enough for some of the following objects, it allocates more memory by default.

The second module consists of functions related to the rank of a system of codewords, information set, orthogonal code, construction of different generator matrices (with non-intersecting information sets), etc.

The following module is related to functions for generating some or all codewords. They give minimum distance, weight spectrum, sufficient set of codewords, coset leaders, etc. We use two general approaches for calculating the weight characteristics of linear codes. One of them is exhaustive search (for small dimensions only) and the other is based on Brouwer-Zimmerman algorithm. Many of the functions check if the minimum distance, weight spectrum and other distance parameters are suitable.

A very important part of the package is the module for canonical form and automorphism group. The central object here is the $(0,1)$-matrix or bipartite graph. The main function in this module obtains canonical form, generators of the automorphism group and orbits of rows and columns (and their ordering) of a given binary matrix. For a linear code C, we use sufficient set $M(C)$ of codewords and invertible mapping of this set to a binary matrix $T(C)$ (see [7]). If two codes C_1 and C_2 are equivalent their corresponding binary matrices $T(C_1)$ and $T(C_2)$ are isomorphic. Moreover, the automorphism groups of C and $T(C)$ are isomorphic, too.

5   How Can We Use the Program for Code Classification?

In this section, we show how the program GENERATION can be used with examples. Let us consider the binary codes with parameters $[24, 7, 10]$. It is known that there are 6 inequivalent codes with these parameters [12].

After starting, GENERATION gives us the following by default:

```
Generating Linear Codes (Generation v1.1 QextNewEdition first module)
                Generate [24,12,8;2]  Linear codes
                With weights:
  wt1= 8, wt2= 12, wt3= 16, wt4= 20, wt5= 24,
                Proportional columns:
d2->800, d3->800, d4->800, d5->800, d6->800, d7->800, d8->800,
d9->800, d10->800, d11->800, d12->800,
        1. Start
        2. Change input parameters
        3. Restriction on weights
        4. Restriction on proportional coordinates
        5. Dual distance 1
        6. Brute generation
        7. About QextNewEdition
        8. Exit
        Choose:
```

To obtain the generator matrices of all 6 inequivalent binary [24,7,10] codes in the file with name 24_7_10.2 we just have to choose point 2, enter the parameters and start the calculations choosing 1. The generator matrices of all inequivalent codes obtained in the generation process (157 in our case) will be written in a file with name 24_7_10.2h. They correspond to all real solutions for $k \neq 1, 7$. The table of optimal codes [10] indicates that binary linear codes with parameters $[22, 6, 10]$ do not exist. Therefore binary $[24, 7, 10]$ codes do not have two proportional coordinates and can be obtained from [23,6,10] codes. That is why we can use restrictions for proportional coordinates (point 4) as follows: up to 4 proportional coordinates in dimension five, 2 in dimension six and no (enter 1) in dimension 7. In this case, the calculation time is 25% less.

If we are interested only in codes with dual distance 4, we can use point 5. The program looks for the codes with dual distance 2 in dimension five and 3 in dimension six. The number of inequivalent codes in the file 24_7_10.2h becomes smaller - 146.

The program has two options for restrictions on possible weights of the codes under investigation. With point 2 we can set an integer w which divides all the weights. After that with point 3 we can choose only some of the weights between d and n, divisible by w. The restriction for self orthogonality works only for codes over fields with 2, 3 and 4 elements.

In the general case, when the program have to generate $[n, k, d]$ codes, the codes with parameters $[n - t, k - i, d]$ are in the search tree, where $1 \leq i \leq k - 1$, $i \leq t \leq n - k + i$.

For optimal search of all codes with fixed n and d, and dimensions from k_{min} to k_{max}, we use point 6. In that case the results will be written in files with extensions "2b" and "2bh". For example, the search trees for constructing $[25, 6, 10]$, $[25, 5, 10]$ and $[25, 4, 10]$ self-orthogonal binary codes have 226, 289 and 99 nodes, respectively (614 summary). If we look for the self-orthogonal codes with the same parameters simultaneously by point 6, the nodes of the corresponding search tree are only 430.

6   Computational Results

In this section we present some examples. To obtain the results, we use one thread of INTEL XEON E5-2620 V4 processor. For natural reasons, the calculational time in the case of relatively small parameters depends on the size of the search tree. For given n, k and q the search tree strictly depends on the restrictions for minimum distance, self-orthogonality, possible weights, dual distance and proportional columns.

In the case of codes with large length, the number of objects that need to be checked for acceptability increases exponentially. That is why, even with a small search tree, the computational time grows.

The following table contains classification results for linear codes with different parameters and restrictions. The first and second columns show the parameters and the used restrictions, respectively. Column 3 contains the execution times in seconds, and the number of equivalent codes in each case is given in the fourth column.

Parameters	Restrictions	Time	#
$[109,5,56]_2$	Weights: 56 64 72	1145.38 s	1
$[34,12,12]_2$	Weights:12 16 20 24 28 32	19404.67 s	11
$[18,6,4]_2$	Even	2337.91 s	434906
$[19,7,9]_3$		72.01 s	61
$[22,6,12]_3$		114.52 s	701
$[24,12,9]_3$	Self-orthogonal	148.73 s	2
$[28,8,15]_3$		47.17 s	1
$[24,5,16]_4$	Weights: 16 18 20 22 24	472.49 s	1

7   Conclusion

In this paper, we present the first interface program GENERATION of the software package QEXTNEWEDITION. There are freely available versions for WINDOWS and LINUX on the webpage
http://www.moi.math.bas.bg/moiuser/~data/Software/QextNewEdition
The package contains two more programs, namely LENGTHEXTENSION and DIMEXTENSION which will be available on the same webpage.

The package QEXTNEWEDITION is a successor of Q-EXTENSION [5]. The aim of both systems is classification of linear codes with different properties and restrictions. They share some ideas in the development of algorithms and have similar interface. The package Q-EXTENSION is written in PASCAL (DELPHI) with static variables depending on the size of the field. QEXTNEWEDITION is a new software system, written in C/C++, designed to be widely portable and suitable for parallelization. All basic functions are rewritten, looking for optimal implementation. The main concept and used methods for classification are different. The classification here is based on canonical augmentation as opposed to Q-EXTENSION where the used method is isomorph-free generation via recorded objects [13].

There are many differences between QEXTNEWEDITION and Q-EXTENSION. We list some new points:

- Programming language is C/C++ which make program portable and proper for MPI parallelization.
- Dynamically allocated variables are used. This means that the size of the input data depends only on the hardware and the range of the program can easily be extended to larger fields.

- The implementation of the generating part presented in Sect. 3 is different. In the beginning, it was by nested loops and now it is based on specific integer partitions [8].
- The algorithm for canonical form is optimized by additional invariants for the partitioning process.
- The representation of the sufficient set as a binary matrix now is much more flexible [7].

With these features, the program GENERATION is a powerful tool for classifying linear codes.

Acknowledgements. We are greatly indebted to the unknown referees for their useful suggestions.

References

1. Baart, R., et al.: GAP package GUAVA. https://www.gap-system.org/Packages/guava.html
2. Betten, A.: Classifying discrete objects with orbiter. ACM Commun. Comput. Algebra **47**(3/4), 183–186 (2014)
3. Betten, A., Braun, M., Fripertinger, H., Kerber, A., Kohnert, A., Wassermann, A.: Error-Correcting Linear Codes - Classification by Isometry and Applications. Springer, Heidelberg (2006). https://doi.org/10.1007/3-540-31703-1
4. Bosma, W., Cannon, J., Playoust, C.: The Magma algebra system I: The user language. J. Symbolic Comput. **24**, 235–265 (1997)
5. Bouyukliev, I.: What is Q-EXTENSION? Serdica J. Comput. **1**, 115–130 (2007)
6. Bouyukliev, I., Bouyuklieva, S., Kurz, S.: Computer classification of linear codes. arXiv:2002.07826 [cs.IT] (2020)
7. Bouyukliev, I., Dzhumalieva-Stoeva, M.: Representing equivalence problems for combinatorial objects. Serdica J. Comput. **8**(4), 327–354 (2014)
8. Bouyukliev, I., Hristova, M.: About an approach for constructing combinatorial objects. Cybernetics Inf. Technol. **18**, 44–53 (2018)
9. Dodunekov, S., Simonis, J.: Codes and projective multisets. Electron. J. Comb. **5**, R37 (1998)
10. Grassl, M.: Bounds on the minimum distance of linear codes and quantum codes. http://www.codetables.de. Accessed 10 March 2020
11. Huffman, W.C., Pless, V.: Fundamentals of Error-Correcting Codes. Cambridge University Press, Cambridge (2003)
12. Jaffe, D.B.: Optimal binary linear codes of length ≤ 30. Discret. Math. **223**, 135–155 (2000)
13. Kaski, P., Östergård, P.R.: Classification Algorithms for Codes and Designs. Springer, Heidelberg (2006). https://doi.org/10.1007/3-540-28991-7
14. McKay, B.: Isomorph-free exhaustive generation. J. Algorithms **26**, 306–324 (1998)

Polyhedral Methods in Geometry and Optimization

Algebraic Polytopes in Normaliz

Winfried Bruns$^{(\boxtimes)}$ (iD)

Institut für Mathematik, Universität Osnabrück, 49069 Osnabrück, Germany
wbrunas@uos.de
http://www.home.uni-osnabrueck.de/wbruns/

Abstract. We describe the implementation of algebraic polyhedra in Normaliz. In addition to convex hull computation/vertex enumeration, Normaliz computes triangulations, volumes, lattice points, face lattices and automorphism groups. The arithmetic is based on the package *e-antic* by V. Delecroix.

Keywords: Polyhedron · Real algebraic number field · Computation

Algebraic polytopes lacking a rational realization are among the first geometric objects encountered in high school geometry: at least one vertex of an equilateral triangle in the plane has non-rational coordinates. Three of the five Platonic solids, namely the tetrahedron, the icosahedron and the dodecahedron are non-rational, and, among the 4-dimensional regular polytopes, the 120-cell and the 600-cell live outside the rational world.

But algebraic polytopes do not only appear in connection with Coxeter groups. Other contexts include enumerative combinatorics [17], Dirichlet domains of hyperbolic group actions [8], $SL(2,\mathbb{R})$-orbit closures in the moduli space of translation surfaces, and parameter spaces and perturbation polyhedra of cut-generating functions in integer programming.

1 Real Embedded Algebraic Number Fields

The notion of convexity is defined over any ordered field, not only over the rationals $\mathbb{Q}$ or the reals $\mathbb{R}$. *Real embedded algebraic number fields* are subfields of the real numbers (and therefore ordered) that have finite dimension as a $\mathbb{Q}$-vector space. It is well known that such a field $\mathbb{A}$ has a primitive element, i.e., an element a such that no proper subfield of $\mathbb{A}$ contains a. The minimal polynomial of a is the least degree monic polynomial μ with coefficients in $\mathbb{Q}$ such that $\mu(a) = 0$. It is an irreducible polynomial, and $\dim_{\mathbb{Q}} \mathbb{A} = \deg \mu$. In particular, every element b of $\mathbb{A}$ has a unique representation $b = \alpha_{n-1}a^{n-1} + \cdots + \alpha_1 a + \alpha_0$ with $\alpha_{n-1}, \ldots, \alpha_0 \in \mathbb{Q}$, $n = \deg \mu$. The arithmetic in $\mathbb{A}$ is completely determined by μ: addition is the addition of polynomials and multiplication is that of polynomials followed by reduction modulo μ. The multiplicative inverse can be computed by the extended Euclidean algorithm. The unique determination of the coefficients α_i allows one

© Springer Nature Switzerland AG 2020
A. M. Bigatti et al. (Eds.): ICMS 2020, LNCS 12097, pp. 193–201, 2020.
https://doi.org/10.1007/978-3-030-52200-1_19

to decide whether $b = 0$. Every element of $\mathbb{A}$ can be written as the quotient of a polynomial expression $\alpha_{n-1}a^{n-1} + \cdots + \alpha_1 a + \alpha_0$ with $\alpha_i \in \mathbb{Z}$ for all i and an integer denominator; this representation is used in the implementation.

However, the algebraic structure alone does not define an ordering of $\mathbb{A}$. For example, $\sqrt{2}$ and $\sqrt{-2}$ cannot be distinguished algebraically: there exists an automorphism of $\mathbb{Q}[\sqrt{2}]$ that exchanges them. For the ordering we must fix a real number a whose minimal polynomial is μ. (Note that not every algebraic number field has an embedding into $\mathbb{R}$.) In order to decide whether $b > 0$ for some $b \in \mathbb{A}$ we need a floating point approximation to b of controlled precision.

Normaliz [4] uses the package *e-antic* of V. Delecroix [7] for the arithmetic and ordering in real algebraic number fields. The algebraic operations are realized by functions taken from the package *antic* of W. Hart and F. Johansson [11] (imported to *e-antic*) while the controlled floating point arithmetic is delivered by the package *arb* of F. Johansson [13]. Both packages are based on W. Hart's *Flint* [12].

In order to specify an algebraic number field, one chooses the minimal polynomial μ of a and an interval I in $\mathbb{R}$ such that μ has a unique zero in I, namely a. An initial approximation to a is computed at the start. Whenever the current precision of b does not allow to decide whether $b > 0$, first the approximation of b is improved, and if the precision of a is not sufficient, it is replaced by one with twice the number of correct digits.

2 Polyhedra

A subset $P \subset \mathbb{R}^d$ is a *polyhedron* if it is the intersection of finitely many affine halfspaces:

$$P = \bigcap_{i=0}^{s} H_i^+, \qquad H_i^+ = \{x : \lambda_i(x) \geq \beta_i\}, \qquad i = 1, \ldots, s,$$

where λ_i is a linear form and $\beta_i \in \mathbb{R}$. It is a *cone* if one can choose $\beta_i = 0$ for all i, and it is a *polytope* if it is bounded.

By the theorem of Minkowski-Weyl-Motzkin [2, 1.C] one can equivalently describe polyhedra by "generators": there exist $c_1, \ldots, c_t \in \mathbb{R}^d$ and $v_1, \ldots, v_u \in \mathbb{R}^d$ such that

$$P = C + Q$$

where $C = \{c\gamma_1 c_1 + \cdots + \gamma_t c_t : \gamma_i \in \mathbb{R}, \gamma_i \geq 0\}$ is the *recession cone* and $Q = \{\kappa_1 v_1 + \cdots + \kappa_u v_u : \kappa_i \in \mathbb{R}, \kappa_i \geq 0, \sum \kappa_i = 1\}$ is a polytope. These two descriptions are often called *H-representation* and *V-representation*. The conversion from H to V is *vertex enumeration* and the opposite conversion is *convex hull computation*.

For theoretical and computational reasons it is advisable to present a polyhedron P as the intersection of a cone and a hyperplane. Let $C(P)$ be the *cone over* P, i.e., the smallest cone containing $P \times \{1\}$, and $D = \{x : x_{d+1} = 1\}$ the *dehomogenizing hyperplane*. Then P can be identified with $C(P) \cap D$. After this step, convex hull computation and vertex enumeration are two sides of the same coin, namely the dualization of cones.

In the definition of polyhedra and all statements following it, the field $\mathbb{R}$ can be replaced by an arbitrary subfield (and even by an arbitrary ordered field), for example a real algebraic number field $\mathbb{A}$. The smallest choice for $\mathbb{A}$ is $\mathbb{Q}$: for it we obtain the class of *rational polyhedra*. For general $\mathbb{A}$ we get *algebraic polyhedra*.

For the terminology related to polyhedra and further details we refer the treader to [2].

3 Normaliz

Normaliz tackles many computational problems for rational and algebraic polyhedra:

- dual cones: convex hulls and vertex enumeration
- projections of cones and polyhedra
- triangulations, disjoint decompositions and Stanley decompositions
- Hilbert bases of rational, not necessarily pointed cones
- normalizations of affine monoids (hence the name)
- lattice points of polytopes and (unbounded) polyhedra
- automorphisms (euclidean, integral, rational/algebraic, combinatorial)
- face lattices and f-vectors
- Euclidean and lattice normalized volumes of polytopes
- Hilbert (or Ehrhart) series and (quasi) polynomials under $\mathbb{Z}$-gradings
- generalized (or weighted) Ehrhart series and Lebesgue integrals of polynomials over rational polytopes

Of course, not all of these computation goals make sense for algebraic polyhedra. The main difference between the rational and the non-rational case can be described as follows: the monoid of lattice points in a full dimensional cone is finitely generated if and only if the cone is rational.

Normaliz is based on a templated C++ library. The template allows one to choose the arithmetic, and so it would be possible to extend Normaliz to more general ordered fields. The main condition is that the arithmetic of the field has been coded in a C++ class library. There is no restriction on the real algebraic number fields that Normaliz can use.

Normaliz has a library as well as a file interface. It can be reached from CoCoA, GAP [9], Macaulay2, Singular, Python [10] and SageMath. The full functionality is reached on Linux and Mac OS platforms, but the basic functionality for rational polyhedra is also available on MS Windows systems.

Its history goes back to the mid 90ies. For recent developments see [3] and [6]. The extension to algebraic polytopes was done in several steps since 2016. We are grateful to Matthias Köppe for suggesting it.

The work on algebraic polytopes has been done in cooperation with Vincent Delecroix (*e-antic*), Sebastian Gutsche (PyNormaliz), Matthias Köppe and Jean-Phiippe Labbé (integration into SageMath). A comprehensive article with these coauthors is in preparation.

4 The Icosahedron

Let us specify the icosahedron, a Platonic solid, by its vertices:

```
amb_space 3
number_field min_poly (a^2 - 5) embedding [2 +/- 1]
vertices 12
0 2 (a + 1) 4
0 -2 (a + 1) 4
2 (a + 1) 0 4
...
(-a - 1) 0 -2 4
Volume
LatticePoints
FVector
EuclideanAutomorphisms
```

The first line specifies the dimension of the affine space. The second defines the unique positive sqare root of 5 as the generator of the number field. It is followed by the 12 vertices. Each of them is given as a vector with 4 components for which the fourth component acts as a common denominator of the first three. Expressions involving a are enclosed in round brackets. The last lines list the computation goals for Normaliz. (Picture by J.-P. Labbé)

Normaliz has a wide variety of input data types. For example, it would be equally possible to define the icosahedron by inequalities. Now we have a look into the output file. (We indicate omitted lines by ...)

```
Real embedded number field:
min_poly (a^2 - 5) embedding [2.23606797...835961152572 +/- 5.14e-54]

1 lattice points in polytope
12 vertices of polyhedron
0 extreme rays of recession cone
20 support hyperplanes of polyhedron (homogenized)
f-vector:
1 12 30 20 1
embedding dimension = 4
affine dimension of the polyhedron = 3 (maximal)
rank of recession cone = 0 (polyhedron is polytope)
...
volume (lattice normalized) = (5/2*a+15/2 ~ 13.090170)
volume (Euclidean) = 2.18169499062
Euclidean automorphism group has order 120
*********************************************************************
1 lattice points in polytope:
0 0 0 1
12 vertices of polyhedron:
...
0 extreme rays of recession cone:
20 support hyperplanes of polyhedron (homogenized):
(-a+1 ~ -1.236068) (-2*a+4 ~ -0.472136)                    0 1
```

```
...
  (a-1 ~ 1.236068)   (2*a-4 ~ 0.472136)                      0 1
```

The output (in homogenized coordinates) is self-explanatory. Note that non-integral numbers in the output are printed as polynomials in a together with a rational approximation. At the top we can see to what precision $\sqrt{5}$ had to be computed. The automorphism group is described in another output file:

```
Euclidean automorphism group of order 120
***********************************************************************
3 permutations of 12 vertices of polyhedron
Perm 1: 1 2 4 3 7 8 5 6 10 9 11 12
Perm 2: 1 3 2 5 4 6 7 9 8 11 10 12
Perm 3: 2 1 3 4 6 5 8 7 9 10 12 11
Cycle decompositions
Perm 1: (3 4) (5 7) (6 8) (9 10) --
Perm 2: (2 3) (4 5) (8 9) (10 11) --
Perm 3: (1 2) (5 6) (7 8) (11 12) --
1 orbits of vertices of polyhedron
Orbit 1 , length 12:  1 2 3 4 5 6 7 8 9 10 11 12
***********************************************************************
3 permutations of 20 support hyperplanes
Perm 1: 2 1 5 6 3 4 7 8 11 12 9 10 13 14 17 18 15 16 20 19
...
Cycle decompositions
Perm 1: (1 2) (3 5) (4 6) (9 11) (10 12) (15 17) (16 18) (19 20) --
...
1 orbits of support hyperplanes
Orbit 1 , length 20:  1 2 3 4 5 6 7 8 9 10 11 12 13 14 15 16 17 18 19 20
```

5   Computation Goals for Algebraic Polyhedra

The basic computation in linear convex geometry is the dualization of cones. We start from a cone $C \subset \mathbb{R}^d$, given by generators $x_1, \ldots, x_n$. The first (easy) step is to find a coordinate transformation that replaces $\mathbb{R}^d$ by the vector subspace generated by $x_1, \ldots, x_n$. In other words, we can assume $\dim C = d$.

The goal is to find a minimal generating set $\sigma_1, \ldots, \sigma_s \in (\mathbb{R}^d)^*$ of the dual cone $C^* = \{\lambda : \lambda(x_i) \geq 0, \ i = 1, \ldots, n\}$. Because of $\dim C = d$, the linear forms $\sigma_1, \ldots, \sigma_s$ are uniquely determined up to positive scalars: they are the extreme rays of C^*. By a slight abuse of terminology we call the hyperplanes $S_i = \{x : \sigma_i(x) = 0\}$ the *support hyperplanes* of C.

Let C_k be the cone generated by $x_1, \ldots, x_k$. Normaliz proceeds as follows:

1. It finds a basis of $\mathbb{R}^d$ among the generators $x_1, \ldots, x_n$, say $x_1, \ldots, x_d$. Computing C_d^* amounts to a matrix inversion.
2. Iteratively it extends the cone C_k to C_{k+1}, and shrinks C_k^* to C_{k+1}^*, $k = d \ldots, n-1$.

Step 2 is done by *Fourier-Motzkin elimination*: if $\sigma_1, \ldots, \sigma_t$ generate C_k^*, then C_{k+1}^* is generated by

$$\left\{\sigma_i : \sigma_i(x_{k+1}) \geq 0\right\} \cup \left\{\sigma_i(x_{k+1})\sigma_j - \sigma_j(x_{k+1})\sigma_i : \sigma_i(x_{k+1}) > 0, \sigma_j(x_{k+1}) < 0\right\}.$$

From this generating set of C_{k+1}^* the extreme rays of C_{k+1}^* must be selected.

This step is of critical complexity. Normaliz has a sophisticated implementation in which *pyramid decomposition* is a crucial tool; see [5]. It competes very well with dedicated packages (see [14]). The implementation is independent of the field of coefficients. As said above, $\mathbb{R}$ can be replaced by an algebraic number field $\mathbb{A}$. In this case Normaliz uses the arithmetic over the field $\mathbb{A}$ realized by *e-antic*, whereas arithmetic over $\mathbb{Q}$ is avoided in favor of arithmetic over $\mathbb{Z}$.

In addition to the critical complexity caused by the combinatorics of cones, one must tame the coordinates of the linear combination $\lambda = \sigma_i(x_{k+1})\sigma_j - \sigma_j(x_{k+1})\sigma_i$. For example, if, over $\mathbb{Z}$, both σ_i and σ_j are divisible by 2, then λ is divisible by 4. If this observation is ignored, a doubly exponential explosion of coefficients will happen. One therefore extracts the gcd of the coordinates. But there is usually no well-defined gcd of algebraic integers, and even if one has unique decomposition into prime elements, there is in general no Euclidean algorithm. Normaliz therefore applies two steps:

1. λ is divided by the absolute value of the last nonzero component (or by another "norm").
2. All integral denominators are cleared by multiplication with their lcm.

Computational experience has shown that these two steps together are a very good choice.

Normaliz tries to measure the complexity of the arithmetic in $\mathbb{A}$ and to control the algorithmic alternatives of the dualization by the measurements. There are several "screws" that can be turned, and it is difficult to find the optimal tuning beforehand.

Normaliz computes lexicographic triangulations of algebraic cones in the same way as triangulations of rational cones. Their construction is interleaved with the extension from C_k to C_{k+1}: the already computed triangulation of C_k is extended by the simplicial cones generated by x_{k+1} and those subcones in the triangulation of C_k that are "visible" from x_{k+1}.

An algebraic polytope P contains only finitely many integral points. They are computed by Normaliz' project-and-lift algorithm. The truncated Hilbert basis approaches, which Normaliz can also use for rational polytopes, are not applicable in the algebraic case. Once the lattice points are known, one can compute their convex hull, called the *integer hull* of P.

At present Normaliz computes volumes only for full-dimensional algebraic polytopes. The volume is the sum of the volumes of the simplices in a triangulation, and these are simply (absolute values of) determinants. We do not see any reasonable definition of "algebraic volume" for lower dimensional polytopes that could replace the lattice normalized volume. The latter is defined for all rational polytopes and is a rational number that can be computed precisely.

It would certainly be possible to extend the computation of the approximate Euclidean volume to all algebraic polytopes, and this extension may be included in future Normaliz versions. Note that the Euclidean volume does in general not belong to $\mathbb{A}$ if P is lower dimensional. Its precise computation would require an extension of $\mathbb{A}$ by square roots.

The computation of automorphism groups follows the suggestions in [1]. First one transforms the defining data into a graph, and then computes the automorphism group of this graph by *nauty* [15]. For algebraic polytopes the Euclidean and the algebraic automorphism groups can be computed, and the combinatorial automorphism group is accessible for all polyhedra.

The Euclidean automorphism group is the group of rigid motions of the ambient space that map the polytope to itself, and the algebraic automorphism group is the group of affine transformations over $\mathbb{A}$ stabilizing the polytope. Both groups are finite, as well as the combinatorial automorphism group, the automorphism group of the face lattice, which can be computed from the facet-vertex incidence vectors, just as in the rational case.

We do not try to define the algebraic (or Euclidean) automorphism group for unbounded polyhedra. First of all, the algebraic automorphism group is infinite in general. Second, it would have to be realized as the permutation group of a vector configuration, and there seems to be no reasonable way to norm the involved vectors. But for polytopes we can and must use the vertices.

6 Scaled Convex Hull Computations

We illustrate the influence of the algebraic number field on the computation time by some examples. For each of them we start from a cone (over a polyhedron) that is originally defined over the integers. Then we scale some coordinates by elements of the field $\mathbb{A}$. This transformation preserves the combinatorial structure throughout. It helps to isolate the complexity of the arithmetic operations. The types of arithmetic that we compare are

int: original input, computation with machine integers,
mpz: same input as int, but computation with GMP mpz_class integers,
rat: same input as int, but computation in $\mathbb{Q}[\sqrt{5}]$,
sc2: scaled input in $\mathbb{Q}[\sqrt{5}]$,
sc8: scaled input in $\mathbb{Q}[\sqrt[8]{5}]$,
p12: scaled input in $\mathbb{Q}[a]$, $a^{12} + a^6 + a^5 + a^2 - 5 = 0$, $a > 1$.

The test candidates are A553 (from the Ohsugi-Hibi classification of contingency tables [16]), the cone q27f1 from [14], the linear order polytope for S_6, and the cyclic polytope of dimension 15 with 30 vertices. The last two are classical polytopes. While the other three cones are given by their extreme rays, q27f1 is defined by 406 equations and inequalities (Table 1).

The Normaliz version is 3.8.4, compiled into a static binary with gcc 5.4 under Ubuntu 16-04. The computations use 8 parallel threads (the default choice of Normaliz). They were taken on the author's PC with an AMD Ryzen 7 1700X

Table 1. Combinatorial data of the test candidates

	amb_space	dim	ext rays	supp hyps
A553	55	43	75	306, 955
q27f1	30	13	68, 216	92
lo6	16	16	720	910
cyc15-30	16	16	30	341088

at 3.2 GHz. Table 2 lists wall times in seconds. As a rule of thumb, for a single thread the times must be multiplied by 6.

Table 2. Wall times of scaled convex hull computations in seconds

Coeff	A553	q27f1	lo6	cyc15-30
int	57	16	5	–
mpz	299	58	5	7
rat	277	40	5	7
sc2	783	166	4	14
sc8	1272	475	15	28
p12	2908	905	31	42

The cyclic polytope and all intermediate polytopes coming up in its computation are simplicial. Therefore it profits from Normaliz' special treatment of simplicial facets—almost everything can be done by set theoretic operations. Also lo6 is combinatorially not complicated. That lo6 is fastest with sc2, is caused by the fine tuning of the pyramid decomposition, which is not always optimal.

Surprisingly, rat is faster than mpz for A553 and q27f1. This can be explained by the fact that linear algebra over $\mathbb{Z}$ must use the Euclidean algorithm, and therefore needs more steps than the true rational arithmetic of rat.

References

1. Bremner, D., Sikirić, M.D., Pasechnik, D.V., Rehn, T., Schürmann, A.: Computing symmetry groups of polyhedra. LMS J. Comput. Math. **17**, 565–581 (2014)
2. Bruns, W., Gubeladze, J.: Polytopes, Rings, and K-Theory. Springer, Dordrecht (2009)
3. Bruns, W., Ichim, B.: Polytope volume by descent in the face lattice and applications in social choice. Preprint arXiv:1807.02835
4. Bruns, W., Ichim, B., Römer, T., Sieg, R., Söger, C.: Normaliz. Algorithms for rational cones and affine monoids. http://normaliz.uos.de
5. Bruns, W., Ichim, B., Söger, C.: The power of pyramid decomposition in Normaliz. J. Symb. Comp. **74**, 513–536 (2016)

6. Bruns, W., Sieg, R., Söger, C.: Normaliz 2013–2016. In: Böckle, G., Decker, W., Malle, G. (eds.) Algorithmic and Experimental Methods in Algebra, Geometry, and Number Theory, pp. 123–146. Springer (2018)
7. Delecroix, V.: Embedded algebraic number fields (on top of antic). Package available at https://github.com/videlec/e-antic
8. Delecroix, V., Page, A.: Dirichlet fundamental domains for real hyperbolic spaces. Work in progress
9. Gutsche, S., Horn, M., Söger, C.: NormalizInterface for GAP. https://github.com/gap-packages/NormalizInterface
10. Gutsche, S., Sieg, R.: PyNormaliz - an interface to Normaliz from python. https://github.com/Normaliz/PyNormaliz
11. Hart, W., Johansson, F.: Algebraic Number Theory In C. Package available at https://github.com/wbhart/antic
12. Hart, W., Johansson, F., Pancratz, S.: FLINT: Fast Library for Number Theory. http://flintlib.org
13. Johansson, F.: Arb - a C library for arbitrary-precision ball arithmetic. http://arblib.org/
14. Köppe, M., Zhou, Y.: New computer-based search strategies for extreme functions of the Gomory-Johnson infinite group problem. Math. Program. Comput. 9, 419–469 (2017)
15. McKay, B.D., Piperno, A.: Practical graph isomorphism. II. J. Symbolic Comput. 60, 94–112 (2014)
16. Ohsugi, H., Hibi, T.: Toric ideals arising from contingency tables. In: Bruns, W. (ed.) Commutative Algebra and Combinatorics. Ramanujan Mathematical Society Lecture Note Series 4, pp. 87–111 (2006)
17. Rote, G.: The maximum number of minimal dominating sets in a tree. In: Proceedings of the Thirtieth Annual ACM-SIAM Symposium on Discrete Algorithms, pp. 1201–1214. SIAM, Philadelphia (2019)

Real Tropical Hyperfaces by Patchworking in polymake

Michael Joswig[1,2(✉)] and Paul Vater[2]

[1] Chair of Discrete Mathematics/Geometry, TU Berlin, Berlin, Germany
joswig@math.tu-berlin.de
[2] MPI MiS, Leipzig, Germany
vater@mis.mpg.de

Keywords: Hilbert's 16th problem · Real algebraic hypersurfaces ·
Viro's patchworking · Tropical hypersurfaces

1 Introduction

Hilbert's 16th problem asks about topological constraints for real algebraic hypersurfaces in projective space. In the 1980s Viro developed patchworking as a combinatorial method to construct real algebraic hypersurfaces with unusually large $\mathbb{Z}_2$-Betti numbers [14–17]. A major breakthrough of this idea was Itenberg's refutation of Ragsdale's Conjecture [9]. Today patchworking is most naturally interpreted within the larger framework of tropical geometry [12]. In this way patchworking is a combinatorial avenue to real tropical hypersurfaces.

Here we report on a recent implementation of patchworking and real tropical hypersurfaces in polymake [1], version 4.1 of June 2020. The first software for patchworking that we are aware of is the "Combinatorial Patchworking Tool" [4], which works web-based and is restricted to the planar case. A second implementation is Viro.sage [18] which is capable of patchworking in arbitrary dimension and degree. Our implementation has the same scope as Viro.sage but it is superior in two ways. First, it naturally ties in with a comprehensive hierarchy of polyhedral objects in polymake; e.g., this allows for a rich choice of constructions of real tropical hypersurfaces. Second, our implementation is more efficient. This is demonstrated by several experiments with curves and surfaces of various degrees. As a new mathematical contribution we provide a census of Betti numbers of real tropical surfaces.

1.1 Tropical Hypersurfaces in $\mathbb{TP}^{n-1}$

Let $f = \bigoplus_{v \in V} c_v \odot x^v \in \mathbb{T}[x_1, \ldots, x_n]$ be a tropical polynomial where V is a finite subset of $\mathbb{Z}^n$. We use the multi-index notation $x^v = x_1^{v_1} \cdots x_n^{v_n}$, and $\mathbb{T} = \mathbb{R} \cup \{\infty\}$, $\oplus = \min$ and $\odot = +$. The *tropical hypersurface* $\mathcal{T}(f)$ is the tropical vanishing locus of f, i.e., the set of points in $\mathbb{R}^n$, where the minimum of the evaluation function $x \mapsto f(x)$ is attained at least twice. Throughout we will

A. M. Bigatti et al. (Eds.): ICMS 2020, LNCS 12097, pp. 202–211, 2020.
https://doi.org/10.1007/978-3-030-52200-1_20

assume that f is homogeneous of degree d, i.e., for each $v \in V$ we have $v_1 + \cdots + v_n = d$. In that case $\mathcal{T}(f)$ descends to the *tropical projective torus* $\mathbb{R}^n/\mathbb{R}\mathbf{1}$, where $\mathbf{1} = (1, \ldots, 1)$. The *Newton polytope* of f is $\mathcal{N}(f) = \operatorname{conv} V$, and the coefficients of f induce a regular subdivision, $\mathcal{S}(f)$. The latter is dual to $\mathcal{T}(f)$. We refer to [12] and [3] for further details.

The tropical projective space $\mathbb{TP}^{n-1} = (\mathbb{T}^n - \{\infty\mathbf{1}\})/\mathbb{R}\mathbf{1}$ compactifies $\mathbb{R}^n/\mathbb{R}\mathbf{1}$. It is naturally stratified into lower dimensional tropical projective tori, marked by those coordinates which are finite. In this way the pair $(\mathbb{TP}^{n-1}, \mathbb{R}^n/\mathbb{R}\mathbf{1})$ is naturally homeomorphic with an $(n-1)$-simplex and its interior. Often we will identify the tropical hypersurface $\mathcal{T}(f)$ with its compactification in $\mathbb{TP}^{n-1}$.

1.2 Viro's Patchworking

The following is essentially a condensed version of [13, §3.1], with minor variations. A *sign distribution* $\epsilon \in \mathbb{Z}_2^V$ can be *symmetrized* to the function

$$s_\epsilon : \mathbb{Z}_2^n \to \mathbb{Z}_2^V, \quad s_\epsilon(z)(v) := \epsilon(v) + \langle z, v \rangle \bmod 2.$$

As in [6] we choose our signs in $\mathbb{Z}_2 = \{0, 1\}$, which corresponds to ± 1 via $z \mapsto (-1)^z$. Further, the elements $z \in \mathbb{Z}_2^n$ are in bijection with the 2^n orthants of $\mathbb{R}^n$ via $z \mapsto \operatorname{pos}\{(-1)^{z_1}e_1, \ldots, (-1)^{z_n}e_n\}$, where $e_1, \ldots, e_n$ are the standard basis vectors of $\mathbb{R}^n$, and $\operatorname{pos}(\cdot)$ denotes the nonnegative hull. We will use this identification throughout and, consequently, we call z itself an *orthant*.

The tropical hypersurface $\mathcal{T}(f)$ is a polyhedral complex in $\mathbb{TP}^{n-1}$, and its k-dimensional cells are dual to the $(n-1-k)$-cells of $\mathcal{S}(f)$. In particular, each maximal cell F of $\mathcal{T}(f)$ corresponds to an edge, $V(F) \subset V$, of $\mathcal{S}(f)$. We write $\mathcal{T}_{n-2}$ for the set of maximal cells (which are $(n-2)$-dimensional polyhedra) and denote powersets as $\mathcal{P}(\cdot)$.

Note that there are no $(n-2)$-cells of $\mathcal{T}(f)$ in the boundary $\mathbb{TP}^{n-1} - \mathbb{R}^n/\mathbb{R}\mathbf{1}$. The *real phase structure* on $\mathcal{T}(f)$ induced by ϵ is the map

$$\phi_\epsilon : \mathcal{T}_{n-2} \to \mathcal{P}(\mathbb{Z}_2^n), \quad F \mapsto \{z \in \mathbb{Z}_2^n \mid s_\epsilon(z)(v) \neq s_\epsilon(z)(w)\} \text{ for } \{v, w\} = V(F).$$

That is, for each maximal cell F of $\mathcal{T}(f)$ this describes the set of orthants, in which the symmetrized sign distribution takes distinct values on the two vertices of the dual edge $V(F)$ in $\mathcal{S}(f)$. This extends to all cells G of $\mathcal{T}(f)$ by setting $\phi_\epsilon(G) := \bigcup \phi_\epsilon(F)$, where the union is taken over all maximal cells $F \in \mathcal{T}_{n-2}$ containing G. The pair $\mathcal{T}_\epsilon(f) = (\mathcal{T}(f), \epsilon)$ is a *real tropical hypersurface*.

Let $\bar{z}$, defined by $\bar{z}_i = 1 - z_i$, be the *antipode* of $z \in \mathbb{Z}_2^n$. We define an equivalence relation $\sim$ on $\mathbb{Z}_2^n \times \mathbb{TP}^{n-1}$, which identifies copies of $\mathbb{TP}^{n-1}$ along common strata, by letting

$$(z, x) \sim (z', y) \; :\Longleftrightarrow \; x = y \text{ and } \big(\bar{z} = z' \text{ or } (x_i = \infty = y_i \Leftrightarrow z_i = 1 = z_i')\big).$$

This identifies $\{z\} \times \mathbb{TP}^{n-1}$ and $\{\bar{z}\} \times \mathbb{TP}^{n-1}$ one to one for each z. It follows that the quotient $(\mathbb{Z}_2^n \times \mathbb{TP}^{n-1})/\sim$ is homeomorphic to the real projective space $\mathbb{RP}^{n-1}$. Combinatorially that construction can be seen as follows: the union of

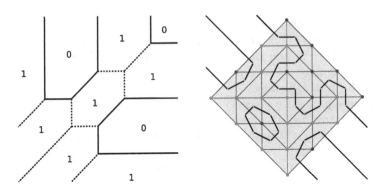

Fig. 1. Real tropical elliptic curve (left) and its real part (right)

the 2^n simplices $\operatorname{conv}\{(-1)^{z_1}e_1, \ldots, (-1)^{z_n}e_n\}$, where z ranges over all orthants, gives the boundary of the regular cross polytope $\operatorname{conv}\{\pm e_1, \ldots, \pm e_n\}$ in $\mathbb{R}^n$. Taking the quotient modulo antipodes yields $\mathbb{RP}^{n-1}$.

The *real part* of the real tropical hypersurface $\mathcal{T}_\epsilon(f) = (\mathcal{T}(f), \epsilon)$, denoted $\mathbb{R}\mathcal{T}_\epsilon(f)$, is now defined as the collection of polyhedral complexes in $\mathbb{Z}_2^n \times \mathbb{TP}^{n-1}$ consisting of the polyhedra

$$\{\{z\} \times F \mid F \in \mathcal{T}_{n-2} \text{ and } z \in \phi_\epsilon(F)\}$$

and their faces. Note that $\{z\} \times F \in \mathbb{R}\mathcal{T}_\epsilon(f)$ if and only if $\{\bar{z}\} \times F \in \mathbb{R}\mathcal{T}_\epsilon(f)$, and hence we may restrict to the part of $\mathbb{R}\mathcal{T}_\epsilon(f)$ in $(\{0\} \times \mathbb{Z}_2^{n-1}) \times \mathbb{TP}^{n-1}$.

To avoid cumbersome notation and language we call the quotient of $\mathbb{R}\mathcal{T}_\epsilon(f)$ by $\sim$ also the *real part* of $\mathcal{T}_\epsilon(f)$ and use the same symbol, $\mathbb{R}\mathcal{T}_\epsilon(f)$. In this way $\mathbb{R}\mathcal{T}_\epsilon(f)$ becomes a piecewise linear hypersurface in $\mathbb{RP}^{n-1} \approx \mathbb{Z}_2^n \times \mathbb{TP}^{n-1}/\sim$.

The above construction is relevant for its connection with real algebraic geometry. To simplify the exposition we now consider a special case: Setting $\Delta_{n-1} = \operatorname{conv}\{e_1, \ldots, e_n\}$, we assume that the set $V = d \cdot \Delta_{n-1} \cap \mathbb{Z}^n$ is the set of lattice points in the dilated unit simplex. This entails that the projective toric variety generated from V is the (complex) projective space $\mathbb{CP}^{n-1}$. The following result comes in various guises; this version occurs in [15] and [8, Proposition 2.6].

Theorem 1 (Viro's combinatorial patchworking theorem). *Let f be a homogeneous tropical polynomial of degree d with support $V = d \cdot \Delta_{n-1} \cap \mathbb{Z}^n$. Then, for each sign distribution $\epsilon \in \mathbb{Z}_2^n$, there exists a nonsingular real algebraic hypersurface X in $\mathbb{CP}^{n-1}$, also with Newton polytope $\mathcal{N}(f) = d \cdot \Delta_{n-1}$, such that*

$$(\mathbb{Z}_2^n \times \mathbb{TP}^{n-1}/\sim, \mathbb{R}\mathcal{T}_\epsilon(f)) \text{ is } \mathbb{Z}_2\text{-homologous to } (\mathbb{RP}^{n-1}, \mathbb{R}X).$$

If additionally $\mathcal{S}(f)$ is *unimodular*, i.e., each simplex has normalized volume one, this is "primitive patchworking". In the primitive case stronger conclusions hold [13, 16]. The notion "combinatorial patchworking" refers to the condition

$\mathcal{N}(f) = d \cdot \Delta_{n-1}$. This is what our implementation supports, for arbitrary degrees and dimensions. More general results require to carefully take into account the toric geometry of $\mathcal{N}(f)$.

Example 2. With $n = d = 3$ we consider the tropical polynomial

$$f = x^3 \oplus 1x^2 y \oplus 1x^2 z \oplus 4xy^2 \oplus 3xyz \oplus 4xz^2 \oplus 9y^3 \oplus 7y^2 z \oplus 7yz^2 \oplus 9z^3$$

in $\mathbb{T}[x, y, z]$, where we omit '$\odot$' for improved readability. The tropical hypersurface $\mathcal{T}(f)$ is the tropical elliptic curve in $\mathbb{R}^3 / \mathbb{R}\mathbf{1}$ in Fig. 1 (left). The sign distribution $\epsilon = (0, 1, 0, 1, 1, 1, 1, 0, 1, 1)$ yields a real tropical curve with real part in $\mathbb{Z}_2^3 \times \mathbb{TP}^2 / \sim$ which has two components; cf. Fig. 1 (right). This primitive patchwork corresponds to a classical Harnack curve of degree 3; cf. [9, Sec. 5].

2   Betti Numbers from Combinatorial Patchworking

Our goal is to exhibit a census of Betti numbers of real tropical surfaces in $\mathbb{Z}_2^4 \times \mathbb{TP}^3 / \sim$. Throughout the following let f be a tropical polynomial of degree d in $n = 4$ homogeneous variables; we will assume that $\mathcal{S}(f)$ is a regular and full triangulation of $V = d \cdot \Delta_3 \cap \mathbb{Z}^4$. That is, we focus on combinatorial patchworks. A triangulation of V is *full* if it uses all points in V; a unimodular triangulation is necessarily full. While the converse holds in the plane, there are many more full triangulations of $d \cdot \Delta_3$ than unimodular ones if $d \geq 3$. Further, with

$$k := \frac{1}{6}d^3 + d^2 + \frac{11}{6}d + 1, \tag{1}$$

which is the cardinality of V, we pick a sign vector $\epsilon \in \mathbb{Z}_2^k$. This gives rise to a real algebraic surface X in $\mathbb{CP}^3$ whose real part $\mathbb{R}X$ is "near the tropical limit" $\mathbb{R}\mathcal{T}_\epsilon(f)$ in the sense of [13]. Itenberg [6, Theorems 3.2/3.3] showed that the Euler characteristic satisfies

$$\chi(\mathbb{R}X) \geq \frac{4d - d^3}{3}, \tag{2}$$

with equality attained in the primitive/unimodular case. Moreover, by [6, Theorem 4.2],

$$b_1(\mathbb{R}X) \leq \frac{2d^3 - 6d^2 + 7d}{3}, \tag{3}$$

where $b_q(\cdot)$ are $\mathbb{Z}_2$-Betti numbers; see also [7] for bounds without the fullness assumption. However, if $\mathcal{S}(f)$ is even unimodular then, by [6, Theorem 4.1],

$$b_0(\mathbb{R}X) \leq \binom{d - 1}{3} + 1. \tag{4}$$

See Table 1 for explicit numbers in the range which is relevant for our experiments. The main result of [13] furnishes a vast generalization of (4) to arbitrary dimensions.

Table 1. Bounds for Euler characteristic and Betti numbers, depending on the degree d. The values k, χ', b_0' and b_1' are the right hand sides of (1), (2), (4) and (3), respectively.

d	k	χ'	b_0'	b_1'
3	20	-5	1	7
4	35	-16	2	20
5	56	-35	5	45
6	84	-64	11	86

Example 3. The subdivision $\mathcal{S}(f)$ induced by the tropical polynomial

$$f = 5x^3 \oplus 1x^2y \oplus 1xy^2 \oplus 5y^3 \oplus 2x^2z \oplus 0xyz \oplus 2y^2z$$
$$\oplus\ 0xz^2 \oplus 0yz^2 \oplus 1z^3 \oplus 2x^2w \oplus 0xyw \oplus 2y^2w \oplus 1xzw$$
$$\oplus\ 1yzw \oplus 1z^2w \oplus 3xw^2 \oplus 3yw^2 \oplus 4zw^2 \oplus 8w^3$$

is a full triangulation of $3 \cdot \Delta_3$ which is not unimodular. Its f-vector reads $(20, 60, 64, 23)$, and its automorphism group is of order 6. The sign distribution

$$\epsilon = (0,0,0,0,0,0,0,1,1,0,1,1,1,1,1,0,1,1,1,0)$$

yields a real tropical surface $\mathbb{R}\mathcal{T}_\epsilon(f)$ whose real part has Betti vector $(2,1,2)$ (Fig. 2).

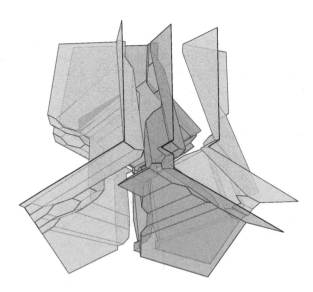

Fig. 2. The real part of a cubic surface with Betti vector $(2,1,2)$. There are three affine sheets, of which the outer two account for one connected component in $\mathbb{RP}^3$, which is homeomorphic to $\mathbb{S}^2$; the middle sheet forms a component homeomorphic to $\mathbb{RP}^2$.

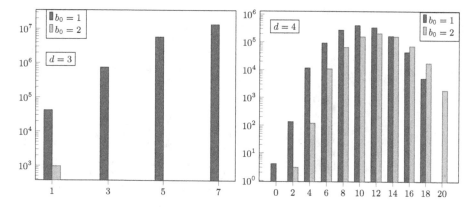

Fig. 3. Distribution of Betti vectors for surfaces of degrees 3 and 4. The colors indicate values for $b_0 = b_2$, the values on the x-axis indicate values for b_1. For $d = 3$ the most frequent vector is $(1, 7, 1)$ with 67.52%. For $d = 4$ it is $(1, 10, 1)$ with 19.86%.

2.1 Combinatorial Description of the Homology

The polyhedral description of $\mathbb{R}\mathcal{T}_\epsilon(f)$ directly gives a combinatorial description of the homology; see also [13, Proposition 3.17]. The cellular chain modules read

$$C_q(\mathbb{R}\mathcal{T}_\epsilon(f); \mathbb{Z}_2) = \bigoplus_{\sigma \text{ cell of } \mathcal{T}_\epsilon(f), \dim \sigma = q} \left(\bigoplus_{z \in \phi_\epsilon(\sigma)} \mathbb{Z}_2^{\{\sigma \times \{z\}\}} \right) \qquad (5)$$

and $\partial(\sigma \times \{z\}) = \partial(\sigma) \times \{z\}$ defines the boundary maps. In fact this construction is a special case of a cellular (co-)sheaf [11]. Algorithmically it is beneficial that this does *not* require the geometric construction of $\mathbb{R}\mathcal{T}_\epsilon(f)$.

2.2 A Census of Betti Numbers of Real Tropical Surfaces

We used `mptopcom` [10] to compute regular and full triangulations of $d \cdot \Delta_3$ for $3 \leq d \leq 6$, which are not necessarily unimodular. For $d = 3$ the total number of such triangulations is known to be 21 125 102 [10, Table 3], up to the natural action of the symmetric group S_4. For higher degrees the corresponding numbers are unknown and probably out of reach for current hard- and software. Still we can compute some of those triangulations, for each degree.

   Our experiments suggest that, in order to see many different Betti vectors (b_0, b_1, b_0), it is preferable to look at many different triangulations. This is feasible for degrees 3 and 4, where we created 1 000 000 and 100 000 orbits of triangulations, respectively. Each of them was equipped with 20 sign distributions which were picked uniformly at random; cf. Fig. 3. For $d = 3$ we obtain all values for b_1 which are allowed by (3) if the surface is connected (i.e., $b_0 = 1$). Additionally, 965 times we saw the Betti vector $(2, 1, 2)$; cf. Example 3. In view of (4) this occurs for non-unimodular triangulations only; all our examples of this kind

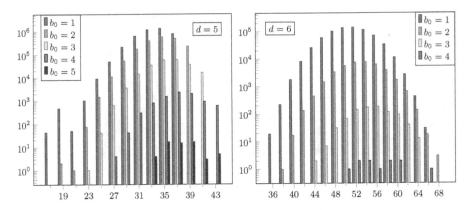

Fig. 4. Distribution of Betti vectors for surfaces of degrees 5 and 6. The colors indicate values for $b_0 = b_2$, the values on the x-axis indicate values for b_1. For $d = 5$ the most frequent vector is $(1, 35, 1)$ with 21.9%. For $d = 6$ it is $(1, 52, 1)$ with 18.97%.

share the same f-vector $(20, 60, 64, 23)$. For $d = 4$ all the possible Betti vectors occur; cf. (2) and (3).

The case of $d = 5$ turned out to be surprisingly difficult. In our standard setup `mptopcom` quickly produced about a hundred full and regular triangulations before it stalled. `mptopcom`'s algorithm employs a very special search through the flip graph of the point configuration, and it finds all regular triangulations plus some non-regular ones connected by a sequence of flips. Apparently, most neighbors to our first 100 triangulations of $5 \cdot \Delta_3$ are not regular or not full. As we were interested in exploring many different Betti vectors, we created a second sample of triangulations; to this end we employed a random walk on the flip graph of $5 \cdot \Delta_3$. After eliminating multiples, this gave an additional 13 000 regular and full triangulations. On each of the resulting 13 100 triangulations we tried 500 random sign distributions; cf. Fig. 4 (left) for the combined statistic. For $d = 6$ we checked 1 500 triangulations with 500 sign distributions each; cf. Fig. 4 (right).

No matter how hard we try we will only see a tiny fraction of all possible real tropical surfaces of higher degrees. So the distributions for $d = 5$ and $d = 6$ may not even be close to the "truth". Yet for $d = 5$ we observed $b_1 = 43$, whereas $b'_1 = 45$; cf. Table 1. We found 61 triangulations of $5 \cdot \Delta_3$ with five components, none of which were unimodular. The maximal number of components in the unimodular case was four. For $d = 6$ our census is way off the theoretical bounds.

3   Implementation in `polymake`

`polymake` is a comprehensive software system for polyhedral geometry and related areas of mathematics [1]. Mathematical objects like tropical hypersurfaces are determined by their *properties*. Upon a user query the system directly returns a property (e.g., a tropical polynomial or the dual polyhedral subdivision) if it is known, or it computes it by applying a sequence of *rules*. Subsequently, the property asked for becomes known, along with any intermediate results. Throughout the life of such a *big object* the number of properties grows; objects, with their properties, can be saved and loaded again. The latter is useful, e.g., for processing data on a cluster and examining them on a laptop later.

The computation which is relevant here takes a tropical polynomial f (such that the Newton polytope $\mathcal{N}(f)$ is a dilated simplex) and a sign distribution ϵ as input and computes the $\mathbb{Z}_2$-Betti numbers of the real part $\mathbb{R}\mathcal{T}_\epsilon(f)$ of the real tropical hypersurface $\mathcal{T}_\epsilon(f)$. The individual steps are: (i) find the maximal cells of $\mathcal{T}(f)$ via a dual convex hull computation; (ii) compute the Hasse diagram of the entire face lattice of $\mathcal{T}(f)$; (iii) construct the chain complex (5) from that Hasse diagram; (iv) compute ranks of the boundary matrices mod 2. Each step is implemented as a separate rule, which makes the code highly modular and reusable. In particular, the only nontrivial implementation which is really new is step (iii).

We wish to give some details about the first two steps. Often the dual convex hull computation is the most expensive part. For this `polymake` has interfaces to several algorithms and implementations, the default being PPL [2] which is also used here. In general, it is difficult to predict which algorithm performs best; see [1] for extensive convex hull experiments. The computation of the Hasse diagram uses a combinatorial procedure whose complexity is linear in the size of the output, i.e., the total number of cells of the tropical hypersurface; cf. [5].

3.1 Running Times

To compare the running times of `Viro.sage` and `polymake` for computing the Betti numbers of patchworked hypersurfaces we conducted two experiments, one for Harnack curves and one for surfaces. All computations were carried out on an AMD Phenom II X6 1090T (3.2 GHz, 38528 bmips).

For the Harnack curves, where we have just one curve per degree (the cubic case is Example 2), we repeated the same computation ten times each. Figure 5 (left) shows the mean running time depending on the degree. The `Viro.sage` code showed a rather wide variety, while the `polymake` computations gave almost identical running times for each test.

The experiment for the surfaces is slightly different in that both the tropical polynomials (and triangulations) and the sign distributions were varied. For degrees 3, 4, 5, and 6 we took the first 2000, 1000, 100, and 75 triangulations (as enumerated by `mptopcom`), respectively, and measured the running time for 10 random sign distributions each. Figure 5 (right) shows a box plot for each degree. The boxes indicate the 2nd and 3rd quartiles, the whiskers mark the minimum and maximum time measurements, excluding outliers (i.e., measurements whose ratio to the median is either bigger than 4, or smaller than 0.25), which are marked separately. Again `Viro.sage` exhibits a much greater variety of running times than `polymake`.

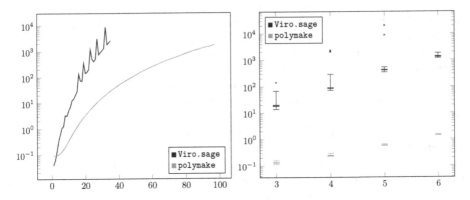

Fig. 5. Time taken to compute Betti numbers (in seconds). Left: Harnack curves, average time by degree. Right: various surfaces, boxplots for each degree.

4   Conclusion

We have shown that our new implementation is capable of determining the $\mathbb{Z}_2$-Betti numbers of a patchworked surface of moderate degree within a few seconds. This allows for providing a rich census.

One major reason for `polymake` being faster than `Viro.sage` [18] is that we avoid the explicit construction of a simplicial complex model of $\mathbb{R}\mathcal{T}_\epsilon(f)$. Moreover, `polymake` computes $\mathbb{Z}_2$ Betti numbers directly, while `Viro.sage` goes through a standard homology computation with integer coefficients. `polymake` provides geometric realizations (and integral homology), too, but this is unnecessary here.

Acknowledgments. We are indebted to Ilia Itenberg, Johannes Rau and Kristin Shaw for valuable comments on a previous version of this text. Moreover, we are grateful to Lars Kastner and Benjamin Lorenz for helping with the experiments. M. Joswig has been supported by DFG (EXC 2046: "MATH^{+}", SFB-TRR 195: "Symbolic Tools in Mathematics and their Application", and GRK 2434: "Facets of Complexity").

References

1. Assarf, B., et al.: Computing convex hulls and counting integer points with `polymake`. Math. Program. Comput. **9**(1), 1–38 (2017)
2. Bagnara, R., Hill, P.M., Zaffanella, E.: The Parma Polyhedra Library: toward a complete set of numerical abstractions for the analysis and verification of hardware and software systems. Sci. Comput. Programm. **72**(1–2), 3–21 (2008)
3. De Loera, J.A., Rambau, J., Santos, F.: Triangulations, Algorithms and Computation in Mathematics, vol. 25. Springer, Berlin (2010). structures for algorithms and applications
4. El-Hilany, B., Rau, J., Reneaudineau, A.: Combinatorial patchworking tool. https://www.math.uni-tuebingen.de/user/jora/patchworking/patchworking.html

5. Hampe, S., Joswig, M., Schröter, B.: Algorithms for tight spans and tropical linear spaces. J. Symbolic Comput. **91**, 116–128 (2019)
6. Itenberg, I.: Topology of real algebraic T-surfaces. Rev. Mat. Univ. Complut. Madrid **10**(Special Issue, suppl.), 131–152 (1997). real algebraic and analytic geometry (Segovia, 1995)
7. Itenberg, I., Shustin, E.: Critical points of real polynomials and topology of real algebraic T-surfaces. Geom. Dedicata. **101**, 61–91 (2003)
8. Itenberg, I., Shustin, E.: Viro theorem and topology of real and complex combinatorial hypersurfaces. Israel J. Math. **133**, 189–238 (2003)
9. Itenberg, I., Viro, O.: Patchworking algebraic curves disproves the Ragsdale conjecture. Math. Intelligencer **18**(4), 19–28 (1996)
10. Jordan, C., Joswig, M., Kastner, L.: Parallel enumeration of triangulations. Electron. J. Combin. **25**(3), Paper 3.6, 27 (2018)
11. Kastner, L., Shaw, K., Winz, A.L.: Cellular sheaf cohomology of `polymake`. In: Combinatorial Algebraic Geometry, Fields Inst. Commun., vol. 80, pp. 369–385. Fields Inst. Res. Math. Sci., Toronto, ON (2017)
12. Maclagan, D., Sturmfels, B.: Introduction to Tropical Geometry, Graduate Studies in Mathematics, vol. 161. American Mathematical Society, Providence (2015)
13. Renaudineau, A., Shaw, K.: Bounding the Betti numbers of real hypersurfaces near the tropical limit (2019). arXiv:1805.02030
14. Viro, O.: Curves of degree 7, curves of degree 8 and the Ragsdale conjecture. Dokl. Akad. Nauk SSSR **254**(6), 1306–1310 (1980)
15. Viro, O.: Gluing algebraic hypersurfaces, removing of singularities and constructions of curves. Topology conference, Proc., Collect. Rep., Leningrad 1982, pp. 149–197 (1983). English translation arXiv:0611382
16. Viro, O.Y.: Gluing of plane real algebraic curves and constructions of curves of degrees 6 and 7. In: Faddeev, L.D., Mal'cev, A.A. (eds.) Topology. LNM, vol. 1060, pp. 187–200. Springer, Heidelberg (1984). https://doi.org/10.1007/BFb0099934
17. Viro, O.: From the sixteenth Hilbert problem to tropical geometry. Jpn. J. Math. **3**(2), 185–214 (2008)
18. de Wolff, T., Kwaakwah, E.O., O'Neill, C.: Viro.sage (2018). https://cdoneill.sdsu. edu/viro/. version 0.4b, posted May 9, 2018

Practical Volume Estimation of Zonotopes by a New Annealing Schedule for Cooling Convex Bodies

Apostolos Chalkis[1]([⊠]) [iD], Ioannis Z. Emiris[1,2] [iD], and Vissarion Fisikopoulos[1] [iD]

[1] Department of Informatics and Telecommunications, National and Kapodistrian University of Athens, Athens, Greece
{achalkis,emiris,vfisikop}@di.uoa.gr
[2] Athena Research and Innovation Center, Maroussi, Greece

Abstract. We study the problem of estimating the volume of convex polytopes, focusing on zonotopes. Although a lot of effort is devoted to practical algorithms for polytopes given as an intersection of halfspaces, there is no such method for zonotopes. Our algorithm is based on Multiphase Monte Carlo (MMC) methods, and our main contributions include: (i) a new uniform sampler employing Billiard Walk for the first time in volume computation, (ii) a new simulated annealing generalizing existing MMC by making use of adaptive convex bodies which fit to the input, thus drastically reducing the number of phases. Extensive experiments on zonotopes show our algorithm requires sub-linear number of oracle calls in the dimension, while the best theoretical bound is cubic. Moreover, our algorithm can be easily generalized to any convex body. We offer an open-source, optimized C++ implementation, and analyze its performance. Our code tackles problems intractable so far, offering the first efficient algorithm for zonotopes which scales to high dimensions (e.g. one hundred dimensions in less than 1 h).

Keywords: Volume approximation · Zonotope · Simulated annealing · Billiard Walk · Mathematical software

1 Introduction

Volume computation is a fundamental problem with many applications. It is #P-hard for explicit polytopes [7,11], and APX-hard [9] for convex bodies in the oracle model. Therefore, a significant effort has been devoted to randomized approximation algorithms, starting with the celebrated result in [8] with complexity $O^*(d^{23})$ oracle calls, where $O^*(\cdot)$ suppresses polylog factors and dependence on error parameters, and d is the dimension. Improved algorithms reduced the exponent to 5 [13] and further results [5,14] reduced the exponent to 3. Current theoretical results consider either the general oracle model or polytopes given as an intersection of halfspaces (i.e. H-polytopes). Regarding implementations, the approach of [13] led to the first practical implementation in [10]

© Springer Nature Switzerland AG 2020
A. M. Bigatti et al. (Eds.): ICMS 2020, LNCS 12097, pp. 212–221, 2020.
https://doi.org/10.1007/978-3-030-52200-1_21

for high dimensions, followed by another practical implementation [6] based on [5,14]. However, both implementations can handle only H-polytopes.

An important class of convex polytopes are zonotopes [15]. A zonotope is the Minkowski sum of k d-dimensional segments. Equivalently, given a matrix $G \in \mathbb{R}^{d \times k}$ a zonotope can be seen as the affine projection of the hypercube $[-1, 1]^k$ to $\mathbb{R}^d$ using the matrix G, while the columns of G are the corresponding segments (or generators). Zonotopes are centrally symmetric and each of their faces are again zonotopes. We call the *order* of a zonotope P the ratio between the number of generators of P over the dimension. For a nice introduction to zonotopes we refer to [20].

Volume approximation for zonotopes is of special interest in several applications in smart grids [1], in autonomous driving [2] or human-robot collaboration [16]. The complexity of algorithms that work on zonotopes strongly depends on their order. Thus, to achieve efficient computations, a solution that is common in practice is to over-approximate P, as tight as possible, with a second zonotope P_{red} of smaller order, while vol(P_{red}) is given by, an easy to compute, closed formula. A good measure for the quality of the approximation is the ratio of fitness, $\rho = (\text{vol}(P_{red})/\text{vol}(P))^{1/d}$, which involves a volume computation problem [3]. Existing work (e.g. in [12]) uses exact - deterministic volume computation [11], and thus ρ can not be computed for $d > 10$ in certain applications.

A typical randomized algorithm uses a Multiphase Monte Carlo (MMC) technique, which reduces volume approximation of convex P to computing a telescoping product of ratios of integrals. Then each ratio is estimated by means of random walks sampling from a proper multivariate distribution. In this paper we rely on MMC of [13] which specifies a sequence of convex bodies $P_m \subseteq \cdots \subseteq P_0 = P$, assuming P is well-rounded, i.e. $B_d \subseteq P \subseteq C\sqrt{d}B_d$, where C is constant and B_d is the unit ball. We define a sequence of scaled copies of B_d, and let $P_i = (2^{(m-i)/d}B_d) \cap P$, $i = 0, \ldots, m$. One computes vol(P_m) and applies:

$$\text{vol}(P) = \text{vol}(P_m)\frac{\text{vol}(P_{m-1})}{\text{vol}(P_m)} \cdots \frac{\text{vol}(P_0)}{\text{vol}(P_1)},$$

$$m = O(d \lg d), \ P_0 = C\sqrt{d}B_d \cap P. \tag{1}$$

There is a closed-form expression to compute vol(P_m) = vol(B_d) . Each ratio $r_i = \text{vol}(P_{i+1})/\text{vol}(P_i)$ in Eq. (1) can be estimated within arbitrary small error ϵ_i by sampling uniformly distributed points in P_i and accept/reject points in P_{i+1} so vol(P) can be derived after m multiplications. The estimation of r_i shows how sampling comes into the picture. In [10], assuming $rB_d \subseteq P \subseteq RB_d$ for $r < R$, they get $m = \lceil d \lg(R/r) \rceil$. The issue is to minimize m while each ratio remains bounded by a constant, and to use a random walk that converges, after a minimum number of steps, to the uniform distribution. The first would permit a larger approximation error per ratio without compromising overall error, while it would require a smaller uniform sample to estimate each ratio. The second would reduce the cost per sample point. Total complexity is determined by the number of ratios, or phases, multiplied by the number of points, or steps, to estimate each ratio, multiplied by the cost to generate a point. The first two factors are determined by the MMC and the third by the random walk.

Previous Work. Exact volume computation for zonotopes can be reduced to a sum of absolute values of determinants, with an exponential number of summands in d [11]. Practical algorithms for volume computation of zonotopes are limited to low dimensions (typically ≤ 10 in [6]). This is due to two main reasons: current algorithms create a long sequence of phases in MMC for zonotopes, and the boundary and membership oracles are costlier than for H-polytope, as they both reduce to Linear Programs (LP). In [6], they consider low dimensional zonotopes: in $\mathbb{R}^{10}$ with $k = 20$ generators, the algorithm performs 1.92×10^5 Boundary Oracle Calls (BOC), whereas our algorithm requires only 8.50×10^3 BOCs, and for $d = 100, k = 200$ it performs 6.51×10^4 BOCs (see Table 1). In [19] they present an implementation of an efficient algorithm that computes Minkowski sums of polytopes (generalization of zonotopes). In [18] they propose a randomized algorithm for enumerating the vertices of a zonotope.

Our Contribution. We focus on zonotopes and introduce crucial algorithmic innovations to overcome the existing barriers, by reducing significantly the number of oracle calls. Thus, our method scales to high dimensions ($d = 100$ in ≤ 1 h), performing computations which were intractable till now.

We use a new simulated annealing method in order to define a sequence of appropriate convex bodies, instead of balls, in MMC, and we exploit the fast convergence of Billiard Walk (BW) [17] to the uniform distribution. We experimentally analyze complexity by counting the number of BOCs, since BW uses boundary reflections.

The new simulated annealing specifies the P_i's by exploiting the statistical properties of the telescoping ratios to drastically reduce the number of phases. In particular, we bound each ratio $r_i = \text{vol}(P_{i+1})/\text{vol}(P_i)$ to a given interval $[r, r + \delta]$ with high probability, for some real r. Moreover, our MMC generalizes balls, used in [13] and previous papers, by taking as input any convex body C and constructing the sequence by only scaling C. It does not need an enclosing body of P nor an inscribed ball (or body), unlike [10,13].

Most of the previous algorithms use a rounding step before volume computation, as preprocessing, to reduce the number of phases in MMC. However, rounding requires uniform sampling from P which makes it costly for zonotopes because of the expensive oracle calls. Our approach is to exploit the fact that the schedule uses any body C and skip rounding by letting C be an H-polytope that fits well to P. The idea is to construct C fast and reduce the number of phases and the total runtime more than a rounding preprocessing would do in practice.

We prove that the number of bodies defined in MMC is, with high probability, $m = O(\lg(\text{vol}(P)/\text{vol}(P_m)))$, where $P_m = qC \cap P$, for some $q \in \mathbb{R}$, is the body with minimum volume, and $\frac{\text{vol}(qC \cap P)}{\text{vol}(qC)} \in [r, r + \delta]$. The bound on m is not surprising, as it does not improve worst-case complexity [5], if C is a ball, but offers crucial advantages in practice. First, the hidden constant is small. More importantly, if C is a good fit to P, $\text{vol}(P_m)$ increases and m decreases (Fig. 1).

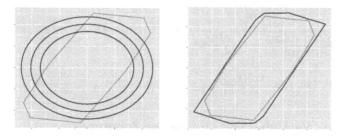

Fig. 1. Different selection of body in our algorithm's MMC; $r = 0.8$ and $r + \delta = 0.85$. Body C: left is the unit ball; right is the centrally symmetric H-polytope of Sect. 2.3.

We also show that, for constant d, and k (number of generators) increasing, m decreases to 1, when we use ball in MMC, since the schedule constructs an enclosing ball of P. Intuitively, while order increases for constant d, a random zonotope approximates the hypersphere. The latter can be approximated up to ϵ in the Hausdorff metric by a zonotope with $k \leq c(d)(\epsilon^2|\lg \epsilon|)^{(d-1)/(d+2)}$, $c(d)$ being a constant [4]. This does not directly prove our claim on m but strengthens it intuitively. So, in our experiments, the number of phases is $m \leq 3$ for any order, without rounding for $d \leq 100$.

Considering uniform sampling, BW defines a linear trajectory starting at the current point, using boundary reflections [17]. No theoretical mixing time exists. We show that with the right selection of parameters, BW behaves like an almost perfect uniform sampler even if the walk length is 1. In particular, for this walk length, it generates just $O^*(1)$ points per phase, with sub-linear number of reflections per point, and provides the desired accuracy. To stop sampling when estimating ratio r_i we modify the binomial proportion confidence interval. We use the standard deviation of a sliding window of the last l ratios, thus defining a new empirical convergence criterion; $l = O(1)$ suffices with BW.

Our software contributions build upon and enhance `volesti`[1] a C++ open source library for high dimensional sampling and volume computation with an R interface. We experimentally show that the total number of oracle calls grows as $o^*(d)$ for random zonotopes; the best available theoretical bound is $O^*(d^3)$ [5].

2   Volume Algorithm

The algorithm first constructs a sequence of convex bodies $C_1 \supseteq \cdots \supseteq C_m$ intersecting the zonotope P; the C_i's are determined by simulated annealing.

---

[1] https://github.com/GeomScale/volume_approximation.

A typical choice of C_i's in this paper is co-centric balls, or centrally symmetric H-polytopes. C_m is chosen for its volume to be computed faster than $\text{vol}(P)$ and easily sampled. Then,

$$\text{vol}(P) = \frac{\frac{\text{vol}(P_m)}{\text{vol}(C_m)}}{\frac{\text{vol}(P_1)}{\text{vol}(P_0)}\frac{\text{vol}(P_2)}{\text{vol}(P_1)} \cdots \frac{\text{vol}(P_m)}{\text{vol}(P_{m-1})}} \text{vol}(C_m), \qquad P_i = C_i \cap P,\, i = 1,\dots,m,$$

where $P_0 = P$. Let $r_i = \dfrac{\text{vol}(P_{i+1})}{\text{vol}(P_i)}$, $i = 0,\dots,m-1$, $r_m = \dfrac{\text{vol}(P_m)}{\text{vol}(C_m)}$.

2.1   Uniform Sampling and Oracles for Zonotopes

We use BW to sample approximately uniform points in P_i at each phase i. BW picks a uniformly distributed line ℓ through the current point. It walks on a linear trajectory of length $L = -\tau \ln \eta$, $\eta \sim \mathcal{U}(0,1)$, reflecting at the boundary. BW can be used to sample only uniform points; in [17] they experimentally show that BW converges fast to the uniform distribution when $\tau \approx \text{diam}(P)$.

The membership oracle is a feasibility problem. A point $p \in P$ iff the following region is feasible: $\sum_{i=1}^{k} x_i g_i = p$, $-1 \le x_i \le 1$, where g_i are the generators of P. Let the uniformly distributed vector on the boundary of the unit ball v define the line ℓ through the current point. The boundary oracle for the intersection $\ell \cap \partial P$ is expressed as a LP. One extreme point of the segment can be computed as follows: $\min -\lambda$, s.t. $p + \lambda v = \sum_{i=1}^{k} x_i g_i$ $-1 \le x_i \le 1$. The second extreme point which corresponds to a negative value of λ is not used by BW. For the BW we need the normal of the facet that intersects ℓ to compute the reflection of the trajectory if needed. We keep the generators that corresponds to $x_i \ne -1, 1$ and then the normal vector is computed straightforwardly.

2.2   Annealing Schedule for Convex Bodies

Given P, the annealing schedule generates the sequence of convex bodies $C_1 \supseteq \cdots \supseteq C_m$ defining $P_i = C_i \cap P$ and $P_0 = P$. The main goal is to restrict each ratio r_i in the interval $[r, r + \delta]$ with high probability. We define the following two statistical tests, which can be reduced to t-tests:

[U-test(P_1, P_2)] H_0: $\text{vol}(P_2)/\text{vol}(P_1) \ge r + \delta$
[L-test (P_1, P_2)] H_0: $\text{vol}(P_2)/\text{vol}(P_1) \le r$

The U-test and L-test are successful iff null hypothesis H_0 is rejected, namely r_i is upper bounded by $r + \delta$ or lower bounded by r, with high probability, respectively. If we sample N uniform points from P_i then r.v. X that counts points in P_{i+1}, follows $X \sim b(N, r_i)$, the binomial distribution, and $Y = X/N \sim \mathcal{N}(r_i, r_i(1 - r_i)/N)$. Then each sample proportion that counts successes in P_{i+1} over N is an unbiased estimator for the mean of Y, which is r_i.

Perform <u>L-test and U-test</u>
Input : convex bodies P_1, P_2, cooling parameters r, δ, s.l. $\alpha, \nu, N \in \mathbb{N}$

Sample νN uniform points from P_1
Partition νN points to lists $S_1, \ldots, S_\nu$, each of length N
Compute ratios $\hat{r}_i = |\{p \in P_2 : p \in S_i\}|/N$, $i = 1, \ldots, \nu$
Compute the mean, $\hat{\mu}$, and st.d., s, of the ν ratios
*if $\hat{\mu} \geq r + t_{\nu-1,\alpha} \frac{s}{\sqrt{\nu}}$ then L-test holds, **otherwise** L-test fails*
*if $\hat{\mu} \leq r + \delta - t_{\nu-1,\alpha} \frac{s}{\sqrt{\nu}}$ then U-test holds, **otherwise** U-test fails*

Let us now describe the annealing schedule: Each C_i in $C_1 \supseteq \cdots \supseteq C_m$ is a scalar multiple of a given body C. Since our algorithm does not use an inscribed body, initialization computes the body with minimum volume, denoted by C' or C_m. This is the last body in the sequence. The algorithm sets $P_0 = P$ and employs C' to decide stopping at the i-th phase.

Initialization. Given C, and interval $[q_{min}, q_{max}]$, one employs binary search to compute $q \in [q_{min}, q_{max}]$ s.t. both U-test$(qC, qC \cap P)$ and L-test$(qC, qC \cap P)$ are successful. Let $q = (q_{min} + q_{max})/2$. If U-test$(qC, qC \cap P)$ succeeds and L-test$(qC, qC \cap P)$ fails, we continue to the left-half of the interval. With inverse outcomes, we continue to the right-half of the interval. If both succeed, stop and set $C' = qC$. The output is C', denoted by C_m at termination.

Regular Iteration. At iteration i, the algorithm determines P_{i+1} s.t. volume ratio $r_i \in [r, r + \delta]$ with high probability. The schedule samples νN points from P_i and binary searches for a q_{i+1} in an updated interval $[q_{min}, q_{max}]$ s.t. both U-test$(P_i, q_{i+1}C \cap P)$ and L-test$(P_i, q_{i+1}C \cap P)$ are successful. Then set $P_{i+1} = q_{i+1}C \cap P$.

Stopping and Termination. The algorithm uses $C' \cap P$ in the i-th iteration for checking whether vol$(C' \cap P)/$vol$(P_i) > r$ with high probability, using only L-test, and then stops if L-test$(P_i, C' \cap P)$ holds. Then, set $m = i + 1$, and $P_m = C' \cap P$.

In the t-tests, errors of different types may occur, thus, binary search may enter intervals that do not contain ratios in $[r, r+\delta]$. Hence, there is a probability that annealing schedule fails to terminate. Let β capture the power of a t-test: $pow = \Pr[\text{reject } H_0 \mid H_0 \text{ false}] = 1 - \beta$.

Theorem 1. *Let J be the minimum number of steps by annealing schedule, corresponding to no errors occurring in the t-tests. Let the algorithm perform $M \geq J$ iterations. Let β_{max}, β_{min} be the maximum and minimum among all β's in the M pairs of t-tests in the U-test and L-test, respectively. Then, annealing schedule terminates with constant probability, namely:*

$$\Pr[\text{an. sched. terminates}] \geq 1 - 2\frac{\alpha(1 - \beta_{min}) + \beta_{max}}{1 - \alpha(1 - \beta_{min}) + \beta_{max}} - \frac{2\beta_{max} - \beta_{min}^2}{1 - 2\beta_{max} - \beta_{min}^2}.$$

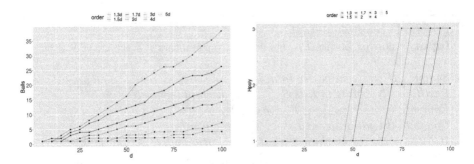

Fig. 2. Number of bodies in MMC. For each dimension we generate 10 random zonotopes and we compute the number of bodies, m, in MMC when C is ball. We keep the zonotope with the larger m and then, for that one, we compute m when C is the P-approx.

2.3 Rounding and Convex Bodies in MMC

The annealing schedule allows as to use any C which must (a) be a good fit to P, (b) allow for more efficient sampling than in P, and (c) for faster volume calculation than of $\mathrm{vol}(P)$. For low order ones C shall be an enclosing H-polytope that fits well to P. Indeed it is possible that with certain choices for C rounding is not needed. We define a centrally symmetric H-polytope with $\leq 2k$ facets:

Construct P-approx
Input : The generator matrix $G \in \mathbb{R}^{d \times k}$ of zonotope P
Output: An H-polytope $C \supset P$

compute the eigenvectors of $G^T G$ (has $k - d$ zero eigenvalues)
let the eigenvectors of $k - d$ zero eigenvalues of $G^T G$ form $E \in \mathbb{R}^{k \times (k-d)}$.
compute an orthonormal basis for E, and the orthogonal complement $W_\perp$
Let $Ay \leq b_0$, $A \in \mathbb{R}^{2k \times k}$ be an H-representation of $[-1,1]^k$
$C := \{x | Mx \leq b_0\}, M = AW_\perp^T (GW_\perp^T)^{-1} \in \mathbb{R}^{2k \times d}$
return C;

2.4 Experimental Complexity

We perform extended experiments analyzing practical complexity. We use `eigen`[2] for linear algebra and `lpSolve`[3] for LPs. All experiments were performed on a PC with Intel® Core™ i7-6700 3.40 GHz 8 CPU and 32 GB RAM. We use three zonotope generators. All of them pick uniformly a direction for each one of the k segments. Then, (a) $Z_\mathcal{U}$-d-k: the length of each segment is uniformly sampled from $[0, 100]$, (b) $Z_\mathcal{N}$-d-k: the length of each segment is random from

---

[2] eigen.tuxfamily.org.
[3] lpsolve.sourceforge.net.

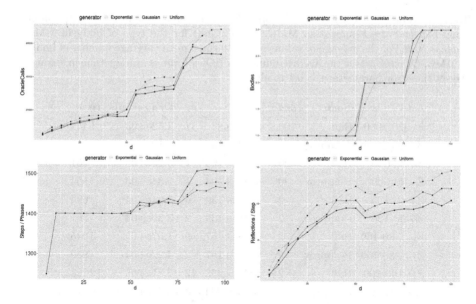

Fig. 3. Experimental complexity for order = 2. Total number of oracle calls is given by the #phases (bodies) × #steps (points) per phase × #reflections per step.

$\mathcal{N}(50, (50/3)^2)$ truncated to $[0, 100]$, (c) Z_{Exp}-d-k: the length of each segment is random from $Exp(1/30)$ truncated to $[0, 100]$. Total number of boundary oracle calls of our algorithm:

> **#BOCs**= **#phases**(bodies)×**#steps**(points)/phase×**#reflections**/point.

Figure 2 denotes the best choice between ball and P-approx in MMC. It moreover shows that for order ≤ 5, the number of phases $m \leq 3$ for $d \leq 100$. In particular, when we use P-approx, m is smaller for order ≤ 4 compared to using balls without rounding. For order equal to 5 the number of balls in MMC is smaller compared to the number of bodies when the choice is the P-approx. Notice than when we use balls in MMC, m decreases for constant d as k increases. Table 1 shows that, for high-order zonotopes, $m = 1$, which implies one or two rejection steps, while the run-time is smaller when we use ball in MMC. It also reports the difference in the run-time for random zonotopes of order = 2 between the cases of using ball and the P-approx in MMC. In all our experiments, BW performs only $O^*(1)$ steps per phase with just a factor of ϵ_i hidden in the complexity. The plot that counts the BW reflections per point in Fig. 3 imply this number grows sub-linearly in d. Hence, the total number of BOCs grows sub-linearly in d.

Table 1. Volume estimation for zonotopes. For each Z-d-k we approximate its volume using ball and the P-approx in MMC. *Body* stands for the type of body in MMC; $order = k/d$, *Vol* the average of volumes over 10 runs; m the average number of bodies in MMC; *OracleCalls* is the average number of BOCs; *time* is average time in seconds. We set the error parameter $\epsilon = 0.1$ in all cases.

Z-d-k	Body	order	Vol	m	OralceCalls	time
$Z_{\mathcal{N}}$-20-2000	Ball	100	3.69e+83	1	3.52e+03	1442
$Z_{\mathcal{N}}$-20-2000	P-approx	100	3.54e+83	1	4.10e+03	1647
$Z_{\mathcal{U}}$-30-600	Ball	20	3.93e+104	1	5.26e+03	451
$Z_{\mathcal{U}}$-30-600	P-approx	20	3.84e+104	1	5.34e+03	554
$Z_{\mathcal{U}}$-60-120	Ball	2	4.31e+139	6	7.94e+04	694
$Z_{\mathcal{U}}$-60-120	P-approx	2	4.18e+139	2	3.39e+04	361
Z_{Exp}-80-160	Ball	2	1.68e+187	9	1.67e+05	3045
Z_{Exp}-80-160	P-approx	2	1.82e+187	2	4.22e+04	950
$Z_{\mathcal{N}}$-100-200	Ball	2	9.77e+233	12	2.81e+05	12223
$Z_{\mathcal{N}}$-100-200	P-approx	2	1.03e+234	3	6.51e+04	2815

References

1. Althoff, M.: Formal and compositional analysis of power systems using reachable sets. IEEE Trans. Power Syst. **29**(5), 2270–2280 (2014)
2. Althoff, M., Dolan, J.M.: Online verification of automated road vehicles using reachability analysis. IEEE Trans. Robotics **30**(4), 903–918 (2014). https://doi.org/10.1109/TRO.2014.2312453
3. Althoff, M.: Reachability Analysis and Its Application to the Safety Assessment of Autonomous Cars (2010). Google-Books-ID: ndN3vgAACAAJ
4. Bourgain, J., Lindenstrauss, J.: Approximating the ball by a Minkowski sum of segments with equal length. Discrete Comput. Geom. **9**(2), 131–144 (1993). https://doi.org/10.1007/BF02189313
5. Cousins, B., Vempala, S.: Bypassing KLS: Gaussian cooling and an $O^*(n^3)$ volume algorithm. In: Proceedings of the ACM STOC, pp. 539–548 (2015)
6. Cousins, B., Vempala, S.: A practical volume algorithm. Math. Prog. Comp. **8**(2), 133–160 (2015). https://doi.org/10.1007/s12532-015-0097-z
7. Dyer, M., Frieze, A.: On the complexity of computing the volume of a polyhedron. SIAM J. Comput. **17**(5), 967–974 (1988). https://doi.org/10.1137/0217060
8. Dyer, M., Frieze, A., Kannan, R.: A random polynomial-time algorithm for approximating the volume of convex bodies. J. ACM **38**(1), 1–17 (1991). http://doi.acm.org/10.1145/102782.102783
9. Elekes, G.: A geometric inequality and the complexity of computing volume. Discrete Comput. Geom. **1**(4), 289–292 (1986). https://doi.org/10.1007/BF02187701
10. Emiris, I., Fisikopoulos, V.: Practical polytope volume approximation. ACM Trans. Math. Soft. **44**(4), 38:1–38:21 (2018). https://doi.org/10.1145/3194656. http://doi.acm.org/10.1145/3194656, Prelim. version: Proc. SoCG 2014
11. Gover, E., Krikorian, N.: Determinants and the volumes of parallelotopes and zonotopes. Linear Algebra Appl. **413**, 28–40 (2010)

12. Kopetzki, A., Schürmann, B., Althoff, M.: Methods for order reduction of zonotopes. In: Proceedings of the IEEE Annual Conference on Decision & Control (CDC), pp. 5626–5633 (2017). https://doi.org/10.1109/CDC.2017.8264508
13. Lovász, L., Kannan, R., Simonovits, M.: Random walks and an $O^*(n^5)$ volume algorithm for convex bodies. Random Struct. Algor. **11**, 1–50 (1997)
14. Lovász, L., Vempala, S.: Simulated annealing in convex bodies and an $O^*(n^4)$ volume algorithms. J. Comput. Syst. Sci. **72**, 392–417 (2006)
15. McMullen, P.: On zonotopes. Trans. Am. Math. Soc. **159**, 91–109 (1971). http://www.jstor.org/stable/1996000
16. Pereira, A., Althoff, M.: Safety control of robots under computed torque control using reachable sets. In: Proceedings of the IEEE International Conference on Robotics Automation (ICRA), pp. 331–338 (2015). https://doi.org/10.1109/ICRA.2015.7139020
17. Polyak, B., Gryazina, E.: Billiard walk - a new sampling algorithm for control and optimization. IFAC Proc. Vol. **47**(3), 6123–6128 (2014). https://doi.org/10.3182/20140824-6-ZA-1003.02312. 19th IFACWorld Congress
18. Stinson, K., Gleich, D.F., Constantine, P.G.: A randomized algorithm for enumerating zonotope vertices. CoRR abs/1602.06620 (2016). http://arxiv.org/abs/1602.06620
19. Weibel, C.: Implementation and parallelization of a reverse-search algorithm for Minkowski sums. In: ALENEX (2010)
20. Ziegler, G.: Lectures on Polytopes. Springer, New York (1995). https://doi.org/10.1007/978-1-4613-8431-1

Slack Ideals in Macaulay2

Antonio Macchia$^{(\boxtimes)}$ and Amy Wiebe

Fachbereich Mathematik und Informatik, Freie Universität Berlin,
Arnimallee 2, 14195 Berlin, Germany
macchia.antonello@gmail.com, w.amy.math@gmail.com

Abstract. Recently Gouveia, Thomas and the authors introduced the slack realization space, a new model for the realization space of a polytope. It represents each polytope by its slack matrix, the matrix obtained by evaluating each facet inequality at each vertex. Unlike the classical model, the slack model naturally mods out projective transformations. It is inherently algebraic, arising as the positive part of a variety of a saturated determinantal ideal, and provides a new computational tool to study classical realizability problems for polytopes. We introduce the package SlackIdeals for *Macaulay2*, that provides methods for creating and manipulating slack matrices and slack ideals of convex polytopes and matroids. Slack ideals are often difficult to compute. To improve the power of the slack model, we develop two strategies to simplify computations: we scale as many entries of the slack matrix as possible to one; we then obtain a reduced slack model combining the slack variety with the more compact Grassmannian realization space model. This allows us to study slack ideals that were previously out of computational reach. As applications, we show that the well-known Perles polytope does not admit rational realizations and prove the non-realizability of a large simplicial sphere.

Keywords: Polytopes · Slack matrices · Slack ideals · Matroids

1 Introduction

Slack matrices of polytopes are nonnegative real matrices whose entries express the slack of a vertex in a facet inequality. In particular, the zero pattern of a slack matrix encodes the vertex-facet incidence structure of the polytope. Slack matrices have found remarkable use in the theory of extended formulations of polytopes: Yannakakis [10] proved that the extension complexity of a polytope is equal to the nonnegative rank of its slack matrix.

More generally, one can define the slack matrix of a matroid by computing the slacks of the ground set vectors in the hyperplanes of the matroid.

If P is d-dimensional polytope, replacing all positive entries in the slack matrix with distinct variables, one obtains a new sparse generic matrix $S_P(\boldsymbol{x})$,

A. Macchia—Supported by the Einstein Foundation Berlin under Francisco Santos grant EVF-2015-230.

A. M. Bigatti et al. (Eds.): ICMS 2020, LNCS 12097, pp. 222–231, 2020.
https://doi.org/10.1007/978-3-030-52200-1_22

called the *symbolic slack matrix* of P. Then we define the *slack ideal* I_P of P as the ideal of all $(d+2)$-minors of $S_P(\boldsymbol{x})$, saturated with respect to the product of all variables in $S_P(\boldsymbol{x})$.

Slack ideals were introduced for polytopes in [5], where it was also noted that they could be used to model the realization space of a polytope. The details of this realization space model and further properties of the slack ideal were studied in [2,3] and [4]. An analogous realization space model for matroids was introduced in [1].

In this paper, we describe the Macaulay2 [6] package SlackIdeals.m2, that is available at https://bitbucket.org/macchia/slackideals/src/master/SlackIdeals. m2. It provides methods to define and manipulate slack matrices of polytopes, matroids, polyhedra, and cones; obtain a slack matrix directly from the Gale transform of a polytope; compute the symbolic slack matrix and the slack ideal from a slack matrix; compute the graphic ideal of a polytope, the cycle ideal and the universal ideal of a matroid.

Slack ideal computations are often out of computational reach. Therefore we develop two techniques to speed up and simplify computations. First, we suitably set to one as many entries of the slack matrix as possible. One can compute the slack ideal of this dehomogenized slack matrix and then rehomogenize the resulting ideal (see Proposition 1). The new ideal coincides with the original slack ideal if the latter is radical. Second, we obtain a reduced slack matrix by keeping the columns of a set of facets F that contains a flag (a maximal chain in the face lattice of P) and such that the facets not in F are simplicial. Combining these two strategies, we have a powerful tool for the study of hard realizability questions. As applications, we show that the well-known Perles polytope does not admit rational realizations and prove the non-realizability of a large simplicial sphere.

2   Slack Matrices and Slack Ideals

Given a collection of points $V = \{\boldsymbol{v}_1, \ldots, \boldsymbol{v}_n\} \subset \mathbb{R}^d$ and a collection of (affine) hyperplanes $H = \{\{\boldsymbol{x} \in \mathbb{R}^d : b_i - \boldsymbol{\alpha}_i^\top \boldsymbol{x} = 0\} : i = 1 \ldots f\}$ we can define a *slack matrix of the pair* (V, H) by

$$
S_{(V,H)} = \begin{bmatrix} \mathbb{1} & \boldsymbol{v}_1 \\ \vdots & \vdots \\ \mathbb{1} & \boldsymbol{v}_n \end{bmatrix} \begin{bmatrix} b_1 & \cdots & b_f \\ \boldsymbol{\alpha}_1 & \cdots & \boldsymbol{\alpha}_f \end{bmatrix} \in \mathbb{R}^{n \times f}.
$$

If P is a d-polytope, we take $V = \mathrm{vert}(P)$ and H to be the set of facet defining hyperplanes. Then $S_P = S_{(V,H)}$. When coordinates V are given for the vectors of a matroid M, they are always assumed to be an affine configuration which gets homogenized to form the matroid; in particular, this means that if $V = \mathrm{vert}(P)$, then the associated matroid is the matroid of the polytope P. The hyperplanes are taken to be all hyperplanes of M, and then $S_M = S_{(V,H)}$.

```
i1 : needsPackage "SlackIdeals";
i2 : V = {{0,0},{0,1},{1,1},{1,0}};
-- Compute the slack matrix of P=conv(V)
i3 : slackMatrix(V)
o3 = | 0 1 0 1 |
     | 1 0 0 1 |
     | 0 1 1 0 |
     | 1 0 1 0 |
-- Compute the slack matrix of matroid of V
i4 : slackMatrix(V, Object=>"matroid")
o4 = | -1 -1  0 -1  0  0  |
     | -1  0  1  0  1  0  |
     |  0  1  1  0  0 -1  |
     |  0  0  0 -1 -1 -1  |
```

The `slackMatrix` command also takes a pre-computed matroid, polyhedron or cone object as input.

Another way to compute the slack matrix of a polytope is from its Gale transform using the command `slackFromGaleCircuits`. Let G be a matrix with real entries whose columns are the vectors of a Gale transform of a polytope P. A slack matrix of P is computed by finding the minimal positive circuits of G, see [7, Section 5.4]. Alternatively, the command `slackFromGalePlucker` applies the maps of [4, Section 5] to fill a slack matrix with Plücker coordinates of the Gale transform.

The slack matrices of a few specific polytopes and matroids of theoretical importance are built-in, using the command `specificSlackMatrix`.

The *symbolic slack matrix* can be obtained by replacing the nonzero entries of a slack matrix by distinct variables; that is,

$$[S_{(V,H)}(\boldsymbol{x})]_{i,j} = \begin{cases} 0 & \text{if } \boldsymbol{v}_i \in H_j \\ x_{i,j} & \text{if } \boldsymbol{v}_i \notin H_j \end{cases}.$$

From this sparse generic matrix we obtain the *slack ideal* as the saturation of the ideal of its $(d+2)$-minors by the product of all variables in $S_{(V,H)}(\boldsymbol{x})$:

$$I_{(V,H)} = \langle (d+2) - \text{minors of } S_{(V,H)}(\boldsymbol{x}) \rangle : \left(\prod_{j=1}^{f} \prod_{i:v_i \notin H_j} x_{i,j} \right)^{\infty}.$$

Given a (symbolic) slack matrix of a d-polytope, $(d+1)$-dimensional cone, or rank $d+1$ matroid, we can compute the associated slack ideal, specifying d as an input. Unless we pass variable names as an option, the function labels the variables consecutively by rows with a single index starting from 1:

```
-- Compute slack ideal of d-polytope P=conv(V)
i10 : V = {{0,0},{0,1},{1,1},{1,0}};
i11 : slackIdeal(2, slackMatrix(V)) -- here d=2
o11 = ideal(x x x x  - x x x x )
            1 4 6 7    2 3 5 8
```

We get the same result if we compute `slackIdeal(2,V)`, giving only the list of vertices of a d-polytope or ground set vectors of a matroid instead of a slack matrix. We also get the same result with `slackIdeal(V)`, but the computation is faster if you provide d as an argument. As optional argument, one can choose the object to be set as `"polytope"`, `"cone"`, or `"matroid"` (default is `Object=>"polytope"`).

To a polytope or matroid we can also associate a specific toric ideal, known as the *graphic* or *cycle ideal*, respectively. These ideals are important in the classification of certain projectively unique polytopes [3] and matroids [1], and can be computed using the commands `graphicIdeal` and `cycleIdeal`.

In [4, Section 4] it is shown that a slack matrix can be filled with Plücker coordinates of a matrix formed from the vertex coordinates of a polytope (or extreme ray generators of a cone or ground set vectors of a matroid). This idea is the basis for the reduction technique described in [4, Section 6] and Sect. 4. The Grassmannian section ideal of a polytope is also defined and shown to cut out exactly a set of representatives of the slack variety that are constructed in this way [4, Section 4.1]. The command `grassmannSectionIdeal` computes this section ideal given a set of vertices of a polytope and the indices of vertices that span each facet.

3   On the Dehomogenization of the Slack Ideal

Let P be a polytope and S_P its slack matrix. We define the *non-incidence graph* G_P as the bipartite graph whose vertices are the vertices and facets of P, and whose edges are the vertex-facet pairs of P such that the vertex is not on the facet. This graphic structure provides a systematic way to scale a maximal number of entries in S_P to 1, as spelled out in [3, Lemma 5.2]. In particular, we may scale the rows and columns of $S_P(x)$ so that it has ones in the entries indexed by the edges in a maximal spanning forest of the graph G_P. This can be done using `setOnesForest`, which outputs a sequence (Y, F) where Y is the scaled symbolic slack matrix and F is the spanning forest used to scale Y.

```
i23 : V = {{0,0,0},{1,0,0},{0,1,0},{0,0,1},{1,0,1},{1,1,0}};
i24 : (Y, F) = setOnesForest(X); Y
o24 = | 0    0 1 0 1   |
      | 1    0 1 0 0   |
      | 0    1 0 0 x_5 |
      | x_6  1 0 0 0   |
      | 0    0 1 1 0   |
      | 0    1 0 1 0   |
```

This leads to a dehomogenized version of the slack ideal defined as follows. Given S_P and a maximal spanning forest F of G_P, let $S_P(x^F)$ be the symbolic slack matrix of P with all the variables corresponding to edges in F set to 1. Then the dehomogenized ideal, I_P^F, is the slack ideal of this scaled slack matrix:

$$I_P^F := \langle (d+2) - \text{minors of } S_P(x^F) \rangle : \left(\prod x^F \right)^\infty .$$

It is natural to ask what is the relation between I_P^F and the original slack ideal I_P. In particular, we might wish to know if we can recover the full slack ideal from I_P^F. From [3, Lemma 5.2] we know that any slack matrix in $\mathcal{V}(I_P)$ (or, in fact, any point in the slack variety with all coordinates that correspond to F being nonzero) can be scaled to a matrix in $\mathcal{V}(I_P^F)$. Conversely, it is clear that any point in $\mathcal{V}(I_P^F)$ can be thought of as a point in $\mathcal{V}(I_P)$. Thus, in terms of the varieties we have $\mathcal{V}(I_P)^* / (\mathbb{R}^v \times \mathbb{R}^f) \cong \mathcal{V}(I_P^F)^*$, where $\mathcal{V}(I)^*$ denotes the part of the variety where all coordinates are nonzero.

To see the algebraic implications of this, let us introduce the following rehomogenization process. Notice that in the proof of [3, Lemma 5.2], we dehomogenize by following the edges of forest F starting from some chosen root(s) and moving toward the leaves. The destination vertex of each edge tells us which row or column to scale, and the edge label is the variable by which we scale. Now, given a polynomial in I_P^F, using the same forest and orientation we proceed in the reverse order: starting at the leaves, for each edge of the forest, we reintroduce the variable corresponding to it in order to rehomogenize the polynomial with respect to the row or column corresponding to the destination vertex of that edge.

Example 1. Consider the slack matrix $S_P(\boldsymbol{x}^F)$ of the triangular prism P scaled according to forest F, pictured in Fig. 1. Then $I_P^F =$ $\langle x_8 - 1, x_{12} - 1 \rangle$. So we can rehomogenize, for example, the element $x_8 - x_{12}$ with respect to forest F as follows.

$$S_P(\boldsymbol{x}^F) = \begin{bmatrix} 0 & 1 & 0 & 0 & 1 \\ 1 & 0 & 0 & 0 & 1 \\ 0 & 1 & 1 & 0 & 0 \\ 1 & 0 & x_8 & 0 & 0 \\ 0 & 1 & 0 & 1 & 0 \\ 1 & 0 & 0 & x_{12} & 0 \end{bmatrix}$$

First, consider the leaf corresponding to column 3. Its edge is labeled with x_6, so we reintroduce that variable to the monomial x_{12} since its degree in column 3 is currently 0, while the degree of x_8 in that column is 1. We continue this process until all the edges of F have been used.

Call the resulting ideal $H(I_P^F)$. By the tree structure, the rehomogenization process does indeed end with a polynomial that is homogeneous, as once we make it homogeneous for a row or column we never add variables in that row or column again. We now consider the effect of this rehomogenization on minors.

Lemma 1. *Let p be a minor of $S_P(\boldsymbol{x})$ and p^F its dehomogenization by F. Then its rehomogenization $H(p^F)$ equals p divided by the product of all variables in F that divide p.*

Proof. Note that all monomials in a minor have degree precisely one on every relevant row and column. In fact they can be interpreted as perfect matchings on the subgraph of G_P corresponding to the $(d+2) \times (d+2)$ submatrix being considered. Let $\boldsymbol{x}^a$ and $\boldsymbol{x}^b$ be two distinct monomials in the minor, then their dehomogenizations are also distinct. To see this, note that if we interpret $\boldsymbol{a}$ and $\boldsymbol{b}$ as matchings, a common dehomogenization would be a common submatching c of both, with all the remaining edges being in F. But $\boldsymbol{a} \setminus c$ and $\boldsymbol{b} \setminus c$ would then be distinct matchings on the same set of variables, hence their union contains a cycle, so they would not be both contained in the forest F.

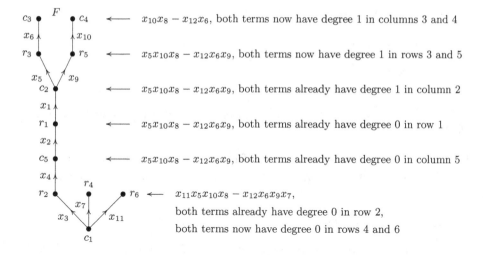

c_3 • $\overset{F}{}$ • c_4 ⟵ $x_{10}x_8 - x_{12}x_6$, both terms now have degree 1 in columns 3 and 4

x_6 ↟ ↟ x_{10}

r_3 • • r_5 ⟵ $x_5x_{10}x_8 - x_{12}x_6x_9$, both terms now have degree 1 in rows 3 and 5

x_5 ↘ ↗ x_9
c_2 • ⟵ $x_5x_{10}x_8 - x_{12}x_6x_9$, both terms already have degree 1 in column 2

x_1 ↟

r_1 • ⟵ $x_5x_{10}x_8 - x_{12}x_6x_9$, both terms already have degree 0 in row 1

x_2 ↟

c_5 • ⟵ $x_5x_{10}x_8 - x_{12}x_6x_9$, both terms already have degree 0 in column 5

x_4 ↟ r_4
r_2 • • • r_6 ⟵ $x_{11}x_5x_{10}x_8 - x_{12}x_6x_9x_7$,

x_3 ↘ ↟ x_7 ↗ x_{11} both terms already have degree 0 in row 2,

c_1 both terms now have degree 0 in rows 4 and 6

Fig. 1. A spanning forest for the triangular prism

Now note that when rehomogenizing a minor, we start with all degrees being zero or one for every row and column, and since we visit each node (corresponding to each of the rows/columns) exactly once by the tree structure, the degree of every row and column is at most one after homogenizing. In the first step of rehomogenizing, we start with a leaf of F, which means the variable x_i labeling its edge is the only variable in the row or column corresponding to that leaf which was set to 1. Thus if any monomial of the minor has degree zero on that row or column, it must be because x_i occurred in that monomial in the original minor.

Hence rehomogenizing will just add that variable to the monomials where it was originally present, with the exception of the case where it was present on all monomials, in which case there will be no need to add it, as the dehomogenized polynomial would be homogeneous (of degree 0) for that particular row/column.

All degrees remain 0 or 1 after this process, and now the node incident to the leaf we just rehomogenized corresponds to a row/column with exactly one variable that is still dehomogenized. Thus we can repeat the argument on the entire forest to find that each monomial rehomogenizes to itself divided by the variables that were originally present in all monomials of the minor.

Remark 1. It is important to note that $H(I_P^F)$ is the ideal of *all elements* of I_P^F rehomogenized. In general, this is different from the ideal generated by the rehomogenized generators of I_P^F. In the package, we rehomogenize the whole ideal by rehomogenizing the generators and saturating the resulting ideal by all the variables we just homogenized by.

For example, let V be the set of vertices of the triangular prism with spanning forest Y as computed before, and let us compute the rehomogenized ideal $H(I_P^F)$.

```
i25 : HIF = rehomogenizeIdeal(3, Y, F)

o25 = ideal (x x x x   - x x x x  , x x x x   - x x x x ,
             3 6 9 10    2 7 8 11   0 5 9 10    1 4 8 11
        x x x x  - x x x x )
        1 3 4 6    0 2 5 7
```

Notice that, in this case the rehomogenized ideal $H(I_P^F)$ equals the slack ideal I_P.

Example 2. Recall that the generators of I_P^F for the triangular prism were $x_8 - 1$ and $x_{12} - 1$, which rehomogenize to $x_2x_3x_5x_8 - x_1x_4x_6x_7$ and $x_2x_3x_9x_{12} - x_1x_4x_{10}x_{11}$, respectively. However,

$$\langle x_2x_3x_5x_8 - x_1x_4x_6x_7, x_2x_3x_9x_{12} - x_1x_4x_{10}x_{11} \rangle \neq H(I_P^F).$$

The relation between the rehomogenized ideal $H(I_P^F)$ and the original slack ideal is given in the following lemma. The proof relies on the key fact that the variety of the rehomogenized ideal is still the same as the slack variety that we started with.

Proposition 1. *Given a spanning forest F for the non-incidence graph of polytope P, the rehomogenization of its scaled slack ideal is an intermediate ideal between the slack ideal and its radical: $I_P \subseteq H(I_P^F) \subseteq \sqrt{I_P}$.*

Proof. To prove the inclusion $I_P \subseteq H(I_P^F)$, note that $p \in I_P$ happens if and only if $x^a p \in J$ for some exponent vector a, where J is the ideal generated by all $(d + 2)$-minors of the symbolic slack matrix of P. Dehomogenizing we get $x^b p^F \in J^F$, which means p^F is in the saturation of J^F by the product of all variables, which is precisely the definition of I_P^F. From Lemma 1 it follows that $p \in H(I_P^F)$.

To prove that $H(I_P^F) \subseteq \sqrt{I_P}$, it is enough to show that any polynomial in $H(I_P^F)$ vanishes in the slack variety. By construction, any such polynomial must vanish on the points of the slack variety where the variables corresponding to the forest F are nonzero, $\mathcal{V}(I_P)\backslash\mathcal{V}(\langle x^F \rangle)$. Thus, they vanish on the Zariski closure of that set. Considering the following containments,

$$\mathcal{V}(I_P)\backslash\mathcal{V}(\langle x \rangle) \subset \mathcal{V}(I_P)\backslash\mathcal{V}(\langle x^F \rangle) \subset \mathcal{V}(I_P),$$

we get that this closure is exactly the slack variety since $\overline{\mathcal{V}(I_P)\backslash\mathcal{V}(\langle x \rangle)} = \mathcal{V}(I_P : \langle x \rangle^\infty) = \mathcal{V}(I_P)$.

Remark 2. One would like to say that $I_P = H(I_P^F)$, and so far we have no counterexample for this equality, since it always holds if I_P is radical, and we also have no examples of non-radical slack ideals.

4 Reduced Slack Matrices

In general, computing the slack ideal may take a long time or be infeasible, especially if the dimension of the polytope is small compared to its number of vertices and facets. In some cases we can speed up this computation combining the slack and the Grassmannian realization space models [4, Section 6]. In fact, we do not need to work with the full slack matrix, since the essential information is contained into a sufficiently large submatrix.

We will see in Examples 3 and 4, that slack ideals which we were not even able to compute (using personal computers) are now able to be calculated in a matter of a few seconds. To give an estimate of the improvement, computing the slack ideal of the full slack matrix in Example 3 requires the computation of about $8.6 \cdot 10^9$ minors, whereas the reduced slack ideal only requires the computation of about $1.9 \cdot 10^4$ minors.

More precisely, let P be a realizable polytope and F be a set of facets of P such that F contains a set of facets that can be intersected to form a flag in the face lattice of P and all facets of P not in F are simplicial. We call a *reduced slack matrix* for P the submatrix, S_F, of S_P consisting of only the columns indexed by F. Set V_F to be the nonzero part of the slack variety $\mathcal{V}(I_F)$.

If $\overline{V_F}$ is irreducible, then $V_F \times \mathbb{C}^h \cong \mathcal{V}(I_P)^*$ are birationally equivalent, where h denotes the number of facets of P outside F [4, Proposition 6.9].

Example 3. Let P be the Perles projectively unique polytope with no rational realization coming from the point configuration in [7, Figure 5.5.1, p. 93]. This is an 8-polytope with 12 vertices and 34 facet and its symbolic slack matrix $S_P(x)$ is a 12×34 matrix with 120 variables.

Let S_F be the following submatrix of S_P whose 13 columns correspond to all the nonsimplicial facets of P:

```
i28 : S = specificSlackMatrix("perles1");
-- Checking that the first 13 columns of S indeed contain a flag
i29 : containsFlag(toList(0..12),S)
o29 = true
i30 : SF = reducedSlackMatrix(8, S, FlagIndices=>toList(0..12));
```

The associated symbolic slack matrix is:

$$S_F(x) = \begin{bmatrix} 0 & 0 & 0 & x_1 & x_2 & x_3 & 0 & 0 & 0 & 0 & 0 & 0 & 0 \\ 0 & 0 & 0 & x_4 & 0 & 0 & x_5 & x_6 & x_7 & 0 & 0 & 0 & 0 \\ 0 & 0 & 0 & 0 & 0 & 0 & x_8 & 0 & 0 & x_9 & x_{10} & 0 & 0 \\ 0 & 0 & 0 & 0 & x_{11} & 0 & 0 & 0 & 0 & 0 & 0 & x_{12} & x_{13} \\ 0 & 0 & 0 & 0 & 0 & 0 & 0 & x_{14} & 0 & x_{15} & 0 & x_{16} & 0 \\ x_{17} & 0 & 0 & 0 & 0 & 0 & 0 & 0 & 0 & 0 & x_{18} & 0 & x_{19} \\ 0 & x_{20} & 0 & 0 & 0 & 0 & 0 & 0 & x_{21} & 0 & 0 & 0 & 0 \\ 0 & 0 & x_{22} & 0 & 0 & x_{23} & 0 & 0 & 0 & 0 & 0 & 0 & 0 \\ x_{24} & 0 & 0 & x_{25} & 0 & 0 & 0 & x_{26} & 0 & 0 & 0 & 0 & 0 \\ 0 & x_{27} & 0 & 0 & x_{28} & 0 & 0 & 0 & 0 & x_{29} & 0 & 0 & 0 \\ 0 & 0 & x_{30} & 0 & 0 & 0 & x_{31} & 0 & 0 & 0 & 0 & 0 & x_{32} \\ 0 & 0 & 0 & 0 & 0 & x_{33} & 0 & 0 & x_{34} & 0 & x_{35} & x_{36} & 0 \end{bmatrix}.$$

Using [3, Lemma 5.2], we first set $x_i = 1$ for $i = 1, 4, 5, 6, 7, 8, 9, 10, 13, 15,$ $16, 17, 18, 21, 22, 26, 27, 28, 29, 30, 31, 32, 33, 35$. The resulting scaled reduced slack ideal is:

$$\langle x_{36}^2 + x_{36} - 1, x_{34} - x_{36} - 1, x_{25} - x_{36}, x_{24} - x_{36}, x_{23} - 1, x_{20} - x_{36},$$
$$x_{19} - x_{36}, x_{14} - x_{36} - 1, x_{12} - x_{36}, x_{11} - 1, x_3 - 1, x_2 - x_{36} - 1 \rangle.$$

It follows that $x_{36} = \frac{-1 \pm \sqrt{5}}{2}$. Hence, P does not admit rational realizations.

Example 4. Let P be the following 3-dimensional simplicial sphere, constructed by Jockusch [8] and studied by Novik and Zheng [9], with 12 vertices labeled by $1, 2, \ldots, 6$ and $-1, -2, \ldots, -6$, and with the following 48 facets:

$$\{1, 2, 5, 6\}, \{2, 3, 5, 6\}, \{3, 4, 5, 6\}, \{-1, -2, 5, 6\}, \{-2, -3, 5, 6\}, \{-3, -4, 5, 6\},$$
$$\{1, -4, 5, 6\}, \{1, -4, -5, 6\}, \{-1, 4, 5, 6\}, \{1, 2, 3, 5\}, \{1, 2, 4, 6\}, \{2, 3, 4, 6\},$$
$$\{-1, -2, 3, 5\}, \{-1, -2, 4, 6\}, \{-2, -3, 4, 6\}, \{1, -2, 3, 5\}, \{1, -3, 4, 6\}, \{2, 3, 4, -5\},$$
$$\{3, 4, 5, -6\}, \{-1, 2, 4, -5\}, \{-1, 3, 5, -6\}, \{1, 2, -3, 4\}, \{1, 2, 3, -4\}, \{1, -2, 3, -4\}.$$

The remaining 24 facets are antipodes of the above ones, i.e., they are of the form $\{-x, -y, -z, -t\}$ for each $\{x, y, z, t\}$ from the above list.

This sphere, denoted by $\Delta_6^{3,2}$ in [9], is centrally symmetric, and is not realizable as the boundary complexes of a centrally symmetric polytope. However, it is not known whether it is realizable as (a non-centrally symmetric) polytope [9, Problem 6.1].

The symbolic slack matrix $S_P(\boldsymbol{x})$ is a 12×48 matrix with 384 variables. A reduced slack matrix (where facets $1, 3, 4, 5, 7$ form a flag) is the matrix $S_F(\boldsymbol{x})$ below, where vertices $1, 3, 5, 6, 9$ form a flag of $S_F(\boldsymbol{x})^\top$. Since row 10 of $S_F(\boldsymbol{x})$ contains four zeros, we can further reduce the above matrix and scale some of the entries according to [3, Lemma 5.2], obtaining the matrix $S_G(\boldsymbol{x})$.

$$S_F(\boldsymbol{x}) = \begin{bmatrix} x_1 & x_2 & x_3 & x_4 & x_5 & x_6 & x_7 \\ 0 & 0 & x_8 & x_9 & x_{10} & x_{11} & x_{12} \\ x_{13} & x_{14} & 0 & 0 & x_{15} & x_{16} & x_{17} \\ 0 & x_{18} & x_{19} & x_{20} & 0 & x_{21} & x_{22} \\ x_{23} & x_{24} & 0 & 0 & 0 & 0 & 0 \\ x_{25} & x_{26} & 0 & x_{27} & x_{28} & 0 & 0 \\ 0 & x_{29} & x_{30} & 0 & x_{31} & x_{32} & x_{33} \\ 0 & 0 & x_{34} & x_{35} & x_{36} & x_{37} & x_{38} \\ x_{39} & 0 & 0 & x_{40} & x_{41} & 0 & x_{42} \\ x_{43} & 0 & x_{44} & x_{45} & x_{46} & x_{47} & x_{48} \\ x_{49} & x_{50} & x_{51} & 0 & 0 & 0 & 0 \\ x_{52} & x_{53} & x_{54} & x_{55} & 0 & x_{56} & 0 \end{bmatrix}$$

$$S_G(\boldsymbol{x}) = \begin{bmatrix} x_1 & x_2 & x_3 & x_4 & x_5 & x_6 & 1 \\ 0 & 0 & 1 & 1 & 1 & 1 & 1 \\ x_{13} & 1 & 0 & 0 & x_{15} & x_{16} & 1 \\ 0 & x_{18} & x_{19} & x_{20} & 0 & x_{21} & 1 \\ 1 & x_{24} & 0 & 0 & 0 & 0 & 0 \\ x_{25} & x_{26} & 0 & 1 & x_{28} & 0 & 0 \\ 0 & x_{29} & x_{30} & 0 & x_{31} & x_{32} & 1 \\ 0 & 0 & x_{34} & x_{35} & x_{36} & x_{37} & 1 \\ x_{39} & 0 & 0 & x_{40} & x_{41} & 0 & 1 \\ 1 & 0 & x_{44} & x_{45} & x_{46} & x_{47} & 1 \\ x_{52} & x_{53} & x_{54} & x_{55} & 0 & 1 & 0 \end{bmatrix}$$

We then reconstruct row 10 by applying the map **GrV** defined in [4, Section 4.1] to $S_G(\boldsymbol{x})^\top$:

```
-- We denote by symbSG the symbolic slack matrix S_G(x) above
i31 : reconstructSlackMatrix(transpose symbSG, {{3,4,5,6}})
o31 = .Macaulay2/local/share/Macaulay2/SlackIdeals.m2:1405:44:(3):
      [4]: error: Cannot extend matrix
```

The above error means that in reconstructing row 10, we get more than four zero entries. Computing explicitly the map $\mathbf{GrV}$, we can see that five entries are zero. This shows that P is not realizable as a polytope.

The previous example shows that the reduction process can be a powerful tool to show nonrealizability of large simplicial spheres.

References

1. Brandt, M., Wiebe, A.: The slack realization space of a matroid. Algebr. Comb. **2**(4), 663–681 (2019)
2. Gouveia, J., Macchia, A., Thomas, R., Wiebe, A.: The slack realization space of a polytope. SIAM J. Discrete Math. **33**(3), 1637–1653 (2019)
3. Gouveia, J., Macchia, A., Thomas, R., Wiebe, A.: Projectively unique polytopes and toric slack ideals. J. Pure Appl. Algebra **224**(5), 14 (2020)
4. Gouveia, J., Macchia, A., Wiebe, A.: Combining realization space models of polytopes (2020). Preprint https://arxiv.org/abs/2001.11999
5. Gouveia, J., Pashkovich, K., Robinson, R., Thomas, R.: Four-dimensional polytopes of minimum positive semidefinite rank. J. Comb. Theory Ser. A **145**, 184–226 (2017)
6. Grayson, D., Stillman, M.: Macaulay 2, a software system for research in algebraic geometry. http://www.math.uiuc.edu/Macaulay2/
7. Grünbaum, B.: Convex Polytopes. Graduate Texts in Mathematics, vol. 221, 2nd edn. Springer, New York (2003). https://doi.org/10.1007/978-1-4613-0019-9
8. Jockusch, W.: An infinite family of nearly neighborly centrally symmetric 3-spheres. J. Comb. Theory Ser. A **72**(2), 318–321 (1995)
9. Novik, I., Zheng, H.: Highly neighborly centrally symmetric spheres (2019). Preprint https://arxiv.org/abs/1907.06115
10. Yannakakis, M.: Expressing combinatorial optimization problems by linear programs. J. Comput. Syst. Sci. **43**(3), 441–466 (1991)

Hyperplane Arrangements in polymake

Lars Kastner and Marta Panizzut[(⊠)]

Chair of Discrete Mathematics/Geometry, Technische Universität Berlin,
Straße des 17. Juni 136, 10623 Berlin, Germany
{kastner,panizzut}@math.tu-berlin.de

Abstract. Hyperplane arrangements form the latest addition to the
zoo of combinatorial objects dealt with by polymake. We report on
their implementation and on a algorithm to compute the associated cell
decomposition. The implemented algorithm performs significantly better
than brute force alternatives, as it requires fewer convex hulls computations. The implementation is included in polymake since release 4.0.

Keywords: Hyperplane arrangements · Cell decomposition

1 Introduction

Hyperplane arrangements are ubiquitous objects appearing in different areas
of mathematics such as discrete geometry, algebraic combinatorics and algebraic
geometry. A common theme is to understand the combinatorics and the topology
of the cells in the complement of the arrangement. Combinatorics and its connections to other areas of mathematics are the focus of the software framework
polymake [GJ00], hence hyperplane arrangements form an almost mandatory
addition to the objects available. We will discuss the implementation, such as
the datatypes and properties, as well as some basic algorithms for analyzing
hyperplane arrangements.

One of the main advantages of polymake are its various interfaces to other
software. This allows keeping the codebase slim, while using powerful software
developed by experts from other fields. Still polymake provides basic algorithms
for many tasks, in case other software is not available. Hence the idea of the
hyperplane arrangements is to provide a datatype with basic functionality as a
basis for future interfaces to other software, e.g. to **ZRAM** [Brü+99] for computing the cell decomposition from the hyperplanes. Nevertheless, the polymake
implementation of hyperplane arrangements comes with a basic algorithm for
computing the associated cell decomposition that performs significantly better
than brute force alternatives. Thus, we will discuss the main ideas of this algorithm in this article as well.

The combinatorics of hyperplane arrangements in real space is linked to zonotopes. Each arrangement endows the support space with a fan structure which

Research by L. Kastner is supported by Deutsche Forschungsgemeinschaft (SFB-TRR
195: "Symbolic Tools in Mathematics and their Application").

A. M. Bigatti et al. (Eds.): ICMS 2020, LNCS 12097, pp. 232–240, 2020.
https://doi.org/10.1007/978-3-030-52200-1_23

is the normal fan of a zonotope. Each hyperplane subdivides the space in two halfspaces. Therefore we can encode relative positions of points with respect to the arrangement. In other words, hyperplane arrangements are examples of (oriented) matroids. Moreover, the hyperplanes in an arrangement can be seen as mirrors hyperplanes of a reflection group.

An interesting application is in Geometric Invariant Theory. GIT constructs quotients of algebraic varieties modulo group actions. The quotients depend on the choice of a linearized ample line bundle. Variation of geometric invariant theory quotients studies how quotients vary when changing the line bundle. Under some hypothesis the classes of equivalent quotients are convex subsets, called chambers. The walls among chambers are defined by certain hyperplane arrangements, see [DH98, Example 3.3.24].

2 Main Definitions

We begin with the basic definitions in the theory of hyperplane arrangements following our implementation in `polymake`.

Definition 1. *A hyperplane arrangement* $H = (H, \mathcal{S}_H)$ *in* $\mathbb{R}^d$ *is given by the following data:*

1. *a finite set of linear forms encoding hyperplanes* $H = \{h \in \mathbb{R}^d \setminus \{0\}\}$ *and*
2. *a polyhedral cone* $\mathcal{S}_H \subseteq \mathbb{R}^d$ *which we call the* support cone.

Given a hyperplane arrangement H*, the induced fan* Σ_H *is a fan with support* $\mathcal{S}_H$ *given by subdividing* $\mathcal{S}_H$ *along all* $\{x \in \mathbb{R}^d \,|\, \langle x, h \rangle = 0\}$ *for* $h \in H$*.*

Every hyperplane in the arrangement subdivides the space into two halfspaces

$$h^+ := \{x \in \mathbb{R}^d \,|\, \langle x, h \rangle > 0\} \text{ and } h^- := \{x \in \mathbb{R}^d \,|\, \langle x, h \rangle < 0\}.$$

We remark that in the definition we allow duplicate hyperplanes, but from each hyperplane arrangement we can construct a reduced one. Let H be a hyperplane arrangement given by the hyperplanes $\{h_1, h_2, \ldots, h_n\}$. The *reduced hyperplane arrangement* H_{red} has the same support cone as H and $h_i \in H_{\mathrm{red}}$ if and only if $h_i \neq \lambda b$, for any $\lambda \in \mathbb{R}$ and any $b \in \{h_1, \ldots, h_{i-1}\}$.

To a hyperplane arrangement $H = \{h_1, \ldots, h_n\} \subseteq \mathbb{R}^d$ we associate the polytope

$$\mathcal{Z}_H := \sum_{i=1}^{n} [-h_i, h_i] + \mathcal{S}_H^\vee,$$

the Minkowski sum of all the line segments $[-h_i, h_i]$ and the dual support cone $\mathcal{S}_H^\vee$. If $\mathcal{S}_H = \mathbb{R}^d$, then $\mathcal{S}_H^\vee = 0$ and $\mathcal{Z}_H$ is a zonotope.

Remark 1. Often hyperplane arrangements are defined without a support cone, i.e. only for the case $\mathcal{S}_H = \mathbb{R}^d$. The connection between intersecting $\Sigma_H \cap \mathcal{S}_H$ is done via taking the Minkowski sum $\mathcal{Z}_H + \mathcal{S}_H^\vee$ on the dual side. The main ingredient is the fact that

$$(\sigma + \tau)^\vee = \sigma^\vee \cap \tau^\vee$$

holds for two cones σ and τ.

Proposition 2.1 *[Zie95, Thm. 7.16] The fan Σ_H is the normal fan of $\mathcal{Z}_H$.*

Definition 2. *To a maximal cone $\sigma \in \Sigma_H$ we associate its signature, which is a set $\mathrm{sig}(\sigma) := \{i \in \{1, \ldots, n\} \mid \sigma \subseteq h_i^+\}$.*

Remark 2. In `polymake` release 4.0 the signature was defined as the set of indices such that $\sigma \subseteq \overline{h_i^-}$. The signature will be automatically updated for data saved in `polymake` 4.0 and loaded in the subsequent releases.

Example 1. Let H be given by

$$H = \{(0,1), (1,1), (-2,1)\} \subseteq \mathbb{R}^2.$$

We will have a look at the induced fans for different support cones $\mathcal{S}_H$. The fan Σ_H and the polytope $\mathcal{Z}_H$ are visualized in Fig. 1 for varying $\mathcal{S}_H$.

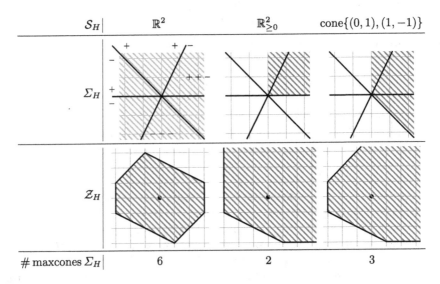

Fig. 1. Visualization of Σ_H and $\mathcal{Z}_H$ for Example 1

In each of the pictures, the support cone is indicated as the shaded area. The structure of the fan Σ_H depends heavily on the support cone $\mathcal{S}_H$. In particular, it is possible for hyperplanes to only intersect $\mathcal{S}_H$ trivially and thereby becoming irrelevant for Σ_H. Thus, one may loose information when going from H to Σ_H.

The labels at the hyperplanes in the first picture indicate which side constitutes h^+, h^- respectively. Using these one can read of the signatures of the single cells, for example the cell σ generated by the rays $(1,0)$ and $(1,2)$ has signature $\mathrm{sig}(\sigma) = \{1, 2\}$.

2.1   Affine Hyperplane Arrangements

An affine hyperplane arrangement is usually given by a finite set of affine hyperplanes:

$$H_{\text{aff}} := \{[a, b] \in \mathbb{R}^d \times \mathbb{R}\}.$$

The whole space $\mathbb{R}^d$ is then subdivided along the hyperplanes

$$\{x \in \mathbb{R}^d \mid \langle a, x \rangle = b\}, \text{ for all } [a, b] \in H,$$

resulting in a polyhedral complex $\mathcal{PC}_{H_{\text{aff}}} \subseteq \mathbb{R}^d$.

Analogously to the connection between polytopes and cones, or polyhedral complexes and fans, every affine hyperplane arrangement gives rise to a (projective) hyperplane arrangement by embedding it at height 1:

$$H_{\text{proj}} := \{[-b, a] \mid [a, b] \in H\}.$$

If we intersect the fan $\Sigma_{H_{\text{proj}}}$ with the affine hyperplane $[x_0 = 1] \subseteq \mathbb{R}^{d+1}$, the resulting polyhedral complex is isomorphic to $\mathcal{PC}_{H_{\text{aff}}}$, via the embedding $\mathbb{R}^d \to \mathbb{R}^{d+1}$, $x \mapsto [1, x]$.

The support cone allows one to deal with affine hyperplanes computationally. Set

$$\mathcal{S}_{H_{\text{proj}}} := \{[x_0, x_1, \dots, x_d] \in \mathbb{R}^{d+1} \mid x_0 \geq 0\},$$

then the maximal cones of $\Sigma_{H_{\text{proj}}}$ are in one-to-one correspondence with the maximal cells of $\mathcal{PC}_{H_{\text{aff}}}$. In particular, `polymake` can interpret $\Sigma_{H_{\text{proj}}}$ as a polyhedral complex via the embedding mentioned above, and this polyhedral complex will be exactly $\mathcal{PC}_{H_{\text{aff}}}$.

Example 2. As a simple example, choose the following hyperplanes in $\mathbb{R}^1$:

$$x_1 = -1, \, x_1 = 0, \, x_1 = 2.$$

The associated hyperplanes of H_{proj} in $\mathbb{R}^2$ are exactly those of the hyperplane arrangement from Example 1. For $\mathcal{S}_H$ we choose the cone $\{x_0 \geq 0\}$, then H_{aff} will be at height one.

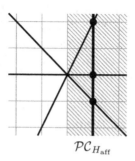

$$\mathcal{PC}_{H_{\text{aff}}}$$

The induced affine hyperplane arrangement is indicated by the dots and thick line. It is one dimensional and the associated polyhedral complex $\mathcal{PC}_{H_{\text{aff}}}$ has four maximal cells.

Example 3. The following is an example of code in `polymake`.

```
fan > $HA = new HyperplaneArrangement(HYPERPLANES
    =>[[0,1],[1,1],[-2,1]],"SUPPORT.INEQUALITIES"=>[[1,0]]);
fan > $HA->CELL_DECOMPOSITION->RAYS; # Force computation
fan > $pc = new PolyhedralComplex($HA->CELL_DECOMPOSITION);
fan > print "(".join("),(",@{$pc->VERTICES}).")\n";
(0 -1),(0 1),(1 -1),(1 0),(1 2)
fan > print join(",",@{$pc->MAXIMAL_POLYTOPES})."\n";
{0 2},{1 4},{2 3},{3 4}
```

3   Implementation

Hyperplane arrangements are implemented in the software `polymake` as a new object `HyperplaneArrangement`, which is derived from the already existing object `VectorConfiguration`. We augment the existing properties of `VectorConfiguration` with the following properties and methods.

1. `HYPERPLANES` A matrix encoding the hyperplanes as rows, this is just an override of the property `VECTORS` of `VectorConfiguration`
2. `SUPPORT` A polymake `Cone`, denoting the support $\mathcal{S}_H$.
3. `CELL_DECOMPOSITION` A polymake `PolyhedralFan`, the cell decomposition Σ_H.
4. `CELL_SIGNATURES` A `Array<Set<Int>>`, the i-th set in the array contains the indices of hyperplanes evaluating positively on the i-th maximal cone of `CELL_DECOMPOSITION`.
5. `signature_to_cell` Given a signature as `Set<Int>`, get the maximal cone with this signature, if it exists.
6. `cell_to_signature` Given a cell, a maximal cone of `CELL_DECOMPOSITION`, determine its signature.

3.1   Cell Decomposition Algorithm

Given $H = \{h_1, \ldots, h_n\}$, we want to compute the subdivision of $\mathcal{S}_H$ induced by the hyperplanes, the induced fan Σ_H. This means, we want to find all the rays and maximal cones of Σ_H. In terms of the zonotope $\mathcal{Z}_H$, this is equivalent to knowing the facets and vertices of $\mathcal{Z}_H$, see [Fuk04, GS93]. The facet directions of $\mathcal{Z}_H$ are the rays of Σ_H. For very vertex of $\mathcal{Z}_H$ we get a maximal cone by determining which facets contain it.

The brute force approach is to loop over all possible signatures in $s \in 2^{\{1,\ldots,n\}}$ and for every signature s to build the cone

$$\bigcap_{i \in s} \overline{h_i^+} \cap \bigcap_{i \notin s} \overline{h_i^-} \cap \mathcal{S}_H.$$

For comparing the different algorithms, we count the number of times they have to perform a convex hull computation for converting a signature to a cone. There are 2^n signatures, so we have to perform 2^n convex hull computations. As we saw in Example 1, it can happen that some hyperplanes are irrelevant, either completely or just for single cells. Furthermore, in Example 1 the fan Σ_H had at most six maximal cones, however we would have to compute eight intersections with the brute force approach regardless.

Remark 3. This brute force approach is in some ways parallel to the brute force approach for computing the Minkowski sum making up $\mathcal{Z}_H$, by taking considering all possible sums of the endpoints of the line segments. One arrives at 2^n points whose convex hull is $\mathcal{Z}_H$. There are several ways to go on: either attempt a massive convex hull computation directly, or check each point individually whether it is a vertex.

Our approach is to first find a full-dimensional cone σ of Σ_H and then to flip hyperplanes in order to compute its neighbors. First take a facet f of σ, then set

$$\text{sig}' := (\text{sig}(\sigma) \setminus \{i \in \text{sig}(\sigma) \mid h_i \| f\}) \cup \{i \notin \text{sig}(\sigma) \mid h_i \| f\}),$$

where $h_i \| f$ denotes that h_i and f are parallel. This is the signature of the cell neighboring σ at facet f, so we can use it to determine the rays of the neighboring cell. Finding the neighbors of a cell allows one to traverse the dual graph of the fan Σ_H. Taking the support cone $\mathcal{S}_H$ into account just requires some minor tweaks, like ignoring facets of σ that are also facets of $\mathcal{S}_H$. By storing signatures one can avoid recomputation of cones.

To find a starting cone, one selects a generic point x from $\mathcal{S}_H$. A generic point will be contained in a maximal cone, this maximal cone will be

$$\sigma(x) := \bigcap_{i \mid x \in h_i^+} \overline{h_i^+} \cap \bigcap_{i \mid x \in h_i^-} \overline{h_i^-}.$$

The point x may be contained in some hyperplanes, but these hyperplanes are exactly those that also contain the entire $\mathcal{S}_H$. Using this approach we would do one convex hull computation per maximal cone, arriving at $\#\text{maxcones}(\Sigma_H)$ convex hull computations.

Remark 4. As the fan Σ_H is polytopal, there is a reverse search structure on it, corresponding to the edge graph of the zonotope $\mathcal{Z}_H$. This has already been exploited by Sleumer in [Sle99] using the software framework [Brü+99]. Reverse search allows for different kinds of parallelisation and it would be interesting to study the performance of budgeted reverse search [AJ18, AJ16] on this particular problem. Note that the dual problem, finding the vertices of $\mathcal{Z}_H$, is equally hard, as it is a Minkowski sum with potentially many summands. We refer the reader to [GS93] for a detailed analysis.

3.2   Sample Code

We conclude with a few examples which illustrate the object `HyperplaneArrangement` and its properties. Example 6 reports the comparison between the running times of the new algorithm implemented in `polymake` and the brute force algorithm to compute cell decompositions.

Example 4. The following examples compute the $4! = 24$ cells in the Coxeter arrangement of type A3. The 6 linear hyperplanes in the arrangements are

$$x_i - x_j = 0, \quad 1 \leq i < j \leq 4.$$

```
fan > $A3 = new HyperplaneArrangement(HYPERPLANES=>
   root_system("A3")->VECTORS->minor(All,~[0]));
fan > $CDA3 = $A3->CELL_DECOMPOSITION;
fan > print $CDA3->N_MAXIMAL_CONES;
24
```

Now we compute the 36 cells in the Linial arrangement [PS00] given by the 6 affine hyperplanes

$$x_i - x_j = 1, \quad 1 \leq i < j \leq 4.$$

As explained in Sect. 2.1, the support cone allows us to deal with affine hyperplanes. We transform the hyperplanes $[a, b] \in \mathbb{R}^4 \times \mathbb{R}$ in the projective arrangement H_{proj} with hyperplanes $[-b, a]$ and then we intersect the latter with the support cone $\mathcal{S}_{H_{\mathrm{proj}}} := \{[x_0, x_1, \ldots, x_5] \in \mathbb{R}^5 \mid x_0 \geq 0\}$ (Fig. 2).

```
fan > $Hyps = new Matrix([[-1,1,-1,0,0],[-1,1,0,-1,0],
[-1,1,0,0,-1],[-1,0,1,-1,0],[-1,0,1,0,-1],[-1,0,0,1,-1]]);
fan > $Lin = new HyperplaneArrangement(HYPERPLANES=>$Hyps,
 "SUPPORT.INEQUALITIES"=>[[1,0,0,0,0]]);
fan > $CDLin = $Lin->CELL_DECOMPOSITION;
fan > print $CDLin->N_MAXIMAL_CONES;
36
```

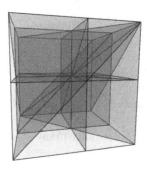

Fig. 2. The arrangement of type A3.

Example 5. This example is based on [Süß19]. Let X be a del Pezzo surface of degree 5 and $[K_X]$ the class of the canonical divisor. The cone of effective divisors $\overline{\mathrm{Eff}}(X)$ is spanned by ten exceptional curves $[C_{ij}]$ indexed by $0 \leq i < j \leq 4$ and characterized by $[C_{ij}]^2 = -1$ and $[C_{ij}] \cdot [K_Y] = -1$. Applying the change of basis $[C_{ij}] = b_i + b_j$, described in [Süß19, Section 3], we see that the polytope P given by points in $\overline{\mathrm{Eff}}(X)$ intersecting $[K_X]$ with multiplicity -1 coincides with the hypersimplex $\Delta(2, 5)$. In the aforementioned article the author considers the cell decomposition of P induced by the hyperplane arrangement defined by

$$\{[D] \in \overline{\mathrm{Eff}}(X) \mid [D] \cdot [C_{ij}] = 0\}.$$

The decomposition is used to study the toric topology of the Grassmannian of planes in complex 5-dimensional space.

The following code allows one to compute the cell decomposition in `polymake`. We first compute the new pairing in the new basis $b_0, b_1, \ldots, b_4$.

```
polytope > $pairing = new Matrix(1/4*ones_matrix(5,5));
polytope > $pairing->row(4) *= -1;
polytope > $pairing->col(4) *= -1;
polytope > for(my $i=0; $i<4; $i++){ $pairing->elem($i,$i) =
    -3/4; }
```

We then introduce the support cone given by the hypersimplex $\Delta(2, 5)$ in the new basis

```
polytope > $R = hypersimplex(2,5)->VERTICES;
polytope > $Z = zero_vector(10);
polytope > $M = hypersimplex(2,5)->VERTICES->minor(All,~[0]);
polytope > $M = $M * $pairing;
polytope > $H = $Z|$M;
```

Finally, we can compute the cell decomposition.

```
fan > $HA = new HyperplaneArrangement(HYPERPLANES=>$H,
"SUPPORT.INPUT_RAYS"=>$R);
fan > print $HA->CELL_DECOMPOSITION->N_RAYS;
15
fan > print $HA->CELL_DECOMPOSITION->N_MAXIMAL_CONES;
27
```

Example 6. Let H be the hyperplane arrangement in $\mathbb{R}^d$ given by the $2^d - 1$ hyperplanes normal to 0/1-vectors. The number of maximal cones in Σ_H are known up to $d = 8$, see entry A034997 in the Online Encyclopedia of Integer Sequences. We run `polymake` implementations of the BFS algorithm described above and the brute force alternative. Our results are reported in Table 1, where we can see that the BFS algorithm performs better than the brute force approach.

Table 1. Results and runtimes for arrangements in Example 6

d	# hyperplanes	# rays	# maximal cones	Time BFS (s)	Time brute force (s)
2	3	6	6	0.10	0.1
3	7	18	32	0.40	1.2
4	15	90	370	4.98	324.1
5	31	1250	11292	209.19	–
6	63	57750	1066044	40517.84	–

References

[AJ18] Avis, D., Jordan, C.: mplrs: a scalable parallel vertex/facet enumeration code. Math. Program. Comput. **10**(2), 267–302 (2017). https://doi.org/10.1007/s12532-017-0129-y

[AJ16] Avis, D., Jordan, C.: A parallel framework for reverse search using mts. Preprint arXiv:1610.07735 (2016)

[Brü+99] Brüngger, A., Marzetta, A., Fukuda, K., Nievergelt, J.: The parallel search bench ZRAM and its applications. Ann. Oper. Res. **90**, 45–63 (1999). https://doi.org/10.1023/A:1018972901171

[DH98] Dolgachev, I.V., Hu, Y.: Variation of geometric invariant theory quotients. Inst. Hautes Études Sci. Publ. Math. **87**, 5–51 (1998). https://doi.org/10.1007/BF02698859. With an appendix by Nicolas Ressayre

[Fuk04] Fukuda, K.: From the zonotope construction to the Minkowski addition of convex polytopes. J. Symb. Comput. **38**(4), 1261–1272 (2004)

[GJ00] Gawrilow, E., Joswig, M.: polymake: a framework for analyzing convex polytopes. In: Kalai, G., Ziegler, G.M. (eds.) Polytopes Combinatorics and Computation (Oberwolfach, 1997). DMV Seminar, vol. 29, pp. 43–73. Birkhäuser, Basel (2000). https://doi.org/10.1007/978-3-0348-8438-9_2

[GS93] Gritzmann, P., Sturmfels, B.: Minkowski addition of polytopes: computational complexity and applications to Gröbner bases. SIAM J. Discrete Math. **6**(2), 246–269 (1993)

[PS00] Postnikov, A., Stanley, R.: Deformations of Coxeter hyperplane arrangements. J. Comb. Theory Ser. A **91**(1–2), 544–597 (2000)

[Sle99] Sleumer, N.H.: Output-sensitive cell enumeration in hyperplane arrangements. Nord. J. Comput. **6**(2), 137–147 (1999)

[Süß19] Süß, H.: Toric topology of the Grassmannian of planes in C5 and the del Pezzo surface of degree 5. arXiv e-prints arxiv.org/abs/1904.13301 (2019)

[Zie95] Ziegler, G.M.: Lectures on Polytopes. GTM, vol. 152. Springer, New York (1995). https://doi.org/10.1007/978-1-4613-8431-1. pp. ix + 370

A Convex Programming Approach
to Solve Posynomial Systems

Marianne Akian, Xavier Allamigeon, Marin Boyet$^{(\boxtimes)}$, and Stéphane Gaubert

INRIA and CMAP, École polytechnique, IP Paris, CNRS, Palaiseau, France
{marianne.akian,xavier.allamigeon,marin.boyet,stephane.gaubert}@inria.fr

Abstract. We exhibit a class of classical or tropical posynomial systems which can be solved by reduction to linear or convex programming problems. This relies on a notion of colorful vectors with respect to a collection of Newton polytopes. This extends the convex programming approach of one player stochastic games.

1 Introduction

A *posynomial* is a function of the form

$$P(x) = \sum_{a \in A} c_a x_1^{a_1} x_2^{a_2} \cdots x_n^{a_n}$$

where the variable $x = (x_1, \ldots, x_n)$ is a vector with real positive entries, A is a finite subset of vectors of $\mathbb{R}^n$, and the c_a are positive real numbers. Here for any $a \in \mathbb{R}^n$, we denote by a_i the i-th coordinate of a. The set A is called the *support* of P, also denoted by S_P, its elements are called the *exponents* of the posynomial and the c_a its *coefficients*.

Unlike polynomials, posynomials can have arbitrary exponents. They arise in convex optimization, especially in geometric and entropic programming [6] and in polynomial optimization [7]. They also arise in the theory of nonnegative tensors [8,11], in risk sensitive control [3] and game theory [1].

A *tropical posynomial* is a function of the form

$$P^{\text{trop}}(x) = \max_{a \in A} (c_a + \langle a, x \rangle)$$

where $\langle \cdot, \cdot \rangle$ is the usual dot product of $\mathbb{R}^n$, the c_a are now real coefficients, and $x = (x_1, \ldots, x_n)$ can take its values in $\mathbb{R}^n$. The terminology used comes from the *tropical* (or max-plus) *semi-field*, whose additive law is the maximum and the multiplicative law is the usual sum.

In this paper, we are interested in solving (square) classical posynomial systems, that are of the form

$$P_i(x) = 1 \quad \text{for all } i \in [n] := \{1, \ldots, n\} \tag{1}$$

© Springer Nature Switzerland AG 2020
A. M. Bigatti et al. (Eds.): ICMS 2020, LNCS 12097, pp. 241–250, 2020.
https://doi.org/10.1007/978-3-030-52200-1_24

with $x \in (\mathbb{R}_{>0})^n$, and the P_i are classical posynomials. We will also study the tropical counterpart,

$$P_i^{\text{trop}}(x) = 0 \quad \text{for all } i \in [n] \tag{2}$$

with now $x \in \mathbb{R}^n$, and the P_i^{trop} are tropical posynomials (hereafter we shall write P_i instead of P_i^{trop}, for brevity). The optimality equations of Markov decision processes [13] are special cases of tropical posynomial systems. More general tropical posynomial systems arise in the performance analysis of timed discrete event systems, see [2].

Solving (square) posynomial systems is in general NP-hard (Sect. 2). However, we identify a tractable subclass. The tropical version can be solved exactly in polynomial time by reduction to a linear program (Sect. 3), whereas the classical version can be solved approximately by reduction to a geometric program (Sect. 4). Our approach is based on a notion of colorful interior of a collection of cones. A point is in the colorful interior if it is a positive linear combination of vectors of these cones, and if at least one vector of every cone is needed in such a linear combination. Our reductions are valid when the colorful interior of the cone generated by the supports of the posynomials is nonempty, and when a point in this interior is known. As special cases, we recover the linear programming formulation of Markov decision processes, and the geometric programming formulation of risk sensitive problems. Properties of the colorful interior and related open problems are discussed in Sect. 5.

2   Solving Posynomial Systems Is NP-hard

The following two results show that the feasibility problems for classical or tropical posynomial systems are NP-hard, even with integer exponents.

Proposition 1. *Solving a square tropical posynomial system is NP-hard.*

Proof. We reduce 3-SAT to the problem (2). Let us consider a Boolean formula in conjunctive normal form $C_1 \wedge \cdots \wedge C_p$ made of p clauses, each one of them using three out of n real variables $x_1, \ldots, x_n$ $(p, n \in \mathbb{N})$.

We introduce the following tropical posynomial system in the $2n + 2p$ variables $(x_1, \ldots, x_n, y_1, \ldots, y_n, z_1 \ldots, z_p, s_1, \ldots, s_p)$, with the same number of equations:

$$\forall i \in [n] \quad \max(x_i - 1, y_i - 1) = 0, \qquad\qquad x_i + y_i - 1 = 0,$$

$$\forall j \in [p] \quad \max\left(\max_{x_i \in C_j} (x_i - z_j), \max_{\neg x_i \in C_j} (y_i - z_j) \right) = 0, \quad \max(\tfrac{1}{2} - z_j, s_j - z_j) = 0.$$

This system can be constructed in polynomial time from the Boolean formula. The first $2n$ equations ensure that for all $i \in [n]$, $x_i \in \{0, 1\}$ and that x_i and y_i have opposite logical values. The next p equations express that for all $j \in [p]$, the variable z_j has the same Boolean value as the clause C_j, with the notation $x_i \in C_j$ (resp. $\neg x_i \in C_j$) if the variable x_i occurs positively (resp. negatively) in the clause C_j. The last equations ensure that $z_j = 1$ for all $j \in [p]$. The instance $C_1 \wedge \cdots \wedge C_p$ is satisfiable if and only if this system admits a solution. □

Theorem 2. *Solving a square classical posynomial system is NP-hard.*

Proof. We modify the previous construction to obtain a square posynomial system over $\mathbb{R}_{>0}^{2n+2p}$, along the lines of Maslov's dequantization principle [12] or Viro's method [14]:

$$\forall i \in [n] \quad \tfrac{2}{5}x_i + \tfrac{2}{5}y_i = 1, \qquad\qquad x_i y_i = 1,$$

$$\forall j \in [p] \quad \sum_{x_i \in C_j} \tfrac{1}{6}x_i z_j^{-1} + \sum_{\neg x_i \in C_j} \tfrac{1}{6}y_i z_j^{-1} = 1 \qquad \tfrac{1}{3}z_j^{-1} + s_j z_j^{-1} = 1.$$

From the first $2n$ equations, the variables x_i and y_i range over $\{2, 1/2\}$, the values 2 and $1/2$ respectively encode the true and false Boolean values. The variable $y_i = 1/x_i$ corresponds to the Boolean negation of x_i. Since each clause has precisely three literals, using the p next equations, we deduce that the variable z_j takes one of the values $\{1/2, 3/4, 1\}$ if the clause C_j is satisfied, and that it takes the value $1/4$ otherwise. The last p equations impose that z_j can take any value in $(1/3, \infty)$. We deduce that the formula $C_1 \wedge \cdots \wedge C_p$ is satisfied if and only if the posynomial system that we have obtained in this way admits a solution in $\mathbb{R}_{>0}^{2n+2p}$. □

3 A Linear Programming Approach to Solve Tropical Posynomial Systems

Given tropical posynomials $P_1, \ldots, P_n$, we write the system (2) as $P(x) = 0$, where $P := (P_1, \ldots, P_n)$. The *support* of this system, denoted $\mathbf{S}$, is defined as the disjoint union $\biguplus_{i \in [n]} S_{P_i}$ of the supports of the posynomials P_i. By *disjoint union*, we mean the coproduct in the category of sets (these supports may have non-empty intersections, and they may even coincide).

Definition 1. We say that a vector y in the (convex) conic hull cone($\mathbf{S}$) is *colorful* if, for all $\mu \in (\mathbb{R}_{\geqslant 0})^{\mathbf{S}}$,

$$y = \sum_{a \in \mathbf{S}} \mu_a\, a \implies \forall i \in [n],\ \exists a \in S_{P_i},\ \mu_a > 0.$$

In other words, a vector $y \in \mathbb{R}^n$ is colorful if it arises as a nonnegative combination of the exponents of P, but also if all such combinations make use of at least one exponent of each of the tropical posynomials $P_1, \ldots, P_n$.

In this way, if we think of $S_{P_1}, \ldots, S_{P_n}$ as colored sets, we need all the colors to decompose a colorful vector y over these. Moreover, by Carathéodory's theorem, every vector in the conic hull cone($\mathbf{S}$) can be written as a positive linear combination of an independent family of vectors of $\mathbf{S}$. Hence, when y is a colorful vector, it is obtained as a positive linear combination of precisely one vector a_i in each color class S_{P_i}, and the family $a_1, \ldots, a_n$ must be a basis. (If not, Carathéodory's Theorem would imply that y is a positive linear combination of a proper subset of $\{a_1, \ldots, a_n\}$, so that y could not be a colorful vector.)

Given a vector y, we consider the following linear program:

$$\text{Maximize} \quad \langle y, x \rangle \quad \text{subject to} \quad \forall a \in \mathbf{S}, \ c_a + \langle a, x \rangle \leqslant 0. \quad \text{(LP}(y))$$

Remark that the feasibility set of this linear program consists of the vectors $x \in \mathbb{R}^n$ satisfying $P(x) \leqslant 0$. In other words, it can be thought of as a relaxation of the system $P(x) = 0$. The following theorem shows that this relaxation provides a solution of $P(x) = 0$ if y is a colorful vector.

Theorem 3. *Assume that y is a colorful vector, and that the linear program (LP(y)) is feasible. Then, the linear program (LP(y)) has an optimal solution, and any optimal solution x satisfies $P(x) = 0$.*

Proof. Since the feasibility set of (LP(y)), $\mathcal{F} := \{x \in \mathbb{R}^n : P(x) \leqslant 0\}$, is nonempty, we can consider its recession cone, which is given by $\mathcal{C} = \{x \in \mathbb{R}^n : \forall a \in \mathbf{S}, \langle a, x \rangle \leqslant 0\}$. As a colorful vector, y belongs to the polyhedral cone generated by the vectors $a \in \mathbf{S}$, so $\langle y, x \rangle \leqslant 0$ for all $x \in \mathcal{C}$. By the Minkowski–Weyl theorem, $\mathcal{F}$ is a Minkowski sum of the form $\mathcal{P} + \mathcal{C}$ where $\mathcal{P}$ is a polytope, i.e., every feasible point x can be written as $x = x' + x''$ with $x' \in \mathcal{P}$ and $x'' \in \mathcal{C}$. Since $\langle y, x'' \rangle \leqslant 0$, the maximum of the objective function $x \mapsto \langle y, x \rangle$ over the polyhedron $\mathcal{F}$ is attained (by an element of $\mathcal{P}$).

Let $x^\star \in \mathbb{R}^n$ be an optimal solution of (LP(y)). From the strong duality theorem, the dual linear program admits an optimal solution $(\mu_a^\star)_{a \in \mathbf{S}} \in (\mathbb{R}_{\geqslant 0})^{\mathbf{S}}$ which satisfies $y = \sum_{a \in \mathbf{S}} \mu_a^\star a$ and $\mu_a^\star(c_a + \langle a, x^\star \rangle) = 0$ for all $a \in \mathbf{S}$. Since y is a colorful vector, for all $i \in [n]$, there is some $a_i \in S_{P_i}$ such that $\mu_{a_i}^\star > 0$. We then get that, for all $i \in [n]$, $P_i(x^\star) \geqslant c_{a_i} + \langle a_i, x^\star \rangle = 0$. As a result, $P(x^\star) = 0$. □

We next provide a geometric condition ensuring that the linear program (LP(y)) is feasible regardless of the coefficients c_a. We say that the tropical posynomial function P has *pointed* exponents if its support is contained in an open halfspace, i.e. there exists $z \in \mathbb{R}^n$ such that $\forall a \in \mathbf{S}, \langle a, z \rangle < 0$. Our interest for pointed systems comes from the following property:

Proposition 4. *The inequality problem $P(x) \leqslant 0$ has a solution $x \in \mathbb{R}^n$ regardless of the coefficients of P if and only if P has pointed exponents.*

Proof. Suppose that for all values of $(c_a)_{a \in \mathbf{S}}$, there exists $x \in \mathbb{R}^n$ such that $P(x) \leqslant 0$. By choosing $c_a \equiv 1$, there exists $x_0 \in \mathbb{R}^n$ that satisfies $\forall a \in \mathbf{S}$, $1 + \langle a, x_0 \rangle \leqslant 0$. Hence, for all $i \in [n]$, the exponents of P_i lie in the open halfspace $\{a \in \mathbb{R}^n \mid \langle a, x_0 \rangle < 0\}$.

Suppose now that P has pointed exponents. Then there is some $z \in \mathbb{R}^n$ such that for all $a \in \mathbf{S}$, we have $\langle a, z \rangle < 0$. We define $\lambda := \max_{a \in \mathbf{S}} (-c_a)/\langle a, z \rangle$ so that $\forall a \in \mathbf{S}, c_a + \langle a, \lambda z \rangle \leqslant 0$ and therefore for all $i \in [n]$, $P_i(\lambda z) \leqslant 0$. □

As a consequence of Theorem 3 and Proposition 4, if the tropical posynomial system $P(x) = 0$ has pointed exponents and there exists a colorful vector, then the system admits a solution which can be found by linear programming.

A remarkable special case consists of Markov decision processes. In this framework, the set $[n]$ represents the state space, and at each state $i \in [n]$, a player has a finite set B_i of available actions included in the n-dimensional simplex $\{p \in \mathbb{R}_{\geqslant 0}^n : \sum_{j=1}^n p_j \leqslant 1\}$. If $p \in B_i$, p_j stands for the probability that the next state is j, given that the current state is i and action p is chosen by the player, so the difference $1 - \sum_{j=1}^n p_j$ is the death probability in state i when this action is picked. To each action p is attached a reward $c_p \in \mathbb{R}$. Given an initial state $i \in [n]$, one looks for the value $v_i \in \mathbb{R}$, which is defined as the maximum over all the strategies of the expectation of the sum of rewards up to the death time, we refer the reader to [13] for background. The value vector $v = (v_i)_{i \in [n]}$ is solution of the tropical posynomial problem

$$v_i = \max_{p \in B_i} (c_p + \langle p, v \rangle), \quad \forall i \in [n].$$

This reduces to the form (2) with $S_{P_i} := B_i - e_i$, where e_i denotes the i-th element of the canonical basis of $\mathbb{R}^n$. We say that a Markov decision process is of *discounted type* if for every state $i \in [n]$ there is at least one action $p \in B_j$ such that $\sum_{j=1}^n p_j < 1$.

Proposition 5. *If a Markov decision process is of discounted type, then any negative vector is colorful with respect to the associated posynomial system.*

Thus, we recover the linear programming approach to Markov decision processes (see [13]), showing that the value is obtained by minimizing the function $v \mapsto \sum_{i \in [n]} v_i$ subject to the constraints $v_i \geqslant c_p + \langle p, v \rangle$ for $i \in [n]$ and $p \in B_i$.

4 Geometric Programming Approach of Posynomials Systems

We refer the reader to [6] for background on geometric programming.

Given a collection $P = (P_1, \ldots, P_n)$ of classical posynomials, we now deal with the system $P_i(x) = 1$ for all $i \in [n]$, which, for brevity, we denote by $P(x) = 1$. We keep the notation of Sect. 3 for the supports of the posynomials. Moreover, the definitions of colorful vectors and pointed exponents, which only depend on these supports, still make sense in the setting of this section.

Lemma 6. *If y is a colorful vector, the polyhedron $\mathcal{P}$ defined by*

$$\mathcal{P} := \{x \in \mathbb{R}^n : \forall a \in \mathbf{S}, \quad \log c_a + \langle a, x \rangle \leqslant 0 \quad \text{and} \quad \langle y, x \rangle \geqslant \mu\}$$

is bounded (possibly empty), regardless of our choice of positive $(c_a)_{a \in \mathbf{S}}$ or $\mu \in \mathbb{R}$.

Proof. If $\mathcal{P}$ is nonempty, let $\mathcal{C} := \{x \in \mathbb{R}^n \mid \forall a \in \mathbf{S}, \langle a, x \rangle \leqslant 0, \langle y, x \rangle \geqslant 0\}$ denote its recession cone, and let $x \in \mathcal{C}$. Since y is a colorful vector, there exists $(\lambda_1, \ldots, \lambda_n) \in \mathbb{R}_{>0}^n$ and a basis $(a_1, \ldots, a_n) \in \prod_{i \in [n]} S_{P_i}$ such that $y = \sum_{i=1}^n \lambda_i a_i$. Thus, $\langle y, x \rangle \leqslant 0$, and so $\langle y, x \rangle = \sum_{i=1}^n \lambda_i \langle a_i, x \rangle = 0$. As a consequence, since $\lambda_i > 0$ for all $i \in [n]$, $\langle a_i, x \rangle = 0$. Since $(a_1, \ldots, a_n)$ is a basis, we get $x = 0$. Thus, $\mathcal{C} = \{0\}$, and $\mathcal{P}$ is bounded by Minkowski–Weyl Theorem. $\qquad\square$

Given $X \in \mathbb{R}^n$, we denote by $\exp X$ the vector with entries $\exp X_i$, $i \in [n]$.

Theorem 7. *Let $P(x) = 1$ be a posynomial system with pointed exponents, and y be a colorful vector. Then, the system has a solution $x = \exp X^* \in (\mathbb{R}_{>0})^n$, where X^* is an arbitrary solution of the following geometric program:*

$$\text{Maximize} \quad \langle y, X \rangle \qquad \text{subject to} \qquad \forall i \in [n] \quad g_i(X) \leqslant 0, \qquad \text{(G)}$$

where $g_i(X) := \log \left(\sum_{a \in S_{P_i}} c_a \, e^{\langle a, X \rangle} \right)$.

Proof. For $x \in \mathbb{R}_{>0}^n$, we define $X = \log(x)$ (component-wise) so that $P(x) = 1$ is equivalent to solving $g_i(X) = 0$ for all $i \in [n]$. By Hölder's inequality, the functions $(g_i)_{i \in [n]}$ are convex. We define $h_i \colon X \mapsto \max_{a \in S_{P_i}} \left(\log(c_a) + \langle a, X \rangle \right)$ for $i \in [n]$ and we observe that $h_i(X) \leqslant g_i(X) \leqslant h_i(X) + \log(|S_{P_i}|)$.

Since the system $P(x) = 1$ has pointed exponents, by Proposition 4, the polyhedron $\{X \in \mathbb{R}^n \colon \forall i \in [n], \ h_i(X) + \log(|S_{P_i}|) \leqslant 0\}$ is nonempty. A fortiori, the feasible set of (G) is nonempty.

Let us now prove that the maximum of (G) is finite and attained, by proving that the μ-superlevel set $\mathcal{S}_\mu = \{X \in \mathbb{R}^n \colon \langle y, X \rangle \geqslant \mu \text{ and } \forall i \in [n], \ g_i(X) \leqslant 0\}$ of the objective function (included in the feasible set) is compact for all $\mu \in \mathbb{R}$. Closedness is direct, and we observe that for $\mu \in \mathbb{R}$, $\mathcal{S}_\mu \subset \{X \in \mathbb{R}^n \colon \langle y, X \rangle \geqslant \mu \text{ and } \forall i \in [n], \ h_i(X) \leqslant 0\}$, but by Lemma 6, this polyhedron is bounded. Hence, (G) admits an optimal solution $X^\star$.

Furthermore, again by Proposition 4, there exists $\overline{X}$ such that for all $i \in [n]$, $h_i(\overline{X}) + \log(|S_{P_i}|) + 1 \leqslant 0$. Therefore, for all $i \in [n]$, $g_i(\overline{X}) < 0$, which means that (G) satisfies Slater's condition. Problem (G) being convex, optimality of $X^\star$ is characterized by the Karush–Kuhn–Tucker conditions (see [4]). Hence, there is a vector of nonnegative multipliers $\lambda^\star = (\lambda_1^\star, \ldots, \lambda_n^\star)$ such that $(X^\star, \lambda^\star)$ is a stationary point of the Lagrangian of (G), and the complementarity slackness conditions hold, i.e. for all $i \in [n]$, $\lambda_i^\star g_i(X^\star) = 0$. Defining $Z_i := \sum_{a \in S_{P_i}} c_a e^{\langle a, X^\star \rangle} > 0$ for $i \in [n]$, the stationarity conditions give

$$y = \sum_{i=1}^n \frac{\lambda_i^\star}{Z_i} \sum_{a \in S_{P_i}} c_a \, e^{\langle a, X^\star \rangle} \, a.$$

Since y is colorful, for all $i \in [n]$, $\lambda_i^\star > 0$. The complementarity slackness conditions yield $g_i(X^\star) = 0$ for all $i \in [n]$. So $x^\star := \exp(X^\star)$ satisfies $P(x^\star) = 1$. $\qquad \square$

5   Properties of the Colorful Interior of Convex Sets

Theorems 3 and 7 rely on the existence of a colorful vector. The purpose of this section is to study the properties of the set of such vectors. In fact, colorful vectors can be defined more generally from a family of n closed convex cones.

Definition 2. Let $\mathcal{C} = (C_1, \ldots, C_n)$ be a collection of n closed convex cones of $\mathbb{R}^n$. A vector $y \in \mathbb{R}^n$ is said to be *colorful* if it belongs to the set

$$\mathrm{cone}(C_1 \cup \cdots \cup C_n) \setminus \bigcup_{i \in [n]} \mathrm{cone}\Big(\bigcup_{j \neq i} C_j\Big).$$

The latter set is referred to as the *colorful interior* of $\mathcal{C}$.

Remark that Definition 1 can be recovered by taking $C_i := \mathrm{cone}(S_{P_i})$ for all $i \in [n]$. In what follows, we restrict to the case where the collection $\mathcal{C}$ is *pointed*, i.e. $\mathrm{cone}(C_1 \cup \cdots \cup C_n)$ is a pointed cone (in the non pointed case, the colorful interior enjoys much less structure than the one proved in Theorem 10, in particular it may not even be connected). Suppose that $\{x \in \mathbb{R}^n : \langle z, x \rangle > 0\}$ is an open halfspace containing the $(C_i)_{i \in [n]}$. Then, as a cone, the colorful interior of $\mathcal{C}$ can be more simply studied from its cross-section with $\{x \in \mathbb{R}^n : \langle x, z \rangle = 1\}$. The latter can be shown to coincide with the set

$$\mathrm{conv}(S_1 \cup \cdots \cup S_n) \setminus \bigcup_{i \in [n]} \mathrm{conv}\Big(\bigcup_{j \neq i} S_j\Big) \tag{3}$$

where for $i \in [n]$, S_i is the cross-section of the cone C_i by $\{x \in \mathbb{R}^n : \langle x, z \rangle = 1\}$. Given a collection $\mathcal{S} = (S_1, \ldots, S_n)$ of closed convex sets of $\mathbb{R}^{n-1}$, we refer to the set (3) as the *colorful interior* of $\mathcal{S}$, and denote it by $\mathrm{colint}\,\mathcal{S}$. We start with a lemma justifying the terminology we have chosen:

Lemma 8. *Let $\mathcal{S} = (S_1, \ldots, S_n)$ be a collection of n closed convex sets of $\mathbb{R}^{n-1}$. Then $\mathrm{colint}\,\mathcal{S}$ is an open set included in $\mathrm{int}\,\mathrm{conv}(S_1 \cup \cdots \cup S_n)$.*

The set $\mathrm{colint}\,\mathcal{S}$ has appeared in a work of Lawrence and Soltan [9], in the proof of the characterization of the intersection of convex transversals to a collection of sets. In more details, Lemma 8 and [9, Lemma 6] imply:

Proposition 9. *Let $\mathcal{S} = (S_1, \ldots, S_n)$ be a collection of n closed convex sets of $\mathbb{R}^{n-1}$. Define $\mathcal{D} := \{\mathrm{conv}(\{x_1, \ldots, x_n\}) : x_1 \in S_1, \ldots, x_n \in S_n\}$, the set of colorful simplices, i.e. with one vertex in each colored set. Then we have*

$$\mathrm{colint}\,\mathcal{S} = \bigcap_{\Delta \in \mathcal{D}} \mathrm{int}\,\Delta = \mathrm{int}\,\bigcap_{\Delta \in \mathcal{D}} \Delta.$$

Remark that Proposition 9 still holds if the colorful simplices $\Delta \in \mathcal{D}$ are replaced by the convex transversals to the sets $S_1, \ldots, S_n$.

Given a hyperplane $H := \{x \in \mathbb{R}^{n-1} : \langle h, x \rangle = b\}$, we shall denote below by $H^>$ (resp. $H^\leqslant$) the open (resp. closed) halfspace $\{x \in \mathbb{R}^{n-1} : \langle h, x \rangle > b\}$ (resp. $\{x \in \mathbb{R}^{n-1} : \langle h, x \rangle \leqslant b\}$). As a corollary of [9, Th. 2], we get the following characterization of the colorful interior:

Theorem 10. *Let $\mathcal{S} = (S_1, \ldots, S_n)$ be a collection of n closed convex sets of $\mathbb{R}^{n-1}$, and assume that $\mathrm{colint}\,\mathcal{S}$ is nonempty. Then, $\mathrm{colint}\,\mathcal{S}$ is the interior of a $(n-1)$-dimensional simplex.*

Moreover, if the sets $(S_i)_{i \in [n]}$ are bounded, then there are n unique hyperplanes $(H_i)_{i \in [n]}$ such that for all $i \in [n]$, $S_i \subset H_i^>$, and for all $j \neq i$, $S_j \subset H_i^\leqslant$ and $S_j \cap H_i \neq \varnothing$. In this case, we have $\mathrm{colint}\,\mathcal{S} = \bigcap_{i \in [n]} H_i^>$.

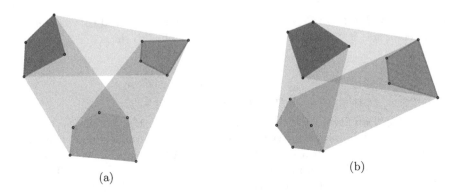

(a)

(b)

Fig. 1. (a) three convex sets S_1 (blue), S_2 (green) and S_3 (orange) in $\mathbb{R}^2$ and their colorful interior (white). The sets $(\widehat{S}_i)_{1 \leqslant i \leqslant 3}$ (resp. $(\overline{S}_i)_{1 \leqslant i \leqslant 3}$) are seen by taking convex hulls of $(S_i)_{1 \leqslant i \leqslant n}$ (resp. intersection of $(\widehat{S}_i)_{1 \leqslant i \leqslant 3}$) pairwise. Observe that the edges of the colorful interior are supported by tangent hyperplanes to two sets of (S_1, S_2, S_3). (b) the colorful interior of (S_1, S_2, S_3) is here empty, although these sets are separated (any three points in each of them are in general position), contrary to the sets $(\overline{S}_1, \overline{S}_2, \overline{S}_3)$, whose intersection is seen in the center of the figure. (Color figure online)

Geometrically, every H_i in Theorem 10 is a tangent hyperplane to the convex sets $(S_j)_{j \neq i}$ which separates them from the set S_i. The existence (and uniqueness) of such tangent hyperplanes follows from the work of Cappell et al. [5], see also the work of Lewis, Klee and von Hohenbalken [10] for a constructive proof. We depict on Fig. 1a three colored sets S_1, S_2 and S_3 in $\mathbb{R}^2$ with nonempty colorful interior colint (S_1, S_2, S_3), illustrating that the latter is the interior of a simplex as claimed in Theorem 10.

Given a collection $\mathcal{S} = (S_1, \ldots, S_n)$ of n closed convex sets of $\mathbb{R}^{n-1}$, we now discuss necessary and sufficient conditions for colint $\mathcal{S}$ to be nonempty. To this purpose we recall that the collection $\mathcal{S}$ is *separated* if for any choice of $k \leqslant n$ points $x_1, \ldots, x_k$ in $S_{i_1} \times \cdots \times S_{i_k}$ (where $i_1, \ldots, i_k$ are pairwise distinct), the points $x_1, \ldots, x_k$ are in general position (spanning a $(k-1)$-dimensional affine space).

Proposition 11. *Let $S_1, \ldots, S_n$ be a collection of n compact convex sets of $\mathbb{R}^{n-1}$, and let us define $\widehat{S}_i := \mathrm{conv}(\bigcup_{j \neq i} S_j)$ for all $i \in [n]$.*
Then, the family $(\overline{S}_i)_{i \in [n]}$ is separated if and only if $\bigcap_{i \in [n]} \widehat{S}_i = \varnothing$.

Proposition 12. *Let $S_1, \ldots, S_n$ be a collection of n compact convex sets of $\mathbb{R}^{n-1}$. Let us define, for all $i \in [n]$,*

$$\overline{S}_i := \bigcap_{j \neq i} \mathrm{conv}\left(\bigcup_{k \neq j} S_k\right).$$

Then, if colint $\mathcal{S}$ is nonempty, the family $(\overline{S}_i)_{i \in [n]}$ is separated.

Proposition 12 provides a necessary condition to ensure that colint $\mathcal{S} \neq \varnothing$. Since for all $i \in [n]$, we have $S_i \subset \overline{S}_i$, we also obtain that the separation of

$(S_i)_{i\in[n]}$ is necessary as well for colint S to be nonempty. However, Fig. 1b shows that this last condition is not sufficient. We conjecture that the necessary condition stated in Proposition 12 is sufficient:

Conjecture 13. Let $S_1,\ldots,S_n$ be a collection of n compact convex sets of $\mathbb{R}^{n-1}$. Then colint S is nonempty if and only if the family $(\overline{S}_i)_{i\in[n]}$ is separated.

We prove this conjecture in the case where $n = 3$ (it is also straightforward to establish for $n = 2$).

Proposition 14. *Let $S = (S_1, S_2, S_3)$ be a collection of three convex compact sets of $\mathbb{R}^2$. Then, colint S is nonempty if and only if $(\overline{S}_1, \overline{S}_2, \overline{S}_3)$ is separated.*

Proof. Suppose that $(\overline{S}_1, \overline{S}_2, \overline{S}_3)$ is separated. We know from [10] that for all $i \in \{1, 2, 3\}$ we have two hyperplanes (in this case affine lines) tangent to sets of the collection $(\overline{S}_j)_{j\neq i}$ and inducing opposite orientation on these. Such lines cannot meet $\overline{S}_i$ by separation property, so one of them, denoted H_i, is such that $\overline{S}_i \subset H_i^>$ and $\overline{S}_j \subset H_i^\leq$ for $j \neq i$. In particular, note that $\mathrm{conv}((S_j)_{j\neq i}) \subset H_i^\leq$. For $i, j \in \{1, 2, 3\}$ and $j \neq i$, the hyperplane H_i is not only tangent to $\overline{S}_j$ but also to S_j: indeed take a support y_i^j of H_i in $\overline{S}_j$, it arises as a convex combination $y_i^j = \sum_{k\neq i} \lambda_k x_k$ with $x_i \in S_k$ for $y_i^j \in \widehat{S}_i$. By $S_i \subset \overline{S}_i$, we derive for all $k \neq i$, $x_k \in H_k$ or $\lambda_k = 0$, the latter being ruled out by separation. Hence, let us denote by x_i^j a support of hyperplane H_i in S_j. Note that once again from the separation of $(\overline{S}_1, \overline{S}_2, \overline{S}_3)$, two supports of a tangent line in two different colors cannot be equal.

If $x := (a, b)^T$ and $y := (a', b')^T$ are two distinct vectors of $\mathbb{R}^2$, we denote $x \wedge y := (ab' - a'b)^{-1}(b - b', a' - a)^T$, the usual cross-product of two vectors in $\mathbb{P}^2$. As is customary, $h_1 := x_2^1 \wedge x_1^3$ (resp. $h_2 := x_3^2 \wedge x_2^1$ and $h_3 := x_1^3 \wedge x_3^2$) is a normal vector to H_1 (resp. H_2 and H_3), and $\langle h_i, x \rangle + 1 = 0$ is an equation defining H_i. Furthermore, the intersection of H_1 and H_2 is given by $s_3 := h_1 \wedge h_2$, or using the triple product formula, by

$$s_3 = h_1 \wedge (x_2^3 \wedge x_2^1) = \frac{(\langle h_1, x_2^1 \rangle + 1)\,x_2^3 - (\langle h_1, x_2^3 \rangle + 1)\,x_2^1}{(\langle h_1, x_2^1 \rangle + 1)\quad - (\langle h_1, x_2^3 \rangle + 1)}. \tag{4}$$

Because $x_2^1 \in \overline{S}_1 \subset H_1^>$ and $x_2^3 \in \overline{S}_3 \subset H_1^\leq$, we have that $\langle h_1, x_2^1 \rangle + 1$ is nonzero and $(\langle h_1, x_2^1 \rangle + 1)(\langle h_1, x_2^3 \rangle + 1) \leq 0$. As a result of (4), s_3 indeed exists and arises as a convex combination of x_2^3 and x_2^1, so $s_3 \in \mathrm{conv}(S_1 \cup S_3)$. By writing $s_3 = (x_1^2 \wedge x_1^3) \wedge h_2$ as in (4), we show likewise that s_3 is a convex combination of x_1^2 and x_1^3, thus $s_3 \in \mathrm{conv}(S_2 \cup S_3)$. This finally entails that $s_3 \in \overline{S}_3$ and therefore $s_3 \in H_3^>$. It now suffices to define $s_1 := h_2 \wedge h_3$ and $s_2 := h_3 \wedge h_1$ in a similar way and consider $y = (s_1 + s_2 + s_3)/3$. It is clear that $y \in \mathrm{conv}(S_1 \cup S_2 \cup S_3)$, and for all $i \in \{1, 2, 3\}$, $y \in H_i^>$, in particular $y \notin \mathrm{conv}((S_j)_{j\neq i})$. As a consequence, y is a colorful vector for S_1, S_2 and S_3. $\square$

To conclude, we point out that another interesting problem is the computational complexity of determining whether the colorful interior is empty or not, in the case where the sets S_i are polytopes. Remark that as a consequence of Proposition 11, if Conjecture 13 holds, then we can determine if colint S is empty in polynomial time using linear programming. Alternatively, the problem could be tackled by studying the complexity of separating a point from the colorful interior. This is tightly linked with the computation of the tangent hyperplanes of Theorem 10, for which the status of the complexity is not well understood.

References

1. Akian, M., Gaubert, S., Grand-Clément, J., Guillaud, J.: The operator approach to entropy games. Theory Comput. Syst. **63**(5), 1089–1130 (2019). https://doi.org/10.1007/s00224-019-09925-z
2. Allamigeon, X., Bœuf, V., Gaubert, S.: Performance evaluation of an emergency call center: tropical polynomial systems applied to timed petri nets. In: Sankaranarayanan, S., Vicario, E. (eds.) FORMATS 2015. LNCS, vol. 9268, pp. 10–26. Springer, Cham (2015). https://doi.org/10.1007/978-3-319-22975-1_2
3. Anantharam, V., Borkar, V.S.: A variational formula for risk-sensitive reward. SIAM J. Control Optim. **55**(2), 961–988 (2017). arXiv:1501.00676
4. Boyd, S., Boyd, S.P., Vandenberghe, L.: Convex Optimization. Cambridge University Press, Cambridge (2004)
5. Cappell, S., Goodman, J., Pach, J., Pollack, R., Sharir, M.: Common tangents and common transversals. Adv. Math. **106**(2), 198–215 (1994)
6. Chandrasekaran, V., Shah, P.: Relative entropy relaxations for signomial optimization. SIAM J. Optim. **26**(2), 1147–1173 (2016)
7. Dressler, M., Iliman, S., de Wolff, T.: A positivstellensatz for sums of nonnegative circuit polynomials. SIAM J. Appl. Algebra Geom. **1**(1), 536–555 (2017)
8. Friedland, S., Gaubert, S.: Spectral inequalities for nonnegative tensors and their tropical analogues (2018). arXiv:1804.00204
9. Lawrence, J., Soltan, V.: The intersection of convex transversals is a convex polytope. Contrib. Algebra Geom. **50**(1), 283–294 (2009)
10. Lewis, T., von Hohenbalken, B., Klee, V.: Common supports as fixed points. Geom. Dedicata. **60**(3), 277–281 (1996)
11. Lim, L.H.: Singular values and eigenvalues of tensors: a variational approach. In: Proceedings of the IEEE International Workshop on Computational Advances in Multi-Sensor Adaptive Processing (CAMSAP 2005), vol. 1, pp. 129–132 (2005)
12. Litvinov, G.L.: Maslov dequantization, idempotent and tropical mathematics: a brief introduction. J. Math. Sci. **140**(3), 426–444 (2007). https://doi.org/10.1007/s10958-007-0450-5
13. Puterman, M.L.: Markov Decision Processes: Discrete Stochastic Dynamic Programming. Wiley, New York (2014)
14. Viro, O.: Dequantization of real algebraic geometry on logarithmic paper. In: Casacuberta, C., Miró-Roig, R.M., Verdera, J., Xambó-Descamps, S. (eds.) European Congress of Mathematics, pp. 135–146. Birkhäuser Basel, Basel (2001). https://doi.org/10.1007/978-3-0348-8268-2_8

Univalent Mathematics: Theory and Implementation

Equality Checking for General Type Theories in Andromeda 2

Andrej Bauer, Philipp G. Haselwarter, and Anja Petković[(⊠)]

University of Ljubljana, Ljubljana, Slovenia
Andrej.Bauer@andrej.com , philipp@haselwarter.org,
Anja.Petkovic@fmf.uni-lj.si

Abstract. We designed a user-extensible judgemental equality checking algorithm for general type theories that supports computation rules and extensionality rules. The user needs only provide the equality rules they wish to use, after which the algorithm devises an appropriate notion of normal form. The algorithm is a generalization of type-directed equality checking for Martin-Löf type theory, and we implemented it in the Andromeda 2 prover.

Keywords: Algorithmic equality checking · Dependent type theory · Proof assistant

1 Introduction

Equality checking algorithms are essential components of proof assistants based on type theories [1,3,7,9,11,13]. They free users from the burden of proving judgemental equalities, and provide computation-by-normalization engines. Indeed, the type theories found in the most popular proof assistants are designed to provide such algorithms. Some systems [6,8] go further by allowing (possibly unsafe) user extensions to the built-in equality checkers.

The situation is less pleasant in a proof assistant that supports arbitrary user-definable theories, such as Andromeda 2 [4,5], where in general no equality checking algorithm may be available. For example, the well-known Martin-Löf "extensional" type theory that includes the equality reflection rule is well-known to have undecidable judgemental equality, and is readily definable in Andromeda 2. Short of implementing exhaustive proof search, the construction of equality proofs must be delegated to the user (and still checked by the trusted nucleus). While some may appreciate the opportunity to tinker with equality checking procedures, they are surely outnumbered by those who prefer good support that automates equality checking with minimal effort, at least for well-behaved type theories that one encounters in practice.

We have designed and implemented in Andromeda 2 an extensible equality checking algorithm that supports user-defined computation rules (β-rules) and

This material is based upon work supported by the Air Force Office of Scientific Research under award number FA9550-17-1-0326.

A. M. Bigatti et al. (Eds.): ICMS 2020, LNCS 12097, pp. 253–259, 2020.
https://doi.org/10.1007/978-3-030-52200-1_25

extensionality rules (inter-derivable with η-rules). The user needs only to provide the equality rules they wish to use, after which the algorithm automatically classifies them either as computation or extensionality rules (and rejects those that are of neither kind), and devises an appropriate notion of weak normal form. For the usual kinds of type theories (simply typed λ-calculus, Martin-Löf type theory), the algorithm behaves like well-known standard equality checkers.

Our algorithm is a variant of a type-directed equality checking [2,14], as outlined below. It is implemented in about 1300 lines of OCaml code, which resides outside the trusted nucleus. The algorithm calls the nucleus to build a trusted certificate of every equality step, and of every term normalization it performs, so all equalities established by the algorithm, including intermediate steps, are verified. It is easy to experiment with different sets of equality rules, and dynamically switch between them depending on the situation at hand. Our initial experiments are encouraging, although many opportunities for optimization and improvements await.

2   Andromeda 2

Andromeda 2 is an experimental LCF-style proof assistant, i.e., it is a meta-level programming language with an abstract datatype of judgements whose constructors are controlled by a trusted nucleus. We review just enough of it to be able to explain the equality checking algorithm.

In Andromeda 2 the user defines their own type theory by declaring the inference rules for types, terms and equalities. For example, formation of dependent products and the successor for natural numbers,

$$\frac{\Gamma \vdash A\ \mathsf{type} \quad \Gamma, x{:}A \vdash B\ \mathsf{type}}{\Gamma \vdash \prod(x{:}A)\,.\,B\ \mathsf{type}} \qquad\qquad \frac{\Gamma \vdash x : \mathbb{N}}{\Gamma \vdash s(x) : \mathbb{N}}$$

are written respectively as

```
rule Π (A type) ({x:A} B type) type
rule s (x : N) : N
```

The typing context Γ is left implicit (henceforth we shall elide Γ from all rules), while the context extension $x{:}A$ in the second premise of the product rule is expressed as an *abstraction*. In Andromeda 2 {x:A} e is a primitive operation that abstracts the variable x in e.

The user may also specify equality rules. For instance, the β-rule for functions is written as

```
rule β (A type) ({x:A} B type) ({x:A} t : B{x}) (a : A)
  : app A B (λ A B t) a ≡ t{a} : B{a}
```

where app and λ are the expected term formers corresponding to application and λ-abstraction, respectively. The notation t{a} instantiates the bound variable x in t with a. Note that all terms are fully annotated with types.

The object type theory has no primitive notion of definition (not to be confused with `let`-binding at the meta-language level). Instead, the user may simply declare an equational rule that serves as a definition, e.g.,

```
rule three  : N
rule three_def  : three ≡ s (s (s z))  : N
```

Structural rules are built into the nucleus. These are reflexivity, symmetry, and transitivity of equality, as well as support for abstraction and substitution. The nucleus automatically generates congruence rules for all term and type formers. For example, the computation

```
congruence (Π A B) (Π C D) α β
```

derives $Π \ A \ B \equiv Π \ C \ D$ by an application of the congruence rule for products. Here $α$ and $β$ are computations that further consult the nucleus to compute equalities $A \equiv C$ and $\{x:A\} \ B\{x\} \equiv D\{x\}$, respectively.

3 Computation and Extensionality Rules

We describe precisely what form computation and extensionality rules take. For this purpose, define an *object judgement* to be one of the form A type or $t : A$, and an *equation judgement* of the form $A \equiv B$ or $s \equiv t : A$. Accordingly, a premise of an inference rule may be either an object or an equation premise.

Term and type *computation rules* respectively have the forms

$$\frac{P_1 \ \cdots \ P_n}{\vdash u \equiv v : A} \qquad \frac{P_1 \ \cdots \ P_n}{\vdash A \equiv B}$$

where the P_i's are object premises. Furthermore, in a term computation rule the left-hand side u must take the form $\mathsf{s}(e_1, \ldots, e_m)$ where s is a term symbol. In other words, u may not be a variable or a meta-variable. Likewise, in an equation computation rule the left-hand side A must take the form $\mathsf{S}(e_1, \ldots, e_m)$ where S is a type symbol. Additionally, all the meta-variables introduced by the premises must appear in the arguments e_j. These conditions ensure that, given a term t, performing simple pattern matching of t against u tells us whether the rule applies to t and how. An example of a computation rule is the usual β-rule for simple products:

$$\frac{\vdash A \ \text{type} \quad \vdash B \ \text{type} \quad \vdash p : A \quad \vdash r : B}{\vdash \mathsf{fst}(A, B, \mathsf{pair}(A, B, p, r)) \equiv p : A}$$

Observe that the left-hand side of the equation mentions all four meta-variables A, B, p, r. In Andromeda 2 the above rule is postulated as

```
rule fst_β (A type) (B type) (p : A) (r : B) :
  fst A B (pair A B p r) ≡ p : A
```

and installed into the equality checker with `eq.add_rule fst_β`. The equality checker automatically determines that `fst_β` is a computation rule.

An *extensionality rule* says, broadly speaking, that two types or terms are equal when their eliminations are equal. Such a rule has the form

$$\frac{P_1 \; \cdots \; P_n \quad \vdash x : A \quad \vdash y : A \quad Q_1 \; \cdots \; Q_m}{\vdash x \equiv y : A},$$

where $P_1, \ldots, P_n$ are object premises and $Q_1, \ldots, Q_m$ are equality premises. We require that every meta-variable introduced by the premises appear in A. To tell whether such a rule applies to $s \equiv t : B$, we pattern match B against A, and recursively check suitably instantiated subsidiary equalities $Q_1, \ldots, Q_m$. Note that both sides of the conclusion of an extensionality rule must be meta-variables, so that the rule applies as soon as the type matches.

As an example we give the extensionality rule for simple products:

$$\frac{\vdash A \text{ type} \quad \vdash B \text{ type} \quad \vdash p : A \times B \quad \vdash q : A \times B}{\vdash \mathsf{fst}(A,B,p) \equiv \mathsf{fst}(A,B,q) : A \quad \vdash \mathsf{snd}(A,B,p) \equiv \mathsf{snd}(A,B,q) : B}{\vdash p \equiv q : A \times B}$$

In Andromeda 2 it is postulated as

```
rule prod_ext (A type) (B type) (p : A × B) (q : A × B)
  (fst A B p ≡ fst A B q : A)
  (snd A B p ≡ snd A B q : B) :
  p ≡ q : A × B
```

Again, the rule is installed with the command `eq.add_rule prod_ext`.

A second example is the extensionality rule for dependent functions (not to be confused with function extensionality):

$$\frac{\vdash A \text{ type} \quad x{:}A \vdash B \text{ type} \quad \vdash f : \prod(x{:}A).B \quad \vdash g : \prod(x{:}A).B}{x{:}A \vdash \mathsf{app}(A,B,f,x) \equiv \mathsf{app}(A,B,g,x) : B(x)}{\vdash f \equiv g : \prod(x{:}A).B}$$

which in Andromeda is written as

```
rule Π_ext (A type) ({x:A} B type)
  (f : Π A B) (g : Π A B)
  ({x:A} app A B f x ≡ app A B g x : B{x}) :
  f ≡ g : Π A B
```

It is easy to see that the `Π_ext` rule is inter-derivable with the η-rule for functions.

4 Normalizing Arguments and Normal Forms

The equality checking algorithm from Sect. 5 requires a notion of normal forms. We define an expression to be *normal* if no computation rule applies to it, and its *normalizing arguments* are in normal form. Thus, our notion of normal form depends on the computation and extensionality rules, as well as on which arguments of term and type symbols are normalizing.

In Andromeda 2 the user may specify the normalizing arguments directly, or let the algorithm determine the normalizing arguments from the computation rules automatically as follows: if $s(u_1, \ldots, u_n)$ appears as a left-hand side of a computation rule, then the normalizing arguments of s are those u_i's that are *not* meta-variables, i.e., matching against them does not automatically succeed, and so they have to be normalized before they are matched.

By varying the notion of normalizing arguments we can control how expressions are normalized. The automatic procedure results in weak head-normal forms, while strong normal forms are obtained if all the arguments are declared to be normalizing.

The normal form of a term t of type A is computed by a call to the command `eq.normalize` t, which outputs a certified equation $t \equiv t' : A$ where t' is the normal form of t. Similarly the command `eq.compute` t provides the strong normal form of t. Normalization of types works analogously.

The user may also verify an equation, say equality of types A and B, by running the command

$$\texttt{eq.prove} \ (A \equiv B \ \texttt{by} \ \texttt{??})$$

The equality checker outputs a certified judgement $A \equiv B$, or reports failure. In the above command, $A \equiv B$ `by` `??` is a *boundary*, which is a primitive notion in Andromeda 2 that expresses a goal. Each judgement form has a corresponding boundary: "`??`:A" is the goal asking that the type A be inhabited, "`??` `type`" that a type be constructed, and "$t \equiv s\colon A$ `by` `??`" that equality of terms s and t be proved.

5 An Overview of the Equality-Checking Algorithm

The equality-checking algorithm has several mutually recursive sub-algorithms:

1. *Normalize a type A:* the user-provided type computation rules are applied to A to give a sequence of (nucleus verified) equalities $A \equiv A_1 \equiv \cdots \equiv A_n$, until no more rules apply. Then the normalizing arguments of A_n are normalized recursively to obtain $A_n \equiv A'_n$, after which the equality $A \equiv A'_n$ is output.
2. *Normalize a term t of type A:* analogously to normalization of types, the user-provided term computation rules are applied to t until no more rules apply, after which the normalizing arguments are normalized.
3. *Check equality of types $A \equiv B$:* the types A and B are normalized and their normal forms are compared.

4. *Check equality of normal types $A \equiv B$:* normal types are compared structurally, i.e., by an application of a suitable congruence rule. The arguments are compared recursively: the normalizing ones by applications of congruence rules, and the non-normalizing ones by applications of the algorithm.
5. *Check equality of terms s and t of type A:*
   (a) *type-directed phase:* normalize the type A and based on its normal form apply user-provided extensionality rules, if any, to reduce the equality to subsidiary equalities,
   (b) *normalization phase:* if no extensionality rules apply, normalize s and t and compare their normal forms.
6. *Check equality of normal terms s and t of type A:* normal terms are compared structurally, analogously to comparison of normal types.

One needs to choose the notions of "computation rule", "extensionality rule" and "normalizing argument" wisely in order to guarantee completeness. In particular, in the type-directed phase the type at which the comparisons are carried out should decrease with respect to a well-founded notion of size, while normalization should be confluent and terminating. These concerns are external to the system, and so the user is allowed to install rules without providing any guarantees of completeness or termination.

6   Related and Future Work

Dedukti [8] is a proof assistant based on $\lambda\Pi$ modulo user-definable equational theories. Its pattern matching and rewriting capabilities are more advanced than ours. It does not have a type-directed phase through which user-defined extensionality rules could be applied, although of course one can reformulate those as η-rules.

Similar in spirit to Andromeda 2 is the equality checking algorithm used in the reconstruction phase of MMT [10,12], a meta-meta-language for description of formal theories. While in Andromeda 2 the user specifies the rules in declarative style that cannot break the trust in the nucleus, in MMT inference rules are implemented directly as executable code. This makes MMT more flexible at the price of importing arbitrary user-code into the trusted part of the system.

Our equality checker is general enough to support a wide range of equality checking algorithms that are based on a type-directed phase followed by normalization. It is easy to use because it automatically classifies equality rules as either computation or extensionality rules, and determines which arguments are normalizing. There are several possible future directions of research, of which we mention three.

First, we have already experimented with *local* equality rules that are installed temporarily. This is sometimes necessary to establish that an object appearing in a rule is well-formed due to an equational premise. More work is needed to design a usable interface for such local rules.

Second, there is no support for checking termination or confluence of the given rules. Consequently, the user may inadvertently install rules that cause the normalization phase to diverge, or experience unpredictable behaviour when the rules are not confluent. It would be worthwhile helping the user in this respect.

Third, combining our equality checker with other kinds of equality-checking algorithms would further facilitate proof development. Even naive proof search could be useful in certain situations. In principle, the user may direct Andromeda 2 to use a specific equality checker in a given situation, but it would be friendlier if the system behaved in an intelligent way with minimal direction from the user.

References

1. Abel, A., Öhman, J., Vezzosi, A.: Decidability of conversion for type theory in type theory. In: Proceedings of the ACM on Programming Languages, vol. 2, no. POPL, December 2017
2. Abel, A., Scherer, G.: On irrelevance and algorithmic equality in predicative type theory. Log. Methods Comput. Sci. **8**(1) (2012). https://lmcs.episciences.org/1045
3. The Agda proof assistant. https://wiki.portal.chalmers.se/agda/
4. The Andromeda proof assistant. http://www.andromeda-prover.org/
5. Bauer, A., Gilbert, G., Haselwarter, P.G., Pretnar, M., Stone, C.A.: Design and implementation of the Andromeda proof assistant. In: 22nd International Conference on Types for Proofs and Programs (TYPES 2016). LIPIcs, vol. 97, pp. 5:1–5:31 (2018)
6. Cockx, J., Abel, A.: Sprinkles of extensionality for your vanilla type theory. In: 22nd International Conference on Types for Proofs and Programs TYPES 2016. University of Novi Sad (2016)
7. The Coq proof assistant. https://coq.inria.fr/
8. The Dedukti logical framework. https://deducteam.github.io
9. Gilbert, G., Cockx, J., Sozeau, M., Tabareau, N.: Definitional proof-irrelevance without K. In: Proceedings of the ACM on Programming Languages, vol. 3, no. POPL, January 2019
10. The MMT language and system. https://uniformal.github.io//
11. de Moura, L., Kong, S., Avigad, J., van Doorn, F., von Raumer, J.: The lean theorem prover (system description). In: Felty, A.P., Middeldorp, A. (eds.) CADE 2015. LNCS (LNAI), vol. 9195, pp. 378–388. Springer, Cham (2015). https://doi.org/10.1007/978-3-319-21401-6_26
12. Rabe, F.: A modular type reconstruction algorithm. ACM Trans. Comput. Log. **19**(4), 24:1–24:43 (2018)
13. Sozeau, M., Boulier, S., Forster, Y., Tabareau, N., Winterhalter, T.: Coq Coq correct! Verification of type checking and erasure for Coq, in Coq. In: Proceedings of the ACM on Programming Languages, vol. 4, no. POPL, December 2019
14. Stone, C.A., Harper, R.: Extensional equivalence and singleton types. ACM Trans. Comput. Log. **7**(4), 676–722 (2006)

Artificial Intelligence and Mathematical Software

GeoLogic – Graphical Interactive Theorem Prover for Euclidean Geometry

Miroslav Olšák[(✉)] [iD]

University of Innsbruck, Innsbruck, Austria
mirek@olsak.net

Abstract. Domain of mathematical logic in computers is dominated by automated theorem provers (ATP) and interactive theorem provers (ITP). Both of these are hard to access by AI from the human-imitation approach: ATPs often use human-unfriendly logical foundations while ITPs are meant for formalizing existing proofs rather than problem solving. We aim to create a simple human-friendly logical system for mathematical problem solving. We picked the case study of Euclidean geometry as it can be easily visualized, has simple logic, and yet potentially offers many high-school problems of various difficulty levels. To make the environment user friendly, we abandoned strict logic required by ITPs, allowing to infer topological facts from pictures. We present our system for Euclidean geometry, together with a graphical application GeoLogic, similar to GeoGebra, which allows users to interactively study and prove properties about the geometrical setup.

Keywords: Euclidean geometry · Logical system

1 Overview

The article discusses GeoLogic 0.2 which can be downloaded from https://github.com/mirefek/geo_logic. It is a logic system for Euclidean geometry together with a graphical application capable of automatic visualization of basic facts (equal angles, equal distances, point being on a line, ...) and allowing user interaction with the logic system. GeoLogic can be used for proving many classical high school geometry problems such as Simson's line, Pascal's theorem, or some problems from International Mathematical Olympiad. Examples of such proofs are available in the package. In this paper, we first explain our motivation, then we describe the underlying logical system, and finally, we present an example of proving the Simson's line to demonstrate GeoLogic's proving and visualization capabilities.

There are many mathematical competitions testing mathematical problem solving capabilities of human beings, presumably most famous of which is the International Mathematical Olympiad (IMO). Writing an automated theorem prover (ATP) that could solve a large portion of IMO problems is a challenge

A. M. Bigatti et al. (Eds.): ICMS 2020, LNCS 12097, pp. 263–271, 2020.
https://doi.org/10.1007/978-3-030-52200-1_26

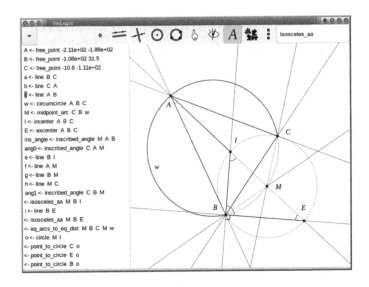

Fig. 1. GeoLogic screenshot

recognized in the field of artificial intelligence [6], and could potentially lead to strong ATPs in general.

IMO, as well as many regional mathematical olympiads divide problems into four categories: algebra, geometry, combinatorics, and number theory. From a human solver's perspective, computers can significantly help with solving geometry problems using an application such as GeoGebra – it allows the user to draw the configuration precisely, and observe how it changes when moving the initial points.

This is one of the reasons why we focused on geometry. Our objective is to capture the steps performed by such human solver in more detail, hoping it could eventually lead to better understanding of human thinking in general.

Therefore, we are building an interactive theorem prover, while preserving usability as an exploration tool. We have implemented a very simple logic, as it is sufficient for Euclidean geometry: most of the geometrical reasoning involves only direct proofs without higher-order logic or case analysis. While some geometrical proofs use case analysis for different topological configurations, we use a different approach. In GeoLogic, we allow inferring topological facts (such as the orientation of a triangle) from the picture (numerical model). This proves only one case of the problem (and its neighborhood), and could potentially lead to inconsistencies caused by numerical errors. However, we believe inconsistency caused by a numerical error is unlikely because we require the fact to be satisfied by a sufficient margin for postulating it.

In the future, we would like to experiment with machine learning agents leading to human-like ATPs for geometry. We would like to also experiment with computer vision components based on the GeoLogic's image output. Another interesting research direction would be adding tools for case analysis, or proving topological facts, so that a solving process of a problem would consist first from finding a solution in the current GeoLogic's flexible logic, and then transforming it into a rigorous one. We believe that such an approach would be very close to the geometrical problem-solving procedure of human beings.

Finally, even though our main motivation was not to make a pedagogical tool, and we do not market GeoLogic as an application for an arbitrary high school student in its current form, we also believe that GeoLogic can be already interesting for talented students. Our objective of making a user-friendly interactive theorem prover for geometry is well-aligned with educational purposes, and if it will get adopted in the future, it can help us with obtaining data for machine learning experiments.

2 Logical System

The logical system of GeoLogic consists of a *logical core* interacting with *tools*. The logical core contains the following data.

- The set of all geometrical objects constructed so far. Every object can be accessed as a reference (for logical manipulation), or as the numerical object (e.g. coordinates of points, for numerical checking).
- The knowledge database. It consists of a disjoint-set data structure for equality checking, equation systems for ratios and angles, and a lookup table for tools.

The logical core also possesses basic automation techniques for angle and ratio calculations, and deductions around equality.

A *tool* is a general concept for construction steps, predicates, or inference rules. It takes a list of geometrical references on an input (and sometimes additional hyper-parameters), possibly adds some objects and some knowledge to the logical core and returns a list of geometrical references on the output, or fails. A tool always fails if the numerical data do not fit.

Besides that, every tool can be executed in a *check mode* or a *postulate mode*. A tool fails in the check mode (and not in the postulate mode) if it requires a fact which is not known by the knowledge database. Otherwise, the outcomes of the two modes are the same.

Most tools are memoized. When they are called, their input is associated with their output in the lookup table of the logical core. In the next call of the same tool on the same input, the tool does not fail (even in check mode) and returns the stored output (the same logical references). This serves three purposes: computation optimization, functional extensionality, and as a database for predicates. In particular, a primitive predicate lies_on is a memoized tool which in postulate mode only checks whether a given point is contained by a given line or circle. If it is not, it fails, otherwise, it returns an empty output. In check mode, however, this tool always fails. It means that the only way how to make this tool executable in the check mode is to have the input already stored in the lookup table by calling it in the postulate mode before. This differs from topological (coexact) predicates such as not_on which in both modes only checks the numerical conditions – whether a given point is not contained by the given line or circle.

By proving a fact (any tool applied to given input) in the logic system, we mean executing certain tools in the check mode (proof), so that in the end the given fact can be also run in the check mode. The graphical interface allows users to run tools in check mode only.

2.1 Composite Tools

A composite tool is a sequence of other tool steps applied to the input objects. More precisely, a composite tool starts with just the input objects, runs several previously defined tools on the objects it has so far, and in the end, it returns some output objects selected from the available created objects. All composite tools are loaded from an external file, so we will explain them together with their format. An example code of the composite tool angle follows.

```
angle 10:L 11:L -> alpha:A
  d0 <- direction_of 10
  d1 <- direction_of 11
  alpha <- angle_compute 0 d0 -1 d1 1
```

The first line of a composite tool is a header specifying the tool name, input, and output objects, the other lines define the individual steps. The header line consists of the name, input objects, forward arrow ->, and output objects separated by space. Every input or output object is given by its label before the colon and its type after the colon. Types are given by letters P (point), L (line), C (circle), A (angle), D (ratio/dimension). Note that the format allows name overloading as long as the input types are different, so there can be an angle tool accepting two lines, and also another angle tool accepting three points. The lines after header describe the tool steps by output objects, backward arrow <-, tool name, and input objects related to the subtool (possibly with numerical hyperparameters) separated by space. Now, we use only labels without types since the parser already knows the input types and it can infer the output types by the used tool. The output labels must be unique unless an anonymous label _ is used. Among the input parameters, there can be also hyperparameters in the form of integers, floats, or fractions. It is not relevant how we mix the hyperparameters with the standard parameters but the order among hyperparameters, and among parameters matters.

The composite tool we described so far is the simplest composite tool (we call it a *macro*) which runs all its tool steps in the same mode as in what the macro is called. If any of the steps fail, the entire macro fails as well. Next to macros, there can be *axioms* and *lemmata*. The axiomatic tool is such a composite tool that contains a single line THEN among the steps. All the steps after THEN are then executed in postulate mode, even if the axiomatic tool is called in a check mode. We call the steps before THEN *assumptions* and the steps after THEN *implications*. Axiomatic tools are used for wrapping up primitive constructions (see direction_of, and line), or formulating real axioms (see isosceles_ss).

```
direction_of l:L -> a:A
  THEN
  a <- prim__direction_of l

line A:P B:P -> p:L
  <- not_eq A B
  THEN
  p <- prim__line A B
  <- lies_on A p
  <- lies_on B p

isosceles_ss A:P B:P C:P ->
  <- not_eq B C
  <- eq_dist A B A C
  THEN
  <- eq_angle A B C B C A
```

Finally, a *lemma* is similar to the axiomatic tool with the exception that there is a third sequence of steps (called *proof*) following a PROOF line. When a lemma is executed in a check-mode, it works the same as an axiomatic tool, but it also calls a *proof check*. The proof check consists of the following steps:

1. opening a new logical core for the following steps,
2. adding the numerical values of input objects as the initial objects,
3. running the assumptions in postulate mode,
4. running the proof in check mode,
5. running the implications in check mode.

If all the tools succeed, the proof check is considered successful. In the following example of isosceles_aa, we have a lemma stating that if the angles β, γ in a triangle ABC are equal, so are the sides b, c. This is proven using an axiom sim_aa_r which takes two indirectly similar triangles CAB and BAC, checks that they are non-degenerated, and their angles are proven to be equal, and infers that the ratios of the sides of the two triangles are equal.

```
isosceles_aa A:P B:P C:P ->
  <- not_collinear A B C
  <- eq_angle A B C B C A
  THEN
  <- eq_dist A B A C
  PROOF
  <- sim_aa_r C A B B A C
```

Adding a macro or a lemma to the toolset creates a conservative extension of the logic – anything that is provable with the usage of lemmata and macros can be proven without them.

3   Example – Simson's Line

We provide an example GeoLogic usage on the example of proving Simson's line. We used Geologic's graphical interface to define the following construction steps written as a code. During the construction, we also directly exported pictures from GeoLogic to show how GeoLogic visualizes known facts.

We start by drawing a triangle ABC, and a point X on its circumcircle.

```
A <- free_point -79.20758056640625 -119.095947265625
B <- free_point -126.97052001953125 23.91351318359375
C <- free_point 108.5352783203125 19.20867919921875
a <- line B C
b <- line C A
c <- line A B
o <- circumcircle A B C
X <- m_point_on 0.6169557687823527 o
```

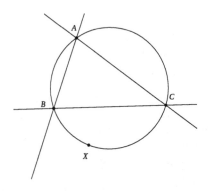

Simson's line is a line passing through feet F_a, F_b, F_c of the point X to the sides of the triangle. However, GeoLogic is not aware (yet) of the fact that these three points are collinear.

```
Fa <- foot X a
Fb <- foot X b
Fc <- foot X c
d <- line Fc Fa
e <- line Fb Fa
```

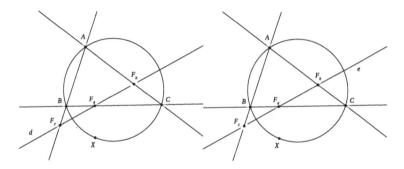

We can use the fact that the angles CF_aX and CF_bX are equal (they are both right angles) to conclude that points C, X, F_a, F_b are concyclic. We consequently use this fact to obtain that the angles F_bF_aC and F_bXC are equal.

```
<- angles_to_concyclic C X Fa Fb
<- concyclic_to_angles Fb C X Fa
```

We can similarly reason that the points B, X, F_a, F_c are concyclic and consequently the angles BF_aF_c and BXF_c are equal.

```
<- angles_to_concyclic B X Fc Fa
<- concyclic_to_angles Fc B Fa X
```

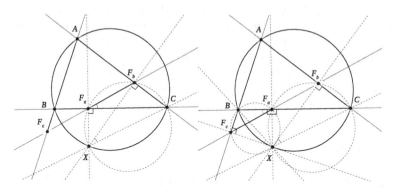

Finally, we use concyclicity of X, A, C, B to conclude that the angle XCA is equal to the complementary angle of ABX.

```
<- concyclic_to_angles X A C B
```

From this point on, GeoLogic's logical core realizes by itself that

$$\angle BF_aF_c = \angle BXF_c = 90° - F_cBX = 90° - F_bCX = CXF_b = CF_aF_b,$$

and since BF_aC are collinear, $F_cF_aF_b$ are collinear as well.

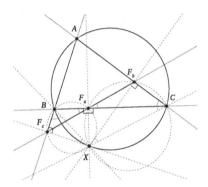

4   Related Work

Jeremy Avigad et al. [1] developed a logical system for formalizing elementary geometrical proofs from Euclid's elements, also distinguishing exact and coexact predicates. Their approach is more formal than ours allowing also proving the coexact statements in the end but it is less extensible by further tools. Michael Beeson et al. [2] connected the interactive theorem prover CoQ with GeoGebra for visualization of the theorem (but not for the proving procedure). Also, note that using a rigid logic system such as in CoQ does not allow numerical checks to be trusted in coexact statements.

The logical core of GeoLogic is partially inspired by General Deduction Database [3] and Full Angle [4] methods for automated synthetic proofs in Euclidean Geometry. These methods are supported by a graphical application Geometry Expert [7] which allows user to state a geometrical problem, run an automated geometrical theorem prover on it, and visualize the proof. Julien Narboux presented a similar graphical interface for construction of geometrical statement translated to CoQ [5]. None of these tools, however, supports constructing and checking proofs in the graphical interface.

5   Conclusion

We designed a semi-formal logic for Euclidean geometry which can be to a great extent controlled with a graphical interface and allows us to prove many standard high school problems. In the future, we would like to perform experiments with machine learning agents.

Acknowledgement. Supported by the ERC starting grant no.714034 SMART.

References

1. Avigad, J., Dean, E., Mumma, J.: A formal system for Euclid's elements. Rev. Symbolic Logic **2**(4), 700–768 (2009). https://doi.org/10.1017/S1755020309990098

2. Beeson, M., Boutry, P., Braun, G., Gries, C., Narboux, J.: GeoCoq (2018). (swh:1:dir:97ce53176b7d5e89d069bc60f49c3fa186831307). (hal-01912024)
3. Chou, S.-C., Gao, X.-S., Zhang, J.-Z.: A deductive database approach to automated geometry theorem proving and discovering. J. Autom. Reasoning **25**, 219–246 (2000). https://doi.org/10.1023/A:1006171315513
4. Chou, S., Gao, X., Zhang, J.: Automated generation of readable proofs with geometric invariants. J Autom. Reasoning **17**, 349–370 (1996). https://doi.org/10.1007/BF00283134
5. Narboux, J.: A graphical user interface for formal proofs in geometry. J. Autom. Reasoning **39**(2), 161–180 (2007). https://doi.org/10.1007/s10817-007-9071-4
6. Selsam, D.: IMO Grand Challenge. https://imo-grand-challenge.github.io/
7. Ye, Z., Chou, S.-C., Gao, X.-S.: An introduction to Java geometry expert. In: Sturm, T., Zengler, C. (eds.) ADG 2008. LNCS (LNAI), vol. 6301, pp. 189–195. Springer, Heidelberg (2011). https://doi.org/10.1007/978-3-642-21046-4_10

A Formalization of Properties
of Continuous Functions
on Closed Intervals

Yaoshun Fu$^{(\boxtimes)}$ and Wensheng Yu$^{(\boxtimes)}$

Beijing Key Laboratory of Space-Ground Interconnection
and Convergence School of Electronic Engineering,
Beijing University of Posts and Telecommunications, Beijing 100876, China
{fuys,wsyu}@bupt.edu.cn

Abstract. Formal mathematics is getting increasing attention in mathematics and computer science. In particular, the formalization of calculus has important applications in engineering design and analysis. In this paper, we present a formal proof of some fundamental theorems of continuous functions on closed intervals based on the Coq proof assistant. In this formalization, we build a real number system referring to Landau's Foundations of Analysis. Then we complete the formalization of the basic definitions of interval, function, and limit and formally prove the theorems including completeness theorem, intermediate value theorem, uniform continuity theorem and others in Coq. The proof process is normalized, rigorous and reliable.

Keywords: Coq · Formalization · Limits · Continuous functions · Closed intervals

1 Introduction

Analysis is one of the greatest achievements in the history of mathematics. The achievement opens a new era of mathematical progress and plays an important role in development of physics, astronomy, signal processing and other disciplines. Analysis which evolved from calculus is a branch of mathematics that studies limits and related theories [12].

At the end of the 19th century, mathematicians deduced many properties of continuous functions on closed intervals, which undoubtedly promoted the development of analytical theory. There are some important properties of continuous functions on closed intervals including Weierstrass second theorem: Boundedness theorem, Weierstrass first theorem: Extreme value theorem, Bolzano-Cauchy second theorem: Intermediate value theorem, Cantor theorem: Uniform continuity theorem.

This research is supported by National Natural Science Foundation (NNSF) of China under Grant 61936008, 61571064.

A. M. Bigatti et al. (Eds.): ICMS 2020, LNCS 12097, pp. 272–280, 2020.
https://doi.org/10.1007/978-3-030-52200-1_27

Bolzano's Function Theory gives the earliest proofs of the Boundedness theorem and the Extreme value theorem (but published some 100 years later) [15], and Weierstrass proved the Extreme value theorem in Berlin lecture. The Intermediate value theorem was first proved in 1817 by Bolzano, and then Cauchy [7] gave a proof in 1821. The definition of uniform continuity is proposed by Heine, and he published a proof of the Uniform continuity theorem.

With the further research of limits by mathematicians, the establishment of a rigorous and complete system of real numbers theory has become a key issue. In 1872, three major real numbers theories appeared in Germany: Dedekind cut theory, Cantor-Henie-Meray "basic sequence" theory, and Weierstrass "bounded monotone sequence" theory. Among them, the Dedekind cut is particularly recognized, and it is called the creation of human intelligence that does not rely on the intuitiveness of space and time. Then Peano established a natural number theory through a set of axioms, thereby solving the core problems of rational number theory and also the basic problems of real number theory.

In recent years, with the rapid development of computer science, especially the emergence of proof assistant Coq, Isabelle and HOL Light and so on [2, 4, 8, 10, 14, 17], formal proof of mathematical theorems has made great progress. In 2005, the international computer experts Gonthier and Werner provided the formal proof in Coq of the famous "four-color theorem" successfully [5]. After years of hard work, Gonthier again achieved the machine proof in Coq of the "odd order theorem" in 2012 [6]. Those progress make Coq more and more popular in academia. Wiedijk pointed out that relevant research teams around the world have completed or plan to formalize the proof of theorems such as Gödel's first incompleteness theorem, Jordan curve theorem, Prime theorem and Fermat last theorem of a hundred well-known mathematical theorems [17].

Based on "Real number theory" formal system, we formalize the properties of continuous functions on closed intervals. Moreover, we give formal proofs of these theorems, which include the Boundedness theorem, the Extreme value theorem, the Intermediate value theorem, the Uniform continuity theorem. It should be noted that the properties of continuous functions on closed intervals is an important theorem in analysis.

The structure of this article is as follows: In Sect. 2, we introduce the "real number theory" machine proof system. In Sect. 3, we present the formal definition of the function limit and related properties. In Sect. 4, we discuss the machine proof of the properties of continuous functions on closed intervals in detail, which are derived by supremum theorem. In Sect. 5, we draw conclusions and discuss some potential further work.

2 Real Number Theory System

Before formally proving the properties of continuous functions on closed intervals, we first need to build a formal system of real number theory. van Benthem Jutting [1] completed the formalization in Automath of Landau's "Foundations of Analysis", which was a significant early progress in formal mathematics.

Harrison [9] presents formalized real numbers and differential calculus on his
HOL Light system. The definition of real numbers in Coq standard library uses
the axiomatic way, and based on this, excellent real analysis library Coquelicot
[3] has been established. This library accomplishes many achievements, but its
definition of real number is non-constructive. Hornung [11] completed the first
four chapters of the "Foundations of Analysis" in Coq, which is closed related
to our work, however our system is closer to the expression of Landau and more
readable. We also completed the complex number part and proved equivalence
between eight completeness theorems of real number.

There are several ways to define natural numbers in Coq. Based on Morse-
Kelley axiomatic set theory, it is designed to give quickly and naturally a foundation for mathematics, and meanwhile deduce the Peano axioms as theorems
[16,18]. If we start from the more higher type rather than set theory, we can
formalize straight Peano axioms as follows:

```
Parameter Nat :Type.
Axiom One :Nat.
Notation "1" := One.
Axiom Successor :Nat -> Nat.
Notation " x ' " := (Successor x)(at level 0).
Axiom AxiomIII : ∀ x, x' <> 1.
Axiom AxiomIV : ∀ x y, x' = y' -> x = y.
Axiom AxiomV : ∀ M, 1 ∈ M /\ (∀ x, x ∈ M -> x' ∈ M) -> ∀ z, z ∈ M.
```

Based on this, we can use "Parameter" and "Axiom" to define natural number
related functions such as addition and multiplication. This way is intuitive but
not elegant. The natural numbers defined by "Inductive" can recursively define
natural number related functions.

Landau's "Foundations of Analysis" [13] is based on naive set theory and
some basic logic. Starting from the Peano axioms, natural numbers (positive
integers), fractions (positive), rational numbers/integer (positive) are defined
in order. The real number, defined by Dedekind cut, defines complex numbers
through real numbers for constructing systematically the whole number system
theory. We have completed the Coq formalization of the system, and the complete source is available online:

https://github.com/coderfys/analysis/

In this system, we can prove Dedekind fundamental theorem, and derive
Supremum theorem. The proof details are not described, and the formalization
is as follows.

2.1   Dedekind Fundamental Theorem

Divide all real numbers into two classes, so that the first class and second class
are not empty, and each number in the first class is less than each number in
the second class. Then there is a unique real number E, so that any number less
than E belongs to the first class, and any number greater than E belongs to the
second class.

```
Section Dedekind.
Variables Fst Snd :Ensemble Real.
Definition R_Divide := ∀ r, r ∈ Fst \/ r ∈ Snd.
Definition ILT_Class := ∀ e f, e ∈ Fst -> f ∈ Snd -> e < f.
Definition Split E := (∀ H, H < E -> H ∈ Fst)
  /\ (∀ G, G > E -> G ∈ Snd).
End Dedekind.
```

```
Theorem DedekindCut_Unique :
  ∀ Fst Snd, R_Divide Fst Snd -> No_Empty Fst -> No_Empty Snd ->
  ILT_Class Fst Snd -> ∀ Z1 Z2, Split Fst Snd Z1 ->
  Split Fst Snd Z2 -> Z1 = Z2.
```

```
Theorem DedekindCut :
  ∀ Fst Snd, R_Divide Fst Snd -> No_Empty Fst -> No_Empty Snd ->
  ILT_Class Fst Snd -> ∃ E, Split Fst Snd E.
```

2.2 Supremum Theorem

If a non-empty real number set has a upper bound, then there must be a least bound (the Supremum as an example).

```
Definition bound_up y A := ∀ z, z ∈ A -> z ≤ y.
Definition supremum y A := bound_up y A /\
  ∀ z, bound_up z -> y ≤ z.
Theorem SupremumT : ∀ R, No_Empty R -> ∃ x, bound_up x R ->
  ∃ y, supremum_s y R.
```

3 Basic Definitions and Properties

The formal definition of functions in this system is as follows:

```
Definition Fun := Real -> Real.
```

Related definitions of function continuity:

```
Definition FunDot_con (f :Fun) x0 := ∀ ε, ε > 0 ->
  ∃ δ, δ > 0 /\ ∀ x, | x - x0 | < δ -> | f x - f x0 | < ε.
Definition FunDot_con (f :Fun) x0 := ∀ ε, ε > 0 ->
  ∃ δ, δ > 0 /\ ∀ x, | x - x0 | < δ -> | f x - f x0 | < ε.
Definition FunDot_conr (f :Fun) x0 := ∀ ε, ε > 0 ->
  ∃ δ, δ > 0 /\ ∀ x, x - x0 < δ -> 0 ≤ x - x0 -> | f x - f x0 | < ε.
Definition FunDot_conl (f :Fun) x0 := ∀ ε, ε > 0 ->
  ∃ δ, δ > 0 /\ ∀ x, x0 - x < δ -> 0 ≤ x0 - x -> | f x - f x0 | < ε.
Definition FunOpen_con f a b := a < b /\
  (∀ z, z ∈ (a,b) -> FunDot_con f z).
Definition FunClose_con f a b := a < b /\ FunDot_conr f a /\
  FunDot_conl f b /\ (∀ z, z ∈ (a,b) -> FunDot_con f z).
```

The function $f(x)$ is continuous at one point implying that

```
Corollary Pr_FunDot : ∀ f x0, FunDot_con f x0 -> ∀ ε, ε > 0 ->
    ∃ δ, δ > 0 /\ ∀ x, | x - x0 | ≤ δ -> | f x - f x0 | < ε.
```

The function $f(x)$ is continuous (left, right) at a point, then the function $f(x)$ is locally bounded at this point (take right continuous for example):

```
Lemma limP1 : ∀ f a, FunDot_conr f a -> f a > 0 -> ∀ r, r > 0 ->
    r < f a -> ∃ δ, δ > 0 /\ ∀ x, x - a < δ -> 0 ≤ x - a -> f x > r.
Lemma limP1' : ∀ f a, FunDot_conr f a -> f a > 0 ->
    ∃ δ, δ > 0 /\ ∀ x, x - a < δ -> 0 ≤ x - a -> f x > ((f a)/2) NoO_N.
```

In this real number system, division function requires three parameters, and the third of which is the proof that the second is not 0. Therefore, the "NoO_N" above means "$2 \neq 0$".

The function $f(x)$ is continuous on $[a, b]$, then $-f(x)$ is continuous on $[a, b]$. The function $f(x)$ is continuous on $[a, b]$ and is not everywhere 0, then $\frac{1}{f(x)}$ is continuous on (take $f(x) > 0$ as an example).

```
Lemma Pr_Fun1 : ∀ f M a b, FunClose_con f a b ->
    FunClose_con (λ x, M - (f x)) a b.
Lemma Pr_Fun2 : ∀ f a b (P: ∀ z, z ∈ [a, b] -> neq_zero (f z))
    (Q:∀ z, z ∈ [a, b] -> (f z) > 0), FunClose_con f a b ->
    FunClose_con (λ x, match classic (x ∈ [a, b]) with
                  | left l => (1 / (f x)) (P _ l)
                  | right _ => f x end) a b.
```

4   Properties of Continuous Functions on Closed Intervals

Continuous functions have four fundamental properties on closed intervals: Boundedness theorem (Weierstrass second theorem), Extreme value theorem (Weierstrass first theorem), Intermediate value theorem (Bolzano-Cauchy second theorem), Uniform continuity theorem (Cantor theorem). These theorems are the basis of mathematical analysis and the direct expression of real number theory in functions. Our formalizations rely on a logical axiom law of excluded middle.

Theorem 1. *Boundedness theorem: A continuous function on a closed interval must be bounded on that interval.*

```
Definition FunClose_boundup f a b := a < b /\
    ∃ up, (∀ z, z ∈ [a,b] -> f z ≤ up).
Definition FunClose_bounddown f a b := a < b /\
    ∃ down, (∀ z, z ∈ [a,b] -> down ≤ f z).
Theorem T1 : ∀ f a b, FunClose_con f a b -> FunClose_boundup f a b.
Theorem T1' : ∀ f a b, FunClose_con f a b -> FunClose_bounddown f a b.
```

First, we prove a lemma L1: if $f(x)$ is continuous on $[a,b]$, then some neighborhood of z is bounded for any $z \in (a,b)$. The notation "$[x0| - \delta]$" below represents $U_{x_0}(\delta)$ in mathematics.

```
Lemma L1 : ∀ f a b, FunClose_con f a b -> ∀ x0, x0 ∈ (a,b) ->
    ∃ δ, δ > 0 /\ (∃ up down, (11∀ z, z ∈ [x0|-δ] -> f z ≤ up) /\
    (∀ z, z ∈ [x0|-δ] -> down ≤ f z)).
```

Upper bound: Construct a real number set $\{t : f(x)$ has an upper bound on $[a,t]\}$. The formal definition is as follows:

```
R:=/{ t | FunClose_boundup f a t /\ t ≤ b /}
```

As $f(x)$ is right continuous at the point a, there exists $\delta > 0$, and $f(x)$ has an upper bound on $(a, a+\delta)$, when $b \le a+\delta$ proves the proposition. When $a+\delta < b$, R is not empty. On the other hand, b is an upper bound of R obviously, so R has supremum ξ, and $\xi \le b$.

Case 1($\xi < b$): According to Lemma L1, there exists $\delta_1 > 0$, and $f(x)$ has an upper bound on $(a, \xi + \delta_1)$. The proposition is proved when $b < \xi + \delta_1$. When $\xi + \delta_1 \le b$, there is $\xi + \delta_1 \in R$, which contradicts ξ is the supremum of R.

Case 2($\xi = b$): The proposition is proved because of $b < \xi + \delta_1$.

Lower bound: According to the lemma Pr_fun1, it can be deduced that $-f(x)$ is continuous on $[a,b]$. From Theorem T1, we know that $-f(x)$ has the upper bound "up", then "-up" is the lower bound of $f(x)$ on $[a,b]$.

Theorem 2. *Extreme value theorem: The continuous function on the closed interval must achieve the maximum and minimum values in this interval.*

```
Theorem T2 : ∀ f a b, FunClose_con f a b ->
    ∃ z, z ∈ [a,b] /\ (∀ w, w ∈ [a,b] -> f w ≤ f z).
Theorem T2' : ∀ f a b, FunClose_con f a b ->
    ∃ z, z ∈ [a,b] /\ (∀ w, w ∈ [a,b] -> f z ≤ f w).
```

Maximum value: Construct a real number set $\{f(x) : x \in [a,b]\}$. The formal definition is as follows:

```
R:=/{ w | ∃ z, z ∈ [a,b] /\ w = f z /}.
```

Obviously, R is not empty and we can deduce R has an upper bound by T1, hence R has a supremum M. If there exists $x \in (a,b)$ and $f(x) = M$, the proposition is proved. Otherwise, $f(x) < M$ for any $x \in [a,b]$. Construct a new function $g(x) = \frac{1}{M-f(x)}$. Since $g(x) > 0$ for any $x \in [a, b]$, so $g(x)$ is continuous on $[a,b]$ by Pr_fun2. From T1, $g(x)$ has an supremum K on $[a,b]$, and $K > 0$. After derivation, $M - \frac{1}{K}$ is the upper bound of $f(x)$ on $[a,b]$, which contradicts M is the supremum of R.

Minimum value: According to the lemma Pr_fun1, it can be deduced that $-f(x)$ is continuous on $[a,b]$. From Theorem T2, we know that $-f(x)$ has a maximum value "max", then "-max" is the minimum value of $f(x)$ on $[a,b]$.

Theorem 3. *Intermediate value theorem: If $f(a) \neq f(b)$, then for any real number C between $f(a)$, $f(b)$, at least one point c on (a, b) satisfies $f(c) = C$.*

```
Theorem T3 : ∀ f a b, FunClose_con f a b -> f a < f b ->
    ∀ C, f a < C -> C < f b -> ∃ ξ, ξ ∈ (a,b) /\ f ξ = C.
Theorem T3' : ∀ f a b, FunClose_con f a b -> f a > f b ->
    ∀ C, f b < C -> C < f a -> ∃ ξ, ξ ∈ (a,b) /\ f ξ = C.
```

First, we prove a lemma L3: if $f(x)$ is left continuous at point b, $a < b$ and $f(b) > C$, then there is z between a, b, satisfies $f(z) > C$.

```
Lemma L3 : ∀ f a b C, FunDot_conl f b -> b > a -> f b > C ->
    ∃ z, a < z /\ z < b /\ f z > C.
```

$f(a) < f(b)$: Construct a real number set $\{t : f(x) < C$ for any x in $[a, t]\}$. The formal definition is as follows:

```
R:=/{ t | (∀ x, x ∈ [a,t] -> f x < C) /\ t < b /}.
```

$f(a) > f(b)$: $f(x)$ is right continuous at point a, then there exists $\delta_1 > 0$, and $|f(x) - f(a)| < C - f(a)$ in $(a, a + \delta_1)$, which can be deduced $f(x) < C$. $(a + \frac{\delta_1}{2}) \in R$ when $\delta_1 < (b - a)$ and $\frac{a+b}{2} \in R$ when $(b - a) \leq \delta_1$. In summary, R is not empty and b is an upper bound of R, so R has a supremum ξ, and $\xi \leq b$. When $\xi = b$, it can deduce contradiction according to Pr_supremum and L3. Therefore $\xi \in (a, b)$.

Case 1 $(f(\xi) < C)$: Because $f(x)$ is continuous at point ξ, by Pr_FunDot there is $\delta > 0$, and $|f(x) - f(\xi)| < \frac{C - f(\xi)}{2}$ for any $x \in [\xi - \delta, \xi + \delta]$, which can be deduced $f(x) < C$. Further, we can conclude that $(\xi + \delta) \in R$ which contradicts ξ is the supremum of R.

Case 2 $(f(\xi) > C)$: $f(x)$ is continuous at point ξ, hence $f(x)$ is left continuous at point ξ. By L3 there must exist $z \in (a, \xi)$ and $f(z) > C$, so z is the upper bound of R, which contradicts ξ is the supremum of R.

$f(a) > f(b)$: Refer to T1', T2' proof.

Theorem 4. *Uniform continuity theorem: A function is continuous on a closed interval then the function is uniformly continuous on that interval.*

```
Definition Un_Con f a b := a < b /\ ∀ ε, ε > 0 ->
    ∃ δ, δ > 0 /\ ∀ x1 x2, x1 ∈ [a,b] -> x2 ∈ [a,b] ->
    |x1 - x2| < δ -> |f x1 - f x2| < ε.
Theorem T4 : ∀ f a b, FunClose_con f a b -> Un_Con f a b.
```

Let $f(x)$ be continuous on $[a, b]$, fix any $\epsilon > 0$. we construct a real number set $\{t : \exists \delta > 0$ and $|f(x_1) - f(x_2)| < \epsilon$ when $x_1, x_2 \in [a, t]$ and $|x_1 - x_2| \leq \delta\}$. The formal definition is as follows:

```
R:=/{ t | (a<t /\ ∃δ, δ > 0 /\ (∀x1 x2,x1 ∈ [a,t] -> x2 ∈ [a,t]
    -> | x1 - x2 | < δ -> | f x1 - f x2 | < ε)) /\ t ≤ b /}.
```

Because $f(x)$ is right continuous at point a, there must exists $\delta_1 > 0$, and $|f(x) - f(a)| < \frac{\epsilon}{2}$ for any $x \in [a, a + \delta_1)$. Let $\delta_2 = min(a + \frac{\delta_1}{2})(b - a)$, we prove that $a + \delta_2 \in R$. Obviously, b is an upper bound of R, So R has a supremum ξ and $\xi \le b$. As $f(x)$ is continuous at point ξ, there exists $\delta > 0$, and $|f(x) - f(\xi)| < \frac{\epsilon}{2}$ for any $x \in [\xi - \delta, \xi + \delta]$. Further, we can deduce $|f(x_1) - f(x_2)| < \epsilon$ for any $x_1, x_2 \in [\xi - \delta, \xi + \delta]$. Therefore, $|f(x_1) - f(x_2)| < \epsilon$ for any $x_1, x_2 \in [a, \xi + \delta]$ and $|x_1 - x_2| \le \delta$.

Case $1(\xi < b)$: When $b < \xi + \delta$, the proposition is proved due to the arbitrariness of ϵ. When $\xi + \delta \le b$, we further prove $\xi + \delta \in R$, which contradicts ξ is the supremum of R.

Case $2(\xi = b)$: $b < \xi + \delta$, the proposition is proved by the arbitrariness of ϵ.

5   Conclusion

This paper formalizes limits, continuous functions and related theorems. These theorems include Boundedness theorem, Extreme value theorem, Intermediate value theorem, and Uniform continuity theorem. We have completed their formal proofs based on the real number theory system developed by ourselves. In the future, we will formalize more theorems of continuous functions and make meaningful attempts for formal work in the fields of real analysis and complex analysis. We are grateful to the anonymous reviewers, whose comments much helped to improve the presentation of the research in this article.

References

1. van Benthem Jutting, L.S.: Checking Landau's "Grundlagen" in the AUTOMATH System. North-Holland, Amsterdam (1994)
2. Bertot, Y., Castéran, P.: Interactive Theorem Proving and Program Development: Coq'Art: The Calculus of Inductive Constructions. Texts in Theoretical Computer Science An EATCS Series. Spring-Verlag, Heidelberg (2004). https://doi.org/10.1007/978-3-662-07964-5
3. Boldo, S., Lelay, C., Melquiond, G.: Coquelicot: a user-friendly library of real analysis for Coq. Math. Comput. Sci. 9(1), 41–62 (2015). https://doi.org/10.1007/s11786-014-0181-1
4. Chlipala, A.: Certified Programming with Dependent Types: A Pragmatic Introduction to the Coq Proof Assistant. MIT Press, Cambridge (2013)
5. Gonthier, G.: Formal proof-the four-color theorem. Notices AMS 55(11), 1382–1393 (2008)
6. Gonthier, G., et al.: A machine-checked proof of the odd order theorem. In: Blazy, S., Paulin-Mohring, C., Pichardie, D. (eds.) ITP 2013. LNCS, vol. 7998, pp. 163–179. Springer, Heidelberg (2013). https://doi.org/10.1007/978-3-642-39634-2_14
7. Grabiner, J.V.: Who gave you the epsilon? Cauchy and the origins of rigorous calculus. Am. Math. Mon. 90(3), 185–194 (1983)
8. Hales, T.C.: Formal proof. Notices AMS 55(11), 1370–1380 (2008)
9. Harrison, J.: Theorem Proving with the Real Numbers. Springer, Heidelberg (1994)

10. Harrison, J.: Formal proof—theory and practice. Notices AMS **55**(11), 1395–1406 (2008)
11. Hornung, C.: Constructing Number Systems in Coq. Saarland University, Saarbrücken (2011)
12. Katz, V.J.: A History of Mathematics. Pearson/Addison-Wesley, Boston (2004)
13. Landau, E.: Foundations of Analysis: The Arithmetic of Whole, Rational, Irrational, and Complex Numbers. Chelsea Publishing Company, New York (1966)
14. Nipkow, T., Wenzel, M., Paulson, L.C. (eds.): Isabelle/HOL. LNCS, vol. 2283. Springer, Heidelberg (2002). https://doi.org/10.1007/3-540-45949-9
15. Rusnock, P., Kerr-Lawson, A.: Bolzano and uniform continuity. Historia Mathematica **32**(3), 303–311 (2005)
16. Sun, T., Yu, W.: A formal system of axiomatic set theory in Coq. IEEE Access **8**, 21510–21523 (2020)
17. Wiedijk, F.: Formal proof-getting started. Notices AMS **55**(11), 1408–1414 (2008)
18. Yu, W., Sun, T., Fu, Y.: Machine Proof System of Axiomatic Set Theory. Science Press, Beijing (2020)

Variable Ordering Selection for Cylindrical Algebraic Decomposition with Artificial Neural Networks

Changbo Chen[1,2(✉)] [iD], Zhangpeng Zhu[1], and Haoyu Chi[1,2]

[1] Chongqing Key Laboratory of Automated Reasoning and Cognition, Chongqing Institute of Green and Intelligent Technology, Chinese Academy of Sciences, Chongqing, China
chenchangbo@cigit.ac.cn
[2] University of Chinese Academy of Sciences, Beijing, China

Abstract. Cylindrical algebraic decomposition (CAD) is a fundamental tool in computational real algebraic geometry. Previous studies have shown that machine learning (ML) based approaches may outperform traditional heuristic ones on selecting the best variable ordering when the number of variables $n \leq 4$. One main challenge for handling the general case is the exponential explosion of number of different orderings when n increases. In this paper, we propose an iterative method for generating candidate variable orderings and an ML approach for selecting the best ordering from them via learning neural network classifiers. Experimentations show that this approach outperforms heuristic ones for $n = 4, 5, 6$.

Keywords: Cylindrical algebraic decomposition · Variable ordering · Machine learning · Neural network

1 Introduction

Cylindrical algebraic decomposition (CAD) was introduced by Collins for solving real quantifier elimination problems [14]. The original framework for computing CAD introduced by Collins is based on a projection and lifting scheme, which has now been gradually improved by many others [1,3,5,19–23]. In 2009, Moreno Maza, Xia, Yang and the first author [13] proposed a new way for computing CAD, which first computes a cylindrical decomposition of complex space and then transforms it into a CAD of real space based on the technique of triangular decompositions and regular chains [13]. Its efficiency was substantially improved in [7] based on an incremental algorithm, which can also take advantage of equational constraints. A complete and efficient algorithm for real quantifier elimination based on it was proposed in [12]. Moreover, it can utilize disjunctive equational constraints via computing a truth table invariant CAD [2].

Today, despite of its doubly exponential complexity [6,14], CAD has been efficiently implemented in many softwares such as QEPCAD, Mathematica, REDLOG and Maple, and found wide applications in geometry theorem proving,

© Springer Nature Switzerland AG 2020
A. M. Bigatti et al. (Eds.): ICMS 2020, LNCS 12097, pp. 281–291, 2020.
https://doi.org/10.1007/978-3-030-52200-1_28

stability analysis of dynamical systems, control system design, verification of hybrid systems, program verification, nonlinear optimization, automatic parallelization, and so on. Recently, it also finds applications on studying quantum nonlocality [8].

The choice of variable ordering has been shown to have a great impact on the performance of CAD, both theoretically [6] and in practice [15]. Several heuristic methods for variable ordering selection have been proposed. In particular, two heuristic strategies [4,9] are implemented in the SuggestVariableOrder (SVO) command of the RegularChains library in Maple. On the other hand, it also becomes a natural option to predict the best variable ordering by the approaches of artificial intelligence, among which machine learning is a natural choice [16–18,24,25].

Existing work for selecting variable ordering by machine learning focus on the trivariate case. For more than three variables, it becomes more difficult to obtain sufficient labelled data due to the doubly exponential behavior of CAD in terms of the number of variables n. Another difficulty is the exponential explosion of number of different orderings when n increases. In this paper, we first propose an iterative approach for generating a better variable ordering starting from the one given by SVO. Then we reduce the potential $n!$ number of classes to predict for the variable ordering problem to n by training a neural network classifier with data generated by the iterative approach. Experiments show that both the iterative approach and the machine learning approach outperform SVO for $n = 4, 5, 6$.

The organization of the paper is as follows. In Sect. 2, we briefly review the concept of CAD and the problem of variable ordering selection. In Sect. 3 and Sect. 4, we present respectively the iterative and the machine learning approaches. In Sect. 5, we show the effectiveness of our approaches by experimentation. Finally, we draw the conclusion in Sect. 6.

2   Cylindrical Algebraic Decomposition

Consider a set of polynomials $F \subset \mathbb{Q}[x_1, \ldots, x_n]$ and a variable ordering $x_{i_1} > \cdots > x_{i_n}$. An F-invariant cylindrical algebraic decomposition (CAD) partitions $\mathbb{R}^n$ into disjoint and cylindrically arranged semi-algebraic subsets (called cells) such that the projection of any two cells onto $\mathbb{R}^k$ (with coordinate variables $x_{i_{n-k+1}}, \ldots, x_{i_n}$), $1 \leq k \leq n-1$, is either disjoint or identically equal. The variable ordering also specifies the order to eliminate variables and the order to construct CAD from projection factors or a complex cylindrical decomposition.

The algorithms presented in [2,10,13] for computing CADs based on regular chains have been implemented in the command CylindricalAlgebraicDecompose in the RegularChains library of Maple [11]. Its latest version is available from http://www.regularchains.org and we use the version from http://www.arcnl.org/cchen/software/cadorder. In the RegularChains library, the command SuggestVariableOrder (SVO for short) implements two different heuristic methods for variable ordering selection, namely the one by Brown [4] (with option "decomposition $=$ cad", SVO(B) for short) and the one by Chen et al. [9] (default option, SVO(C) for short).

The variable ordering plays an important role in the efficiency of computing CAD, as illustrated by the following example.

Example 1. *Let* $F := \{68\,x_1{}^2 - 12\,x_3\,x_2 + 46\,x_3 - 126,\ -54\,x_2\,x_1 + 11\,x_1 + 92\,x_2 - 21,\ -60\,x_3\,x_1{}^2 - 42\,x_3\,x_2\,x_1 + 45\,x_4{}^2 - 35\}$. *Table 1 lists the computation times and number of cells for several variable orders. As we can see, for this example, current heuristic methods avoid picking the worst variable order, but also miss the best variable order.*

Table 1. Impact of different variable orders

Order	Method	Timing (seconds)	#cells
$x_4 \succ x_3 \succ x_2 \succ x_1$	–	5	3373
$x_3 \succ x_1 \succ x_4 \succ x_2$	–	93	43235
$x_2 \succ x_3 \succ x_4 \succ x_1$	SVO(B)	16	11953
$x_3 \succ x_2 \succ x_4 \succ x_1$	SVO(C)	14	9253

3   An Iterative Method

In this section, we present an iterative variable ordering selection method, called IVO, for cylindrical algebraic decomposition. It starts with an initial ordering $x_{i_1} > x_{i_2} > \cdots > x_{i_n}$ provided by SVO. Then it calls a subroutine, called RVO, to generate n orderings in a round-robin manner and picks the best by calculating the shortest time of computing CAD with them. Next, it fixes the largest variable and calls RVO again on the rest ones to select the second largest variable, and so on. Precise description of IVO and RVO are given as below. We denote by $\|$ the concatenation of two sequences.

- **Algorithm** RVO.
- Input: a set of polynomials F; a sequence of variables O defining a descending ordering; the time t for running CAD(F, O), an integer k.
- Output: a new ordering O' and running time t' of CAD(F, O') such that $t' \le t$.
- Steps:
  1. Let $P := O_1, \ldots, O_k$ and $Q := O_{k+1}, \ldots, O_n$.
  2. Let $Q^{(i)} := Q_i, Q \setminus \{Q_i\}$, $i = 1, \ldots, |Q|$.
  3. For each $Q^{(i)} \ne Q$, $i = 1, \ldots, |Q|$, call CAD$(F, P\|Q^{(i)})$, $Q^{(i)} \ne Q$ and record the running times.
  4. Compare these running times with t and return the shortest one and the corresponding order.

- **Algorithm** IVO.
- Input: a set of polynomials F; a sequence of variables X.
- Output: a permutation of X, which defines a descending variable order.
- Steps:
  1. If $X \leq 1$, return X.
  2. Let $O_B := \mathsf{SVO}(F, X, B)$ and $O_C := \mathsf{SVO}(F, X, C)$.
  3. Let t be the shorter running time between $\mathsf{CAD}(F, O_B)$ and $\mathsf{CAD}(F, O_C)$ and let O be the corresponding order with shorter time (if equal, we use O_B).
  4. For k from 0 to $n - 2$ do
     (a) $O, t := \mathsf{RVO}(F, O, t, k)$.
  5. Return O.

It is easy to see that IVO calls CAD at most $2 + \sum_{k=1}^{n}(k - 1) = (n^2 - n + 2)/2$ times. In the rest of this paper, if no confusion arises, we denote by $\mathsf{SVO}(*)$ an oracle that always returns the better ordering between $\mathsf{SVO}(B)$ and $\mathsf{SVO}(C)$ and by SVO either of the three.

4   A Machine Learning Approach

To train a useful machine learning model for predicting the best variable ordering for CAD, it is important to have a dataset of enough labelled examples and the size of the dataset cannot be too small. On the other hand, since computing CAD is expensive when the number of variables is larger than 3, a larger dataset demands more computing resources. To make the learned model useful, it is better that the training dataset contains diverse examples. On the other hand, if the data are too diverse, it will be hard to learn. The following table summarizes the information of the dataset we generate using random polynomials as input to IVO. The whole dataset is divided into three datasets, used respectively for training, validation and testing with ratio 9/1/1. The validation dataset is used for tuning the machine learning model while the testing dataset is treated as unseen data used only once for reporting experimental results in the paper and showing the generalization ability of the ML model.

Table 2. Dataset

n	Degree	#terms	#polynomials	Equations	#valid examples
4	2..3	2..5	2	No	10957
5	2..3	3..6	2	No	6875
6	2..3	4..6	2	No	3751

The data in Table 2 were generated on a cluster (4 compute nodes, each of which has two Intel E5-2620 CPU (6-core each) and 64 GB memory). On each node, 6 Maple sessions were run in parallel. The time limit is set as 15 min. The total time for generating the dataset is about 1 month. In Table 2, we only record the number of valid examples. An example is valid if CAD finishes the computation within the time limit for at least one ordering computed by IVO. Note that it is possible that SVO returns an ordering for which CAD times out.

Next, we recall the features to represent the polynomials introduced in our earlier work [24]. These features are generated based on a graph structure defined for polynomial systems. For a given variable x_i, $i = 1, \ldots, n$, an equivalent description of the features is summarized in the following Table 3. Let $E(i)$ be the features associated with x_i. Then the feature vector $E = \cup_{i=1}^n E(i)$.

Table 3. Features

Feature	Description
$E_1(x_i)$	$|\{x_j : x_j, j \neq i, \text{ appears in the same polynomial as } x_i\}|$
$E_2(x_i)$	$|\{f \in F : x_i \text{ appears in } f\}|$
$E_3(x_i)$	$\max_{f \in F}\{\deg(f, x_i)\}$
$E_4(x_i)$	$\sum_{f \in F} \deg(f, x_i)$
$E_5(x_i)$	$\max_{f \in F}\{\deg(lc(f, x_i))\}$, where lc denotes for leading coefficient
$E_6(x_i)$	$\max_{f \in F}\{|\{M : M \text{ is a monomial of } f \text{ and } x_i|M\}|\}$
$E_7(x_i)$	$\max_{f \in F}\{\deg(M) : M \text{ is a monomial of } f \text{ and } x_i|M\}$
$E_8(x_i)$	$\sum_{f \in F} \sum_M \text{ is a monomial of } f \text{ and } x_i|M \deg(M, x_i)$
$E_9(x_i)$	$\sum_{f \in F}\{\deg(lc(f, x_i))\}$
$E_{10}(x_i)$	$\sum_{f \in F} |\{M : M \text{ is a monomial of } f \text{ and } x_i|M\}|$

We aim to train a model which can predict variable orders for $n = 4, 5, 6$. Instead of treating a variable order as a class, which may lead to huge number of classes for a fixed n, we would like to train a multiclass classifier M_n, which only predicts the largest variable in an ordering. To achieve this, for each example in the dataset, we will call SVO to return an initial ordering, and then call RVO once to get a hopefully better ordering. Then the first variable in the ordering

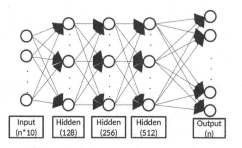

Fig. 1. The neural network classifier

is set as the label of the example. For each $n = 4, 5, 6$, we train an artificial neural network classification model implemented in TensorFlow. The structure and parameters of the neural network are illustrated in Fig. 1. Each M_n is a full-connected neural network with one input layer, three hidden layers and one softmax output layer. The activation function used is ReLu. Hyperparameters of the network are hand-tuned to maximize validation accuracy.

Suppose that we have obtained the well-trained models M_n, $n = 4, 5, 6$. We then employ the following procedure PVO to predict the variable ordering.

- **Algorithm** PVO
- Input: a set of polynomials F; a sequence of n variables X.
- Output: a permutation of X, which defines a descending variable order.
- Steps:
  1. Compute a sequence of feature vectors $E = E(1), \ldots, E(n)$ for F.
  2. Let $O := \mathsf{SVO}(F, X)$.
  3. Let $x_i := M_n(E)$.
  4. Return $O \setminus \{x_i\}$.

The overall training and predicting process is depicted in Fig. 2.

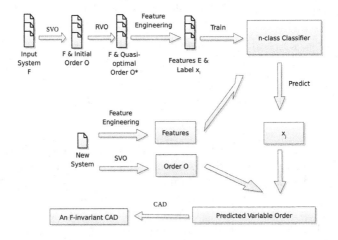

Fig. 2. The flow graph for finding variable order based on an artificial neural network.

5   Experiments

In this section, we report on the experimental results of the iterative method and the machine learning approach for selecting variable orderings.

Note that when we call PVO to predict the variable ordering, we have three options. One can use $\mathsf{SVO}(B)$ or $\mathsf{SVO}(C)$ to get an initial ordering without calling CAD. Or one can use $\mathsf{SVO}(*)$ to get the better one between $\mathsf{SVO}(B)$ and $\mathsf{SVO}(C)$, but this requires calling CAD twice. Nevertheless, for $a = B, C, *$, we use IVO(a), RVO(a) and PVO(a) to denote the iterative method, the first iteration of the iterative method and the ML-based method for variable ordering selection which uses SVO(a) as an initial ordering. As a result, the testing dataset for $a = B, C, *$ is respectively the set of examples, for which SVO(a) provides the initial ordering in the original testing dataset. Table 4 summarizes the size of the datasets. Note that the dataset for $a = C$ does not contain timeout examples. This is because whenever CAD times out with the ordering given by SVO(C), IVO will use the ordering given by SVO(B). Although the testing datasets and inference procedures for $a = B$ and $a = C$ are different, for a given n, both $a = B$ and $a = C$ use the same classifier trained with the dataset in Table 2, where the examples are labelled by RVO(*) with an initial variable ordering provided by SVO(*). Table 5 summarizes the average computation times (in seconds) and timeout rates of IVO(*) and RVO(*) for the whole datasets. Note that the datasets only contain examples on which IVO(*) succeeds.

Table 4. Size of testing datasets

n	B	C	*	n	B	C	*	n	B	C	*
4	593	404	997	5	359	266	625	6	214	127	341

Table 5. Comparison between IVO(*) and RVO(*)

n	RVO(*) (time)	IVO(*) (time)	RVO(*) (timeout)	IVO(*) (timeout)
4	7.46	2.25	0.5225	0
5	38.45	9.91	0.3025	0
6	63.69	18.36	0.1954	0

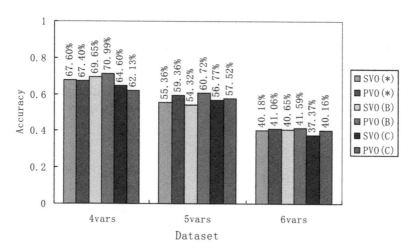

Fig. 3. Accuracies of different order selecting methods

Figure 3 summarizes the accuracies of six ordering selecting methods. Here the ground truth for SVO(a) and PVO(a) is given by RVO(a), for $a = B, C, *$. The accuracy is defined as the percentage of best orderings chosen by each method over the total number of test examples, given in Table 4. We observe that the accuracy of PVO(a) is higher than SVO(a) for $n = 5, 6$, for all $a = B, C, *$. For $n = 4$, the accuracy of PVO(a) is slightly lower than SVO(a) for $a = C, *$ and higher than SVO(a) for $a = B$.

Figure 4 provides the average running time of CAD with the variable orderings predicted by different methods. If there is a timeout, the time is counted as twice the time limit, that is half an hour. For $n = 4, 5, 6$ and $a = *, B$, the average computation time of PVO(a) is considerably less than SVO(a). The average computation time of PVO(C) is less than SVO(C) for $n = 4, 6$ but greater than SVO(C) for $n = 5$.

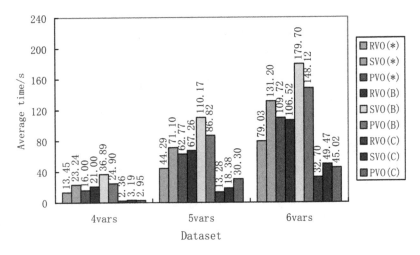

Fig. 4. Average running time of CAD with different variable orderings

Figure 5 gives the percentage of timeout examples for CAD with the variable orderings predicted by different methods. The phenomenon here is similar to the computation time as illustrated in Fig. 4.

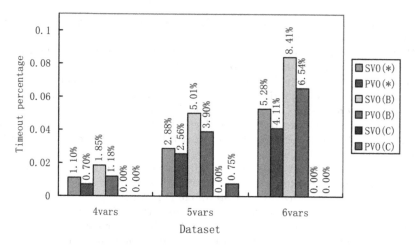

Fig. 5. Percentage of timeout examples for CAD with different variable orderings

6 Conclusion and Future Work

In this paper, we propose a machine learning based approach for predicting the best variable ordering for CAD targeting on $n > 3$. The experiments show that it outperforms traditional heuristic approaches for $n = 4, 5, 6$ on randomly generated datasets. The CylindricalAlgebebraicDecompose command can compute CAD for even larger n, say $n \leq 10$ in reasonable time if there are several equational constraints in the system [7]. It will be interesting to extend the model for $n > 6$, test it on CAD-QE problems from real applications and finally make the ML-based variable ordering selection method a useful option for Suggest-VariableOrder. The data and code used in this paper is available at http://doi.org/10.5281/zenodo.3818086.

Acknowledgments. The authors would like to thank anonymous referees for helpful comments. This research was supported by NSFC (11771421, 11671377, 61572024), CAS "Light of West China" Program, the Key Research Program of Frontier Sciences of CAS (QYZDB-SSW-SYS026), and cstc2018jcyj-yszxX0002 of Chongqing.

References

1. Arnon, D.S., Collins, G.E., McCallum, S.: Cylindrical algebraic decomposition I: the basic algorithm. SIAM J. Comput. **13**(4), 865–877 (1984)
2. Bradford, R., Chen, C., Davenport, J.H., England, M., Moreno Maza, M., Wilson, D.: Truth table invariant cylindrical algebraic decomposition by regular chains. In: Gerdt, V.P., Koepf, W., Seiler, W.M., Vorozhtsov, E.V. (eds.) CASC 2014. LNCS, vol. 8660, pp. 44–58. Springer, Cham (2014). https://doi.org/10.1007/978-3-319-10515-4_4
3. Bradford, R.J., Davenport, J.H., England, M., McCallum, S., Wilson, D.J.: Truth table invariant cylindrical algebraic decomposition. J. Symb. Comput. **76**, 1–35 (2016)

4. Brown, C.: Tutorial: Cylindrical algebraic decomposition, at ISSAC (2004). http://www.usna.edu/Users/cs/wcbrown/research/ISSAC04/handout.pdf

5. Brown, C.W.: Improved projection for cylindrical algebraic decomposition. J. Symb. Comput. **32**(5), 447–465 (2001)

6. Brown, C.W., Davenport, J.H.: The complexity of quantifier elimination and cylindrical algebraic decomposition. In: Proceedings of ISSAC, pp. 54–60 (2007)

7. Chen, C., Moreno Maza, M.: An incremental algorithm for computing cylindrical algebraic decompositions. In: Feng, R., Lee, W., Sato, Y. (eds.) Computer Mathematics, pp. 199–221. Springer, Heidelberg (2014). https://doi.org/10.1007/978-3-662-43799-5_17

8. Chen, C., Ren, C., Ye, X.J., Chen, J.L.: Mapping criteria between nonlocality and steerability in qudit-qubit systems and between steerability and entanglement in qubit-qudit systems. Phys. Rev. A **98**(5), 052114 (2018)

9. Chen, C., et al.: Solving semi-algebraic systems with the RegularChains library in Maple. In: Proceedings of MACIS, pp. 38–51 (2011). in the long version

10. Chen, C., Moreno Maza, M.: Algorithms for computing triangular decomposition of polynomial systems. J. Symb. Comput. **47**(6), 610–642 (2012)

11. Chen, C., Moreno Maza, M.: Cylindrical algebraic decomposition in the RegularChains library. In: Hong, H., Yap, C. (eds.) ICMS 2014. LNCS, vol. 8592, pp. 425–433. Springer, Heidelberg (2014). https://doi.org/10.1007/978-3-662-44199-2_65

12. Chen, C., Moreno Maza, M.: Quantifier elimination by cylindrical algebraic decomposition based on regular chains. J. Symb. Comput. **75**, 74–93 (2016)

13. Chen, C., Moreno Maza, M., Xia, B., Yang, L.: Computing cylindrical algebraic decomposition via triangular decomposition. In: Proceedings of ISSAC, pp. 95–102 (2009)

14. Collins, G.E.: Quantifier elimination for real closed fields by cylindrical algebraic decompostion. In: Brakhage, H. (ed.) GI-Fachtagung 1975. LNCS, vol. 33, pp. 134–183. Springer, Heidelberg (1975). https://doi.org/10.1007/3-540-07407-4_17

15. Dolzmann, A., Seidl, A., Sturm, T.: Efficient projection orders for CAD. In: Proceedings of ISSAC, pp. 111–118. ACM (2004)

16. England, M., Florescu, D.: Comparing machine learning models to choose the variable ordering for cylindrical algebraic decomposition. In: Kaliszyk, C., Brady, E., Kohlhase, A., Sacerdoti Coen, C. (eds.) CICM 2019. LNCS (LNAI), vol. 11617, pp. 93–108. Springer, Cham (2019). https://doi.org/10.1007/978-3-030-23250-4_7

17. Florescu, D., England, M.: Improved cross-validation for classifiers that make algorithmic choices to minimise runtime without compromising output correctness. In: Slamanig, D., Tsigaridas, E., Zafeirakopoulos, Z. (eds.) MACIS 2019. LNCS, vol. 11989, pp. 341–356. Springer, Cham (2020). https://doi.org/10.1007/978-3-030-43120-4_27

18. Huang, Z., England, M., Wilson, D.J., Bridge, J.P., Davenport, J.H., Paulson, L.C.: Using machine learning to improve cylindrical algebraic decomposition. Math. Comput. Sci. **13**(4), 461–488 (2019)

19. Lazard, D.: An improved projection for cylindrical algebraic decomposition. In: Bajaj, C.L. (ed.) Algebraic Geometry and Its Applications, pp. 467–476. Springer, New York (1994). https://doi.org/10.1007/978-1-4612-2628-4_29

20. McCallum, S.: An improved projection operator for cylindrical algebraic decomposition. In: Caviness, B., Johnson, J. (eds.) Quantifier Elimination and Cylindrical Algebraic Decomposition. Texts and Monographs in Symbolic Computation, pp. 242–268. Springer, Vienna (1998)

21. McCallum, S., Parusiński, A., Paunescu, L.: Validity proof of Lazard's method for CAD construction. J. Symb. Comput. **92**, 52–69 (2019)
22. Strzeboński, A.: Solving systems of strict polynomial inequalities. J. Symb. Comput. **29**(3), 471–480 (2000)
23. Strzeboński, A.: Cylindrical algebraic decomposition using validated numerics. J. Symb. Comput. **41**(9), 1021–1038 (2006)
24. Zhu, Z., Chen, C.: Variable order selection for cylindrical algebraic decomposition based on machine learning. J. Syst. Sci. Math. (2018, accepted). (in Chinese)
25. Zhu, Z., Chen, C.: Variable ordering selection for cylindrical algebraic decomposition based on a hierarchical neural network. Comput. Sci. (2020, accepted). (in Chinese)

Applying Machine Learning to Heuristics for Real Polynomial Constraint Solving

Christopher W. Brown$^{(\boxtimes)}$ and Glenn Christopher Daves

United States Naval Academy, Annapolis, USA
{wcbrown,m201362}@usna.edu

Abstract. This paper considers the application of machine learning to automatically generating heuristics for real polynomial constraint solvers. We consider a specific choice-point in the algorithm for constructing an open Non-uniform Cylindrical Algebraic Decomposition (NuCAD) for a conjunction of constraints, and we learn a heuristic for making that choice. Experiments demonstrate the effectiveness of the learned heuristic. We hope that the approach we take to learning this heuristic, which is not a natural fit to machine learning, can be applied effectively to other choices in constraint solving algorithms.

Keywords: Non-linear polynomial constraints · Machine learning

1 Introduction

In [2] the first author proposed Non-uniform Cylindrical Algebraic Decomposition (NuCAD) as an alternative to the well-known Cylindrical Algebraic Decomposition (CAD) as a data-structure for representing sets of points in Euclidean space defined by boolean combinations of real polynomial equalities and inequalities. The process of constructing a NuCAD involves many points at which an arbitrary choice needs to be made—a choice that does not affect correctness, but can have a considerable impact on running time, memory usage, and quality of solution. The purpose of this work is to consider one such choice, and attempt to use machine learning to automatically learn a successful heuristic. This paper will introduce the problem, explain why the application of machine learning to the problem is exceptionally challenging, describe the process we developed to handle not just this heuristic-learning problem, but others with similar challenges, and report experimental results.

2 The Problem

A *semi-algebraic set* is a set of points in Euclidean space defined by a boolean combination of polynomial equalities and inequalities (known as a *Tarski formula*). Non-uniform Cylindrical Algebraic Decomposition (NuCAD) is a data structure providing an explicit representation of semi-algebraic sets. From this

A. M. Bigatti et al. (Eds.): ICMS 2020, LNCS 12097, pp. 292–301, 2020.
https://doi.org/10.1007/978-3-030-52200-1_29

data structure, a variety of questions can be answered. In this article we only address the question of whether the semi-algebraic set is non-empty or, equivalently, whether the associated Tarski formula is satisfiable. For example, to determine the satisfiability of the Tarski formula $[x^2 + y^2 < 1 \wedge y > x^2 \wedge 3x > 2y^2 + 1]$, we would construct a NuCAD data structure representing the decomposition of $\mathbb{R}^2$ depicted in Fig. 1.

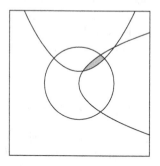 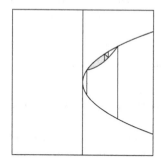

Fig. 1. Depicted on the left is the semi-algebraic set defined by the Tarski formula $[x^2 + y^2 < 1 \wedge y > x^2 \wedge 3x > 2y^2 + 1]$ along with the curves defined by the three polynomials in the formula. On the right is depicted the decomposition described by a NuCAD data structure produced from the formula. This is not unique, as different choices during the construction lead to different NuCADs.

The NuCAD data structure contains a sample point for each cell in the decomposition it represents. In this case, two cells have sample points at which the input formula is satisfied. Thus, not only do we learn that the formula is satisfiable, but we also have witness points to prove it.

NuCADs are constructed by a simple refinement process. At each step a cell is selected, and from amongst the constraints that are violated at its sample point, one is chosen to be the basis for splitting the selected cell into subcells—thus refining the decomposition. The key idea is that the subcell containing the original sample point has the property that the chosen constraint is violated throughout the subcell. To illustrate, suppose that in the previous example our initial sample point was $(0, 0)$. Two constraints (the parabolas) are violated at this point, so we must choose one. It is this choice that we concentrate on in the present paper. We would like to learn a heuristic for it. To be precise, we give the following definition of a *choice function*.

Definition 1. *A choice function maps a set of multivariate polynomials and a variable ordering to an element of the input polynomial set. When a cell has been selected to refine, the set of polynomials from constraints that are violated at the selected cell's sample point is given as input to a choice function, and the output of the choice function is the constraint that will be used for the next refining step. Additional input consisting of state information from the selected* *cell and the NuCAD as a whole (e.g. the polynomials defining the selected cell's boundaries) are optionally allowed.*

It is important to point out that making good choices vs. bad choices can make orders of magnitude differences in computing time and number of cells in the resulting NuCAD. Of course for very easy problems, the difference is not that great. Likewise, there are combinations of problem and initial point for which, at every sample point, there is only one violated constraint. This means that there never is a real choice, and so the heuristic is not needed. However, as our experiments show, there are also problems for which a good choice heuristic vs. bad choice heuristic means the difference between problems that are solvable within a reasonable amount of time and those that are not.

3   ML and and Why Learning a Choice Function Is Hard

Machine Learning (ML) is a huge field, and one very much in the spotlight at the moment. Moreover, machine learning has been applied to related problems (see for example [5–8]), though in a significantly different ways than what we do in this paper. So we will not try to explain what machine learning is, and trust that any reader who is not sufficiently conversant in the subject will have many options for obtaining the necessary background. Machine learning typically centers around trying to learn a function that maps input feature vectors to output result vectors. The dimensions and component types of these vectors is fixed for a given learning problem. Also pertinent to this discussion is the basic categorizations of learning as "supervised" vs. "unsupervised" and "regressions" vs. "classification".

Unfortunately, learning a choice function does not fit these typical machine learning paradigms well. There is no limit on the size of the input space, since there is no bound on the number of polynomials in the input set, nor is there a bound on the number of variables, degrees, or number of terms in the individual polynomials within the set. So it does not even fit the general framework. It does not fit as a regression problem, because we don't have continuous outputs (the output being one polynomial from the set). It doesn't fit as a classification problem, because each polynomial from the input set constitutes a class label, and there is not a fixed number of polynomials in the set (classification usually assuming a fixed set of labels). It does not fit as a supervised learning problem because we have no ground truth, i.e. we do not know what the "right" decisions are, so we do not have labeled data[1]. Unsupervised learning, at first, seems like it might be promising. Specifically, we could try to cast learning a choice function as a reinforcement learning problem, in which a decision made during the construction of a NuCAD for an input is rewarded or penalized by the success of the final result, appropriately discounted by its distance from that result. Unfortunately, there are two problems with this approach. First is that we have no basis for determining an award or penalty. If we end up with a decomposition consisting of 500 cells, is that good? Perhaps a different sequence of choices would have yielded a decomposition into 20 cells, or perhaps 500 is in fact the minimum achievable. With no way of knowing, we cannot determine an award or penalty. Second, the choices made during the construction of a NuCAD are not generally linear. The first choice, for example, produces several sibling cells, and the further refinement of each of those sibling cells is independent of the others. It is then unclear how to split up rewards.

Another substantial impediment to applying ML to the problem of producing an effective choice function is the very common problem of finding good data. In the case of polynomial constraints, we have the issue that the problem space is so vast, that it is not feasible to have anything like reasonable coverage of the space. Moreover, it is typically observed that most symbolic algorithms for dealing with polynomial constraints behave very differently on random problems than on "real-world" problems. This means that if we learn from randomly generated inputs we have to be concerned that the result might not transfer well to real-world problems or, indeed, problems that were not generated in a similar way to the training data. While we do have the SMT-LIB [1] QF_NRA library of problems as a valuable resource, we have to be careful learning from it as well. This is specifically true because most problems come from a large group generated by one of a relatively few applications. Typically all the problems within a group are very similar. This means, that if you learn from some of the problems in a given group and you evaluate on others from the same group, the learned function may test well because it has memorized correct actions for that group. This may result in overfitting that would only get exposed if you were to test on problems that were not from any of the groups that were used in

---

[1] In some other contexts in which machine learning has been applied to learn heuristic choice functions in symbolic computing, for example in work about learning to choose good variable orderings, the approach has been to compute the results of all possible choices, so that one does indeed know ground truth. That is impractical in our context for the following reason: in order to determine the best option for the first choice-point, it is not enough to try each of the options for just that choice; one has to make the optimum decision for each of the follow-on choices until NuCAD construction is complete. This means that one would have to try every option at every choice-point! One can compute that for the three-variable formula $\wedge_{i=1}^{3}(0 < x_i + 1/5 \wedge 0 < x_i + 2/5 \wedge 0 > x_i + 3/5 \wedge 0 > x_i + 4/5)$, which consists solely of linear univariate constraints, there are 5,760 ways that these choices can be made, each resulting in a different NuCAD data structure in the end. By comparison, for three variables there are exactly six variable orders to consider. So this exhaustive approach to finding ground truth to learn from is not feasible for our problem.

training. Finally, the current implementation of NuCAD is for "open" NuCAD only, and only for conjunctions. This means that we are restricted to problems from the SMT-LIB that are conjunctions of strict inequalities.

4 Using ML to Learn Choice Functions

The preceding examination of all the reasons why it might not appear to be promising to apply ML to the problem of learning choice functions is important to understanding why we developed the approach described below.

Reduce to Binary Classification: Trying to learn a choice function directly is problematic in part because the number of classes is not fixed, rather it is determined by the size of the input set of polynomials, which varies from choice-to-choice. In a crucial change of perspective, we redefine our problem to learning a *choice ordering predicate*, i.e. a function that takes two polynomials (and optionally extra information about the underlying cell and NuCAD context), and returns 1 if the first polynomial is a preferable choice to the second, and 0 otherwise. The choice function applied to a set of n polynomials then becomes $n - 1$ applications of the choice ordering predicate, at each step retaining the polynomial preferred by the predicate. If the choice ordering predicate fails to induce a proper total order, the order in which the predicate is applied to the polynomials in the set could affect the result. In the present experiments we accept this fact.

Reduce Inputs to Fixed Size Feature Set: Having restricted the learning problem to learning a choice ordering predicate, we have reduced the dimension of the input vector to the function to be learned, but not yet fixed it to a constant. After all, while there are now always two input polynomials, rather than an arbitrary sized set, those polynomials are still unbounded in the number of variables, degrees, number of terms and coefficient sizes. The solution to this is clear cut from a machine learning perspective: extract from the two polynomials (and optionally from the cell and NuCAD context) a fixed sized set of *features*. Although part of the revolution in deep learning is that the learning algorithm is, in some sense, supposed to do this "feature engineering" for us, in this case our domain forces us to do it. However, because of our restriction to learning choice ordering predicates, there are a number of natural *comparative features*, for example the difference in $level^2$ of the two input polynomials, or the difference in the two polynomials' total degrees, number of terms, etc.

Reduce to Supervised Learning: The biggest hurdle to applying ML to this problem is that, as described above, it fits neither the supervised nor unsupervised paradigms. To address this, we use the fact that we have total control of the executing algorithm. In particular, for a given input problem, we execute

---

[2] Prior to constructing a NuCAD, a variable ordering is fixed, and the level of a polynomial is highest index within that ordering of any variable appearing in the polynomial.

NuCAD construction following the "current" choice ordering predicate, but we stop at a random point in the process at which we have more than one constraint violated at a sample point and thus have a choice to make. At this point, we separately try each of the possible choices (i.e. each polynomial), finishing out the entire construction from that point using the "current" choice ordering predicate for all subsequent choices. This gives us the exact number of cells resulting from each polynomial we could have chosen. Thus, for each pairing of polynomials from the set, we construct the associated feature vector and, since we know the correct result for the choice ordering predicate with that pair as input, we have the correct output to go along with it. By collecting these feature vectors and correct output values over many inputs, we produce a data set that can be used for a supervised, binary classification machine learning problem.

Of course, the "correct" results we gathered are only correct for what was then the "current" choice ordering predicate, and by learning we have derived a new choice ordering predicate. This means we should iterate the process until the performance of NuCAD construction using the learned choice ordering predicate converges. It is not clear that this convergence will happen. For example, the initial choice ordering function is chosen at random, and it could easily happen that NuCADs constructed with it are so far off from NuCADs constructed with a good choice ordering predicate, that the data we try to learn from is garbage. However, our experiments indicate that we get good performance in relatively few iterations.

Learn from Randomly Generated Inputs, Evaluate With "Real" Input: Our approach was to learn from randomly generated formulas—formulas in five variables with many constraints, both linear and non-linear, different levels of sparsity, and different numbers of variables appearing. Each input formula was used to analyze only one choice-point. The rationale for this choice is simply that we need a lot of data to learn from. The justification is that our initial experiments show that the learned choice ordering predicate does indeed transfer well to the "real" problems pulled from SMT-LIB, despite the fact that they often have very different shape and that all the problems we tested against had fewer than five variables.

5 Experiments

The TARSKI[3] system [9] was used in our experiments for its implementation of NuCAD. For our machine learning, we used the Keras [4] package.

Random formulas were generated to consist of 13 "<" constraints, eight quadratic and five linear, with between two and four terms. Below is an example of one of them.

$$
\begin{aligned}
[\,0 < {} & 18v - 9w \wedge 0 < 9vx + 4vy - 3y + 3 \wedge 0 < 5yz + 8y + 2 \wedge 0 < -4x^2 + 6w - 6y - 7 \wedge 0 < -8wz \\
& + 9z^2 + 8w \wedge 0 < -3vz + 6w - 8z - 4 \wedge 0 < v^2 + 22vy + 13y^2 + 4 \wedge 0 < -27w^2 - 72wz - 48z^2 \\
& - 20v + 14w - 8z \wedge 0 < -10w - 10y - 5z \wedge 0 < 9v + 8w - 4x \wedge 0 < -8v + 7w + 7x \wedge 0 < -8v \\
& - 5w + 1 \wedge 0 < -4w - 8x + 1\,]
\end{aligned}
\tag{1}
$$

---

[3] TARSKI is available from https://www.usna.edu/Users/cs/wcbrown/tarski/.

Given a choice-point in a NuCAD construction and two polynomials, we construct a vector of 22 features capturing individual and comparative properties of the two polynomials and the context of the NuCAD construction. These features are described in the table below. The first six are comparative features based on the two polynomials under consideration (p_1 and p_2). Note that "spsize" is a function that measures the size of the internal representation of a polynomial—it is roughly proportional to the print length, but without any dependency on the length of variable names. Features 6–10 are comparative, but based on "p_1^*" and "p_2^*". Here p_1^* is the same as p_1, but with any term removed for which p_2 has a term with the same power product, and p_2^* is defined analogously. For example, if $p_1 = 2xz^2 - 3zy + x + 1$ and $p_2 = -xz^2 - 2y^2 + y - 5x$ then $p_1^* = -3zy + 1$ and $p_2^* = -2y^2 + y$. Features 11–14 involve a process we call "pseudo-projection", which gives estimates of the size of the projection set arising from choosing a particular polynomial with which to refine the current cell. It gives a very rough estimate, as no resultants or discriminants are computed. Features 15–18 are based on geometric information concerning the roots of $p_1(\alpha_1, \ldots, \alpha_{n-1}, z)$ and $p_2(\alpha_1, \ldots, \alpha_{n-1}, z)$, where α is the cell's sample point. Feature 21 is an arbitrary feature—in the context of learning, it is pure noise.

0	$\texttt{ite}\,\text{tdeg}\,p_1 = 1 \land \text{tdeg}\,p_2 > 1, -1, (\texttt{ite}\,\text{tdeg}\,p_1 > 1 \land \text{tdeg}\,p_2 = 1, +1, 0)$
1	$(\text{level}\,p_1 - \text{level}\,p_2)/(n-1)$, where $n =$ number of variables
2	$(\text{tdeg}\,p_1 - \text{tdeg}\,p_2)/5$
3	$\text{spsize}\,p_1 - \text{spsize}\,p_2$, where spsize is the internal data structures size
4	degree of p_1 in its main variable
5	degree of p_2 in its main variable
6–10	same as 1–5 except with p_1^* and p_2^*
11–14	pseudo-projection sizes and weighted sizes for p_1 and p_2
15	number of roots of $p_1(\alpha_1, \ldots, \alpha_{n-1}, z)$ inside cell
16	number of roots of $p_2(\alpha_1, \ldots, \alpha_{n-1}, z)$ inside cell
17	± 1 according to which polynomial gives a weaker lower bound over α
18	± 1 according to which polynomial gives a weaker upper bound over α
19	number of constraint polynomials known to be sign-invariant in cell
20	number of constraint polynomials not known to be sign-invariant in cell
21	± 1 based on hashes of p_1 and p_2

There is an existing hand-crafted heuristic used in TARSKI, called "BPC", against which we will compare our learned heuristic. It is expressible in terms of features zero, one and three of the feature vector F for polynomials p_1 and p_2, but we describe it more directly here as follows: if exactly one of p_1 and p_2 has total degree one, choose it; otherwise choose the polynomial with lower level, breaking ties by choosing the polynomial with smaller "spsize". The first feature in our feature vector was included so that other, less capable, learning paradigms (like decision lists) would still be expressive enough to learn BPC. For neural nets we could have replaced features zero and one with $\text{tdeg}\,p_1$ and $\text{tdeg}\,p_2$, which would have been more natural, but these same feature vectors were used for experiments outside of the scope of this paper.

For our learning model, we use a feed forward neural network that takes a dense network with 22 inputs and dimension $22 \times 22 \times 5 \times 5 \times 1$ with ReLU activation functions for internal nodes and sigmoid activation function for the output node. This network, seeded with random weights initially, represents the "current" choice ordering predicate.

Learning proceeded in rounds. In each round the neural net is used as the choice ordering predicate for producing data, as described in the previous section. From this training data, a new set of network weights was learned (starting from a different set of random weights), and this newly learned function served as the "current" choice ordering predicate for the subsequent round. This process ran for 10 rounds, producing the initial (randomly weighted) choice ordering predicate, and 10 learned predicates. Figure 2 shows the performance of the learned heuristics. Also shown is performance of "BPC" the hand-crafted heuristic that is used by default in our implementation of NuCAD. The performance does not smoothly increase from one iteration to the next, nor show the classic "U" shape often seen in machine learning, which warrants further investigation. The neural net resulting from the 6th iteration, "NN06", is the best performing network on these test problems.

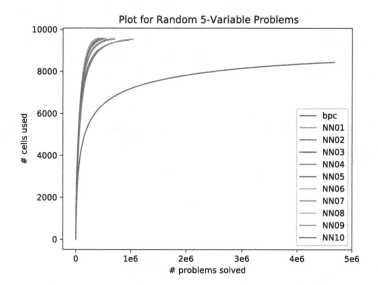

Fig. 2. This plot shows the performance of the trained neural networks for each round on the test data, which consists of problems generated randomly in the same way as the training data. Though difficult to see, "NN06", the network generated after the 6th round performs the best. Also shown is the performance of "BPC" the hand-crafted heuristic that is used by default in our implementation of NuCAD.

The crucial question is whether our learned choice ordering predicate transfers to problems that aren't of the same "shape" as the problems on which we trained. To evaluate this, we considered two data sets. The first is a set of 4,235 problems from [3]. These problems are conjunctions of strict inequalities derived from a subset of problems in the SMT-LIB QF_NRA collection by simplifications and case splittings. Essentially, in these problems "simplification" alone was insufficient, so we need to turn to a solver, like NuCAD. They are a natural choice for us to test with, as they meet the requirements of the current NuCAD implementation, namely that they are conjunctions of strict inequalities. None of these problems are terribly difficult—almost all have three or fewer variables—so we also produced a more challenging set consisting of 1,000 randomly generated four variable problems, generated in a different way than our test problems: each formula consisted of $x > 0 \land y > 0$ and nine conditions generated in Maple as `randpoly([x,y,z,w],degree=2,terms=4,coeffs=rand(-9..9))>0`. These examples differ from the training data in a number of ways. They have four variables instead of five. The only linear constraints are $x > 0$ and $y > 0$, whereas the training examples had five linear constraints, each of which was a multi-term constraint, like $-8v+7w+7x > 0$, that ties variables together. Finally the eight non-linear constraints in the training data were generated differently. Figure 3 shows the performance of BPC, NN06, and NN00, a randomly weighted neural net, which serves to show the performance of a random choice function. These experiments indicate that what gets learned from the training set does transfer over to other, unrelated input formulas. Obviously more testing on a wider range of inputs is required to be very confident of this, but these results are promising.

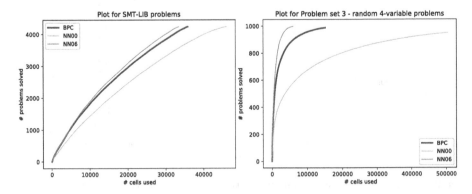

Fig. 3. Plots showing the performance of the hand-crafted heuristic BPC, the randomly weighted network NN00, and NN06, the trained network that performed best on the original set of training data. The left plot shows performance for the problem set stemming from the SMT-LIB, and the right plot shows performance on the four-variable randomly generated problem set. Note that NN00 failed to solve 51 problems within the 60 s timeout, BPC failed 13 problems within the 60 s timeout, and NN06 failed to solve two problems within the 60 s timeout.

6   Future Work and Acknowledgments

There are several avenues for future work that should be considered. The first is to improve upon what we've already done by training on a broader range of input formula "shapes"—different numbers of variables, different distributions for the constraints within formula, etc.—and by evaluating on a wider range of data sets, especially more "real" rather than randomly generated problems. The second avenue is to apply the basic approach outlined here to different problems, for example to the problem of selecting a variable ordering in Cylindrical Algebraic Decomposition, considered in [7], or choosing pivots in parametric Gaussian elimination.

Acknowledgements. This research was supported in part by the DOD High Performance Computing Modernization Program. We recognize and appreciate its support in enhancing our undergraduate education and research. Parts of this work were supported by National Science Foundation Grant 1525896.

References

1. Barrett, C., Fontaine, P., Tinelli, C.: The satisfiability modulo theories library (SMT-LIB) (2016). www.SMT-LIB.org
2. Brown, C.W.: Open non-uniform cylindrical algebraic decompositions. In: Proceedings of the 2015 ACM on International Symposium on Symbolic and Algebraic Computation, ISSAC 2015, pp. 85–92. ACM, New York (2015)
3. Brown, C.W., Vale-Enriquez, F.: From simplification to a partial theory solver for non-linear real polynomial constraints. J. Symb. Comput. **100**, 72–101 (2020). Symbolic Computation and Satisfiability Checking
4. Chollet, F., et al.: Keras (2015). https://keras.io
5. Florescu, D., England, M.: Algorithmically generating new algebraic features of polynomial systems for machine learning (2019)
6. Huang, Z., England, M., Davenport, J.H., Paulson, L.C.: Using machine learning to decide when to precondition cylindrical algebraic decomposition with Groebner bases. In: 18th International Symposium on Symbolic and Numeric Algorithms for Scientific Computing (SYNASC 2016), pp. 45–52, September 2016
7. Huang, Z., England, M., Wilson, D., Davenport, J.H., Paulson, L.C., Bridge, J.: Applying machine learning to the problem of choosing a heuristic to select the variable ordering for cylindrical algebraic decomposition. In: Watt, S.M., Davenport, J.H., Sexton, A.P., Sojka, P., Urban, J. (eds.) CICM 2014. LNCS (LNAI), vol. 8543, pp. 92–107. Springer, Cham (2014). https://doi.org/10.1007/978-3-319-08434-3_8
8. Kobayashi, M., Iwane, H., Matsuzaki, T., Anai, H.: Efficient subformula orders for real quantifier elimination of non-prenex formulas. In: Kotsireas, I.S., Rump, S.M., Yap, C.K. (eds.) MACIS 2015. LNCS, vol. 9582, pp. 236–251. Springer, Cham (2016). https://doi.org/10.1007/978-3-319-32859-1_21
9. Vale-Enriquez, F., Brown, C.W.: Polynomial constraints and unsat cores in TARSKI. In: Davenport, J.H., Kauers, M., Labahn, G., Urban, J. (eds.) ICMS 2018. LNCS, vol. 10931, pp. 466–474. Springer, Cham (2018). https://doi.org/10.1007/978-3-319-96418-8_55

A Machine Learning Based Software Pipeline to Pick the Variable Ordering for Algorithms with Polynomial Inputs

Dorian Florescu and Matthew England[✉]

Faculty of Engineering, Environment and Computing,
Coventry University, Coventry CV1 5FB, UK
fdorian88@gmail.com, Matthew.England@coventry.ac.uk

Abstract. We are interested in the application of Machine Learning (ML) technology to improve mathematical software. It may seem that the probabilistic nature of ML tools would invalidate the exact results prized by such software, however, the algorithms which underpin the software often come with a range of choices which are good candidates for ML application. We refer to choices which have no effect on the mathematical correctness of the software, but do impact its performance.

In the past we experimented with one such choice: the variable ordering to use when building a Cylindrical Algebraic Decomposition (CAD). We used the Python library Scikit-Learn (`sklearn`) to experiment with different ML models, and developed new techniques for feature generation and hyper-parameter selection.

These techniques could easily be adapted for making decisions other than our immediate application of CAD variable ordering. Hence in this paper we present a software pipeline to use `sklearn` to pick the variable ordering for an algorithm that acts on a polynomial system. The code described is freely available online.

Keywords: Machine learning · Scikit-learn · Mathematical software · Cylindrical algebraic decomposition · Variable ordering

1 Introduction and Context

Mathematical Software, i.e. tools for effectively computing mathematical objects, is a broad discipline: the objects in question may be expressions such as polynomials or logical formulae, algebraic structures such as groups, or even mathematical theorems and their proofs. In recent years there have been examples of software that acts on such objects being improved through the use of artificial intelligence techniques. For example, [21] uses a Monte-Carlo tree search to find the representation of polynomials that are most efficient to evaluate; [22] uses a machine learnt branching heuristic in a SAT-solver for formulae in Boolean logic; [18] uses pattern matching to determine whether a pair of elements from a specified group are conjugate; and [1] uses deep neural networks for premise selection

© Springer Nature Switzerland AG 2020
A. M. Bigatti et al. (Eds.): ICMS 2020, LNCS 12097, pp. 302–311, 2020.
https://doi.org/10.1007/978-3-030-52200-1_30

in an automated theorem prover. See the survey article [12] in the proceedings of ICMS 2018 for more examples.

Machine Learning (ML), that is statistical techniques to give computer systems the ability to *learn* rules from data, may seem unsuitable for use in mathematical software since ML tools can only offer probabilistic guidance, when such software prizes exactness. However, none of the examples above risked the correctness of the end-result in their software. They all used ML techniques to make non-critical choices or guide searches: the decisions of the ML carried no risk to correctness, but did offer substantial increases in computational efficiency. All mathematical software, no matter the mathematical domain, will likely involve such choices, and our thesis is that in many cases an ML technique could make a better choice than a human user, so-called magic constants [6], or a traditional human-designed heuristic.

Contribution and Outline

In Sect. 2 we briefly survey our recent work applying ML to improve an algorithm in a computer algebra system which acts on sets of polynomials. We describe how we proposed a more appropriate definition of model accuracy and used this to improve the selection of hyper-parameters for ML models; and a new technique for identifying features of the input polynomials suitable for ML.

These advances can be applied beyond our immediate application: the feature identification to any situation where the input is a set of polynomials, and the hyper-parameter selection to any situation where we are seeking to take a choice that minimises a computation time. Hence we saw value in packaging our techniques into a software pipeline so that they may be used more widely. Here, by pipeline we refer to a succession of computing tasks that can be run as one task. The software is freely available as a Zenodo repository here: https://doi.org/10.5281/zenodo.3731703

We describe the software pipeline and its functionality in Sect. 3. Then in Sect. 4 we describe its application on a dataset we had not previously studied.

2 Brief Survey of Our Recent Work

Our recent work has been using ML to select the variable ordering to use for calculating a cylindrical algebraic decomposition relative to a set of polynomials.

2.1 Cylindrical Algebraic Decomposition

A *Cylindrical Algebraic Decomposition* (CAD) is a *decomposition* of ordered $\mathbb{R}^n$ space into cells arranged *cylindrically*, meaning the projections of cells all lie within cylinders over a CAD of a lower dimensional space. All these cells are (semi)-algebraic meaning each can be described with a finite sequence of polynomial constraints. A CAD is produced for either a set of polynomials, or a logical formula whose atoms are polynomial constraints. It may be used to analyse these objects by finding a finite sample of points to query and thus understand the behaviour over all $\mathbb{R}^n$. The most important application of CAD is to perform Quantifier Elimination (QE) over the reals. I.e. given a quantified formula, a CAD may be used to find an equivalent quantifier free formula[1].

CAD was introduced in 1975 [10] and is still an active area of research. The collection [7] summarises the work up to the mid-90s while the background section of [13], for example, includes a summary of progress since. QE has numerous applications in science [2], engineering [25], and even the social sciences [23].

CAD requires an ordering of the variables. QE imposes that the ordering matches the quantification of variables, but variables in blocks of the same quantifier and the free variables can be swapped[2]. The ordering can have a great effect on the time/memory use of CAD, the number of cells, and even the underlying complexity [5]. Human designed heuristics have been developed to make the choice [3,4,11,14] and are used in most implementations.

The first application of ML to the problem was in 2014 when a support vector machine was trained to choose which of these heuristics to follow [19,20]. The machine learned choice did significantly better than any one heuristic overall.

2.2   Recent Work on ML for CAD Variable Ordering

The present authors revisited these experiments in [15] but this time using ML to predict the ordering directly (because there were many problems where none of the human-made heuristics made good choices and although the number of orderings increases exponentially, the current scope of CAD application means this is not restrictive). We also explored a more diverse selection of ML methods available in the Python library scikit-learn (sklearn) [24]. All the models tested outperformed the human made heuristics.

The ML models learn not from the polynomials directly, but from features: properties which evaluate to a floating point number for a specific polynomial set. In [20] and [15] only a handful of features were used (measures of degree and frequency of occurrence for variables). In [16] we developed a new feature generation procedure which used combinations of basic functions (average, sign, maximum) evaluated on the degrees of the variables in either one polynomial or the whole system. This allowed for substantially more features and improved the performance of all ML models. The new features could be used for any ML application where the input is a set of polynomials.

The natural metric for judging a CAD variable ordering is the corresponding CAD runtime: in the work above models were trained to pick the ordering which minimises this for a given input. However, this meant the training did not distinguish between any non-optimal ordering even though the difference between these could be huge. This led us to a new definition of accuracy in [17]: to picking an ordering which leads to a runtime within $x\%$ of the minimum possible.

---

[1] E.g. QE would transform $\exists x, ax^2 + bx + c = 0 \land a \neq 0$ into the equivalent $b^2 - 4ac \geq 0$.

[2] In Footnote 1 we must decompose (x, a, b, c)-space with x last, but the other variables can be in any order. Using $a \prec b \prec c$ requires 27 cells but $c \prec b \prec a$ requires 115.

We then wrote a new version of the `sklearn` procedure which uses cross-validation to select model hyper-parameters to minimise the total CAD runtime of its choices, rather than maximise the number of times the minimal ordering is chosen. This also improved the performance of all ML models in the experiments of [17]. The new definition and procedure are suitable for any situation where we are seeking to take a choice that minimises a computation time.

3   Software Pipeline

The input to our pipeline is given by two distinct datasets used for training and testing, respectively. An individual entry in the data set is a set of polynomials that represent an input to a symbolic computation algorithms, in our case CAD. The output is a corresponding sequence of variable ordering suggestions for each set of polynomials in the testing dataset.

The pipeline is fully automated: it generates and uses the CAD runtimes for each set of polynomials under each admissible variable ordering; uses the runtimes from the training dataset to select the hyper-parameters with cross-validation and tune the parameters of the model; and evaluates the performance of those classifiers (along with some other heuristics for the problem) for the sets of polynomials in the testing dataset.

We describe these key steps in the pipeline below. Each of the numbered stages can be individually marked for execution or not in a run of the pipeline (avoiding duplication of existing computation). The code for this pipeline, written all in Python, is freely available at: https://doi.org/10.5281/zenodo.3731703.

I. Generating a Model Using the Training Dataset

(a) Measuring the CAD Runtimes: The CAD routine is run for each set of polynomials in the training dataset. The runtimes for all possible variable orderings are stored in a different file for each set of polynomials. If the runtime exceeds a pre-defined timeout, the value of the timeout is stored instead.

(b) Polynomial Data Parsing: The training dataset is first converted to a format that is easier to process into features. For this purpose, we chose the format given by the `terms()` method from the `Poly` class located in the `sympy` package for symbolic computation in Python.

Here, each monomial is defined by a tuple, containing another tuple with the degrees of each variable, and a value defining the monomial coefficient. The polynomials are then defined by lists of monomials given in this format, and a point in the training dataset consists of a list of polynomials. For example, one entry in the dataset is the set $\{235x_1 + 42x_2^2, 2x_1^2x_3 - 1\}$ which is represented as

$$[[((1,0,0),235),((0,2,0),42)],[((2,0,1),2),((0,0,0),-1)]].$$

All the data points in the training dataset are then collected into a single file called `terms_train.txt` after being placed into this format. Subsequently, the file y_train.txt is created storing the index of the variable ordering with the minimum computing times for each set of polynomials, using the runtimes measured in Step I(a).

(c) **Feature Generation:** Here each set of polynomials in the training dataset is processed into a fixed length sequence of floating point numbers, called features, which are the actual data used to train the ML models in sklearn. This is done with the following steps:

i. **Raw feature generation**
   We systematically consider applying all meaningful combinations of the functions average, sign, maximum, and sum to polynomials with a given number of variables. This generates a large set of feature descriptions as proposed in [16]. The new format used to store the data described above allows for an easy evaluation of these features. An example of computing such features is given in Fig. 1. In [16] we described how the method provides 1728 possible features for polynomials constructed with three variables for example. This step generates the full set of feature descriptions, saved in a file called features_descriptions.txt, and the corresponding values of the features on the training dataset, saved in a file called features_train_raw.txt.

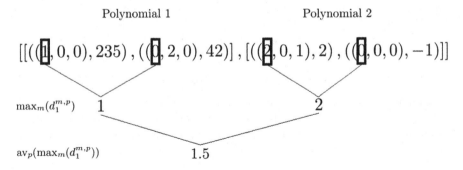

Fig. 1. Generating feature $av_p\left(\max_m\left(d_1^{m,p}\right)\right)$ from data stored in the format of Section I(b). Here $d_1^{m,p}$ denotes the degree of variable x_1 in polynomial number p and monomial number m, and av_p denotes the average function computed for all polynomials [16].

ii. **Feature simplification**
   After computing the numerical values of the features in Step I(c)i this step will remove those features that are constant or repetitive for the dataset in question, as described in [16]. The descriptions of the remaining features are saved in a new file called features_descriptions_final.txt.

iii. **Final feature generation**
   The final set of features is computed by evaluating the descriptions in features_descriptions_final.txt for the training dataset. Even though these were already evaluated in Step I(c)i we repeat the evaluation for the

final set of feature descriptions. This is to allow the possibility of users entering alternative features manually and skipping steps i and ii. As noted above, any of the named steps in the pipeline can be selected or skipped for execution in a given run. The final values of the features are saved in a new file called features_train.txt.

(d) Machine Learning Classifier Training:

i. Fitting the model hyperparameters by cross-validation

The pipeline can apply four of the most commonly used deterministic ML models (see [15] for details), using the implementations in sklearn [24].
- The K-Nearest Neighbors (KNN) classifier
- The Multi-Layer Perceptron (MLP) classifier
- The Decision Tree (DT) classifier
- The Support Vector Machine (SVM) classifier

Of course, additional models in sklearn and its extensions could be included with relative ease. The pipeline can use two different methods for fitting the hyperparameters via a cross-validation procedure on the training set, as described in [17]:
- Standard cross-validation: maximizing the prediction accuracy (i.e. the number of times the model picks the optimum variable ordering).
- Time-based cross-validation: minimizing the CAD runtime (i.e. the time taken to compute CADs with the model's choices).

Both methods tune the hyperparameters with cross-validation using the routine RandomizedSearchCV from the sklearn package in Python (the latter an adapted version we wrote). The cross-validation results (i.e. choice of hyperparameters) are saved in a file hyperpar_D**_**_T**_**.txt, where D**_** is the date and T**_** denotes the time when the file was generated.

ii. Fitting the parameters

The parameters of each model are subsequently fitted using the standard sklearn algorithms for each chosen set of hyperparameters. These are saved in a file called par_D**_**_T**_**.txt.

II. Predicting the CAD Variable Orderings Using the Testing Dataset

The models in Step I are then evaluated according to their choices of variable orderings for the sets of polynomials in the testing dataset. The steps below are listed without detailed description as they are performed similarly to Step I for the testing dataset.

(a) Polynomial Data Parsing: The values generated are saved in a new file called terms_test.txt.

(b) Feature Generation: The final set of features is computed by evaluating the descriptions in Step I(b)ii for the testing dataset. These values are saved in a new file called `features_test.txt`.

(c) Predictions Using ML: Predictions on the testing dataset are generated using the model computed in Step I(c). The model is run with the data in Step II(a)ii, and the predictions are stored in a file called `y_D**_**_T**_**_test.txt`.

(d) Predictions Using Human-Made Heuristics: In our prior papers [15–17] we compared the performance of the ML models with the human-designed heuristics in [4] and [11]. For details on how these are applied see [15]. Their choices are saved in two files entitled `y_brown_test.txt` and `y_sotd_test.txt`, respectively.

(e) Comparative Results: Finally, in order to compare the performance of the proposed pipeline, we must measure the actual CAD runtimes on the testing dataset. The results of the comparison is saved in a file with the template: `comparative_results_D**_**_T**_**.txt`.

Adapting the Pipeline to Other Algorithms

The pipeline above was developed for choosing the variable ordering for the CAD implementation in Maple's Regular Chains Library [8,9]. But it could be used to pick the variable ordering for other procedures which take sets of polynomials as input by changing the calls to CAD in Steps I(a) and II(e) to that of another implementation/algorithm. Step II(d) would also have to be edited to call an appropriate competing heuristic.

4 Application of Pipeline to New Dataset

The pipeline described in the previous section makes it easy for us to repeat our past experiments (described in Sect. 2) for a new dataset. All that is needed to do is replace the files storing the polynomials and run the pipeline.

To demonstrate this we test the proposed pipeline on a new dataset of randomly generated polynomials. We are not suggesting that it is appropriate to test classifiers on random data: we simply mean to demonstrate the ease with which the experiments in [15–17] that originally took many man-hours can be repeated with just a single code execution.

The randomly generated parameters are: the degrees of the three variables in each polynomial term, the coefficient of each term, the number of terms in a polynomial and the number of polynomials in a set. The means and standard deviations of these parameters were extracted from the problems in the `nlsat` dataset[3], which was used in our previous work [15] so that the dataset is of a

---

[3] https://cs.nyu.edu/~dejan/nonlinear/.

Table 1. The comparative performance of DT, KNN, MLP, SVM, the Brown and sotd heuristics on the testing data for our randomly generated dataset. A random prediction, and the virtual best (VB) and virtual worst (VW) predictions are also included.

	DT	KNN	MLP	SVM	Brown	sotd	rand	VB	VW
Prediction time (s)	$4.8 \cdot e^{-4}$	0.68	$2.8 \cdot e^{-4}$	0.99	53.01	15 819			
Total time (s)	6 548	6 610	6 548	6 565	6 614	22 313	16 479	5 610	25 461

comparable scale. We generated 7500 sets of random polynomials, where 5000 were used for training, and the remaining 2500 for testing.

The results of the proposed processing pipeline, including the comparison with the existing human-made heuristics are given in Table 1. The prediction time is the time taken for the classifier or heuristic to make its predictions for the problems in the training set. The total time adds to this the time for the actual CAD computations using the suggested orderings. We do not report the training time of the ML as this is a cost paid only once in advance. The virtual solvers are those which always make the best/worst choice for a problem (in zero prediction time) and are useful to show the range of possible outcomes. We note that further details on our experimental methodology are given in [15–17].

As with the tests on the original dataset [15,16] the ML classifiers outperformed the human made heuristics, but for this dataset the difference compared to the Brown heuristic was marginal. We used a lower CAD timeout which may benefit the Brown heuristic as past analysis shows that when it makes suboptimal choices these tend to be much worse. We also note that the relative performance of the Brown heuristic fell significantly when used on problems with more than three variables in [17]. The results for the sotd heuristic are bad because it had a particularly long prediction time on this random dataset. We note that there is scope to parallelize sotd which may make it more competitive.

5   Conclusions

We presented our software pipeline for training and testing ML classifiers that select the variable ordering to use for CAD, and described the results of an experiment applying it to a new dataset.

The purpose of the experiment in Sect. 4 was to demonstrate that the pipeline can easily train classifiers that are competitive on a new dataset with almost no additional human effort, at least for a dataset of a similar scale (we note that the code is designed to work on higher degree polynomials but has only been tested on datasets of 3 and 4 variables so far). The pipeline makes it possible for a user to easily tune the CAD variable ordering choice classifiers to their particular application area.

Further, with only a little modification, as noted at the end of Sect. 3, the pipeline could be used to select the variable ordering for alternative algorithms that act on sets of polynomials and require a variable ordering. We thus expect the pipeline to be a useful basis for future research and plan to experiment with its use on such alternative algorithms in the near future.

Acknowledgements. This work is funded by EPSRC Project EP/R019622/1: *Embedding Machine Learning within Quantifier Elimination Procedures.* We thank the anonymous referees for their comments.

References

1. Alemi, A., Chollet, F., Een, N., Irving, G., Szegedy, C., Urban, J.: DeepMath - deep sequence models for premise selection. In: Proceedings of the NIPS 2016, pp. 2243–2251 (2016). https://doi.org/10.5555/3157096.3157347
2. Bradford, R., et al.: Identifying the parametric occurrence of multiple steady states for some biological networks. J. Symb. Comput. **98**, 84–119 (2020). https://doi.org/10.1016/j.jsc.2019.07.008
3. Bradford, R., Davenport, J.H., England, M., Wilson, D.: Optimising problem formulation for cylindrical algebraic decomposition. In: Carette, J., Aspinall, D., Lange, C., Sojka, P., Windsteiger, W. (eds.) CICM 2013. LNCS (LNAI), vol. 7961, pp. 19–34. Springer, Heidelberg (2013). https://doi.org/10.1007/978-3-642-39320-4_2
4. Brown, C.: Companion to the tutorial: cylindrical algebraic decomposition. In: Presented at ISSAC 2004 (2004). http://www.usna.edu/Users/cs/wcbrown/research/ISSAC04/handout.pdf
5. Brown, C., Davenport, J.: The complexity of quantifier elimination and cylindrical algebraic decomposition. In: Proceedings of the ISSAC 2007, pp. 54–60. ACM (2007). https://doi.org/10.1145/1277548.1277557
6. Carette, J.: Understanding expression simplification. In: Proceedings of the ISSAC 2004, pp. 72–79. ACM (2004). https://doi.org/10.1145/1005285.1005298
7. Caviness, B., Johnson, J.: Quantifier Elimination and Cylindrical Algebraic Decomposition. Texts & Monographs in Symbolic Computation. Springer, Heidelberg (1998). https://doi.org/10.1007/978-3-7091-9459-1
8. Chen, C., Moreno Maza, M., Xia, B., Yang, L.: Computing cylindrical algebraic decomposition via triangular decomposition. In: Proceedings of the ISSAC 2009, pp. 95–102. ACM (2009). https://doi.org/10.1145/1576702.1576718
9. Chen, C., Moreno Maza, M.: Quantifier elimination by cylindrical algebraic decomposition based on regular chains. J. Symb. Comput. **75**, 74–93 (2016). https://doi.org/10.1016/j.jsc.2015.11.008
10. Collins, G.E.: Quantifier elimination for real closed fields by cylindrical algebraic decompostion. In: Brakhage, H. (ed.) GI-Fachtagung 1975. LNCS, vol. 33, pp. 134–183. Springer, Heidelberg (1975). https://doi.org/10.1007/3-540-07407-4_17. Reprinted in [7]
11. Dolzmann, A., Seidl, A., Sturm, T.: Efficient projection orders for CAD. In: Proceedings of the ISSAC 2004, pp. 111–118. ACM (2004). https://doi.org/10.1145/1005285.1005303
12. England, M.: Machine learning for mathematical software. In: Davenport, J.H., Kauers, M., Labahn, G., Urban, J. (eds.) ICMS 2018. LNCS, vol. 10931, pp. 165–174. Springer, Cham (2018). https://doi.org/10.1007/978-3-319-96418-8_20
13. England, M., Bradford, R., Davenport, J.: Cylindrical algebraic decomposition with equational constraints. J. Symb. Comput. **100**, 38–71 (2020). https://doi.org/10.1016/j.jsc.2019.07.019

14. England, M., Bradford, R., Davenport, J.H., Wilson, D.: Choosing a variable ordering for truth-table invariant cylindrical algebraic decomposition by incremental
triangular decomposition. In: Hong, H., Yap, C. (eds.) ICMS 2014. LNCS, vol.
8592, pp. 450–457. Springer, Heidelberg (2014). https://doi.org/10.1007/978-3-
662-44199-2_68

15. England, M., Florescu, D.: Comparing machine learning models to choose the
variable ordering for cylindrical algebraic decomposition. In: Kaliszyk, C., Brady,
E., Kohlhase, A., Sacerdoti Coen, C. (eds.) CICM 2019. LNCS (LNAI), vol. 11617,
pp. 93–108. Springer, Cham (2019). https://doi.org/10.1007/978-3-030-23250-4_7

16. Florescu, D., England, M.: Algorithmically generating new algebraic features of
polynomial systems for machine learning. In: Proceedings of the SC^2 2019. CEUR-
WS, vol. 2460 (2019). http://ceur-ws.org/Vol-2460/

17. Florescu, D., England, M.: Improved cross-validation for classifiers that make algorithmic choices to minimise runtime without compromising output correctness. In:
Slamanig, D., Tsigaridas, E., Zafeirakopoulos, Z. (eds.) MACIS 2019. LNCS, vol.
11989, pp. 341–356. Springer, Cham (2020). https://doi.org/10.1007/978-3-030-
43120-4_27

18. Gryak, J., Haralick, R., Kahrobaei, D.: Solving the conjugacy decision problem
via machine learning. Exp. Math. **29**(1), 66–78 (2020). https://doi.org/10.1080/
10586458.2018.1434704

19. Huang, Z., England, M., Wilson, D., Bridge, J., Davenport, J., Paulson, L.: Using
machine learning to improve cylindrical algebraic decomposition. Math. Comput.
Sci. **13**(4), 461–488 (2019). https://doi.org/10.1007/s11786-019-00394-8

20. Huang, Z., England, M., Wilson, D., Davenport, J.H., Paulson, L.C., Bridge, J.:
Applying machine learning to the problem of choosing a heuristic to select the
variable ordering for cylindrical algebraic decomposition. In: Watt, S.M., Davenport, J.H., Sexton, A.P., Sojka, P., Urban, J. (eds.) CICM 2014. LNCS (LNAI),
vol. 8543, pp. 92–107. Springer, Cham (2014). https://doi.org/10.1007/978-3-319-
08434-3_8

21. Kuipers, J., Ueda, T., Vermaseren, J.: Code optimization in FORM. Comput. Phys.
Commun. **189**, 1–19 (2015). https://doi.org/10.1016/j.cpc.2014.08.008

22. Liang, J.H., Ganesh, V., Poupart, P., Czarnecki, K.: Learning rate based branching
heuristic for SAT solvers. In: Creignou, N., Le Berre, D. (eds.) SAT 2016. LNCS,
vol. 9710, pp. 123–140. Springer, Cham (2016). https://doi.org/10.1007/978-3-319-
40970-2_9

23. Mulligan, C.B., Davenport, J.H., England, M.: TheoryGuru: a mathematica package to apply quantifier elimination technology to economics. In: Davenport, J.H.,
Kauers, M., Labahn, G., Urban, J. (eds.) ICMS 2018. LNCS, vol. 10931, pp. 369–
378. Springer, Cham (2018). https://doi.org/10.1007/978-3-319-96418-8_44

24. Pedregosa, F., et al.: Scikit-learn: machine learning in Python. J. Mach. Learn.
Res. **12**, 2825–2830 (2011). http://www.jmlr.org/papers/v12/pedregosa11a.html

25. Sturm, T.: New domains for applied quantifier elimination. In: Ganzha, V.G.,
Mayr, E.W., Vorozhtsov, E.V. (eds.) CASC 2006. LNCS, vol. 4194, pp. 295–301.
Springer, Heidelberg (2006). https://doi.org/10.1007/11870814_25

Databases in Mathematics

FunGrim: A Symbolic Library for Special Functions

Fredrik Johansson$^{(\boxtimes)}$ (ID)

LFANT, Inria Bordeaux, Talence, France
fredrik.johansson@gmail.com
http://fredrikj.net

Abstract. We present the Mathematical Functions Grimoire (Fun-Grim), a website and database of formulas and theorems for special functions. We also discuss the symbolic computation library used as the backend and main development tool for FunGrim, and the Grim formula language used in these projects to represent mathematical content semantically.

Keywords: Special functions · Symbolic computation · Mathematical databases · Semantic mathematical markup

1 Introduction

The *Mathematical Functions Grimoire*[1] (FunGrim, http://fungrim.org/) is an open source library of formulas, theorems and data for mathematical functions. It currently contains around 2600 entries. As one example entry, the modular transformation law of the Eisenstein series G_{2k} on the upper half-plane $\mathbb{H}$ is given in http://fungrim.org/entry/0b5b04/ as follows:

$$G_{2k}\left(\frac{a\tau + b}{c\tau + d}\right) = (c\tau + d)^{2k}G_{2k}(\tau)$$

Assumptions: $k \in \mathbb{Z}_{\geq 2}$ and $\tau \in \mathbb{H}$ and $\begin{pmatrix} a & b \\ c & d \end{pmatrix} \in \mathrm{SL}_2(\mathbb{Z})$

FunGrim stores entries as symbolic expressions with metadata, in this case:

```
Entry(ID("0b5b04"),
    Formula(Equal(EisensteinG(2*k, (a*tau+b)/(c*tau+d)),
        (c*tau+d)**(2*k) * EisensteinG(2*k, tau))),
    Variables(k, tau, a, b, c, d),
    Assumptions(And(Element(k, ZZGreaterEqual(2)), Element(tau, HH),
        Element(Matrix2x2(a, b, c, d), SL2Z))))
```

---

[1] A *grimoire* is a book of magic formulas.

© Springer Nature Switzerland AG 2020
A. M. Bigatti et al. (Eds.): ICMS 2020, LNCS 12097, pp. 315–323, 2020.
https://doi.org/10.1007/978-3-030-52200-1_31

Formulas are fully quantified (*assumptions* give conditions for the free variables such that the formula is valid) and context-free (symbols have a globally consistent meaning), giving precise statements of mathematical theorems. The metadata may also include bibliographical references. Being easily computer-readable, the database may be used for automatic term rewriting in symbolic algorithms. This short paper discusses the semantic representation of mathematics in FunGrim and the underlying software.

2   Related Projects

FunGrim is in part a software project and in part a reference work for mathematical functions in the tradition of Abramowitz and Stegun [1] but with updated content and a modern interface. There are many such efforts, notably the NIST Digital Library of Mathematical Functions (DLMF) [4] and the Wolfram Functions Site (WFS) [10], which have two rather different approaches:

- DLMF uses LaTeX together with prose for its content. Since many formulas depend on implicit context and LaTeX is presentation-oriented rather than semantic (although DLMF adds semantic extensions to LaTeX to alleviate this problem), the content is not fully computer-readable and can also sometimes be ambiguous to human readers. DLMF is edited for conciseness, giving an overview of the main concepts and omitting in-depth content.
- WFS represents the content as context-free symbolic expressions written in the Wolfram Language. The formulas can be parsed by Mathematica, whose evaluation semantics provide concrete meaning. Most formulas are computer-generated, sometimes exhaustively (for example, WFS lists tens of thousands of transformations between elementary functions and around 200,000 formulas for special cases of hypergeometric functions).

FunGrim uses a similar approach to that of WFS, but does not depend on the proprietary Wolfram technology. Indeed, one of the central reasons for starting FunGrim is that both DLMF and WFS are not open source (though freely accessible). Another central idea behind FunGrim is to provide even stronger semantic guarantees; this aspect is discussed in a later section.

Part of the motivation is also to offer complementary content: in the author's experience, the DLMF and WFS are strong in some areas and weak in others. For example, both have minimal coverage of some important functions of number theory and they cover *inequalities* far less extensively than *equalities*. At this time, FunGrim has perhaps 10% of the content needed for a good general reference on special functions, but as proof as concept, it has detailed content for some previously-neglected topics. The reader may compare the following:

- http://fungrim.org/topic/Modular_lambda_function/ versus
  http://functions.wolfram.com/EllipticFunctions/ModularLambda/ versus
  formulas for $\lambda(\tau)$ in https://dlmf.nist.gov/23.15 + https://dlmf.nist.gov/23.17.

- http://fungrim.org/topic/Barnes_G-function/ versus
  https://dlmf.nist.gov/5.17. (The Barnes G-function is not covered in WFS.)

Most FunGrim content is hand-written so far; adding computer-generated entries in the same fashion as WFS is a future possibility.

We mention three other related projects:

- FunGrim shares many goals with the NIST Digital Repository of Mathematical Formulas (DRMF) [3], a companion project to the DLMF. We will not attempt to compare the projects in depth since DRMF is not fully developed, but we mention one important difference: DRMF represents formulas using a semantic form of LaTeX which is hard to translate perfectly to symbolic expressions, whereas FunGrim (like WFS) uses symbolic expressions as the source representation and generates LaTeX automatically for presentation.
- The Dynamic Dictionary of Mathematical Functions (DDMF) [2] generates information about mathematical functions algorithmically, starting *ab initio* only from the defining differential equation of each function. This has many advantages: it enables a high degree of reliability (human error is removed from the equation, so to speak), the presentation is uniform, and it is easy to add new functions. The downside is that the approach is limited to a restricted class of properties for a restricted class of functions.
- The LMFDB [7] is a large database of L-functions, modular forms, and related objects. The content largely consists of data tables and does not include "free-form" symbolic formulas and theorems.

3   Grim Formula Language

Grim is the symbolic mathematical language used in FunGrim.[2] Grim is designed to be easy to write and parse and to be embeddable within a host programming language such as Python, Julia or JavaScript using the host language's native syntax (similar to SymPy [8]). The reference implementation is Pygrim, a Python library which implements Grim-to-LaTeX conversion and symbolic evaluation of Grim expressions. Formulas are converted to HTML using KaTeX for display on the FunGrim website; Pygrim also provides hooks to show Grim expressions as LaTeX-rendered formulas in Jupyter notebooks. The FunGrim database itself is currently part of the Pygrim source code.[3]

Grim has a minimal core language, similar to Lisp S-expressions and Wolfram language M-expressions. The only data structure is an expression tree composed of function calls $f(x, y, \ldots)$ and atoms (integer literals, string literals, alphanumerical symbol names). For example, $\texttt{Mul(2, Add(a, b))}$ represents $2(a + b)$. For convenience, Pygrim uses operator overloading in Python so that the same expression may be written more simply as $\texttt{2*(a+b)}$.

---

[2] Documentation of the Grim language is available at http://fungrim.org/grim/.

[3] Pygrim is currently in early development and does not have an official release. The source code is publicly available at https://github.com/fredrik-johansson/fungrim.

On top of the core language, Grim provides a vocabulary of hundreds of builtin symbols (`For`, `Exists`, `Matrix`, `Sin`, `Integral`, etc.) for variable-binding, logical operations, structures, mathematical functions, calculus operations, etc.
The following dummy formula is a more elaborate example:

```
Where(Sum(1/f(n), For(n, -N, N), NotEqual(n, 0)), Def(f(n),
    Cases(Tuple(n**2, CongruentMod(n, 0, 3)), Tuple(1, Otherwise))))
```

$$\sum_{\substack{n=-N \\ n\neq 0}}^{N} \frac{1}{f(n)} \text{ where } f(n) = \begin{cases} n^2, & n \equiv 0 \pmod{3} \\ 1, & \text{otherwise} \end{cases}$$

Grim can be used both as a mathematical markup language and as a simple functional programming language. Its design is deliberately constrained:

- Grim is not intended to be a typesetting language: the Grim-to-LaTeX converter takes care of most presentation details automatically. (The results are not always perfect, and Grim does allow including typesetting hints where the default rendering is inadequate.)
- Grim is not intended to be a general-purpose programming language. Unlike full-blown Lisp-like programming languages, Grim is not meant to be used to manipulate symbolic expressions from within, and it lacks concrete data structures for programming, being mainly concerned with representing immutable mathematical objects. Grim is rather meant to be embedded in a host programming language where the host language can be used to traverse expression trees or implement complex algorithms.

Grim formulas entered in Pygrim are preserved verbatim until explicitly evaluated. This contrasts with most computer algebra systems, which automatically convert expressions to "canonical" form. For example, SymPy automatically rewrites $2(b + a)$ as $2a + 2b$ (distributing the numerical coefficient and sorting the terms). SymPy's behavior can be overridden with a special "hold" command, but this can be a hassle to use and might not be recognized by all functions.

4   Evaluation Semantics

FunGrim and the Grim language have the following fundamental semantic rules:

- Every mathematical object or operator must have an unambiguous interpretation, which cannot vary with context. In principle, every syntactically valid constant expression should represent a definitive mathematical object (possibly the special object undefined when a function is evaluated outside its domain of definition). This means, for example, that multivalued functions have fixed branch cuts (analytic continuation must be expressed explicitly), and removable singularities do not cancel automatically. Many symbols which have an overloaded meaning in standard mathematical notation require disambiguation; for example, Grim provides separate `SequenceLimit`, `RealLimit` and `ComplexLimit` operators to express $\lim_{x \to c} f(x)$, depending on whether the set of approach is meant as $\mathbb{Z}$, $\mathbb{R}$ or $\mathbb{C}$.

- The standard logical and set operators ($=$ and $\in$, etc.) compare identity of mathematical objects, not equivalence under morphisms. The mathematical universe is constructed to have few, orthogonal "types": for example, the integer 1 and the complex number 1 are the same object, with $\mathbb{Z} \subset \mathbb{C}$.
- Symbolic evaluation (rewriting an expression as a simpler expression, e.g. $2 + 2 \to 4$) must preserve the exact value of the input expression. Formulas containing free variables are implicitly quantified over the whole universe unless explicit assumptions are provided, and may only be rewritten in ways that preserve the value for all admissible values of the free variables. For example, $yx \to xy$ is not a valid rewrite operation *a priori* since the universe contains noncommutative objects such as matrices, but it is valid when quantified with assumptions that make x and y commute, e.g. $x, y \in \mathbb{C}$.

These semantics are stronger than in most symbolic computing environments. Computer algebra systems traditionally ignore "exceptional cases" when rewriting expressions. For example, many computer algebra systems automatically simplify x/x to 1, ignoring the exceptional case $x = 0$ where a division by zero occurs.[4] A more extreme example is to blindly simplify $\sqrt{x^2} \to x$ (invalid for negative numbers), and more generally to ignore branch cuts or complex values.

Indeed, one section of the Wolfram Mathematica documentation helpfully warns users: "The answer might not be valid for certain exceptional values of the parameters." As a concrete illustration, we can use Mathematica to "prove" that $e = 2$ by evaluating the hypergeometric function ${}_1F_1(a, b, 1)$ at $a = b = -1$ using two different sequences of substitutions:

- ${}_1F_1(a, b, 1) \to [a = b] \to e \to [b = -1] \to e$
- ${}_1F_1(a, b, 1) \to [a = -1] \to 1 - \frac{1}{b} \to [b = -1] \to 2$

The contradiction happens because Mathematica uses two different rules to rewrite the ${}_1F_1$ function, and the rules are inconsistent with each other in the exceptional case $a = b \in \mathbb{Z}_{\leq 0}$).[5] (SymPy has the same issue.)

Our aspiration for the Grim formula language and the FunGrim database is to make such contradictions impossible through strong semantics and pedantic use of assumptions. This should aid human understanding (a user can inspect the source code of a formula and look up the definitions of the symbols) and help support symbolic computation, automated testing, and possibly formal theorem-proving efforts. Perfect consistency is particularly important for working with multivariate functions, where corner cases can be extremely difficult to spot.

---

[4] The simplification is valid if x is viewed as a formal indeterminate generating $\mathbb{C}[x]$ rather than a free variable representing a complex number. The point remains that some computer algebra systems overload variables to serve both purposes, and this ambiguity is a frequent source of bugs. In Grim, the distinction is explicit.

[5] In WFS, corresponding contradictory formulas are http://functions.wolfram.com/07.20.03.0002.01 and http://functions.wolfram.com/07.20.03.0118.01.

In reality, eliminating inconsistencies is an asymptotic goal: there are certainly present and future mathematical errors in the FunGrim database and bugs in the Pygrim reference implementation. We believe that such errors can be minimized through randomized testing (ideally combined with formal verification in the future, where such methods are applicable).

5   Evaluation with Pygrim

Pygrim has rudimentary support for evaluating and simplifying Grim expressions. It is able to perform basic logical and arithmetic operations, expand special cases of mathematical functions, perform simple domain inferences, partially simplify symbolic arithmetic expressions, evaluate and compare algebraic numbers using an exact implementation of $\overline{\mathbb{Q}}$ arithmetic, and compare real or complex numbers using Arb enclosures [5] (only comparisons of unequal numbers can be decided in this way; equal numbers have overlapping enclosures and can only be compared conclusively when an algebraic or symbolic simplification is possible).

Calling the .eval() method in Pygrim returns an evaluated expression:

```
>>> Element(Pi, SetMinus(OpenInterval(3, 4), QQ)).eval()
True_

>>> Zeros(x**5 - x**4 - 4*x**3 + 4*x**2 + 2*x - 2,
...            ForElement(x, CC), Greater(Re(x), 0)).eval()
...
Set(Sqrt(Add(2, Sqrt(2))), 1, Sqrt(Sub(2, Sqrt(2))))

>>> ((DedekindEta(1 + Sqrt(-1)) / Gamma(Div(5, 4))) ** 12).eval()
Div(-4096, Pow(Pi, 9))
```

To simplify formulas involving free variables, the user needs to supply sufficient assumptions:

```
>>> (x / x).eval()
Div(x, x)
>>> (x / x).eval(assumptions=Element(x, CC))
Div(x, x)
>>> (x / x).eval(assumptions=And(Element(x, CC), NotEqual(x, 0)))
1
>>> Sin(Pi * n).eval()
Sin(Mul(Pi, n))
>>> Sin(Pi * n).eval(assumptions=Element(n, ZZ))
0
```

In some cases, Pygrim can output conditional expressions: for example, the evaluation $_2F_1(1, 1, 2, x) = -\log(1-x)/x$ is made with an explicit case distinction for the removable singularity at $x = 0$ (the singularity at $x = 1$ is consistent with $\log(0) = -\infty$ and does not require a case distinction).

```
>>> f = Hypergeometric2F1(1, 1, 2, x); f.eval()
Hypergeometric2F1(1, 1, 2, x)              # no domain -- no evaluation
>>> f.eval(assumptions=Element(x, CC))
Cases(Tuple(Div(Neg(Log(Sub(1, x))), x), NotEqual(x, 0)),
    Tuple(1, Equal(x, 0)))                  # separate case for x = 0
>>> f.eval(assumptions=Element(x, SetMinus(CC, Set(0))))
Div(Neg(Log(Sub(1, x))), x)                 # no case distinction needed
```

Pygrim is not a complete computer algebra system; its features are tailored to developing FunGrim and exploring special function identities. Users may also find it interesting as a symbolic interface to Arb (the `.n()` method returns an arbitrary-precision enclosure of a constant expression).

6 Testing Formulas

To test a formula $P(x_1, \ldots, x_n)$ with free variables $x_1, \ldots, x_n$ and corresponding assumptions $Q(x_1, \ldots, x_n)$, we generate pseudorandom values $x_1, \ldots, x_n$ satisfying $Q(x_1, \ldots, x_n)$, and for each such assignment we evaluate the constant expression $P(x_1, \ldots, x_n)$. If P evaluates to False, the test fails (a counterexample has been found). If P evaluates to True or cannot be simplified to True/False (the truth value is unknown), the test instance passes.

As an example, we test $P(x) = [\sqrt{x^2} = x]$ with assumptions $Q(x) = [x \in \mathbb{R}]$:

```
>>> formula = Equal(Sqrt(x**2), x)
>>> formula.test(variables=[x], assumptions=Element(x, RR))
{x: 0}     ... True
{x: Div(1, 2)}    ... True
{x: Sqrt(2)}    ... True
{x: Pi}    ... True
{x: 1}    ... True
{x: Neg(Div(1, 2))}     ... False
```

The test passes for $x = 0, \frac{1}{2}, \sqrt{2}, \pi, 1$, but $x = -\frac{1}{2}$ is a counterexample. With correct assumptions $x \in \mathbb{C} \wedge (\mathrm{Re}(x) > 0 \vee (\mathrm{Re}(x) = 0 \wedge \mathrm{Im}(x) > 0))$, it passes:

```
>>> formula.test(variables=[x], assumptions=And(Element(x, CC),
...     Or(Greater(Re(x), 0), And(Equal(Re(x), 0), Greater(Im(x), 0)))))
...
Passed 91 instances (77 True, 14 Unknown, 0 False)
```

It currently takes two CPU hours to test the FunGrim database with up to 100 test instances (assignments $x_1, \ldots, x_n$ that satisfy the assumptions) per entry. We estimate that around 75% of the entries are effectively testable. For the other 25%, either the symbolic evaluation code in Pygrim is not powerful enough to generate any admissible values (for which Q is provably True), or P contains constructs for which Pygrim does not yet support symbolic or numerical evaluation. For 30% of the entries, Pygrim is able to symbolically simplify P

to True in at least one test instance (in the majority of cases, it is only able to check consistency via Arb). We aim to improve all these statistics in the future.

The test strategy is effective: the first run to test the FunGrim database found errors in 24 out of 2618 entries. Of these, 4 were mathematically wrong formulas (for example, the Bernoulli number inequality $(-1)^n B_{2n+2} > 0$ had the prefactor negated as $(-1)^{n+1}$), 6 had incorrect assumptions (for example, the Lambert W-function identity $W_0(x \log(x)) = \log(x)$ was given with assumptions $x \in [-e^{-1}, \infty)$ instead of the correct $x \in [e^{-1}, \infty)$); the remaining errors were due to incorrect metadata or improperly constructed symbolic expressions.

A similar number of additional errors were found and corrected after improving Pygrim's evaluation code further. An error rate near 5% seems plausible for untested formulas entered by hand (by this author!). We did not specifically search for errors in the literature used as reference material for FunGrim; however, many corrections were naturally made when the entries were first added, prior to the development of the test framework.

7   Formulas as Rewrite Rules

The FunGrim database can be used for term rewriting, most easily by applying a specific entry as a rewrite rule. For example, FunGrim entry ad6c1c is the trigonometric identity $\sin(a) \sin(b) = \frac{1}{2} (\cos(a - b) - \cos(a + b))$:

```
>>> (Sin(2) * Sin(Sqrt(2))).rewrite_fungrim("ad6c1c")
Div(Sub(Cos(Sub(2, Sqrt(2))), Cos(Add(2, Sqrt(2)))), 2)
```

This depends on pattern matching. To ensure correctness, a match is only made if parameters in the input expression satisfy the assumptions for free variables listed in the FunGrim entry. The pattern matching is currently implemented naively and will fail to match expressions that are mathematically equivalent but structurally different (better implementations are possible [6]).

A rather interesting idea is to search the whole database automatically for rules to apply to simplify a given formula. We have used this successfully on toy examples, but much more work is needed to develop a useful general-purpose simplification engine; this would require stronger pattern matching as well as heuristics for applying sequences of rewrite rules. Rewriting using a database is perhaps most likely to be successful for specific tasks and in combination with advanced hand-written search heuristics (or heuristics generated via machine learning). A prominent example of the hand-written approach is Rubi [9] which uses a decision tree of thousands of rewrite rules to simplify indefinite integrals.

References

1. Abramowitz, M., Stegun, I.A.: Handbook of Mathematical Functions with Formulas, Graphs, and Mathematical Tables. Dover, New York (1964)

2. Benoit, A., Chyzak, F., Darrasse, A., Gerhold, S., Mezzarobba, M., Salvy, B.: The dynamic dictionary of mathematical functions (DDMF). In: Fukuda, K., Hoeven, J., Joswig, M., Takayama, N. (eds.) ICMS 2010. LNCS, vol. 6327, pp. 35–41. Springer, Heidelberg (2010). https://doi.org/10.1007/978-3-642-15582-6_7
3. Cohl, H.S., McClain, M.A., Saunders, B.V., Schubotz, M., Williams, J.C.: Digital Repository of mathematical formulae. In: Watt, S.M., Davenport, J.H., Sexton, A.P., Sojka, P., Urban, J. (eds.) CICM 2014. LNCS (LNAI), vol. 8543, pp. 419–422. Springer, Cham (2014). https://doi.org/10.1007/978-3-319-08434-3_30
4. NIST. Digital Library of Mathematical Functions (2019). http://dlmf.nist.gov/
5. Johansson, F.: Arb: efficient arbitrary-precision midpoint-radius interval arithmetic. IEEE Trans. Comput. **66**, 1281–1292 (2017). https://doi.org/10.1109/TC.2017.2690633
6. Krebber, M., Barthels, H.: MatchPy: pattern matching in python. J. Open Source Softw. **3**(26), 670 (2018). https://doi.org/10.21105/joss.00670
7. LMFDB: The L-functions and modular forms database (2020). http://lmfdb.org
8. Meurer, A., et al.: SymPy: symbolic computing in Python. PeerJ Comput. Sci. **3**, e103 (2017). https://doi.org/10.7717/peerj-cs.103
9. Rich, A., Scheibe, P., Abbasi, N.: Rule-based integration: an extensive system of symbolic integration rules. J. Open Source Softw. **3**(32), 1073 (2018). https://doi.org/10.21105/joss.01073
10. The Wolfram Functions Site (2020). http://functions.wolfram.com/

Accelerating Innovation Speed in Mathematics by Trading Mathematical Research Data

Operational Research Literature as a Use Case for the Open Research Knowledge Graph

Mila Runnwerth[(⊠)], Markus Stocker[ⓘ], and Sören Auer[ⓘ]

TIB - Leibniz Information Centre for Science and Technology,
Welfengarten 1B, 30167 Hannover, Germany
{mila.runnwerth,markus.stocker,soren.auer}@tib.eu
http://www.tib.eu

Abstract. The Open Research Knowledge Graph (ORKG) provides machine-actionable access to scholarly literature that habitually is written in prose. Following the FAIR principles, the ORKG makes traditional, human-coded knowledge findable, accessible, interoperable, and reusable in a structured manner in accordance with the Linked Open Data paradigm. At the moment, in ORKG papers are described manually, but in the long run the semantic depth of the literature at scale needs automation. Operational Research is a suitable test case for this vision because the mathematical field and, hence, its publication habits are highly structured: A mundane problem is formulated as a mathematical model, solved or approximated numerically, and evaluated systematically. We study the existing literature with respect to the Assembly Line Balancing Problem and derive a semantic description in accordance with the ORKG. Eventually, selected papers are ingested to test the semantic description and refine it further.

Keywords: Knowledge graph · Mathematical knowledge management · Operational research literature · Operations research literature

1 Introduction

Today's scholarly communication behaviour and logistics is still defined by centuries of printed document culture. Although there is progress by transforming journals into digital article repositories that, in principle, provide access to the content at all times and irrespective of a researcher's location, the nature of an article itself has not changed: The investigated hypothesis, the used methodology, the experiment, and the outcome are written in prosaic form; the final document is usually published for no other purposes than reading, seemingly optimised for human cognition.

The Open Research Knowledge Graph (ORKG) [8] questions the "paradigm of document-centric scholarly information communication" [2]. It aims at transforming research literature into structured, machine-actionable data in order to

© Springer Nature Switzerland AG 2020
A. M. Bigatti et al. (Eds.): ICMS 2020, LNCS 12097, pp. 327–334, 2020.
https://doi.org/10.1007/978-3-030-52200-1_32

represent and express information through semantically rich, interlinked knowledge graphs. Similarly to DBpedia, a prosaic knowledge source is transformed according to Linked Open Data standards [1]. Users are enabled to compare papers, discover patterns across methods or disciplines, or get a structured overview in a chosen context.

The main use cases of the ORKG's beta version[1] are article search, a machine-actionable, semantic representation, and especially paper comparison as introduced in [10]. To date, it indexes about 400 research articles. More than half are assigned to the subject cluster *Physical Sciences & Mathematics*.

1.1 Operational Research as a Use Case from Mathematics

The structural science mathematics provides particularly suitable content for the ORKG: Its published prose is clear and dense from a linguistic point of view. However, as [4] have shown the high degree of abstraction in mathematics makes a conceptualisation consisting of the categories *process*, *method*, *material*, and *data*, which have been adapted to empirical sciences, inexpedient. The ORKG is not limited to this model but its feature, the abstract annotator, has been shown to be out of its depth with regard to mathematics. The applied mathematical science of operational research (OR) combines the rather abstract fields of combinatorics and numerical analysis with mundane research questions from economics. In favour of this study, we narrowed the topic down to the optimisation problem of *Assembly Line Balancing*. Its name derives from mass production where the intricate logistics for paced manufacturing assembly lines have to be organised efficiently, i.e. optimally. The Assembly Line Balancing Problem (ALBP) and its variations are not only well-covered in scholarly literature but also provide an abundance of structured overviews of exact and heuristic algorithms or benchmarks thereof. Thus, the research literature about ALBPs is an appropriate use case for the ORKG. In a first step, we choose literature reviews and articles that suggest minor optimisations to existing methods, which are compared to each other. Then, we suggest a semantic description that covers the content of the collection. It will serve as a prospective template for the ORKG. Furthermore, we will look for elements and patterns in the papers that are suited for automatic extraction in the future. Third, we ingest those literature reviews or articles that are published under an eligible licence, i.e. a CC-BY, CC-BY-SA, or arXiv's *Non-exclusive licence to distribute* into the ORKG. During this intellectual step, we will test and refine the proposed data model. Finally, we consider the representations and comparisons of the scholarly contributions in the ORKG and discuss its added value for researchers.

2   A Template for the Assembly Line Balancing Problem

Scholarly research of the ALBP can be traced back to the 1960s when it was shown to be a NP-hard combinatorial optimisation problem [7]. Since then scientists work on the sophistication of the mathematical model, exact algorithms for

---

[1] https://orkg.org/.

defined special cases or heuristic algorithms in order to find optimal solutions in adequate time. Recently, several reviews have been published to benchmark the stated mathematical model, exemplary data scenarios [5] or performances of the selected methods. We chose to set a focus on these reviews at first, but most publications were not openly accessible or free to be reused in the ORKG according to their respective licences. Eventually, articles introducing a new model statement for the ALBP or a heuristic to solve it were also considered. The collection comprised 28 topically relevant papers of which eight provide openly accessible preprints on *arXiv*[2]. These were manually ingested into the ORKG with varying degrees of thoroughness (cf. Sect. 2.2): From the single statement of the research problem to detailed descriptions of the algorithms and data sets that were applied[3].

The collection of research articles was organised in the open source reference management system Zotero[4], also including documented experiences of the whole process.

2.1 A Semantic Model to Reflect ALBP Research

The ORKG's performance depends on a data model that is well-tuned to the content it is supposed to represent. That means expert knowledge in both the considered field and data modelling is required. Authors who possess the domain knowledge may not be able to squeeze it into the RDF scheme of the ORKG because there is no or little expertise in knowledge engineering. Data curators on the other hand may struggle with the proper in-depth indexing of the latest research knowledge. The ORKG's flexibility is an advantage because it allows almost limitless adaptions to describing papers by reusing existing concepts (mostly entries by former contributors) and relations but also by introducing new ones. The default schema stems from the comprehension of empirical sciences: A method is applied to a defined research question. This application causes a process that involves material to be observed or changed. Meanwhile observational data is collected and eventually evaluated in order to prove or disprove a hypothesis constructed prior to the experiment.

In operational research in general and with respect to the ALBP in particular, there is also a rather standardised development that can be represented by a data model: The practical problem is formulated as a mathematical model or programme. Depending on the choice of the model, there is a toolbox of direct or heuristic algorithms to yield an exact (or approximate) solution to the model. Usually, in scholarly literature either a new variant of the ALBP is stated and the derived model is traced back to established methods or a new or rather slightly modified method is tested against known methods to solve the same problem. Thus, we conclude that most research papers about the ALBP are comprised of the elements listed in Table 1.

---

[2] arxiv.org.

[3] An exemplary comparison of three selected papers can be found at https://www. orkg.org/orkg/comparison?contributions=R12018,R12059,R12193.

[4] https://www.zotero.org/.

Table 1. A semantic model translating the OR research process into the ORKG scheme. The arrows connote a hierarchical descent, the asterisks connote a newly introduced property.

OR Term	ORKG Properties	Example
Name of the optimisation problem	Has research	ALBP
Model/programme	Has approach → Has model*	MIP
Exact method	Has method → Has exact solution method*	Branch & Bound
Heuristic	Has method → Has heuristic*	Tabu search
Instance data set	Has instance*	Roszieg
Programming language	Has implementation → Has programming language*	C, GCC 3.4.0
System specifications	Has implementation → Has system specification*	Athlon 64 X2 4400
Performance	Has performance*	$\mathcal{O}(n \log n)$, 0,2 ms

Has Performance contains the results that depend on the method that is applied, the graph the algorithm is applied on, and the specification of the implementation and system. Thus, it is semantically interlinked with other elements.

If the suggested structure in Table 1 proves valid, it can be cast into a topic-specific template on its own in order to facilitate highly consistent knowledge graphs of further relevant papers independent of the curator.

2.2   Entering ALBP Literature into the ORKG

After careful study and annotation of the eight papers from arXiv, we entered the data into the ORKG. In the first of three steps of the procedure, the formal metadata can be automatically ingested via DOI[5] or a BibTeX entry. There is an additional fallback option to enter the formal metadata manually. Since the preprint repository arXiv does not provide a DOI for its documents, we chose BibTeX entries for the import.

In the second step, the document is classified by subject. The ORKG's specified, hierarchical classification does currently not allow for several attributions. Hence, when assigning a single subject, a multidisciplinary field such as operational research is prone to inconsistencies with respect to its main focus in the respective paper or the curator. We chose to consistently assign the collection to Applied Mathematics → Numerical Analysis & Computation, although several other closely related fields would have been adequate as well, for example Engineering → Operations Research (and more). However, OR being predominantly

---

[5] https://www.doi.org/.

a multidisciplinary subject involving mathematics, computer science, and economics, engineering seemed too misleading for a semantically sound assignment.

The curator may choose between several templates; the template called *Research Problem* is closest to the data model as suggested in Sect. 2.1. The template provides the field *Has research* where keywords can be chosen from the suggested list or entered manually. Each entry is added to a bag-of-words and, thus, will be provided for autocompletion further on. This unrestricted freedom leads to a number of challenges. We struggled with typos (e.g. 'optimisation') as it was not immediately obvious how to correct these. Moreover, the same word was included in its American and British form, respectively, i.e. 'optimization' and 'optimisation'. We plan to add some functionality to ORKG to semi-automatically interlink such surface forms as they are describing the same concept. An underlying controlled vocabulary with an additional feature to enter free text would avoid wreaking havoc in the bag-of-words. A user entering 'optimisation problem' may thus be faced with four versions of which one contains a typo and two are identical.

Further predefined fields are *Has evaluation, Has approach, Has method, Has implementation, Has result, Has value,* and *Has metric.* Not all of these semantic relations make sense for describing a mathematical paper, or rather, they lack distinctive accuracy, e.g. when does an approach become a method; or do we mean the outcome of the algorithm or its performance when stating the result? Yet, the relevant semantic units of an OR paper can be transferred and amended easily.

Each field contains further fields in turn that may be annotated and indexed. And as a last resort, new relations can be introduced on every hierarchical level. The OR terms introduced in Table 1 were mapped by employing existing relations and introducing new ones (marked by an asterisk in the table). After leaving the hierarchical top level which is edited in the main browser window, every edit thereafter is conducted in a small overlay window. So while modelling, there is no visual aid where the description process is hierarchically taking place at the moment. However, we made it a habit to describe the top level first, save the description, such that the visualisation of the graph is available. From there, refinement is more accessible.

The eight papers are not consistently described in this fashion because each paper gave reason to a refinement iteration of previous graphs. Thus, after each paper, there are (or should be) well-documented, retroactive modifications to each graph representing a paper. Again, this inevitably leads to inconsistencies even among papers that are ingested by the same curator. Another critical observation is our choice of terms: General denominations such as *Has model, Has instance,* or *Has performance* could mean completely different things in another context. Even between OR researchers these terms might not be semantically tight enough to guarantee frictionless communication. Hence, the relations are prone to cross-contextual use that might make the otherwise carefully created model fuzzy.

In theory, a paper can be thoroughly represented by modelling each sentence as Linked Data, at arbitrary granularity. Again, in agreement with another field expert from biochemistry, we concluded: when to finish the indexing procedure is at the margin of discretion. Of course, the ORKG's crowd-sourcing philosophy allows and even demands for further refinement by others or at a later stage. Thus, a knowledge graph is never truly complete, especially if dynamic data such as citations will be taken into account in the future. An exemplary paper description is shown in Fig. 1.

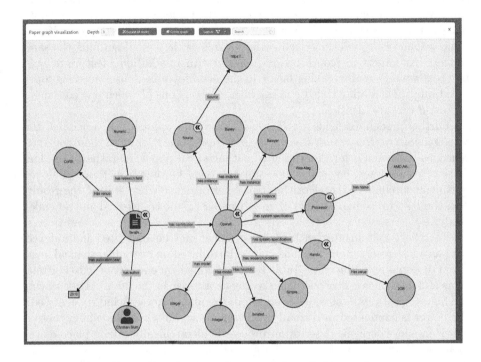

Fig. 1. Visualisation of a paper's knowledge graph.

3   Conclusion and Further Work

OR is a suitable test case for the ORKG, because the topic itself and the appropriate publication habits are highly structured and can easily be mapped to the default data schema already provided by the ORKG. However, on the basis of this study we tackle several general and subject-specific further improvements in the future:

– Creating checklists and guidelines to define both minimum requirements and a gold standard for a paper's knowledge graph.

- Underlying a general, and for templates a subject-specific, vocabulary with moderated editing workflows. Also, the resources should be displayed alphabetically or by assigned relevance instead of a last-in-first-out fashion in the tabular view.
- Linear user guidance for generating a skeleton data set and visual support for the refinement.
- Similarly described papers with identical or semantically close descriptions should yield similarity.
- The selected papers met our expectations of being highly structured and easy to parse for the defined information patterns. They are well suited for a pilot study of automated extraction for information framing a basic knowledge graph. Automated indexing where scientific literature is indexed with terms of well-maintained thesauri like the German Authority File or automatically classified with the Mathematics Subject Classification (MSC)[6] may provide a first draft to be ingested into the ORKG [9].
- The aforementioned MSC would provide the obvious classification backbone for contributions from the mathematical sphere. The extremely confined example of the ALBP suggests that this would not only call for 63 template schemas for each top level class but at least 5.000 refinements accounting for MSC's subclasses. However, with this first experiment we cannot estimate the structural synergies between classes. We rather expect, given a wisely chosen sample that future work might result in a manageable number of mathematical templates with few extensions for the subclass topics. An example are the MSC classes 44 and 45 covering ordinary and partial differential equations, respectively. Even if they turn out to differ minutely in their ORKG template, these differences will be provided for in other templates, e.g. (numerical) analysis.
- Since the ORKG follows a crowdsourcing philosophy, seeking support from and collaborate with further projects in the field of mathematical knowledge engineering guarantees high quality and integrity of the data and its community-curated modelling. Critical exchange with the researchers of *MathDataHub* is established [3], but projects like swMATH, a database for mathematical software, should be considered more closely [6].

References

1. Auer, S., Bizer, C., Kobilarov, G., Lehmann, J., Cyganiak, R., Ives, Z.: DBpedia: a nucleus for a web of open data. In: Aberer, K., et al. (eds.) ASWC/ISWC -2007. LNCS, vol. 4825, pp. 722–735. Springer, Heidelberg (2007). https://doi.org/10.1007/978-3-540-76298-0_52
2. Auer, S.; Kovtun, V.; Prinz, M.; Kasprzik, A.; Stocker, M.: Towards a knowledge graph for science. In: Proceedings of the 8th International Conference on Web Intelligence, Mining and Semantics (WIMS 2018). https://doi.org/10.15488/3401

---

[6] http://msc2010.org/mediawiki/index.php?title=MSC2010.

3. Berčič, K., Kohlhase, M., Rabe, F.: Towards a unified mathematical data infrastructure: database and interface generation. In: Proceedings of the Twelfth Conference on Intelligent Computer Mathematics (CICM 2019) (2019). https://doi.org/10.1007/978-3-030-23250-4_3

4. Brack, A., D'Souza, J., Hoppe, A., Auer, S., Ewerth, R.: Domain-independent extraction of scientific concepts from research articles. CoRR (2020). http://arxiv.org/abs/2001.03067

5. Chaves, A.A., Miralles, C., Lorena, L.A.N.: Clustering search approach for the assembly line worker assignment and balancing problem. In: Proceedings of the 37th International Conference On Computers And Industrial Engineering (2007)

6. Chrapary, H., Dalitz, W., Neun, W., Sperber, W.: Design, concepts, and state of the art of the swMATH service. Math. Comput. Sci. 469–481 (2017). https://doi.org/10.1007/s11786-017-0305-5

7. Gutjahr, A.L., Nemhauser, G.L.: An algorithm for the line balancing problem. In: Management science (1964). https://doi.org/10.1287/mnsc.11.2.308

8. Jaradeh, M.Y., et al.: Open research knowledge graph: next generation infrastructure for semantic scholarly knowledge. In: Proceedings of the 10th International Conference on Knowledge Capture - K-CAP 2019 (2019). https://doi.org/10.1145/3360901.3364435

9. Kasprzik, A.: Putting research-based machine learning solutions for subject indexing into practice. In: Proceedings of the Conference on Digital Curation Technologies (Qurator 2020) (2020). http://ceur-ws.org/Vol-2535/paper_1.pdf

10. Oelen, A., Jaradeh, M.Y., Farfar, K.E., Stocker, M., Auer, S.: Comparing research contributions in a scholarly knowledge graph. In: Proceedings of the Third International Workshop on Capturing Scientific Knowledge Co-located with the 10th International Conference on Knowledge Capture (K-CAP 2019) (2019). http://ceur-ws.org/Vol-2526/paper3.pdf

Making Presentation Math Computable: Proposing a Context Sensitive Approach for Translating LaTeX to Computer Algebra Systems

André Greiner-Petter[1]([✉])[ID], Moritz Schubotz[1,2][ID], Akiko Aizawa[3], and Bela Gipp[1][ID]

[1] University of Wuppertal, Wuppertal, Germany
`andre.greiner-petter@zbmath.org`, `gipp@uni-wuppertal.de`
[2] FIZ-Karlsruhe, Berlin, Germany
`moritz.schubotz@fiz-karlsruhe.de`
[3] National Institute of Informatics, Tokyo, Japan
`aizawa@nii.ac.jp`

Abstract. Scientists increasingly rely on computer algebra systems and digital mathematical libraries to compute, validate, or experiment with mathematical formulae. However, the focus in digital mathematical libraries and scientific documents often lies more on an accurate presentation of the formulae rather than providing uniform access to the semantic information. But, presentational math formats do not provide exclusive access to the underlying semantic meanings. One has to derive the semantic information from the context. As a consequence, the workflow of experimenting and publishing in the Sciences often includes time-consuming, error-prone manual conversions between presentational and computational math formats. As a contribution to improve this workflow, we propose a context-sensitive approach that extracts semantic information from a given context, embeds the information into the given input, and converts the semantically enhanced expressions to computer algebra systems.

Keywords: Presentation to computation · Translation · Computer algebra systems · Mathematical information retrieval

1 Introduction

The document preparation system LaTeX has become a de facto standard[1] for writing scientific papers in STEM disciplines over the last 30 years [1]. Numerous other editors, such as the editor for Wikipedia articles[2] or Microsoft Word [11], entirely or partially support LaTeX expressions. LaTeX provides a syntax for printing mathematical formulae that is similar to the way a person would write

---

[1] https://www.latex-project.org/ [Accessed 03-24-2020].
[2] https://en.wikipedia.org/wiki/Help:Displaying_a_formula [Accessed 03-24-2020].

© Springer Nature Switzerland AG 2020
A. M. Bigatti et al. (Eds.): ICMS 2020, LNCS 12097, pp. 335–341, 2020.
https://doi.org/10.1007/978-3-030-52200-1_33

the math by hand. Thus, LaTeX focuses on the presentation of formulae but does not explicitly carry their semantic information.

For a human reader, LaTeX's focus on formulae presentation is typically not a problem since readers can deduce the semantics of the formulae from the surrounding context and the reader's prior knowledge. Consider the Euler-Mascheroni constant represented by the Greek letter γ. Without further information, γ is just a Greek letter, often used to describe this mathematical constant but can also be used to represent curve parametrization, among other things. Based on the context, a human reader can interpret γ correctly and connect the letter with the semantic background. Computational systems, however, have issues identifying the correct semantics of formulae if the formulae do not provide enough context. For example, in LaTeX, γ is represented as \gamma.

Explicitly given semantic information in mathematical expressions becomes increasingly relevant in computational mathematics. Nowadays, many scientists also compute formulae from their papers [2,3]. They evaluate specific values, create diagrams, and search or calculate practical solutions. Computer Algebra Systems (CAS) are software tools that allow for such computations and visualizations of mathematical expressions. CAS create their representations (hereafter referred to as CAS input) with the intent of creating an input syntax that is intuitive and easy to type. CAS input must be unambiguous to CAS. Otherwise, a CAS is unable to perform computations and visualizations. CAS input is not standardized; instead, each CAS provider has created its own syntax that differs from other systems [10]. The workflow of writing a paper, therefore, leads to the problem of continually transforming mathematical expressions from LaTeX to CAS input and back. Since LaTeX does not carry the semantic information explicitly, the CAS is unable to parse complex input directly. Thus, the author must perform the transformation manually, which is time-consuming and error-prone.

Transformations between CAS input and LaTeX are not straightforward and require substantial knowledge of the internal processes for the CAS [10]. Table 1 illustrates the differences in representations exemplified for a Jacobi polynomial [5]. The expression in generic LaTeX, i.e., general LaTeX without custom macros, sharply differs from the semantically unique terms in CAS inputs. To overcome the issue of missing explicit semantic information in LaTeX expressions, the National Institute of Standards and Technology (NIST) has developed a unique set of semantic LaTeX macros. NIST uses these macros for the Digital Library of Mathematical Functions (DLMF) [13] and the Digital Repository of Mathematical Formulae (DRMF) [4]. Both DLMF and DRMF macros enhance the search capabilities on the DLMF and DRMF websites and establish info boxes that provide short descriptions of the symbols, link to their definitions, and further literature. Table 1 shows that the semantically enhanced LaTeX is closer to the syntax supported by a CAS. In the following, we will refer to semantically enhanced LaTeX as semantic LaTeX, and general LaTeX expressions as generic LaTeX, respectively. In the following, we will propose a context-sensitive approach to convert the generic LaTeX expressions to CAS. The approach will take advantage of existing tools and datasets.

Table 1. Representations of a Jacobi polynomial in different systems.

Systems	Representations
Rendered Version	$P_n^{(\alpha,\beta)}(\cos(a\Theta))$
Generic LATEX	`P_n^{(\alpha,\beta)}(\cos(a\Theta))`
Semantic LATEX	`\JacobiP{\alpha}{\beta}{n}@{\cos@{a\Theta}}`
CAS Maple	`JacobiP(n,alpha,beta,cos(a*Theta))`
CAS Mathematica	`JacobiP[n,\[Alpha],\[Beta],Cos[a \[CapitalTheta]]]`

1.1 Related Work

To the best of our knowledge, there is no system nor a theoretical concept yet that allows for translating LATEX expressions to CAS and taking the context of the expression into account. Existing tools, such as the inbuild import/export functions of CAS, ignore context information and are therefore limited to simple, unambiguous cases (e.g., `\frac{1}{2}` or `\cos x`) [10].

We previously developed a system called LACAST, that converts semantic LATEX expressions to the CAS Maple and Mathematica [10]. LACAST is essentially a rule-based engine that performs translations based on manually crafted patterns. The engine follows a modular concept, which allows for extending the system without additional coding, e.g., by extending or creating new lists of translation patterns. Cohl et al. [8] have shown that LACAST is able to identify errors in digital mathematical libraries and CAS. However, LACAST also does not consider the context of math formulae, since the necessary semantic information is encoded in the semantic macros. Moreover, the use of the semantic LATEX dialect is currently limited to the DLMF and DRMF. Hence, the next step is to extend the system to work with generic LATEX inputs.

2 Towards a Context-Sensitive Approach

LACAST performs the translation based on parse trees, which are generated by the Part-of-Math (POM) tagger [7]. Similar to the Part-of-Speech (POS) taggers in natural language processing (NLP), the POM tagger also tags tokens with additional information. In its current state, the POM tagger does not consider context information. Thus, the parse tree generated by the POM tagger should not be misunderstood as a syntax tree of equations. Since semantic LATEX is an extension of generic LATEX, the POM tagger is also able to parse semantic LATEX expressions. The POM tagger stores the information about tokens in a manually crafted database, called lexicons. The lexicons contain possible semantic information for symbols. For example, the lexicon entry for ζ contains twelve different meanings [7,10]. Three of the twelve entries are special functions: the Weierstrass zeta function, the Riemann zeta function, and the Hurwitz zeta function. Each meaning also provides information about the structure of the function. For example, the Hurwitz zeta function $\zeta(s,a)$ has two arguments. The first argument is a complex variable, while the second is a complex parameter.

The semantic information of a mathematical formula is either given in the context or can be derived from the structure of the formula (e.g., when the notation of an expression is unambiguous). The lexicons of the POM tagger and the definitions of the semantic LaTeX macros provide a database of standardized notations of mathematical functions. Hence, this knowledgebase can be used to derive semantic information from the structure of an expression. To analyze the textual context, we can use the Mathematical Language Processor (MLP) [6]. The MLP aims to extract the textual descriptions, called definiens, from the context of a mathematical expression. The MLP focuses on single mathematical symbols, named identifiers. An identifier might also include the subscript since a symbol with a subscript is often interpreted as one mathematical object. The basic approach of the MLP is that candidates of definiens and identifiers are connected when the distance between them is small, i.e., fewer words appear between the identifier and its definiens. The score also considers the distance of identifier-definiens pairs to complex mathematical expressions that contain the identifier. Schubotz et al. [6] also presented ten patterns of phrases, defined by domain experts, that introduce a new pair of definiens and identifier, such as <identifier> (is|are) <definiens>. The authors reported the precision of $p = 0.4860$ and the recall of $r = 0.2806$ for their new machine learning approach. The concept of the MLP is implemented in a publicly available Java framework called mathosphere[3].

For the Jacobi polynomial from Table 1, $P_n^{(\alpha,\beta)}(x)$, mathosphere extracts four identifier P_n, α, β, and x rather than groups of tokens, such as $P_n^{(\alpha,\beta)}(x)$. Without considering $P_n^{(\alpha,\beta)}(x)$ as one mathematical object, it is challenging to identify α, β, and n as parameters and x as the variable. We addressed this issue in [12] by identifying so-called Mathematical Objects of Interest (MOI). MOI represent meaningful groups of tokens rather than single identifiers. In [12], we developed a search engine to find MOI by a given textual query. For example, the top-3 results for the search query '*Jacobi Polynomial*' were $P_n^{(\alpha,\beta)}(x)$, $P_n^{(\alpha,\beta)}$, and $\beta > -1$ (which is one of the constraints of Jacobi polynomials). The search engine allows for linking mathematical expressions with textual queries. The retrieved MOIs are based on the distributions of mathematical formulae in the corpus of arXiv[4] and zbMATH[5]. Hence, they represent common relevant expressions for a given textual query.

3   Conversion and Evaluation Pipeline

Figure 1 illustrates the pipeline of the proposed system to convert generic LaTeX expressions to CAS. The figure contains numbered badges that represent the different steps in the system. Steps 2–5 represent the conversion pipeline, while steps 1, 6, and 7 are different ways to evaluate the system. Mathosphere [6] will

---

[3] https://github.com/ag-gipp/mathosphere [Accessed 03-24-2020].
[4] https://arxiv.org [Accessed 03-24-2020].
[5] https://zbmath.org [Accessed 03-24-2020].

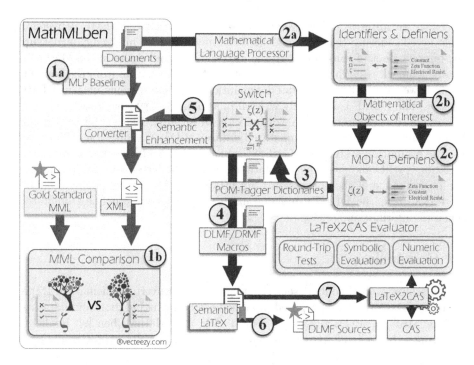

Fig. 1. Pipeline of the proposed context-sensitive conversion process. The project extracts semantic information from real-world documents (2), enhances the mathematical input expressions with the extracted information (3–4), and transforms the math into CAS representations in the final step (6–7).

serve as the baseline. With MathMLben [9], a benchmark for MathML, we tested the performance of several LaTeX to MathML conversion tools. MathMLben provides a manually crafted semantically annotated dataset for 300 mathematical formulae. We evaluate mathosphere on this annotated dataset in step 1a.

The conversion pipeline starts with mathosphere (step 2a) to extract identifier-definiens pairs from the given context. Since mathosphere only considers single identifiers, we will use the developed search engine in [12] to derive MOIs for the extracted definiens (step 2b). The identified MOIs can be matched against complex expressions in the context. Therefore, we end up with MOI-definiens pairs in step 2c, where the scores are calculated based on the relevance of MOIs and the original scores generated by mathosphere.

Once we extracted the MOI-definiens pairs, we replace the generic LaTeX expressions by their semantic counterparts (steps 3–4). This can be done based on the lexicons of the POM tagger and the DLMF Macro definition files, which both provide information about the argument layout of functions. This information is important to identify fixed notations, i.e., P in $P_n^{(\alpha,\beta)}(x)$, and the variables/parameters, i.e., α, β, n, and x in $P_n^{(\alpha,\beta)}(x)$. After these steps, we have the option to evaluate the system in three different ways.

First, we improve the conversion process of LaTeX to MathML conversion tools by considering the extracted MOI-definiens pairs. Thus, we can measure the improvement of considering the context against the results in the MathML-ben benchmark tests in [9], which did not use the information from the context. Second, we evaluate the generated semantic LaTeX expressions on the DLMF dataset. The DLMF is internally written in semantic LaTeX, but provides external access to the generic LaTeX version of each formula. Hence, the DLMF can be interpreted as a manually annotated dataset of LaTeX expressions. Third, we use the evaluation system of LACaST [8], which uses CAS to check if a translated equation is still valid after the translation system. The latter is useful to compare the performance of the conversion from LaTeX to CAS with manually (semantic LaTeX from the DLMF) and automatically (proposed pipeline) annotated semantic information.

4   Conclusion

We presented a novel context-sensitive approach to convert mathematical LaTeX expressions to CAS. The proposed pipeline based on existing tools and datasets, such as MLP [6], POM tagger [7], LACaST [10], and MathMLben [9]. Realizing the proposed pipeline is part of our current research.

Acknowledgments. This work was supported by the German Research Foundation (DFG grant GI-1259-1).

References

1. Gaudeul, A.: Do open source developers respond to competition?: the (La)TeX case study. Rev. Netw. Econ. **6**(2), 9 (2006). https://doi.org/10.2139/ssrn.908946
2. Karampetakis, N.P., Vardulakis, A.I.G.: Special issue on the use of computer algebra systems for computer aided control system design. Int. J. Control **79**(11), 1313–1320 (2006). https://doi.org/10.1080/00207170600882346
3. von zur Gathen, J., Gerhard, J.: Modern Computer Algebra, 3rd edn. Cambridge University Press, Cambridge (2013)
4. Cohl, H.S., et al.: Growing the digital repository of mathematical formulae with generic LaTeX sources. In: Kerber, M., Carette, J., Kaliszyk, C., Rabe, F., Sorge, V. (eds.) CICM 2015. LNCS (LNAI), vol. 9150, pp. 280–287. Springer, Cham (2015). https://doi.org/10.1007/978-3-319-20615-8_18
5. Cohl, H.S., et al.: Semantic preserving bijective mappings of mathematical formulae between document preparation systems and computer algebra systems. In: Geuvers, H., England, M., Hasan, O., Rabe, F., Teschke, O. (eds.) CICM 2017. LNCS (LNAI), vol. 10383, pp. 115–131. Springer, Cham (2017). https://doi.org/10.1007/978-3-319-62075-6_9
6. Schubotz, M., Krämer, L., Meuschke, N., Hamborg, F., Gipp, B.: Evaluating and improving the extraction of mathematical identifier definitions. In: Jones, G.J.F., et al. (eds.) CLEF 2017. LNCS, vol. 10456, pp. 82–94. Springer, Cham (2017). https://doi.org/10.1007/978-3-319-65813-1_7

7. Youssef, A.: Part-of-math tagging and applications. In: Geuvers, H., England, M., Hasan, O., Rabe, F., Teschke, O. (eds.) CICM 2017. LNCS (LNAI), vol. 10383, pp. 356–374. Springer, Cham (2017). https://doi.org/10.1007/978-3-319-62075-6_25

8. Cohl, H.S., Greiner-Petter, A., Schubotz, M.: Automated symbolic and numerical testing of DLMF formulae using computer algebra systems. In: Rabe, F., Farmer, W.M., Passmore, G.O., Youssef, A. (eds.) CICM 2018. LNCS (LNAI), vol. 11006, pp. 39–52. Springer, Cham (2018). https://doi.org/10.1007/978-3-319-96812-4_4

9. Schubotz, M., et al.: Improving the representation and conversion of mathematical formulae by considering their textual context. In: Chen, J. et al. (eds.) Proceedings of ACM IEEE JCDL, Fort Worth, USA, pp. 233–242. ACM (2018). https://doi.org/10.1145/3197026.3197058

10. Greiner-Petter, A., et al.: Semantic preserving bijective mappings for expressions involving special functions in computer algebra systems and document preparation systems. Aslib J. Inform. Manage. **71**(3), 415–439 (2019). https://doi.org/10.1108/AJIM-08-2018-0185

11. Matthews, D.: Craft beautiful equations in word with LaTeX. Nature **570**(7760), 263–264 (2019). https://doi.org/10.1038/d41586-019-01796-1

12. Greiner-Petter, A., et al.: Discovering mathematical objects of interest - a study of mathematical notations. In: Proceedings of The Web Conference 2020 (WWW 2020), 20–24 April 2020, Taipei, Taiwan, April 2020. https://doi.org/10.1145/3366423.3380218

13. Olver, F.W.J., et al. (eds.) NIST Digital Library of Mathematical Functions, Release 1.0.25 of 15 December 2019. http://dlmf.nist.gov/

Employing C++ Templates in the Design of a Computer Algebra Library

Alexander Brandt$^{(\boxtimes)}$ ⓘ, Robert H.C. Moir ⓘ, and Marc Moreno Maza

Department of Computer Science, The University of Western Ontario,
London, Canada
{abrandt5,rmoir3}@uwo.ca
moreno@csd.uwo.ca

Abstract. We discuss design aspects of the open-source Basic Polynomial Algebra Subprograms (BPAS) library. We build on standard C++11 template mechanisms to improve ease of use and accessibility. The BPAS computer algebra library looks to enable end-users to do work more easily and efficiently through optimized C code wrapped in an object-oriented and user-friendly C++ interface. Two key aspects of this interface to be discussed are the encoding of the algebraic hierarchy as a class hierarchy and a mechanism to support the combination of algebraic types as a new type. Existing libraries, if encoding the algebraic hierarchy at all, use run-time value checks to determine if two elements belong to the same ring for an incorrect false sense of type safety in an otherwise statically-typed language. On the contrary, our template metaprogramming mechanism provides true compile-time type safety and compile-time code generation. The details of this mechanism are transparent to end-users, providing a very natural interface for an end-user mathematician.

Keywords: Algebraic hierarchy · C++ templates · Type safety

1 Introduction

In the world of computer algebra software there are two main categories. The first is computer algebra systems, self-contained environments providing an interactive user-interface and usually their own programming language. Custom interpreters and languages yield powerful functionality and expressibility, however, obstacles remain. For a basic user, they must learn yet another programming language. For an advanced user, interoperability and obtaining fine control of hardware resources is challenging. AXIOM [12] is a classic example of such a system. Moreover, these problems are exacerbated by systems being proprietary and closed-source, such as MAPLE [5], MAGMA [6], and MATHEMATICA [18]. The second category is computer algebra libraries, which add support for symbolic computation to an existing programming environment. Since such libraries extend existing environments, and are often free (as in free software), they can have a lower barrier to entry and better accessibility. Some examples are NTL [14], FLINT [11], and CoCoALIB [1].

© Springer Nature Switzerland AG 2020
A. M. Bigatti et al. (Eds.): ICMS 2020, LNCS 12097, pp. 342–352, 2020.
https://doi.org/10.1007/978-3-030-52200-1_34

The Basic Polynomial Algebra Subprograms (BPAS) library [2] is a free and open-source computer algebra library for polynomial algebra, and is the subject of this paper. The BPAS library looks to improve the efficiency of end-users through both usability and performance, providing high-performance code along with an interface which incorporates some of the expressibility of a custom computer algebra system. The library's core is implemented in C for performance and wrapped in a C++ interface for usability. Like any computer algebra software, functionality is highly important, yet usability makes the software practical.

The implementation of BPAS is focused on performance for modern computer architectures by optimizing for data locality and through the effective use of parallelization. These techniques have been applied to our implementations of multi-dimensional FFTs, real root isolation, dense modular polynomial arithmetic, and dense integer polynomial multiplications; see [7] and references therein. Recent works have extended BPAS to include arithmetic over large prime fields [8] and sparse multivariate polynomial arithmetic [3]. Experimentation presented in those works indicates that the performance of BPAS surpasses other existing works. All of this functionality culminates into a high-performance and parallel polynomial system solver (currently under development) based on the theory of regular chains [4]. However, in the present discussion, we look to describe our efforts to make these existing high-performance implementations accessible and practical through user-interface design and improved usability.

Usability includes many things: ease of use in interfaces, syntax, and semantics; mathematical correctness; accessibility and extensibility for end-users; and maintainability for developers. The BPAS library follows two driving principles in its design. The first is to encapsulate as much complexity as possible on the developer's side, where the developer's intimacy with the code allows her to bear such a burden, in order to leave the end-user's code as clean as possible. The second can be described by a common phrase in user experience design: "make it hard to do the wrong thing."

The object-oriented nature of C++, along with its automatic memory management, provides a very natural environment for a user-interface. While C++ is notoriously difficulty to learn, it remains ubiquitous in industry and scientific computing, making it reasonably accessible, and particularly so, if complexity can be well-encapsualted. Moreover, C++ being a compiled, statically- and strongly-typed language, further aids the end-user. The compilation process itself provides the user with checks on their code before it even runs. Meanwhile, statically-typed languages have been shown to be beneficial to usability, and decreases development time, compared to dynamic languages [10].

In the present work, we discuss our early efforts to use C++ metaprogramming to aid in the usability of our interface, for which we hope that BPAS will be easily adopted by other practitioners. Our discussion focuses on two aspects relating to type safety and expressibility. First, encoding the algebraic hierarchy as a class hierarchy is discussed in Sect. 2. Doing so while maintaining type safety is difficult; syntactically valid operations may yield mathematically invalid operations between incompatible rings. Secondly, we examine a mechanism to automatically adjust the definition of a class created from the composition of other classes. In particular, we look at polynomials adapting to different ground rings in Sect. 3. Our techniques are discussed and contrasted with existing works in Sect. 4. We conclude and present future work in Section 5.

We note that our techniques are not entirely new; the underlying template metaprogramming constructs have been adopted into the C++ standard since as early as C++11. Nevertheless, it remains useful to explore how these advanced concepts can be employed in the context of computer algebra. For details on C++, templates, and their capabilities, see [17].

2   Algebraic Hierarchy as a Class Hierarchy

In object-oriented programming (OOP) classes form a fundamental part of software design. A class defines a type and how all instances of that type should behave. Through a class hierarchy, or a tree of inheritance, classes have increasing specialization while maintaining all of the functionality of their superclasses. The benefits of a class hierarchy are numerous, including providing a common interface to which all objects should adhere, minimizing code duplication, facilitating incremental design, and of course, polymorphism. All of this provides better maintainability of the software and a more natural use of the classes themselves since they directly model their real-world counterparts.

For algebraic structures, the chain of class inclusions naturally admits an encoding as a class hierarchy. For example, the class inclusions of some rings[1],

$$\text{field} \subset \text{Euclidean domain} \subset \text{GCD domain} \subset \text{integral domain} \subset \text{ring},$$

would allow rings as the topmost superclass with an incremental design down to fields. Let us call such an encoding of algebraic types as a class hierarchy the *algebraic class hierarchy*. Particularly, we look to implement this hierarchy as a collection of abstract classes for the benefits of code re-use and enforcing a uniform interface across all concrete types (e.g. integers, rational numbers).

Unfortunately, an encoding of algebraic structures as classes in this way yields incorrect type safety. Through polymorphism, two objects sharing a superclass interact and behave in a uniform way, without regard to if they are mathematically compatible. Consider the C++ function declaration which could appear in the topmost `Ring` class: `Ring add(Ring x, Ring y)`. By polymorphism, any two Ring objects could be passed to this function to produce valid code, but, if those objects are from mathematically incompatible rings, this will certainly lead to errors. A more robust system is needed to facilitate strict type safety.

Some libraries (see Sect. 4) solve this by checking runtime values to ensure compatibility, throwing an error otherwise. Instead, our main idea is to define the interface of a ring (or a particular subclass, e.g. integral domain) in such a way where a function declaration itself restricts its parameters to be from compatible rings.

---

[1] Throughout this paper we assume commutative rings with unity.

In our algebraic class hierarchy, function declarations themselves restrict their parameters to be from compatible rings through the use of template parameters. Particularly, our algebraic class hierarchy is a hierarchy of class templates with the template parameter `Derived`. This template parameter identifies the concrete ring(s) with which the one being defined is compatible. In this design, all abstract classes in the hierarchy have the template parameter `Derived` while the concrete classes instantiate this template parameter of their superclass with that concrete class itself being defined. This yields the C++ idiom, the Curiously Recurring Template Pattern (CRTP) (see [17, Ch. 16]).

While CRTP has several functions, it is used here to facilitate *static polymorphism*. That is to say, it forces function resolution to occur at compile-time, instead of dynamically at runtime via virtual tables, providing compile-time errors for incompatibility. For example, the topmost `Ring` class would become a class template `Ring<Derived>` and the `add` function would become `Derived add(Derived x, Derived y)`.

This process works from a key observation when considering simultaneously templates and class inheritance: different template parameter specializations produce distinct classes and thus distinct inheritance hierarchies. Recall that template instantiation in fact causes code generation at compile-time. Thus, each concrete ring defined via CRTP exists in its own class hierarchy, and dynamic dispatch via polymorphism cannot cause runtime inconsistencies. This concept is illustrated in Listing 1 where the abstract classes for ring and Euclidean domain are shown, as well as the concrete class for the ring of integers. The `Integer` class uses template instantiation where it defines its superclass, specializing the `Derived` parameter of `BPASEuclideanDomain` to be `Integer`, following CRTP.

```
1   template <class Derived>
2   class BPASRing;
3
4   //... more abstract algebraic classes, e.g. BPASGCDDomain, BPASField
5
6   template <class Derived>
7   class BPASEuclideanDomain : BPASGCDDomain<Derived>;
8
9   class Integer : BPASEuclideanDomain<Integer>;
```

Listing 1. A subset of the algebraic class hierarchy, using CRTP to declare the integers.

While this design provides the desired compile-time type safety, it may be viewed as too strict, since each concrete ring exists in an independent class hierarchy. For example, arithmetic between integers and rational numbers would be restricted. More generally, natural ring embeddings are neglected. However, we can make use of *implicit conversion* in C++. Where a constructor exists for type `A` taking an object of type `B` as input, an object of type `B` can be implicitly converted to an object of type `A`, and used anywhere type `A` is expected. A `RationalNumber` constructor taking an `Integer` parameter thus allows for automatic and implicit conversion, allowing integers to be used as rational numbers.

This design via implicit conversion can be seen as giving permission for compatibility between rings by defining such a constructor. Errors are then discovered at compile-time where implicit conversion fails. This is in opposition to other methods which act in a restrictive manner, allowing everything at compile-time and then throwing errors at runtime if incompatible.

We now look to extend the abstract algebraic class hierarchy to include polynomials. For genericity and a common structured interface we wish to parameterize polynomials by their ground ring. This can be accomplished with a secondary template parameter in addition to the `Derived` parameter already included by virtue of polynomials existing in the algebraic class hierarchy (see Listing 2).

However, this is not fully sufficient, and two issues arise. First, while polynomials do form a ring, they often form more specialized algebraic structures, e.g. a GCD domain. We leave that discussion to Sect. 3. Secondly, there is no restriction on the types which can be used as template parameter specializations of the ground ring. Any type used as a specialization of this ground ring template parameter should truly be a ring and not any other nonsense type. Recall, it should be hard to do the wrong thing.

Leveraging another template trick along with multiple inheritance, this can be solved with the so-called `Derived_from` class[2] which determines at compile-time if one class is the subclass of another. `Derived_from` is a template class with two parameters: one a potential subclass, and the other a superclass. This class defines a function converting the apparent subclass type to the superclass. If the conversion is valid via implicit up-casting, then the function is well-formed, otherwise, a compiler error occurs.

To make use of `Derived_from`, a class template inherits from `Derived_from`, passing its own template parameter to `Derived_from` as the potential subclass, along with a statically defined superclass type. This enforces that the template parameter be a subclass of that superclass. In our implementation, shown in Listing 2, polynomial classes enforce that their ground ring should be a `BPASRing`, our abstract class for rings (recall the declaration of `BPASRing` from Listing 1).

```
1  // If T is not a subclass of Base, a compiler error occurs
2  template <class T, class Base> class Derived_from {
3      static void constraints(T* p) { Base* pb = p; }
4      Derived_from() { void(*p)(T*) = constraints; }
5  };
6
7  template <class Ring, class Derived>
8  class BPASPolynomial : BPASRing<Derived>, Derived_from<Ring, BPASRing<Ring
        >>;
```

Listing 2. An implementation of a polynomial interface using CRTP and `Derived_from`.

All of these functionalities together create an algebraic hierarchy as a class hierarchy while maintaining strict type safety. Yet, our scheme remains flexible enough to support implicit conversions, such as natural ring embeddings, and generic enough to allow, for example, polynomials over user-defined classes, as long as those classes inherit from `BPASRing`. What remains now is to address the

---

[2] `Derived_from` is a long-known trick, but is now adopted into the C++20 standard.

issue of polynomial rings sometimes forming different algebraic types depending on their particular ground ring.

3 "Dynamic" Type Creation, Conditional Export

In object-oriented design, the combination of types to create another type is known as composition. In this section, let us consider univariate polynomial rings; one can always work recursively for multivariate polynomials. Viewing a polynomial ring as a ring extension of its ground ring, polynomials can be seen as the composition of some finite number of elements of that ground ring. Moreover, we know that the properties of a polynomial ring depend on the properties of the ground ring. For example, the ring of univariate polynomials over a field is a Euclidean domain while the ring of polynomials over a ring is itself only a ring. Recall from the previous section that our implementation of polynomials are templated by their ground ring. Our goal then is to capture the idea that the position of a polynomial ring in the algebraic class hierarchy changes depending on the particular specialization of this template parameter.

More generally, we would like that the type resulting from the composition of another type depends on the type being composed. Hence, a sort-of "dynamic" type creation. This is not truly dynamic, since it is a compile-time operation, but it nonetheless feels dynamic since it is an automatic process by the compiler via template instantiation. In fact, having this occur at compile-time is actually a benefit where errors can be determined preemptively. One can also view this mechanism as a way of controlling the methods which the newly created type exports. That is, conditionally exposing methods (or other attributes) in its interface depending on the particular template parameter specialization. This technique relies on compile-time introspection and SFINAE.

3.1 SFINAE and Compile-Time Introspection

Substitution Failure Is Not An Error (SFINAE), coined by Vandevoorde in [17], refers to a fundamental part of C++ templating. The invalid substitution of a type as a template parameter is itself not an error. Such a principle is required for templates to be practical. Where two or more template specializations exist, it is not required that the substitution of the template parameter fit all of the specializations, but only one. This principle, combined with compile-time function overload resolution, provides template metaprogramming its power. In particular, *compile-time introspection* is possible: using templates, truth values about a type can be determined and then made use of within the program.

Consider the typical example, adapted from [17, Section 8.3], shown in Listing 3. type_has_X determines if a type has a member X by checking the size of the return type of a function. By function overload resolution, if T has a member X the test<T> function chosen will be the first, whose return type has size 1. Otherwise the second function is chosen with return type of size (at least) 2.

```
1   template<typename T> char test(typename T::X const*);
2   template<typename T> int  test(...);
3   #define type_has_X(T) (sizeof(test<T>(NULL)) == 1);
```

Listing 3. A simple compile-time introspection to determine if type T has member X.

This idea can be generalized to many introspective metaprogramming techniques. For example, is_base_of, a standard feature in C++11, is much like Derived_from. However, instead of creating a compiler error, is_base_of determines a Boolean value representing if one type is derived from another.

Using introspection, one may think that enable_if, another standard C++11 template construct, is sufficient. The enable_if struct template conditionally compiles and exposes a function template based on the value of a Boolean known at compile-time. This Boolean value can of course be determined by introspection. Unfortunately, function templates cannot be virtual, thus this solution cannot be used within a class hierarchy. Conditionally exposing methods in our algebraic class hierarchy requires a different solution.

3.2   Conditional Inheritance for Polynomials

Defining new types dependent on the value of another type, as well as conditionally exposing member functions, can both be fulfilled by *conditional inheritance*. Specifically, we implement a compile-time case discussion for inheritance based on introspective values. In the context of polynomials in our algebraic class hierarchy, that case discussion works as a cascade of type checks on the ground ring, say R, when forming the polynomial ring $R[x]$. For example: if R is a field, then $R[x]$ is a Euclidean domain; else if R is a GCD domain, so is $R[x]$; else if R is an integral domain, so is $R[x]$; else $R[x]$ is a ring. This case discussion can be extended to include as much granularity as needed.

To perform this case discussion, we use the C++11 metaprogramming feature conditional, which uses a Boolean value known at compile-time to choose between two types. This is much like the ternary conditional operator which uses a Boolean to choose between two statements. Using is_base_of to determine the Boolean, conditional chooses one of two types to use as a class's superclass.

As a simple example, consider Listing 4. The definition of BPASPolynomial tests if the Ring template parameter is a subclass of BPASField. If so, conditional chooses BPASEuclideanDomain as the the superclass of BPASPolynomial, otherwise BPASRing is chosen. Additionally, a concrete class SparseUnivarPoly is shown, still parameterized by a coefficient ring. In this concrete class, the interface of the class will adapt "dynamically" to a particular template specialization via the conditional in its superclass. Notice also that the template parameter of SparseUnivarPoly is enforced to be a subclass of BPASRing on specialization via the Derived_from of its superclass.

```
1  template <class Ring, class Derived>
2  class BPASPolynomial : conditional< is_base_of<Ring, BPASField<Ring>>::
       value,
3                                   BPASEuclideanDomain<Derived>,
4                                   BPASRing<Derived> >::type,
5                          Derived_from< Ring, BPASRing<Ring> >;
6
7  template <class CoefRing>
8  class SparseUnivarPoly : BPASPolynomial<CoefRing,SparseUnivarPoly<CoefRing
       >>;
```

Listing 4. A simple use of `conditional` to choose between Euclidean domain or ring as the algebraic type of a polynomial based on its template parameter.

The presented code for `BPASPolynomial` in Listing 4 is rather simple, showing only a single type check. To implement a chain of type checks, the "else" branch of a `conditional` should simply be another `conditional`. To improve the readability of this case discussion, we avoid directly implementing nested if-else chains, and thus avoid using one `conditional` inside another. Instead, we create two symmetric class hierarchies, one representing the true algebraic class inclusions while the other is a "tester" hierarchy.

This tester hierarchy uses one `conditional` to determine if a property holds and, if so, chooses the corresponding class from the algebraic hierarchy as superclass. Otherwise, the next tester in the hierarchy is chosen as superclass to trigger the evaluation of the next `conditional`. Finally, all concrete polynomial classes inherit from `BPASPolynomial` to automatically determine their correct interface based on their ground ring. This structure is shown in Fig. 1, with the algebraic hierarchy on the left, and the tester hierarchy on the right.

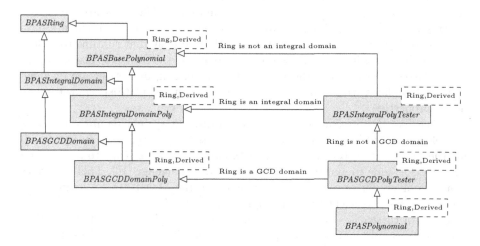

Fig. 1. UML diagram for a subset of the polynomial abstract class hierarchy. Recall in UML that template parameters are shown in dashed boxes. Template parameters for non-polynomial classes are omitted for clarity. Note also that the multiple inheritance diamond problem is easily solved using virtual inheritance.

This technique of conditional inheritance is a powerful tool in any class template hierarchy. By understanding the properties of a type via introspection, it can automatically be incorporated into an existing class hierarchy either as itself or when used in composition to create a new type. For example, based on the specialization of a template parameter, the definition of a class template can be changed automatically and dynamically. Not only does this enforce a proper class interface, but it allows the possibility of choosing between several different abstract implementations in order to best support the new type (i.e. the result of a composition).

4   Discussion and Related Work

For decades, computer algebra systems have worked towards type safety. Axiom [12] is a pioneering work on that front, but has grown out of popularity. Functional languages, like Scala and Haskell, have seen some progress in developing computer algebra systems thanks to type classes (see, e.g., [13] and references therein). These languages and their type classes provide a very suitable environment to define algebraic structures. However, while powerful, functional languages can be seen as an obscure and inaccessible programming paradigm compared to the mainstream imperative paradigm.

Considering other C/C++ computer algebra libraries, there are many examples with interesting mechanisms for handling algebraic structures. The SINGULAR library [9] perhaps has the most simple mechanism: a single class represents all rings, using a number of enum and Boolean variables to determine properties of instances at runtime. In COCOALIB [1] an abstract base class `RingBase` declares many functions returning Boolean values. Concrete subclasses define these functions to determine properties at runtime. While rings are subclasses of `RingBase`, elements of a ring are an entirely different class. Elements have pointers to the ring they belong, which are then compared at runtime to ensure compatibility in arithmetic between two elements. LINBOX [15] also has separate classes for rings and their elements. There, ring properties are encoded as class templates where concrete rings use explicit template specialization to define properties.

Much like the previous cases, the MATHEMAGIX system requires instances (i.e. elements) of a ring to have a specific reference to a separate entity encoding the ring itself. Notably, MATHEMAGIX also includes a scheme to import and export C++ code to and from the MATHEMAGIX language [16]. This uses templates to allow, for example, a ring specified in the MATHEMAGIX language to be used as the coefficient ring for polynomials defined in C++.

In all of these cases there is some limiting factor. Most often, mathematical type safety is only a runtime property maintained by checking values. In some cases this is implemented by separating rings themselves from elements of a ring, a process counterintuitive to object-oriented design where one class should define the behaviour of all instances of that type.

On the contrary, our scheme does not rely on runtime checks. Instead, a function declaration itself restricts its arguments to be mathematically compatible at compile-time via the use of template parameters and the Curiously Recurring Template Pattern. By using an abstract class hierarchy many such function declarations are combined through consecutive inheritances to build up an interface incrementally. This closely follows the chain of class inclusions for algebraic types, where each type adds properties to the previous. The symmetry between the algebraic hierarchy and our class hierarchy hopes to make our interfaces natural and approachable to an end-user. This symmetry comes at the price of creating a deep class hierarchy, and thus strong coupling within the class hierarchy. Yet, this price is worth the symmetry and comprehensibility of the class hierarchy with the algebraic hierarchy.

In contrast with our class hierarchy solution to type safety, a different compile-time solution could be crafted through further use of type traits (see, e.g., [17, Ch. 15, 17]). Type traits are template metaprogramming constructs for type introspection and modification, some of which have already been seen, such as is_base_of, and conditional. Type traits are arguably more flexible, however, template metaprogramming is already rather difficult, and is essentially limited to C++. Class hierarchies, on the other hand, are present in every object-oriented language and should therefore be more accessible to end-users. The use of class hierarchies, in addition to encapsulating much of the template metaprogramming in our design, should provide better extensibility to end-users in general.

5   Conclusion and Future Work

In this work we have explored part of the implementation and design of the C++ interface of the BPAS library. Through the use of template metaprogramming we have devised a so-called algebraic class hierarchy which directly models the algebraic hierarchy while providing compile-time type safety. This hierarchy is type-safe both in the programming language sense and the mathematical sense.

Using inheritance throughout the algebraic abstract class hierarchy, the interface of algebraic types is constructed incrementally. Therefore, a concrete type's properties and interface is determined by its particular abstract superclass from this hierarchy. Through additional templating techniques, we can automatically infer, at compile-time, the correct superclass (and thus interface) of new types created by template parameter specialization (e.g. polynomials). The result is a consistent and enforced interface for all classes modelling algebraic types.

We are currently working to extend our algebraic class hierarchy to include multivariate power series, polynomials with power series coefficients, and polynomials in prime characteristic. This more capable hierarchy will be used within our library to implement a sophisticated solver for systems of polynomial equations, a prototype of which is already available in recent releases of BPAS. Finally, we hope to create a Python interface to the BPAS library (i.e. an extension module) to further improve the accessibility and ease of use of our library.

Acknowledgements. The authors would like to thank IBM Canada Ltd (CAS project 880) and NSERC of Canada (CRD grant CRDPJ500717-16, award CGSD3-535362-2019).

References

1. Abbott, J., Bigatti, A.M.: CoCoALib: a C++ library for doing Computations in Commutative Algebra. http://cocoa.dima.unige.it/cocoalib
2. Asadi, M., et al.: Basic polynomial algebra subprograms (BPAS) (2020). http://bpaslib.org
3. Asadi, M., Brandt, A., Moir, R.H.C., Moreno Maza, M.: Algorithms and data structures for sparse polynomial arithmetic. Mathematics **7**(5), 441 (2019)
4. Asadi, M., Brandt, A., Moir, R.H.C., Moreno Maza, M., Xie, Y.: On the parallelization of triangular decomposition of polynomial systems. In: CoRR abs/1906.00039 (2019)
5. Bernardin, L., et al.: Maple programming guide (2018). www.maplesoft.com/documentation_center/maple2018/ProgrammingGuide.pdf
6. Bosma, W., Cannon, J., Playoust, C.: The Magma algebra system. I. The user language. J. Symbolic Comput. **24**(3–4), 235–265 (1997). Computational algebra and number theory (London, 1993)
7. Chen, C., Covanov, S., Mansouri, F., Maza, M.M., Xie, N., Xie, Y.: The Basic polynomial algebra subprograms. In: Hong, H., Yap, C. (eds.) ICMS 2014. LNCS, vol. 8592, pp. 669–676. Springer, Heidelberg (2014). https://doi.org/10.1007/978-3-662-44199-2_100
8. Covanov, S., Mohajerani, D., Moreno Maza, M., Wang, L.: Big prime field FFT on multi-core processors. In : 2019 Proceedings of the International Symposium on Symbolic and Algebraic Computation, pp. 106–113 (2019)
9. Decker, W., Greuel, G.-M., Pfister, G., Schönemann, H.: Singular 4-1-1 – a computer algebra system for polynomial computations (2018). http://www.singular.uni-kl.de
10. Endrikat, S., Hanenberg, S., Robbes, R., Stefik, A.: How do API documentation and static typing affect API usability? In: Proceedings of the 36th International Conference on Software Engineering, pp. 632–642. ACM (2014)
11. Hart, W., Johansson, F., Pancratz, S.: FLINT: fast library for number theory. Version 2.5.2 (2015). http://flintlib.org
12. Jenks, R.D., Sutor, R.S.: Axiom, the scientific computation system (1992)
13. Jolly, R.: Categories as type classes in the scala algebra system. In: Gerdt, V.P., Koepf, W., Mayr, E.W., Vorozhtsov, E.V. (eds.) CASC 2013. LNCS, vol. 8136, pp. 209–218. Springer, Cham (2013). https://doi.org/10.1007/978-3-319-02297-0_18
14. Shoup, V., et al.: NTL: a library for doing number theory. www.shoup.net/ntl/
15. The LinBox group. LinBox. v1.6.3. 2019. url: http://github.com/linbox-team/linbox
16. van der Hoeven, J., Lecerf, G.: Interfacing mathemagix with C++. In: Proceedings of the 2013 International Symposium on Symbolic and Algebraic Computation, pp. 363–370 (2013)
17. Vandevoorde, D., Josuttis, N.M.: C++ Templates. Addison-Wesley Longman Publishing Co., Inc (2002)
18. Wolfram Research, Inc., Mathematica, Version 11.3. Champaign, IL (2018)

Mathematical World Knowledge Contained in the Multilingual Wikipedia Project

Dennis Tobias Halbach[✉][iD]

University of Wuppertal, Wuppertal, Germany
dennis.halbach@uni-wuppertal.de

Abstract. The purpose of this project is to test and evaluate an approach for Formula Concept Discovery (FCD). FCD aims at retrieving a formula concept (in the form of a Wikidata item) together with its defining formula within documents, in this case 100 English Wikipedia articles. To correctly identify the defining formula of a Wikipedia article, this approach searches for shared formulae across Wikipedia articles available in different languages. The formula shared in the most languages is then assumed to be the defining formula. The results show that neither this approach alone nor a combination with an existing approach that considers the order of the formulae inside an article leads to satisfying results. It is thus concluded that the number of times a formula is shared across a Wikipedia article in different languages is not a good indicator to determine the defining formula with the current approach. Consequently, several ideas for further research are proposed which could improve the results.

1 Introduction

For many generations mathematical textbooks were the primary source of information for pupils and laypersons to acquire mathematical knowledge. However, since the beginning of the 21st century and the rise of collaborative online encyclopaediae such as Wikipedia, this situation is changing. Wikipedia can basically be seen as a digital book organized in classical articles with cross-references. This format is similar to printed textbooks and not designed to be machine-readable. Thus, the automated retrieval of properties (like a formula) describing a related topic is a non-trivial task. To assist this task, the Wikidata knowledge graph was established in 2012. Wikidata connects different language versions of Wikipedia and stores data related to Wikipedia as triples, linking a data item (via its unique identifier called a 'QID') to one or multiple properties and their respective value. One such property can be the so called *defining formula* of a Wikidata item, which can be stored in the Wikidata knowledge graph since 2016. For example the Wikidata item on Schwarz's theorem (Q1503239) connects Wikipedia articles in 15 languages on the topic. Here the formula

$$\frac{\partial}{\partial x}\left(\frac{\partial}{\partial y}f(x,y)\right) = \frac{\partial}{\partial y}\left(\frac{\partial}{\partial x}f(x,y)\right)$$

© Springer Nature Switzerland AG 2020
A. M. Bigatti et al. (Eds.): ICMS 2020, LNCS 12097, pp. 353–361, 2020.
https://doi.org/10.1007/978-3-030-52200-1_35

is the defining formula of this concept and thus included in most of the 15 articles, although partly in different mathematical notations (see Fig. 1). As the data format, Wikidata uses *Presentation MathML* as the exchange format and the LaTeX dialect *texvc* as the input format.

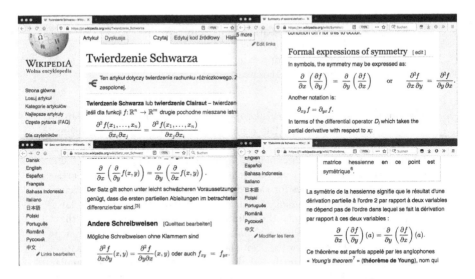

Fig. 1. The Wikipedia articles on Schwarz's theorem in the languages Polish, English, German and French. Accessed on 28th of March, 2020.

While some versions of Wikipedia like German and Portuguese include the exact form of the formula, the French and Spanish versions use a rather than x for the function argument and the English article lacks the function argument. Moreover, the Russian and Polish versions use numeric indices for the variables, i.e, x_1, x_2 instead of x, y. Still, judging from this one example, it seems possible to infer the defining formula from the reoccurrence of a formula across different Wikipedias, although advanced techniques might be needed, e.g. to recognize slightly different formulae as representing the same mathematical concept.

In this paper, we aim at improving the automatic extraction of defining formulae over an already existing approach from Schubotz et al. [2], who chose to extract the first formula included in the English Wikipedia article after a manual investigation showed that the first formula is often the most relevant one for that article [2] as it is frequently included in the introductory part of an article. Knowing the formula with the highest probability to be the defining formula can then be used to suggest formula edits to Wikidata editors.

2   Method

The Wikipedia articles used are obtained from a collection of Wikipedia article dumps[1] for all 309 official Wikipedia languages[2]. Specifically, we use the 100 articles defined as QIDs in the dataset from Schubotz et al. and their respective articles in other Wikipedia languages available. We infer the articles titles from the 100 QID using the MediaWiki API and use these titles to filter the dumps for all pages containing one of the titles in their title-tag. These pages will then be filtered to extract the formulae. Schubotz et al. consider a string a *formula* if it fulfills the following two conditions: Firstly, it has to be enclosed in a wikitext tag, namely 'math', 'ce', or 'chem'. Secondly, it needs to include (at least) one formula-indicator [1,4], in our case '$=$', '$<$', '$>$', '$\leq$', '$\geq$', '$\approx$' and/or '$\equiv$' were used. These formula-indicators prevent that variables are recognized as formulae since a formula typically relates the definiens and the definiendum using formula-indicators.

After filtering all articles for formulae, the extracted strings[3] of all articles with the same QID are compared to determine the string shared by the most articles corresponding to each QID. The resulting 100 most common strings are then compared to a gold standard dataset derived from [2] to evaluate the results. This dataset has been built by randomly choosing 100 English Wikipedia articles, each containing at least one math-tag, and manually determining the correct defining formula for each article.[4] Thus, the gold standard consists of 100 defining formulae and their respective QID of the corresponding article(s). Instead of using the Latex notations for the 100 defining formulae provided by the gold standard dataset, we copied the current Latex notations from the dumps. This decreases the probability that a most common formula does not match an equivalent defining formula simply due to slightly different Latex notations: The notations were found to have changed since the publication of the dataset of [2], e.g. optional brackets were added in the formulae. Thus, this approach ensures better comparability of our results with the results from [2]: While they

---

[1] The downloaded Wikipedia data dump files were created on 2nd & 3rd of March, 2020, and are available on https://dumps.wikimedia.org/.

[2] https://meta.wikimedia.org/wiki/List_of_Wikipedias. Accessed on 6th of March, 2020.

[3] Note: While the word 'string' typically refers to a formula in this paper, it can also mean an empty string (if no formula exists in the article or the string shared most often across all articles is 'none').

[4] Note: While our definition of a 'formula' means a string containing a formula-indicator, the term 'defining formula' references an arbitrary, possibly empty string in the gold standard. This definition is in accordance with Wikidatas *defining formula* property, which does allow strings without a formula-indicator. Thus, it is obvious that our filtering approach cannot find the four defining formulae without a formula-indicator (e.g. $\pi \int_a^b [R(x)]^2 \, dx$ is the defining formula of the article about Disc integration (Q3825524)).

manually confirmed if each extracted formula visually[5] matches the defining formula - an approach that does not depend on exactly matching Latex notations - we automatically check for matching strings. To ensure that we recognize most formulae that visually match their defining formula as a true positive, we check if they are similar: Two mathematical expressions are considered similar if they only differ due to whitespaces, irrelevant characters at the end (like a comma or dot that are part of the sentence surrounding the formula) or optional brackets around a sub- or superscript. These factors were found to be the cause for most different, despite visually matching formulae in a small manual investigation. We rectified the entry for 'plastic number' in the gold standard dataset by using $\rho = \sqrt[3]{\frac{9+\sqrt{69}}{18}} + \sqrt[3]{\frac{9-\sqrt{69}}{18}}$ instead of an empty string (no defining formula).

We classified a result as relevant if and only if its defining formula is not an empty string and is included in (at least) one of the articles of the corresponding QID. To make sure we correctly identify relevant results as such, a defining formula is considered 'included' in an article if (at least) one mathematical expression is similar to it. If a result is relevant and gets retrieved, i.e. the most common formula is the defining formula, it is counted as a true positive (TP). If a result is relevant, but the defining formula is not retrieved, this is classified as a false negative (FN). Non-relevant results are counted as true negatives (TN) if the most common string matches the defining formula, otherwise as false positives (FP). These definitions are in accordance with Schubotz et al. in order to ensure the comparability of the results.

We first investigate the results of our approach of counting the occurrences of formulae as well as a combined approach that also considers the order of the formulae in the articles. Afterwards, we inspect the findings of the combined approach with regard to the number of Wikipedia languages used. When filtering only a number of all 309 Wikipedia languages, we choose to filter the biggest language (English) as well as the biggest five and 20 Wikipedias, while excluding Cebuano and Waray-Waray, since both have a high number of bot-generated articles[6] and low number of community members (see footnote 2).[7] We determine the size of the Wikipedias by the number of articles in its respective language according to a list of all Wikipedias (see footnote 2). Afterwards we use our definition of similarity to determine the most common formula and investigate the number of true positives.

---

[5] Two mathematical expressions are considered visually matching if the expressions generated from the (possibly different) Latex notations look the same, e.g. x_i and $x\_\{i\}$ generate the same expression.

[6] https://stats.wikimedia.org/EN/BotActivityMatrixCreates.htm. Accessed on 22nd of March, 2020.

[7] As it turns out, this measure was unnecessary since neither language included an article corresponding to one of the 100 QIDs.

3   Evaluation

As a first, simple approach we filter one, five, 20, and all 309 Wikipedias while counting the number of articles a formula occurs in. If more than one formula is the most common one, the extracted formula is chosen randomly among them. The results show that while we do get better results by using more Wikipedias, the number of false results is always higher than 70 (mostly due to FN), irrespective of the number of Wikipedias used, and thus too high for this approach to reliably work. An investigation of the results when using 309 languages shows that more than half of the most common strings only occur in one or two languages, thus 54 most common strings have at least one other string with the same number of occurrences. Consequently, ~80% of those are falsely identified — in comparison to ~61% for the more common formulae. This shows an obvious problem in the data: A lot of strings only reoccur very rarely across articles, mostly because they occur in a similar mathematical form or depend on a different Latex notation to generate a visually equivalent formula. Before trying to solve this problem by recognizing similar formulae when determining the most common formula, we will focus on another point: Randomly choosing the extracted string among multiple most common strings is a simple but unsophisticated approach. Instead, we now use the order of the formulae as a measure in case two formulae have the same number of occurrences.

As it turns out, this allows us to easily reproduce the findings of Schubotz et al. when using English as the sole language to filter: Since we only count every formula in an article once, we essentially disregard the occurrences of the formulae when using only one language; instead only the order of the formulae will be taken into account, as is the case in [2].

The results in Table 1 reveal that we find nine TP less, while getting five FP and six FN more than Schubotz et al. The higher number of FN is in about four cases attributed to the fact that we — in contrast to Schubotz et al. — automatized the comparison of the extracted formulae with the defining formulae. As a consequence, four extracted formulae could not be identified as equal to their visually equivalent defining formula since they were not similar. The remaining five missing TP are probably attributable to the time-conditioned changes of the Wikipedia articles since the publication date of [2]: The gold standard depends on the defining formulae that were based on mathematical expressions of former Wikipedia sites. As such, today some Wikipedia articles only include a mathematically equivalent, but different formulation, which does not match our defining formula, e.g. $a = b$ and $b = a$. Thus, such a result is falsely recognized as 'non-relevant' and classified as a FP instead of TP.

Altogether, we can verify the findings of Schubotz et al. The investigation of the results revealed that an automated classification of results in 'relevant' and 'non-relevant' is not perfectly accomplishable with the current approach and a more sophisticated method is needed to determine if a formula matches its defining formula.

Table 1. Contingency table comparison of (a) our results and (b) the findings from Schubotz et al. when using one Wikipedia language (English)

	(a) relevant	not relevant		(b) relevant	not relevant
retrieved	62 (TP)	22 (FP)	retrieved	71 (TP)	17 (FP)
not retrieved	16 (FN)	0 (TN)	not retrieved	10 (FN)	2 (TN)

Next, we take a look at the impact the combination of the approach of Schubotz et al. with our approach has on the results when using more than one language. The results in Fig. 2 show that we do get less TP as the number of languages increases and that we get the best results with one language. In other words, as the influence of the order of the formulae gets smaller and, consequently, as the influence of the reoccurrences of formulae gets bigger, the results worsen. This suggests that the order is significantly more important and thus, that the approach of using only the order of the formulae is most probably better than only choosing the most common formula. Note, however, that we cannot verify this: A direct comparison of both methods is not possible as the method of simply counting the reoccurrences always needs an accompanying measure in case multiple most common strings exist. While we could generate an arbitrary dataset such that in no case multiple most common strings exist — simply by excluding all QIDs whose article(s) contain more than one most common string — such a dataset would probably be biased: The number of most common strings might correlate with the number of formulae in the article and thus the length and quality of the article, consequently influencing the results. This was not further investigated.

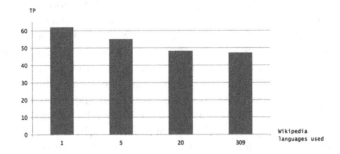

Fig. 2. Number of true positives (TP) depending on the number of Wikipedia languages used.

In the following, we examine the impact that checking for similarity has on the results when we check this not only when comparing a formula and a defining formula as before, but also when determining the most common formula. This allows us to recognize ~13% of the formulae as similar to another formula found and thus increases our number of occurrences per formula. Figure 3 shows that the number of TP negatively correlates with the number of languages used, as is the case in the last approach (see Fig. 2). Contrary to our initial expectations, the current approach could not improve our results compared to Fig. 2. The reason is most probably that the average number of articles containing the most common string increased from 4.0 to ~4.7 (for 309 languages), thus the number of cases where only one most common string exists increased from 46 to 54. As a consequence, in eight fewer cases the order of the formulae is considered. This is another indicator that, as the impact of the occurrences of formulae on our results gets bigger, our results worsen.

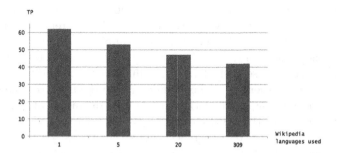

Fig. 3. Number of true positives (TP) depending on the number of Wikipedia languages used when checking for similarity in every comparison between strings.

4   Future Work

The current approach only takes the number of occurrences of a formula and the order of the formulae into consideration, which leaves out a lot of information like the quality of the article, the formula-indicator used in each formula or whether a formula is visually highlighted by placing it in a separate line. Thus, we propose a score-based system using all the information to determine the defining formula more accurately. The information should also include the number of occurrences of a formula, even though it might not improve the results as seen in this project. It is still believed that knowing how often a formula occurs across multiple articles is important information that can improve the detection-rate if used correctly. As the investigation shows, it cannot be the only information used in conjunction with the order of the formulae, although no advanced techniques like unification [3] were used to verify more similar formulae as actually being similar, which might better the results, although the results indicate otherwise. We suggest that in this proposed approach the occurrences of formulae should probably not be weighted heavily as this might negatively impact the results considering our findings.

To build the proposed score, it is necessary to find an optimal weighting of the different pieces of information. To do so, a bigger dataset is needed. We suggest to use Wikidata, which contains a manual assignment of the QIDs of more than 4,300 Wikipedia articles to their respective defining formula.

5   Conclusion

Our findings verify the results of Schubotz et al. who extracted the first formula of a Wikipedia article as an approach to obtain the defining formula related to an article. Nevertheless, it was not possible to achieve the same amount of true positives as Schubotz et al., most probably due to the lack of advanced techniques used to determine whether two formulae are equivalent.

Furthermore, our results were negatively impacted when considering the order of the formulae in their respective article together with the number of languages it occurs in. This suggests that the order of the formulae is a much more important indicator to determine the defining formulae than the number of its occurrences across multiple languages. Thus, reducing the influence the order has on the results in favor of the number of occurrences decreases the number of extracted defining formulae. This assumption is further supported by the fact that the number of true positives negatively correlates with the number of Wikipedia languages used, which in turn influences the number of languages a formula occurs in. Furthermore, when we determine the most common formula by regarding formulae as equal if they match our definition of being similar, the number of true positives further decreases. This is reasoned to be another indication that the number of occurrences of a string across articles is a bad factor for determining the most common formula. Consequently, other indicators are proposed that should be able to improve the current approach. It is worth including the number of occurrences across articles as one of the factors, as it cannot be said with certainty that the number of occurrences is an inherently bad indicator. It might be possible that much more sophisticated measures are needed to determine if two formulae are similar, though our findings suggest otherwise. Improving the results of the current approach will be a focus of future work.

Acknowledgments. The project is based on our contribution to the seminar 'Selected Topics in Data Science' from the Data and Knowledge Engineering group of the University of Wuppertal headed by Bela Gipp [ORCID: 0000 − 0001 − 6522 − 3019]. The author thanks his seminar-advisor Moritz Schubotz [ORCID: 0000−0001−7141−4997].

References

1. Schubotz, M., et al.: Evaluation of similarity-measure factors for formulae based on the NTCIR-11 math task. In: Kando, N., Joho, H., Kishida, K. (eds.) Evaluation of Similarity-Measure Factors for Formulae. Proceedings of the NTCIR National Institute of Informatics (NII) (2014)
2. Schubotz, M., et al.: Introducing MathQA – a math-aware question answering system. Inf. Discov. Deliv. **46**(4), 214–224 (2018). https://doi.org/10.1108/IDD-06-2018-0022

3. Sojka, P., Ruzicka, M., Novotný, V.: MIaS: math-aware retrieval in digital mathematical libraries. In: Cuzzocrea, A., et al. (eds.) Proceedings of the 27th ACM International Conference on Information and Knowledge Management, CIKM 2018, Torino, Italy, 22–26 October 2018, pp. 1923–1926. ACM (2018). https://doi.org/10.1145/3269206.3269233
4. Zhang, Q., Youssef, A.: An approach to math-similarity search. In: Watt, S.M., Davenport, J.H., Sexton, A.P., Sojka, P., Urban, J. (eds.) CICM 2014. LNCS (LNAI), vol. 8543, pp. 404–418. Springer, Cham (2014). https://doi.org/10.1007/978-3-319-08434-3_29

Archiving and Referencing Source Code with Software Heritage

Roberto Di Cosmo[⊠]

Software Heritage, Inria and University of Paris, Paris, France
roberto@dicosmo.org

Abstract. Software, and software source code in particular, is widely used in modern research. It must be properly archived, referenced, described and cited in order to build a stable and long lasting corpus of scientific knowledge. In this article we show how the Software Heritage universal source code archive provides a means to fully address the first two concerns, by archiving seamlessly all publicly available software source code, and by providing *intrinsic persistent identifiers* that allow to reference it at various granularities in a way that is at the same time convenient and effective.

We call upon the research community to adopt widely this approach.

Keywords: Software source code · Archival · Reference · Reproducibility

1 Introduction

Software source code is *an essential research output*, and there is a growing general awareness of its importance for supporting the research process [6,20,27]. Many research communities focus on the issue of *scientific reproducibility* and strongly encourage making the source code of the artefact available by archiving it in publicly-accessible long-term archives; some have even put in place mechanisms to assess research software, like the *Artefact Evaluation* process introduced in 2011 and now widely adopted by many computer science conferences [7], and the *Artifact Review and Badging* program of the ACM [4]. Other raise the complementary issues of making it easier to discover existing research software, and giving academic credit to authors [21,22,25].

These are important issues that are similar in spirit to those that led to the current FAIR data movement [28], and as a first step it is important to clearly identify the different concerns that come into play when addressing software, and in particular its source code, as a research output. They can be classified as follows:

Archival: software artifacts must be properly **archived**, to ensure we can *retrieve* them at a later time;

Reference: software artifacts must be properly **referenced** to ensure we can *identify* the exact code, among many potentially archived copies, used for reproducing a specific experiment;

A. M. Bigatti et al. (Eds.): ICMS 2020, LNCS 12097, pp. 362–373, 2020.
https://doi.org/10.1007/978-3-030-52200-1_36

Description: software artifacts must be equipped with proper **metadata** to make it easy to *find* them in a catalog or through a search engine;

Citation: research software must be properly **cited** in research articles in order to give *credit* to the people that contributed to it.

As already pointed out in the literature, these are not only different concerns, but also *separate* ones. Establishing proper *credit* for contributors via *citations* or providing proper metadata to *describe* the artifacts requires a *curation* process [2,5,18] and is way more complex than simply providing stable, intrinsic identifiers to *reference* a precise version of a software source code for reproducibility purposes [3,16,21]. Also, as remarked in [3,20], resrach software is often a thin layer on top of a large number of software dependencies that are developed and maintained outside of academia, so the usual approach based on institutional archives is not sufficient to cover all the software that is relevant for reproducibility of research.

In this article, we focus on the first two concerns, *archival* and *reference*, showing how they can be addressed fully by leveraging the Software Heritage universal archive [1], and also mention some recent evolutions in best practices for embedding *metadata* in software development repositories.

In Sect. 2 we briefly recall what is Software Heritage and what makes it special; in Sect. 3 we show how researchers can easily ensure that any relevant source code is archived; in Sect. 4 we explain how to use the *intrinsic identifiers* provided by Software Heritage to enrich research articles, making them more useful and appealing for the readers, and providing stable links between articles and source code in the web of scientific knowledge we are all building. Finally, we point to ongoing collaborations and future perspectives in Sect. 5.

2 Software Heritage: The Universal Archive of Software Source Code

Software Heritage [1,17] is a non profit initiative started by Inria in partnership with UNESCO, to build a long term universal archive specifically designed for software source code, and able to store not only a software artifact, but *also its full development history.*

Software Heritage's mission is to collect, preserve, and make easily accessible the source code of *all* publicly available software. Among the strategies designed for collecting the source code there is the development of a large scale automated crawler for source code, whose architecture is shown in Fig. 1.

The sustainability plan is based on several pillars. The first one is the support of Inria, a national research institution that is involved for the long term. A second one is the fact that Software Heritage provides a common infrastructure catering to the needs of a variety of stakeholders, ranging from industry to academia, from cultural heritage to public administrations. As a consequence, funding comes from a diverse group of sponsors, ranging from IT companies to public institutions.

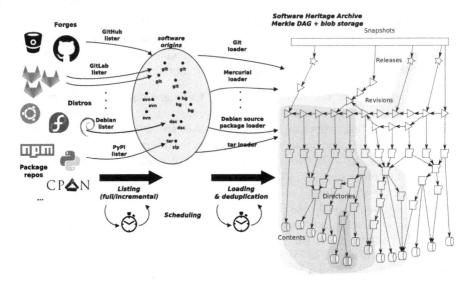

Fig. 1. Architecture of the Software Heritage crawler

Finally, an extra layer of security is provided by a network of independent, international mirrors that maintain a full copy of the archive[1].

We recall here a few key properties that set Software Heritage apart from all other scholarly infrastructures:

- it *proactively* archives *all software*, making it possible to store and reference any piece of publicly available software relevant to a research result, independently from any specific field of endeavour, and even when the author(s) did not take any step to have it archived [1,17];
- it stores the source code with its development history in a uniform data structure, a Merkle DAG [23], that allows to provide uniform, *intrinsic* identifiers for the billions of software artifacts of the archive, independently of the version control system or package format used [16].

At the time of writing this article, the Software Heritage archive contains over 7 billions unique source code files, from more than 100 million different software origins[2].

It provides the ideal place to *preserve research software artifacts*, and offers powerful mechanisms to *enhance research articles* with precise references to relevant fragments of your source code. Using Software Heritage is straightforward and involves very simple steps, that we detail in the following sections.

---

[1] More details can be found at https://www.softwareheritage.org/support/sponsors and https://www.softwareheritage.org/mirrors.

[2] See https://www.archive.softwareheritage.org for the up to date figures.

3   Archiving and Self Archiving

In a research article one may want to reference different kinds of source code artifacts: some may be popular open source components, some may be general purpose libraries developed by others, and some may be one's own software projects.

All these different kinds of software artifacts can be archived extremely easily in Software Heritage: it's enough that their source code is hosted on a publicly accessible repository (Github, Bitbucket, any GitLab instance, an institutional software forge, etc.) using one of the version control systems supported by Software Heritage, currently Subversion, Mercurial and Git[3].

For source code developed on popular development platforms, chances are that the code one wants to reference is already archived in Software Heritage, but one can make sure that the archived version history is fully up to date, as follows:

- go to https://save.softwareheritage.org,
- pick the right version control system in the drop-down list, enter the code repository url[4],
- click on the Submit button (see Fig. 2).

Fig. 2. The save code now form

That's all. No need to create an account or disclose personal information of any kind. If the provided URL is correct, Software Heritage will archive the repository shortly after, with its full development history. If it is hosted on one of the major forges we already know, this process will take just a few hours; if it is in a location we never saw before, it can take longer, as it will need to be manually screened[5].

---

[3] For up to date information, see https://archive.softwareheritage.org/browse/origin/save/.

[4] Make sure to use the clone/checkout url as given by the development platform hosting your code. It can easily be found in the web interface of the development platform.

[5] It is also possible to request archival programmatically, using the Software Heritage API, which can be quite handy to integrate in a Makefile; see https://archive.softwareheritage.org/api/1/origin/save/ for details.

3.1   Preparing Source Code for Self Archiving

In case the source code is one own's, before requesting its archival it is important
to structure the software repository following well established good practices for
release management [24]. In particular one should add README and AUTHORS files
as well as licence information following industry standard terminology [19,26].

Future users that find the artifact useful might want to give credit by citing
it. To this end, one might want to provide instructions on how one prefers the
artifact to be cited. We would recommend to also provide structured metadata information in machine readable formats. While practices in this area
are still evolving, one can use the CodeMeta generator available at https://
codemeta.github.io/codemeta-generator/ to produces metadata conformant to
the CodeMeta schema: the JSON-LD output can be put at the root of the
project in a **codemeta.json** file. Another option is to use the Citation File
Format, CFF (usually in a file named **citation.cff**).

4   Referencing

Once the source code has been archived, the Software Heritage *intrinsic identifiers*, called SWH-ID, fully documented online and shown in Fig. 3, can be used
to reference with great ease any version of it.

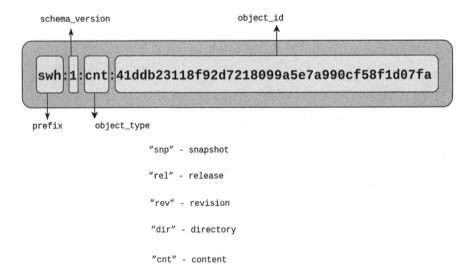

Fig. 3. Schema of the core Software Heritage identifiers

SWH-IDs are URIs with a very simple schema: the swh prefix makes explicit
that these identifiers are related to Software Heritage; the colon (:) is used as
separator between the logical parts of identifiers; the schema version (currently 1)

is the current version of this identifier schema; then follows the type of the objects identified and finally comes a hex-encoded (using lowercase ASCII characters) cryptographic signature of this object, computed in a standard way, as detailed in [15,16].

These core identifiers may be equipped with the *qualifiers* that carry contextual *extrinsic* information about the object:

origin : the *software origin* where an object has been found or observed in the wild, as an URI;

visit : persistent identifier of a *snapshot* corresponding to a specific *visit* of a repository containing the designated object;

anchor : a *designated node* in the Merkle DAG relative to which a *path to the object* is specified;

path : the *absolute file path*, from the *root directory* associated to the *anchor node*, to the object;

lines : *line number(s)* of interest, usually within a content object

The combination of the core SWH-IDs with these qualifiers provides a very powerful means of referring in a research article to all the software artefacts of interest.

To make this concrete, in what follows we use as a running example the article *A "minimal disruption" skeleton experiment: seamless map and reduce embedding in OCaml* by Marco Danelutto and Roberto Di Cosmo [9] published in 2012. This article introduced Parmap [12], an elegant library for multicore parallel programming that was distributed via the gitorious.org collaborative development platform, at gitorious.org/parmap. Since Gitorious has been shut down a few years ago, like Google Code and CodePlex, this example is particularly fit to show why pointing to an *archive* of the code is better than pointing to the collaborative development platform where it is developed.

4.1 Specific Version

The Parmap article describes a *specific version* of the Parmap library, the one that was used for the experiments reported in the article, so in order to support reproducibility of these results, we need to be able to pinpoint precisely the state(s) of the source code used in the article.

The exact revision of the source code of the library used in the article has the following SWH-ID:

```
swh:1:rev:0064fbd0ad69de205ea6ec6999f3d3895e9442c2;
origin=https://gitorious.org/parmap/parmap.git;
visit=swh:1:snp:78209702559384ee1b5586df13eca84a5123aa82
```

This identifier can be turned into a clickable URL by prepending to it the prefix https://archive.softwareheritage.org/ (one can try it by clicking on this link).

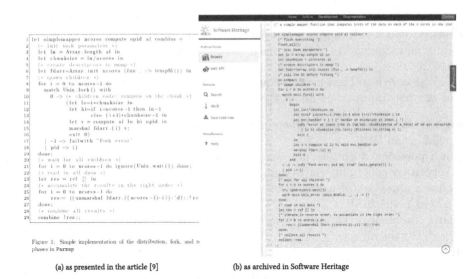

(a) as presented in the article [9]   (b) as archived in Software Heritage

Fig. 4. Code fragment from the published article compared to the content in the Software Heritage archive

4.2 Code Fragment

Having a link to the exact archived revision of a software project is important in all research articles that use software, and the core SWH-IDs allow to drill down and point to a given directory or even a file content, but sometimes, like in our running example, one would like to do more, and pinpoint a fragment of code inside a specific version of a file. This is possible using the `lines=` qualifier available for identifiers that point to file content.

Let's see this feature at work in our running example, showing how the experience of studying or reviewing an article can be greatly enhanced by providing pointers to code fragments.

In Figure 1 of [9], which is shown here as Fig. 4a, the authors want to present the core part of the code implementing the parallel functionality that constitutes the main contribution of their article. The usual approach is to typeset in the article itself *an excerpt of the source code*, and let the reader try to find it by delving into the code repository, which may have evolved in the mean time. Finding the exact matching code can be quite difficult, as the code excerpt is *often edited* a bit with respect to the original, sometimes to drop details that are not relevant for the discussion, and sometimes due to space limitations.

In our case, the article presented 29 lines of code, slightly edited from the 43 actual lines of code in the Parmap library: looking at Fig. 4a, one can easily see that some lines have been dropped (102–103, 118–121), one line has been split (117) and several lines simplified (127, 132–133, 137–142).

Using Software Heritage, the authors can do a much better job, because the original code fragment can now be precisely identified by the following Software Heritage identifier:

> swh:1:cnt:d5214ff9562a1fe78db51944506ba48c20de3379;
> origin=https://gitorious.org/parmap/parmap.git;
> visit=swh:1:snp:78209702559384ee1b5586df13eca84a5123aa82;
> anchor=swh:1:rev:0064fbd0ad69de205ea6ec6999f3d3895e9442c2;
> path=/parmap.ml;
> lines=101-143

This identifier will **always** point to the code fragment shown in Fig. 4b.

The caption of the original article shown in Fig. 4a can then be significantly enhanced by incorporating a clickable link containing the SWH-ID shown above: it's all is needed to point to the exact source code fragment that has been edited for inclusion in the article, as shown in Fig. 5. The link contains, thanks to the SWH-ID qualifiers, all the contextual information necessary to identify the context in which this code fragment is intended to be seen.

> Simple implementation of the distribution, fork, and recollection phases in Parmap (slightly simplified from the the actual code in the version of Parmap used for this article)

Fig. 5. A caption text with the link to the code fragment and its contextual information

When clicking on the hyperlinked text in the caption shown above, the reader is brought seamlessly to the Software Heritage archive on a page showing the corresponding source code archived in Software Heritage, with the relevant lines highlighted (see Fig. 4b).

4.3   Software Bibliographies with `biblatex-software`

Another way to enrich an article with precise pointers to software source code is by adding entries for it in the bibliography. Unfortunately, standard bibliography styles do not treat software as a first class citizen, and for example BibTeX users often resort to the `@misc` entry to this end, which is really unsatisfactory.

Since April 2020, users of the BibLaTeX package can leverage the `biblatex-software` package [10], available on CTAN [8], to produce rich software bibliographies.

This package support four kind of different entries:

- `@software` for describing the general information about a software project
- `@softwareversion` for describing a specific version or release of a software project

- @softwaremodule for describing a module that is part of a larger software project
- @codefragment for describing a fragment of code (full file, or selected lines of a file)

Using these special BibTeX entries, the various examples presented in the previous sections above can be described as follows

```
@software {parmap,
  title = {The Parmap library},
  author = {Di Cosmo, Roberto and Marco Danelutto},
  year = {2012},
  institution = {{University Paris Diderot} and {University of Pisa}},
  url = {https://rdicosmo.github.io/parmap/},
  license = {LGPL-2.0},
}


@softwareversion {parmap-0.9.8,
  title = {The Parmap library},
  author = {Di Cosmo, Roberto and Marco Danelutto},
  version = {0.9.8},
  swhid = {swh:1:rev:0064fbd0ad69de205ea6ec6999f3d3895e9442c2;
    origin=https://gitorious.org/parmap/parmap.git;
    visit=swh:1:snp:78209702559384ee1b5586df13eca84a5123aa82},
  crossref = {parmap}
}


@codefragment {simplemapper,
  subtitle = {Core mapping routine},
  swhid = {
  swh:1:cnt:d5214ff9562a1fe78db51944506ba48c20de3379;
  origin=https://gitorious.org/parmap/parmap.git;
  visit=swh:1:snp:78209702559384ee1b5586df13eca84a5123aa82;
  anchor=swh:1:rev:0064fbd0ad69de205ea6ec6999f3d3895e9442c2;
  path=/parmap.ml;
  lines=101-143},
  crossref = {parmap-0.9.8}
}
```

The result can be seen in the bibliography of this article as [13,14].

4.4   Getting the SWH-ID

A fully qualified SWH-ID is rather long, and it needs to be, as it contains quite a lot of information that is essential to convey. In order to make it easy to use SWH-IDs, we provide a very simple way of getting the right SWH-ID without having to type it by hand. Just browse the archived code in Software Heritage and navigate to the software artifact of interest. Clicking on the *permalinks vertical red tab* that is present on all pages of the archive, opens up a tab that allows to select the identifier for the object of interest: an example is shown in Fig. 6.

The two buttons on the bottom right allow to copy the identifier or the full permalink in the clipboard, and to paste it in an article as needed.

Fig. 6. Obtaining a Software Heritage identifier using the permalink box on the archive Web user interface

4.5   Generating and Verifying SWH-IDs

An important consequence of the fact that SWH-IDs are *intrinsic identifiers* is that they can be generated and verified *independently* of Software Heritage, using `swh-identify`, an open source tool developed by Software Heritage, and distributed via PyPI as `swh.model`, with the stable version at the time of writing being this one.

Version 1 of the SWH-IDs uses git-compatible hashes, so if the source code that one wants to reference uses git as a version control system, one can create the right SWH-ID by just prepending `swh:1:rev:` to the commit hash. This comes handy to automate the generation of the identifiers to be included in an article, as one will always have code and article in sync.

5   Perspectives for the Scholarly World

We have shown how Software Heritage and the associated SWH-IDs enables the seamless archival of all publicly available source code. It provides for all kind of software artifacts the *intrinsic identifiers* that are needed to establish long lasting, resilient links between research articles and the software they use or describe.

All researchers can use *right now* the mechanisms presented here to produce improved and enhanced research articles. More can be achieved by establishing collaborations with academic journals, registries and institutional repositories and registries, in particular in terms of description and support for software citation. Among the initial collaborations that have been already established, we are happy to mention the cross linking with the curated mathematical software descriptions maintained by the swMath.org portal [5], and the curated deposit of software artefacts into the HAL french national open access portal [18], which is performed via a standard SWORD protocol inteface, an approach that is currently being explored by other academic journals.

We believe that the time has come to see software become a first class citizen in the scholarly world, and Software Heritage provides a unique infrastructure to support an open, non profit, long term and resilient web of scientific knowledge.

Acknowledgements. This article is a major evolution of the research software archival and reference guidelines available on the Software Heritage website [11] resulting from extensive discussions that took place over several years with many people. Special thanks to Alain Girault, Morane Gruenpeter, Antoine Lambert, Julia Lawall, Arnaud Legrand, Nicolas Rougier and Stefano Zacchiroli for their precious feedback on these issues and/or earlier versions of this document.

References

1. Abramatic, J.-F., Di Cosmo, R., Zacchiroli, S.: Building the universal archive of source code. Commun. ACM **61**(10), 29–31 (2018)
2. Allen, A., Schmidt, J.: Looking before leaping: creating a software registry. J. Open Res. Softw., 3.e15 (2015)
3. Alliez, P., et al.: Attributing and referencing (research) software: best practices and outlook from Inria. Comput. Sci. Eng. **22**(1), 39–52 (2020). https://hal.archives-ouvertes.fr/hal-02135891
4. Association for Computing Machinery. Artifact Review and Badging, April 2018. https://www.acm.org/publications/policies/artifact-review-badging. Accessed 27 April 2019
5. Bönisch, S., Brickenstein, M., Chrapary, H., Greuel, G.-M., Sperber, W.: swMATH – a new information service for mathematical software. In: Carette, J., Aspinall, D., Lange, C., Sojka, P., Windsteiger, W. (eds.) CICM 2013. LNCS (LNAI), vol. 7961, pp. 369–373. Springer, Heidelberg (2013). https://doi.org/10.1007/978-3-642-39320-4_31
6. Borgman, C.L., Wallis, J.C., Mayernik, M.S.: Who's got the data? Interdependencies in science and technology collaborations. Comput. Support. Coop. Work **21**(6), 485–523 (2012)
7. Childers, B.R., et al.: Artifact evaluation for publications (Dagstuhl Perspectives Workshop 15452). In: Childers, B.R., et al. (eds.) Dagstuhl Reports, vol. 5, no. 11, pp. 29–35 (2016)
8. CTAN: the Comprehensive TeX Archive Network. http://www.ctan.org/. Visited on 29 April 2020
9. Danelutto, M., Di Cosmo, R.: A "Minimal Disruption" skeleton experiment: seamless map & reduce embedding in OCaml. Procedia CS **9**, 1837–1846 (2012)
10. [SW] Roberto Di Cosmo, BibLaTeX stylefiles for software products (2020). https://ctan.org/tex-archive/macros/latex/contrib/biblatex-contrib/biblatex-software
11. Di Cosmo, R.: How to use software heritage for archiving and referencing your source code: guidelines and walkthrough, April 2019. https://hal.archives-ouvertes.fr/hal-02263344

12. [SW] Di Cosmo, R., Danelutto, M.: The Parmap library. University Paris Diderot and University of Pisa. LIC: LGPL-2.0 (2012). https://rdicosmo.github.io/parmap/
13. [SW REL.] Di Cosmo, R., Danelutto, M.: The Parmap library version 0.9.8. University Paris Diderot and University of Pisa. LIC: LGPL-2.0 (2012). SWHID: ⟨ swh:1:rev:0064fbd0ad69de205ea6ec6999f3d3895e9442c2;origin=https://gitorious. org/parmap/parmap.git;visit=swh:1:snp:78209702559384ee1b5586df13eca84a51 23aa82 ⟩
14. [SW EXC.] Di Cosmo, R., Danelutto, M.: "Core mapping routine", from The Parmap library version 0.9.8. University Paris Diderot and University of Pisa (2012). LIC: LGPL-2.0. SWHID: ⟨swh:1:cnt:d5214ff9562a1fe78db51944506ba48c20de 3379;origin=https://gitorious.org/parmap/parmap.git;visit=swh:1:snp:782097025 59384ee1b5586df13eca84a5123aa82;anchor=swh:1:rev:0064fbd0ad69de205ea6ec699 9f3d3895e9442c2;path=/parmap.ml;lines=101-143⟩
15. Di Cosmo, R., Gruenpeter, M., Zacchiroli, S.: Identifiers for digital objects: the case of software source code preservation. In: Proceedings of the 15th International Conference on Digital Preservation, iPRES: Boston, p. 2018, September 2018
16. Di Cosmo, R., Gruenpeter, M., Zacchiroli, S.: Referencing source code artifacts: a separate concern in software citation. Comput. Sci. Eng. **22**(2), 33–43 (2020)
17. Di Cosmo, R., Zacchiroli, S.: Software heritage: why and how to preserve software source code. In: Proceedings of the 14th International Conference on Digital Preservation, iPRES 2017, September 2017
18. Di Cosmo, R., et al.: Curated archiving of research software artifacts: lessons learned from the French open archive (HAL). Presented at the International Digital Curation Conference, submitted to IJDC, December 2019
19. Free Software Foundation Europe. REUSE Software, September 2019. https:// reuse.software. Accessed 24 Sept 2019
20. Hinsen, K.: Software development for reproducible research. Comput. Sci. Eng. **15**(4), 60–63 (2013)
21. Howison, J., Bullard, J.: Software in the scientific literature: Problems with seeing, finding, and using software mentioned in the biology literature. J. Assoc. Inf. Sci. Technol. **67**(9), 2137–2155 (2016). https://onlinelibrary.wiley.com/doi/pdf/10. 1002/asi.23538
22. Lamprecht, A.-L., et al.: Towards FAIR principles for research software, pp. 1–23 (2019)
23. Merkle, R.C.: A digital signature based on a conventional encryption function. In: Pomerance, C. (ed.) CRYPTO 1987. LNCS, vol. 293, pp. 369–378. Springer, Heidelberg (1988). https://doi.org/10.1007/3-540-48184-2_32
24. Raymond, E.S.: Software release practice HOWTO, January 2013. https://www. tldp.org/HOWTO/html_single/Software-Release-Practice-HOWTO/. Accessed 5 June 2019
25. Smith, A.M., Katz, D.S., Niemeyer, K.E.: Software citation principles. PeerJ Comput. Sci. **2**, e86 (2016)
26. SPDX Workgroup. Software Package Data Exchange Licence List (2019). https:// spdx.org/license-list. Accessed 30 Mar 2020
27. Stodden, V., LeVeque, R.J., Mitchell, I.: Reproducible research for scientific computing: tools and strategies for changing the culture. Comput. Sci. Eng. **14**(4), 13–17 (2012)
28. Wilkinson, M.D., et al.: The FAIR guiding principles for scientific data management and stewardship. Sci. Data **3**(1), 160018 (2016)

The Jupyter Environment for Computational Mathematics

Polymake.jl: A New Interface to `polymake`

Marek Kaluba[1,2], Benjamin Lorenz[1], and Sascha Timme[1][(✉)]

[1] Chair of Discrete Mathematics/Geometry, Technische Universität Berlin,
Berlin, Germany
`timme@math.tu-berlin.de`
[2] Adam Mickiewicz University in Poznań, Poznań, Poland

Abstract. We present the Julia interface `Polymake.jl` to `polymake`, a
software for research in polyhedral geometry. We describe the technical
design and how the integration into Julia makes it possible to combine
`polymake` with state-of-the-art numerical software.

Keywords: Polymake · Julia

1 Introduction

`polymake` is an open source software system for computing with a wide range
of objects from polyhedral geometry and related areas [4]. This includes convex
polytopes and polyhedral fans as well as matroids, finite permutation groups,
ideals in polynomial rings and tropical varieties. The user is interfacing the
`polymake` library through Perl language.

In this note we provide a brief overview of a new interface `Polymake.jl`, which
allows the use of `polymake` in Julia [1]. Julia is a high-level, dynamic programming
language. Distinctive aspects of Julia's design include a type system with parametric polymorphism and multiple dispatch as its core programming paradigm.
The package `Polymake.jl` can be installed, without any preparations, using the
build-in package manager that comes with Julia. The source code is available at
https://github.com/oscar-system/Polymake.jl.

2 Functionality

In `polymake` the objects that a user encounters can be roughly divided into the
following three classes

M. Kaluba—The author was supported by the National Science Center, Poland grant
2017/26/D/ST1/00103. This research is carried out in the framework of the DFG
funded Cluster of Excellence EXC 2046 MATH+: *The Berlin Mathematics Research
Center* within the Emerging Fields area.
S. Timme—The author was supported by the Deutsche Forschungsgemeinschaft (German Research Foundation) Graduiertenkolleg *Facets of Complexity* (GRK 2434).

© Springer Nature Switzerland AG 2020
A. M. Bigatti et al. (Eds.): ICMS 2020, LNCS 12097, pp. 377–385, 2020.
https://doi.org/10.1007/978-3-030-52200-1_37

- *big objects* (e.g., cones, polytopes, simplicial complexes),
- *small objects* (e.g., matrices, polynomials, tropical numbers),
- *user functions.*

Broadly speaking, *big objects* correspond to mathematical concepts with well defined semantics. These can be queried, accumulate information (e.g., a polytope defined by a set of points can "learn" its hyperplane representation), and are constructed usually in Perl. Big objects implement *methods*, i.e., functions which operate on, and perform computations specific to the corresponding object. *Small objects* correspond to types or data structures which are implemented in C++. Standalone *user functions* are exposed to the user via the Perl interpreter.

These entities are mapped to Julia in the following way:

- big objects are exposed as opaque Perl objects that can be queried for their properties (they are only computed when queried for the first time),
- small objects are wrapped through an intermediate C++ layer between Julia and `libpolymake` generated by `CxxWrap.jl`,
- methods and user functions are mapped to Julia functions, in the case of methods, the parent object being the first argument.

A unique feature of `Polymake.jl` is based on the affinity of Julia to C and C++ programming languages. As Julia provides the possibility to call functions from dynamic libraries directly, one can call any function from the `polymake` library as long as the function symbol is exported. In `polymake`, due to extensive use of templates in the C++ library, the precise definition of a function needs to be often explicitly instantiated. Such instantiaton can be easily added to the `Polymake.jl` C++ wrapper. An example of such functionality is

```
Polymake.solve_LP(inequalities, equalities, objective; sense=max)
```

a function, which directly taps into the `polymake` framework for linear programming. It is worth pointing that the signature of the exposed `solve_LP` will accept any instances of Julias **AbstractMatrix** or **AbstractVector** (where appropriate) in the paradigm of generic programming.

3   Technical Contribution

The `Polymake.jl` interface is based on `CxxWrap.jl`[1], a Julia package which aims to provide a seamless interoperability between C++ and Julia. The interface is separated into two parts: a C++ wrapper library and a Julia package. The former, `libpolymake`, a dynamic library, wraps the data structures (*small types*) in a Julia-compatible way and exposes functions from the callable C++ `polymake` library. It is then loaded through `CxxWrap.jl` where the Julia part of the package generates functions accessible from Julia.

The installation of `Polymake.jl` is performed through Julia's package manager with the help of `BinaryBuilder.jl` infrastructure. Thanks to this infrastructure

---

[1] Available at https://github.com/JuliaInterop/CxxWrap.jl.

it is not necessary for the user to perform any preparations except for installing Julia itself. All dependencies of `Polymake.jl` (including the `polymake` library, the Perl interpreter and supplementary libraries) are installed in a binary form. The complete installation of `Polymake.jl` should take no longer than 5 minutes on modern hardware

Due to extensive use of metaprogramming, relatively little code was necessary to make most of the functionality of `polymake` available in Julia: as of version 0.4.1 `Polymake.jl` consists of about 1200 lines of C++ code and 1600 lines of Julia code. In particular, only the small objects need to be manually wrapped, while functions, constructors for big objects and their methods are generated automatically from the information provided by `polymake` itself. This automatic code generation takes place during precompilation which is done only once during the installation. Loading `Polymake.jl` brings the familiar `polymake` welcome banner.

```
julia> using Pkg; Pkg.add("Polymake")
  Updating registry at ~/.julia/registries/General
  [ ... ]
  Building Polymake → ~/.julia/packages/Polymake/[...]/deps/build.log
julia> using Polymake
[ Info: Precompiling Polymake [d720cf60-89b5-51f5-aff5-213f193123e7]
[ Info: Generating module common
[ Info: Generating module ideal
[ Info: Generating module graph
[ Info: Generating module fulton
[ Info: Generating module fan
[ Info: Generating module group
[ Info: Generating module polytope
[ Info: Generating module topaz
[ Info: Generating module tropical
[ Info: Generating module matroid

polymake version 4.0
Copyright (c) 1997-2020
Ewgenij Gawrilow, Michael Joswig, and the polymake team
Technische Universität Berlin, Germany
https://polymake.org

This is free software licensed under GPL; see the source for
copying conditions.
There is NO warranty; not even for MERCHANTABILITY or
FITNESS FOR A PARTICULAR PURPOSE.
```

The latest version of `Polymake.jl` is 0.4.1 which is compatible with at Julia 1.3 and newer. The latest `polymake` version is available in `Polymake.jl` within two weeks of release (currently `polymake` 4.0).

Big Objects

All big objects are constructed by direct calls to their constructors, e.g.

```
polytope.Polytope(POINTS=[1 1 2; 1 3 4])
```

constructs a rational polytope from (homogeneous) coordinates of points given row-wise. We attach the `polymake` docstring to the structure such that the documentation is readily available in Julia.

```
help?> polytope.Polytope
Not necessarily bounded convex polyhedron, i.e., the feasible region of a linear
 ↪   program.
 [...]
```

Template parameters can also be passed to big objects, e.g., to construct a polytope with floating point precision it is sufficient to call

$$\texttt{polytope.Polytope\{Float64\}(...)}\ .$$

A caveat is that all parameters must be valid Julia objects.[2] The properties of big objects are accessible through the `bigobject.property` syntax which mirrors the `$bigobject->property` syntax in `polymake`. Note that some properties of big objects in `polymake` are indeed methods with no arguments and therefore in Julia they are only available as such.

Small Objects

The list of small objects available in `Polymake.jl` includes basic types such as arbitrary size integers (subtypes **Integer**, rationals (subtypes **Real**, vectors and matrices (subtypes of **AbstractArray**), and many more. These data types can be converted to appropriate Julia types, but are also subtypes of the corresponding Julia abstract types (as indicated above). This allows to use `Polymake.jl` types in generic methods, which is the paradigm of Julia programming.

As already mentioned, these small objects need to be manually wrapped in the C++ part of `Polymake.jl`. In particular, all possible combinations of such types, e.g., an array of sets of rationals, need to be explicitly wrapped. Note that `polymake` is able to generate dynamically any combination of small objects. Thus, we cannot guarantee that all small objects a user will encounter is covered. However, the small objects available in `Polymake.jl` are sufficient for the most common use cases.

Functions

A function in `Polymake.jl` calling `polymake` may return either a big or a small object, and the generic return type (`PropertyValue`, a container opaque to Julia) is transparently converted to one of the known (small) data types[3]. If the data type of the returned function value is not known to `Polymake.jl`, the conversion fails and an instance of `PropertyValue` is returned. It can be either passed back as an argument to a `Polymake.jl` function, or converted to a known type using the `@convert_to` macro.

---

[2] For advanced use (when this is not the case) we provide the `@pm` macro.

[3] This conversion can be deactivated by passing `PropertyValue` type as the first argument to function/method call.

```
julia> K5 = graph.complete(5);
julia> K5.MAX_CLIQUES
PropertyValue wrapping pm::PowerSet<long, pm::operations::cmp>
{{0 1 2 3 4}}
julia> @convert_to Array{Set} K5.MAX_CLIQUES
pm::Array<pm::Set<long, pm::operations::cmp> >
{0 1 2 3 4}
```

All user functions from `polymake` are available in modules corresponding to their applications, e.g. `homology` functions from the application `topaz` can be called as `topaz.homology(...)` in Julia. Moreover `polymake` docstrings for functions are available in Julia to allow for easy help[4]:

```
julia> ?topaz.homology
  homology(complex, co; Options)

  Calculate the reduced (co-)homology groups of a simplicial complex.

  Arguments:
    Array<Set<Int>> complex
    Bool co set to true for cohomology

  Options:
    dim_low => Int narrows the dimension range of interest, with negative values being
 ↪   treated as co-dimensions
    dim_high => Int see dim_low

    [ ... ]
```

Function Arguments. Functions in `Polymake.jl` accept the following as their arguments: simple data types (bools, machine integers, floats), wrapped native types, or objects returned by `polymake` (e.g., **BigObject**, or `PropertyValue`). Due to the easy extendability of methods in Julia, a foreign type could be passed seamlessly to `Polymake.jl` function if an appropriate `Base.convert` method, which return one of the above types, is defined:

```
Base.convert(::Type{Polymake.PolymakeType}, x::ForeignType)
```

`Polymake.jl` also wraps the extensive visualization methods of `polymake` which can be used to produce images and animations of geometric objects. These include the interactive visualizations using **three.js**. Due to the convenient extendability of Julia, the visualization also integrates seamlessly with Jupyter notebooks.

4   Example

This section demonstrates the interface of `Polymake.jl` on a concrete example. An advantage of the package is that it allows effortless combination of computations in polyhedral geometry with e.g., state-of-the-art numerical software. Here we combine `Polymake.jl` with **HomotopyContinuation.jl** [3], a Julia package for

---

[4] The documentation currently uses the Perl syntax.

numerically solving systems of polynomial equations. In particular, we test a theoretical result from Soprunova and Sottile [8] on non-trivial lower bounds for the number of real solutions to sparse polynomial systems.

The results show how we can construct a sparse polynomial system that has a non-trivial lower bound on the number of real solutions starting from an integral point configuration. We start with the 10 lattice points $A = \{a_1, \ldots, a_{10}\} \subset \mathbb{Z}^2$ of the scaled two-dimensional simplex $3\Delta_2$ and look at the regular triangulation $\mathcal{T}$ induced by the lifting $\lambda = [12, 3, 0, 0, 8, 1, 0, 9, 5, 15]$.

```
julia> A = polytope.lattice_points(polytope.simplex(2,3));
julia> λ = [12, 3, 0, 0, 8, 1, 0, 9, 5, 15];
julia> F = polytope.regular_subdivision(A, λ);
julia> T = topaz.GeometricSimplicialComplex(COORDINATES = A[:,2:end], FACETS = F)
type: GeometricSimplicialComplex<Rational>

COORDINATES
      0 0
      0 1
      [ ... ]

FACETS
  pm::Set<long, pm::operations::cmp>
{5 6 8}
  pm::Set<long, pm::operations::cmp>
{5 7 8}
   [ ... ]
```

Fig. 1. Foldable subdivision of $3\Delta_2$.

The triangulation $\mathcal{T}$ is very special in that it is *foldable* (or "balanced"), i.e., the dual graph is bipartite. This means that the triangles can be colored, say, black and white such that no two triangles of the same color share an edge. See Fig. 1 for an illustration. The *signature* $\sigma(\mathcal{T})$ of a balanced triangulation of a polygon is the absolute value of the difference of the number of black triangles and the number of the white triangles whose normalized volume is odd. The vertices of a foldable triangulation can be colored by $d + 1$ colors [6] (such that vertices of the same color do not share an edge), where d is the dimension. Here $d = 2$, so 3 colors suffice. We can check both properties with `polymake`.

```
julia> (foldable = T.FOLDABLE, signature = T.SIGNATURE)
  (foldable = true, signature = 3)
```

Now, a *Wroński polynomial* $W_{\mathcal{T},s}(x)$ has the lifted lattice points as exponents, and only one non-zero coefficient $c_i \in \mathbb{R}$ per color class of vertices of the triangulation

$$W_{\mathcal{T},s}(x) = \sum_{i=0}^{d} c_i \left(\sum_{j:\, \mathrm{color}(a_j)=i} s^{\lambda_i} x^{a_j} \right).$$

A *Wroński system* consists of d Wroński polynomials with respect to the same lattice points A and lifting λ such that for general $s = s_0 \in [0, 1]$ it has precisely $d!\,\mathrm{vol}(\mathrm{conv}(A))$ distinct complex solutions, which is the highest possible number by Kushnirenko's Theorem [7].

Soprunova and Sottile showed that a Wroński system has at least $\sigma(\mathcal{T})$ distinct real solutions if two conditions are satisfied. First, a certain double cover of the real toric variety associated with A must be orientable. This is the case here. Second, the *Wroński center ideal*, a zero-dimensional ideal in coordinates x_1, x_2 and s depending on $\mathcal{T}$, has no real roots with s coordinate between 0 and 1. Let us verify this condition using `HomotopyContinuation.jl`. Luckily for us, polymake already has an implementation of the Wroński center ideal. However, we have to convert the ideal returned by `Polymake.jl` to a polynomial system which `HomotopyContinuation.jl` understands. This can be accomplished with a simple routine.

```julia
julia> using HomotopyContinuation
julia> function hc_poly(f, vars)
           M = Polymake.monomials_as_matrix(f)
           monomials = [prod(vars.^m) for m in eachrow(M)]
           coeffs = Int.(Polymake.coefficients_as_vector(f))
           sum(map(*, coeffs, monomials))
       end;
julia> I = polytope.wronski_center_ideal(A, λ)
julia> @polyvar x[1:2] s;
julia> HC_I = [hc_poly(f, [x;s]) for f in I.GENERATORS]
3-element Array{Polynomial{true,Int},1}:
 x_2^3 s^{15} + s^{12} + x_1 x_2 s + x_2^3
 x_1^2 s^9 + x_2 s^3 + x_1 x_2^2
 x_1 s^8 + x_1^2 x_2 s^5 + x_2^2
```

Since we are only interested in solutions in the algebraic torus $(\mathbb{C}^*)^3$ we can use polyhedral homotopy [5] to efficiently compute the solutions.

```julia
julia> @time res = solve(HC_I; start_system = :polyhedral, only_torus = true)
  0.010595 seconds (3.03 k allocations: 215.766 KiB)
Result{Array{Complex{Float64},1}} with 54 solutions
=======================================================
* 54 non-singular solutions (2 real)
* 0 singular solutions (0 real)
* 54 paths tracked
* random seed: 782949
```

Out of the 54 complex roots only two solutions are real. Strictly speaking, this is here only checked heuristically by looking at the size of the imaginary parts.

However, a certified version can be obtained by using **alphaCertified**[2]. By closer inspection, we see that no solution has the s-coordinate in $(0, 1)$.

```julia
julia> HomotopyContinuation.real_solutions(res)
2-element Array{Array{Float64,1},1}:
 [-0.2117580095433453, -215.72260079314424, 4.411470567441922]
 [-0.6943590430596768, -0.41424188458258815, -0.8952189506082179]
```

Therefore, the Wroński system with respect to A and λ for $s = 1$ has at least $\sigma(\mathcal{T}) = 3$ real solutions. Let us verify this on an example.

```julia
julia> c = Vector{Polymake.Rational}[[19,8,-19], [39,7,42]];
julia> W = polytope.wronski_system(A, λ, c, 1)
julia> HC_W = [hc_poly(f, x) for f in W.GENERATORS];
julia> W_res = HomotopyContinuation.solve(HC_W)
Result{Array{Complex{Float64},1}} with 9 solutions
=====================================================
* 9 non-singular solutions (3 real)
* 0 singular solutions (0 real)
* 9 paths tracked
* random seed: 813729
```

Finally, we can use the **ImplicitPlots.jl** package to visualize the real solutions of the Wroński system W (Fig. 2).

```julia
julia> W_real = HomotopyContinuation.real_solutions(W_res)
julia> using ImplicitPlots, Plots;
julia> p = plot(aspect_ratio = :equal);
julia> implicit_plot!(p, HC_W[1]);
julia> implicit_plot!(p, HC_W[2]; color=:indianred);
julia> scatter!(first.(r), last.(r), markercolor=:black)
```

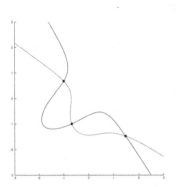

Fig. 2. Visualization of the Wroński system W and its 3 solutions.

Acknowledgements. We would like to express our thanks to Alexej Jordan and Sebastian Gutsche for all their help during the development of **Polymake.jl**.

References

1. Bezanson, J., Edelman, A., Karpinski, S., Shah, V.B.: Julia: a fresh approach to numerical computing. SIAM Rev. **59**(1), 65–98 (2017)
2. Hauenstein, J.D., Sottile, F.: Algorithm 921: alphaCertified: certifying solutions to polynomial systems. ACM Trans. Math. Softw. (TOMS) **38**(4), 1–20 (2012)
3. Breiding, P., Timme, S.: HomotopyContinuation.jl: a package for homotopy continuation in Julia. In: International Congress on Mathematical Software, pp. 458–465. Springer (2018)
4. Gawrilow, E., Joswig, M.: polymake: a framework for analyzing convex polytopes. Polytopes–combinatorics and computation (Oberwolfach, 1997). DMV Sem. **29**, 43–73 (2000)
5. Huber, B., Sturmfels, B.: A polyhedral method for solving sparse polynomial systems. Math. Comput. **64**(212), 1541–1555 (1995)
6. Joswig, M.: Projectivities in simplicial complexes and colorings of simple polytopes. Math. Z. **240**, 243–259 (2002)
7. Kushnirenko, A.G.: A Newton polyhedron and the number of solutions of a system of k equations in k unknowns. Usp. Math. Nauk. **30**(1975), 266–267 (1975)
8. Soprunova, E., Sottile, F.: Lower bounds for real solutions to sparse polynomial systems. Adv. Math. **204**, 116–151 (2006)

Web Based Notebooks for Teaching, an Experience at Universidad de Zaragoza

Miguel Angel Marco Buzunariz$^{(\boxtimes)}$ (iD)

Universidad de Zaragoza-IUMA, 50003 Zaragoza, Spain
`mmarco@unizar.es`

Abstract. Since 2012, web based notebooks have been used as interface for computer algebra systems in teaching mathematics courses at Universidad de Zaragoza. We present an overview of the experience, detailing the advantages and problems that have been noticed during this time.

1 Introduction

Computer algebra systems, both proprietary and free, have been used as part of some mathematics related courses since several decades ago. In particular, the Universidad de Zaragoza used to have campus licenses for Mathematica, Maple and Matlab. Some professors were not happy with the use of proprietary software, and opted for free systems such as CoCoA [1] or GAP [2].

However, the user interfaces of these systems used to be less friendly to the user than the ones of the mentioned proprietary systems, which constituted a problem for the adoption. Moreover, many of the free systems were oriented to a very specific area of mathematics (e.g. CoCoA is focused specifically in computations in polynomial rings, GAP is focused in group theory and so on).

By 2010, SageMath [6] had reached a level of maturity that made its use for these tasks viable. It was feature rich enough for the courses, and it had a user interface that was much more user friendly than the ones of other free systems. An aspect that was very unusual at that time was that this user interface was web-based (that is pretty common nowadays, but at that point SageMath was a pioneer). But the main drawback was that it couldn't be installed over Windows operating systems, which is what most students used in their personal computers.

We then decided to take advantage of the fact that the user interface was web based to use a server-centered approach, setting up a server that run SageMath's web interface, students could log in and use SageMath without installing any software in their computers.

This approach has been used for nine years. In the 2019–2020 academic year we started a migration process from the classic SageMath notebook to a Jupyter [3] based one. In the rest of the paper we will describe the technical details of the setups we have used, and analyze their differences. We will also mention the advantages and drawbacks that we have experienced in each case.

Partially supported by MTM2016-76868-C2-2-P and Grupo "Investigación en Educación Matemática" of Gobierno de Aragón/Fondo Social Europeo.

A. M. Bigatti et al. (Eds.): ICMS 2020, LNCS 12097, pp. 386–392, 2020.
https://doi.org/10.1007/978-3-030-52200-1_38

2   The SageNB Notebook

SageMath included from the very beginning a web based notebook GUI. It included some options that were useful for our use case:

- It allowed using user accounts, so each instructor and student could log in, and keep his/her worksheets in the server.
- It allowed publishing worksheets, so everybody could see them, and logged in users could make a copy to work in them.
- It allowed to share a worksheet with other specific users, so they could both see and edit it.

Our usual workflow was the following:

1. The instructor of each course creates a worksheet with explanations, examples, and exercises; and published it.
2. The students create a copy of it, work on their copy solving the exercises, and share it with the instructor.
3. The instructor sees the work done by each student, and modifies the student's worksheet to grade it or add further comments or explanations. In some cases, they can include further work or corrections for the students to do.
4. The student does the extra corrections requested by the instructor.

This approach worked reasonably well, although there were some problems that were solved on the way. We will now mention some of them, and how they were dealt with.

The server we used couldn't handle the peaks of work. It was solved by adding another machine and using the ability of SageNB to create sessions in remote machines through ssh.

The number of published worksheets, and the list of worksheets shared to each user kept growing to the point of making it hard for the user to find a specific one. Sadly, SageNB never included a proper method to organize them with folders or tags. We partially mitigated this by restarting the database of worksheets each academic year. Besides, we kept different instances of the notebook server for different degrees. All of them run in the same server, with a proxy server that filtered and redirected connections by domain name. That is, for example, requests to `sage-mtm.unizar.es` were redirected to the port were one of the notebook servers was listening; whereas `sage-inf.unizar.es` was redirected to another port.

The problems of lack of CPU and RAM were specially usual during exams. We suspect that it was due to some students trying to sabotage the exams by overloading the server. We tried limiting the amount of RAM and CPU that each session could use, but this remained a problem.

Also, users were created manually (theoretically, SageNB allowed LDAP identification, but we were never able to make it work with our university's directory). So we had to create scripts to create the users each academic year, from a list of students registered in each course.

The tasks of installing, updating and managing the server were done mainly by one person, with another one helping in minor issues. Both admins were professors that dedicated part of their spare time for this task. Most of the time this wasn't a problem, but at some specific times they couldn't handle the issues that appeared as quickly as it would be desirable.

Overall, it was a viable option, although far from perfect.

3   JupyterHub and JupyterLab

The SageNB notebook was deprecated in SageMath version 9 (although its development was virtually abandoned long ago). Instead, SageMath had been moving since several years ago towards the Jupyter notebook. In these circumstances, we had to start migrating our infrastructure to this new model. However, we kept the old SageNB server running for one more year to allow a grace period for the instructors that had a hard time making the switch.

When we started considering the options to switch to, our desired features were the following:

1. It should allow different users to log in, ideally using the university's directory.
2. It should allow users to organize their worksheets in folders or similar.
3. It should allow a persistent storage of the work of each user.
4. It should allow users to share worksheets with other users, at least in a similar fashion than the workflow we had in SageNB.
5. It should be able to scale to many users.
6. It should be able to isolate the computations of each users, in such a way that one user cannot exhaust the available resources.

The obvious answer for several of the requirement was to use the Jupyter ecosystem.

The Jupyter notebook [3] is a web based application that allows to combine text, executable code and graphics in the same document. The code can be executed interactively. It uses a specific protocol to communicate with the different kernels, which are programs that execute the corresponding code and return the result. That way, one can run sessions of different programming languages (the name was chosen as a combination of Julia, Python and R, but many more kernels have been added since then). JupyterLab is a redesign of Jupyter, including a desktop-like environment, with a file browser, an embedded tiling window (and tabs) manager for notebooks, interactive consoles and file editors.

Both Jupyter and JupyterLab are single-user applications, but there is a front-end that can handle Jupyter sessions for several users called JupyterHub [4]. It has different modules for authenticating users and spawning sessions. In the authentication side, we used the CAS authenticator, that worked seamlessly with our university's Single Sign On system.

As for the spawner choice, one of the most popular solutions to accomplish the isolation requirement is to use Docker. However, the Docker approach is hard to mix with the requirement of having persistent storage for each user. So we opted for the Systemd spawner that runs each session as the corresponding system's user, under a Systemd container that allows to limit the CPU and RAM available.

To better handle the requirement of organizing the files in folders, we opted for JupyterLab instead of plain Jupyter (since it allows, for example, moving files to folders by just dragging them). Moreover, the Jupyter project has already stated that their long term plan is to move to JupyterLab and deprecate plain Jupyter.

As for the requirement to share worksheets with other users, we looked into nbgrader [5], which is a tool designed for this specific purpose, but it wasn't a good fit since it assumes that the user logs in a session that is dedicated to each course; whereas we envisioned a global system for all the university, where each student could log in and have access to his/her files related to all courses. We started working on a JupyterLab extension that would allow instructors to send files to the students enrolled in a certain course, and students to send them back to instructors. However, that work is no ready yet, so for the moment we are using external channels (Moodle and email) to send files back and forth. Luckily, the JupyterLab interface makes uploading and downloading files easier than the SageNB one.

To achieve the scalability, we contacted a research institute in our university that provides cloud services. They were kind to provide us with some virtual machines to make a test deployment. In those machines, we deployed the following design:

- A HAProxy server acts as a web frontal and https terminator. It redirects each http session to one of the computation nodes.
- A database node provides a persistent NFS volume to the computing nodes, and also a user database to make sure that usernames and uid's are kept in sync in all computing nodes. We also planned to keep here the database of teachers/students/courses for the extension that would allow to share files between users, but it is not ready yet.
- In each computing node, JupyterHub and JupyterLab are installed. Each request is authenticated by the CASauthenticator against the university's SSO. Then the system user is created (if it doesn't exist already) and the session is started under a Systemd container.

As a backup option in case of failures, a single instance was also installed in a regular computer (so it is not scalable).

The work of making the systems design, installing and configuring the software was done by a student as part of his degree thesis, together with the professor that managed the older server. This professor has also been the one that has worked in the JupyterLab extension to send files back and forth between students and instructors. Both that and the task of administering the server and deal with the problems that have arisen have proven to be too much load for the time he can dedicate to this task, which is the main reason why the extension is still in early development phase.

This approach has worked during the current academic year, where we have had troubles with the peak capacity. A limit of 1 GB RAM per user seemed insufficient for some tasks (partly due to the fact that most of them didn't properly shut down the worksheets that they don't use anymore, the UI allows to do so, but it is not evident, and most assume that just closing the tab is enough). However, if we raised that limit to 1.5 GB or 2 GB, we encountered that the whole computing node got its RAM exhausted, and didn't respond until the service was restarted. This contrasts with the peaks that the SageNB server was able to handle, which, we suspect, shows that the Jupyter server introduces a non-negligible overhead.

4   Impact on Teaching and Learning

We started with a test experience with only three instructors on the 2011–2012 academic year. After that, it was offered to all instructors that wanted to use it in their courses. The adoption was modest the first year, but then it growed quickly and has been stable since then. At the current academic year, the JupyterLab servers were used by 24 instructors and 442 students. Most of them used SageMath notebooks, but some used other ones, such as Python and R.

The typical way to use it was during problem/exercises sessions in the computer lab: the students got a notebook document that combined the theoretical explanations, code examples, and the questions they should answer by running the corresponding computations (either by using the CAS as a calculator or writing some actual code to solve the problems). Maybe the instructor could combine it with a general explanation, and/or give specific hints to the students that got stuck at some point.

As we can see, adoption increased quickly when the first notebook was introduced; and then it has remained stable. The introduction of the Jupyter environment resulted in some instructors migrating to it, while others preferred to keep using the legacy one (under the warning that it would be eventually deprecated). An undesired result of this way of switching is that some students are forced to use the legacy system for some courses, and the new one for others.

Although we didn't make a general survey about the satisfaction of students, we did receive some comments (that should only be taken as anecdotal). Some of those comments were:

- Some engineering students felt overwhelmed by having to learn one programming language (C++) for their programming courses, and a different one (Sage/Python) for their mathematics courses.
- Several complained about the punctual availability and stability problems of the server.
- Several students of science degrees without a big mathematical content (Optics, Biotech...) had a very hard time grasping the general ideas of a programming language.

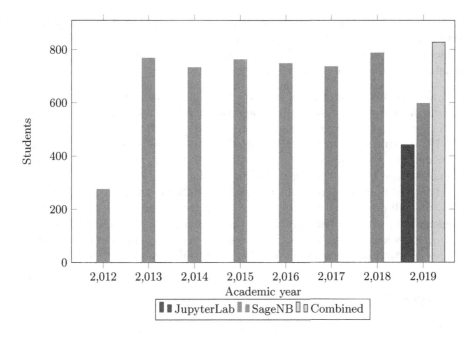

Fig. 1. Number of students that used the notebooks each year.

- In general, students of mathematics and physics degree showed a more positive attitude towards these tools. Some of them used these tools also as an assistance for doing exercises in courses where it was not formally used (Fig. 1).

During the confinement declared in the COVID-19 pandemic, the university switched to an on-line model with virtually no time to redesign the courses and workflows. It was very hard in general, and some aspects could not really be adapted. However, in particular, the computer lab lessons were one of the aspects that was more easily adapted, since it didn't require the students to install any software on their computers, and the instructors could use the same exercises and documents that were planned.

5   Conclusions

The combination of JupyterHub/JupyterLab has the potential to be a very useful tool in teaching, and maybe some day it might be considered as a standard service provided by universities (just like email or virtual campus). Its modularity and flexibility makes it adaptable to each institution's needs. However, most of the development, modules and documentation seems to be focused to different use cases, which makes it non obvious how to proceed for the university wide approach.

To find a setup that fits our requirements is still a challenge. Significant computing resources would be necessary to provide this kind of services at a whole university level. It would also require dedicated personnel to install and maintain it. An estimate of this requirements would be:

- A minimum of 2 GB of RAM per expected simultaneous user at peak time. A conservative security threshold should be added to that to prevent problems in case of an unexpectedly high peak.
- CPU doesn't seem to be a big limitation, unless the students are expected to do very CPU intensive computations. And even in that case, the only expected problem is that they would take longer.
- About personnel requirements, the initial installation and setup could be done by a small group (one or two persons) with the appropriate skills and documentation, within a few days. Proper maintenance would require a sysadmin with knowledge of the system (although, as many sysadmin work, the difference in workload between normal days and peak times would be very big).

References

1. Abbott, J., Bigatti, A.M.: CoCoA: a system for doing Computations in Commutative Algebra. http://cocoa.dima.unige.it
2. The GAP Group, GAP - Groups, Algorithms, and Programming, Version 4.11.0; 2020. https://www.gap-system.org
3. Project Jupyter. https://Jupyter.org/
4. JupyterHub. https://JupyterHub.readthedocs.io/en/stable/
5. nbgrader. https://nbgrader.readthedocs.io
6. The Sage Developers, SageMath, the Sage Mathematics Software System (Version 9.0) (2020). http://www.sagemath.org

Phase Portraits of Bi-dimensional Zeta Values

Olivier Bouillot[(✉)]

Gustave Eiffel University, Marne-la-Vallée, France
olivier.bouillot@u-pem.fr
http://www-igm.univ-mlv.fr/~bouillot/

Abstract. In this extended abstract, we present how to compute and visualize phase portraits of bi-dimensional Zeta Values. Such technology is useful to explore bi-dimensional Zeta Values and in long-term quest to discover a 2D-Riemann hypothesis.

To reach this goal, we need two preliminary steps:
- the notion of phase portraits and a general tool to visualize phase portrait based on interactive Jupyter widgets.
- the ability to compute numerical approximations of bi-dimensional Zeta values, using mpmath, a Python library for arbitrary-precision floating-point arithmetic. To this end, we develop a theory to numerically compute double sums and produce the first algorithm to compute bi-dimensional Zeta Values with complex parameters.

Keywords: Phase portrait · Lindelöf formula · double Zeta Values

1 Introduction

The Riemann Zeta function ζ and the bi-dimensional Zeta Values, also called 2D-Zeta Value, bizetas or double Zeta Values, are respectively defined by:

$$\zeta(s) = \sum_{k>0} \frac{1}{k^s}, \text{ for all } s \in \mathbb{C} \text{ , } \Re e\ s > 1. \tag{1}$$

$$\mathcal{Z}e^{s_1,s_2} = \sum_{k>l>0} \frac{1}{k^{s_1} l^{s_2}}, \text{ for all } s_1, s_2 \in \mathbb{C} \text{ , } \begin{cases} \Re e\ s_1 > 1. \\ \Re e\ (s_1 + s_2) > 2. \end{cases} \tag{2}$$

It is well-known that these functions can be respectively meromorphically extended to $\mathbb{C}$ and $\mathbb{C}^2$ (see [9]). It is also conjectured that the zeros of the Riemann Zeta function be complex numbers with real part $\frac{1}{2}$: this is the Riemann hypothesis, stated by Riemann himself in 1859 in [10]. A nice long term quest is to discover the location of zeros of bi-dimensional Zeta Values.

This quest needs first a tool to visualize a representation of a function f defined over (a part of) $\mathbb{C}$ and valued in $\mathbb{C}$, such that looking for zeros of f becomes easy.

A. M. Bigatti et al. (Eds.): ICMS 2020, LNCS 12097, pp. 393–405, 2020.
https://doi.org/10.1007/978-3-030-52200-1_39

Constructing a graphical representation of a function $f : \mathbb{R} \longrightarrow \mathbb{R}$ is quite easy, we just need a 2-dimensional space. However, constructing the graph of a function $f : \mathbb{C} \longrightarrow \mathbb{C}$ is not so simple because we need a 4-dimensional space. In particular, this implies that most of us have difficulties visualizing such a graph. The first objective of this note is to review how to read and construct phase portraits, which are heat maps where colors encode simultaneously phase and modulus (see Sect. 2).

Let us notice that there already exists a tool, developed by Elias Wegert in Matlab (see [12]), to represent phase portraits of complex functions. In particular, it has been used to explore many functions and illustrate a first course on complex analysis (see [13]). Unfortunately, this tool computes a lot of values to produce the desired representation and does not save them. So, research the localization of the zeros of a function with it is necessarily time-consuming. In other words, it is not an efficient tool for our purpose.

Nevertheless, Jupyter widgets (see [14]) are wonderful tools that easily enable us to construct a general phase portrait visualization tool, where the input is just a complex function. The author has implemented such a widget (see [1]) where, with interactivity, the user can change the visualization window, the number of pixels used in a unit square, reuse already computed values or store new ones in a database, save pictures or see information on how the computations progress.

This widget can be extended to visualize phase portraits of functions with two complex variables. Visualizing such a function is nothing else than drawing a representation of the partial functions and move inside them. According to this, the author has implemented a second interactive Jupyter widget to realize this walk.

To benefit from the power of the second widget, we need to have an efficient algorithm to numerically compute bi-dimensional Zeta Value. This is the second preliminary step to look for a 2D-Riemann hypothesis.

Nowadays, researchers can compute multiple zeta values (and in particular bizetas), with integer parameters, with as many digits as we wanted (see [3]). But, to the best of the author's knowledge, nothing is known to compute these numbers with complex parameters in any length.

Therefore, the last objective of this note is to present how a double sum can be numerically computed, and apply the developed method to bizetas. This method is based on a generalization of a poorly known Lindelöf formula explaining how to compute the sum of the values at integers of a class of holomorphic functions. Section 3 contains a presentation of Lindelöf formula, as well as a comparison between three methods to compute evaluations of the Riemann Zeta function.

Lindelöf formula generalization to double sum, and therefore to bizetas, is discussed in Sect. 4. One could notice that Lindelöf formula can be written as an integral. Therefore, we generalize the process to compute double sums using double integrals and, then, expansions of these integrals. Of course, this technic is still valid for multiple sums and gives, in theory, an algorithm to compute multiple Zeta Value in any length, with complex parameters!

Finally, once the Jupyter visualization widget of a two complex variables function is available and the computational machinery is developed, we can explore phase portraits of convergent 2D-Zeta Values, or linear combinations of these numbers, to find out some zeros. Therefore, the last necessary point is to be able to numerically compute the analytic extension of bizetas. Only then, using our Jupyter widget, a Riemann hypothesis for bizetas could be conjectured.

2   Drawing Function from $\mathbb{C}$ to $\mathbb{C}$: Phase Portraits

In this section, we review how to construct and read phase portraits. We will see how to find zeros and poles of a function, but also points (called *color saddle*) where the derivative becomes null. References for this section are [11] and [13].

2.1   Analytic Landscapes vs Phase Portraits

Drawing a function $\mathbb{R}^2 \longmapsto \mathbb{R}$ is typically achieved through 2D heat maps (see [15]) or 3D plots. For a function $f : \mathbb{C} \longmapsto \mathbb{C}$, these approaches may be used to visualize the modulus; that is *analytic landscapes*. We could also use such plots to visualize the phase of f.

However, many applications like ours require to visualize the modulus and the phase at the same time to visualize poles and zeroes. In addition, such visualization should highlight simultaneously very small modulus (near zeroes) and huge modulus (near poles).

Here come into play *phase portraits*: heat maps where the color of a pixel z encodes simultaneously the phase of $f(z)$ (by giving color to its values) and the (logarithm of the) modulus (by using different brightness). Eventually, the modulus can then be emphasized by adding contours.

2.2   HSL-representation of Colors

The HSL-representation of colors is an alternative representation of the RGB color model and can be seen as a cylindrical geometry (see Fig. 1). Its angular dimension, the hue, starts at the red color at $0°$, passing through the green color at $120°$ and the blue one at $240°$ to finally come back to red at $360°$. The lightness, *i.e.* the central vertical axis, describes the gray colors, rang-ing from black at the bottom of the cylinder with

Fig. 1. The HSL cylinder <small>Source: Wikipedia, "HSL and HSV" (Color figure online)</small>

lightness or value 0 to white at the top of the cylinder with lightness or value 1.

2.3 Principle of Phase Portraits

A pixel of coordinates (x, y) is associated with a complex number z. Then, we find a color c related to the value $f(z) = \rho e^{i\theta}$ using the HLS representation of colors. Therefore, the pixel (x, y) is colored by the color c. (See Fig. 2).

Let us mention that zeros and poles play an important role when visualizing a phase portrait. Using the logarithm of the modulus, instead of the modulus itself, allows us to have a smoother graphic representation and also to have symmetrical behavior between zeros and poles, as well as symmetric visualization of small and big values.

Moreover, to encounter all the possible values of the modulus, we compactify the extended real number line $[-\infty; +\infty]$ to $[-1; 1]$, using $x \longmapsto \dfrac{x}{1 + |x|}$.

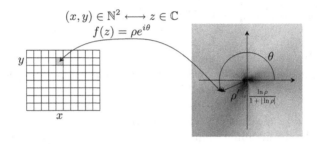

Fig. 2. Color coding of pixels to draw phase portraits of a complex function f

Consequently, a pixel of coordinates (x, y), via its associated complex number z, is associated with the color obtained in the HSL model by:

$$H = \text{phase}\big(f(z)\big) \qquad L = \frac{1}{2}\left(\frac{\ln|f(z)|}{1 + |\ln|f(z)||} + 1 \right) \qquad S = 1. \qquad (3)$$

In particular, positive real numbers appear reddish in a phase portrait when negative real numbers appear cyan. Moreover, a zero is a black point while a pole is a white point.

Let us notice that the color map can be, in principle, adapted to colorblind people provided that the inherent periodicity of color be fulfilled.

2.4 Easy Properties to Read on Phase Portraits

The phase portraits not only show the location of zeros and poles of a function but also reveal their multiplicity. As an example, Fig. 3a shows a zero with multiplicity three. It is easily recognizable: z^3 travels 3 times around 0 when z moves once around 0 along a small circle. We emphasize that the colors met are in the reverse order for zeros and poles (compare the pole 0 to the three zeros located in cube root of -2 in Fig. 3b).

Moreover, it can be shown that the points where $f'(z) = 0$, with $f(z) \neq 0$, are where the isochromatic lines meet. Such points are called *color saddles*. See Fig. 3b where the color saddles are located at cube roots of -2. See also Fig. 3c.

2.5 Necessary Precision to Choose Exact Color

In most cases, only a few significant digits (typically five), need to be known to plot phase portraits of a function $f : \mathbb{C} \longrightarrow \mathbb{C}$. In particular, we do not assume that the underlying system can do arbitrary precision computations.

We review here how to assess the required precision and how to achieve that precision.

If ε_L and ε_H denote the absolute error of L and H, the RGB components of the associated color are fixed as soon as we have: $\dfrac{\varepsilon_H}{60} + 6\varepsilon_L \leq \dfrac{1}{255}$. More precisely, we seek to have: $\varepsilon_L \leq 2040^{-1}$ and $\varepsilon_H \leq 20^{-1}$.

If we assume that Re $f(z)$ and Im $f(z)$ are known up to $\varepsilon > 0$, then, $|f(z)|$ is known up to 2ε. If $|f(z)| \geq 1$, then we can prove that $\varepsilon_L \leq 2\varepsilon$. If $|f(z)| < 1$, then this time $\varepsilon_L \leq \dfrac{k+2}{\ln^2(k+2)}\varepsilon$, where $k \in \mathbb{N}^*$ is such that $|f(z)| \in \left[\dfrac{1}{k+1}; \dfrac{1}{k}\right[$.

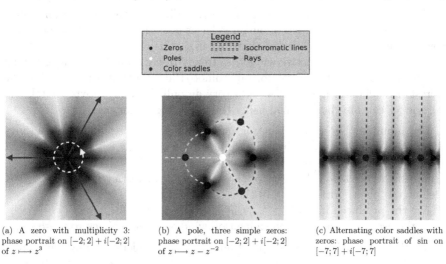

(a) A zero with multiplicity 3: phase portrait on $[-2; 2] + i[-2; 2]$ of $z \longmapsto z^3$

(b) A pole, three simple zeros: phase portrait on $[-2; 2] + i[-2; 2]$ of $z \longmapsto z - z^{-2}$

(c) Alternating color saddles with zeros: phase portrait of sin on $[-7; 7] + i[-7; 7]$

Fig. 3. Examples of easy properties seen on phase portraits

Consequently, if $f(z)$ is computed up to $\varepsilon = 10^{-4}$ and satisfied $|f(z)| \geq 0.01$, we have $\varepsilon_L \leq 2040^{-1}$ for all cases and the parameter L is well-approximated.

On the H-side, if we assume that $R = \text{Re } f(z)$ and $I = \text{Im } f(z)$ are known up to $\varepsilon > 0$ and satisfied $|R|, |I| \geq \alpha$ for a positive real number α, then, we can show that $\varepsilon_H \leq \varepsilon\left(\alpha^{-1} + (\alpha - \varepsilon)^{-1}\right)$. In particular, for $\alpha = 5 \cdot 10^{-3}$ and $\varepsilon = 10^{-4}$, we thus have $\varepsilon_H \leq 20^{-1}$ and the parameter H is well-approximated.

Consequently, in most cases, we compute the function $f : \mathbb{C} \longrightarrow \mathbb{C}$ up to 10^{-4}. However, in the other cases, *i.e.* when $|f(z)| < 0.01$, or $|\mathrm{Re}\ f(z)| < 5 \cdot 10^{-3}$ or $|\mathrm{Im}\ f(z)| < 5 \cdot 10^{-3}$, then we will compute again $f(z)$ up to a smaller precision, *e.g.* $\varepsilon = 10^{-5}$ or 10^{-6}, in order to finaly have $\varepsilon_L \leq 2040^{-1}$ and $\varepsilon_H \leq 20^{-1}$.

2.6 Interactive Visualization of Phase Portraits for Jupyter Notebook

The next step is to create a tool to visualize phase portraits of a given complex function $\mathbb{C} \longrightarrow \mathbb{C}$.

Let us remind such a tool was already available in Matlab (see [12]). However, for our long-term quest of a 2D Riemann hypothesis, we need a tool where the visualization window of our phase portraits can be easily modified.

According to the official website, a "Jupyter Notebook is an open-source web application that allows you to create and share documents that contain live code, equations, visualizations and narrative text" (see also [7] about Jupyter). Moreover, interactive widgets like Sliders, Buttons or Images are now available in Jupyter (see [14]).

Consequently, the author has implemented an interactive Jupyter widget for general visualization of phase portraits (see [1]). Figure 4 is a graphic representation of its usage. For examples of outputs, let us note that all the phase portraits shown in this paper have been constructed from this widget.

The code of this interactive widget is already freely available in the following GitHub repository: https://github.com/tolliob/PhasePortrait. In particular, the interested reader will find in this repository an online usable Jupyter notebook. The widget can also be installed using the `pip` Unix command.

The input of this widget is a complex function $f : \mathbb{C} \longrightarrow \mathbb{C}$. Interactivity allows the user to fix a visualization window by setting the left below and right upper corners. Once these corners fixed, the user can:
⤳ launch the needed evaluations of the function f, or retrieve them;
⤳ show the phase portrait of f in the desired window;
⤳ save the produced image
Then, the user can change the visualization window by going back to these steps.

Another interactive Jupyter widget has also been implemented by the author. It aims to visualize phase portraits of two variables complex functions from its partial functions phase portraits. The input is a function $f : \mathbb{C}^2 \longrightarrow \mathbb{C}$, outputs are again phase portraits, alternatively of $f(s_0, \cdot)$ and $f(\cdot, s_0)$. Interactivity allows the user to move a partial function to another one by moving the point s_0 to a close complex point, or to switch from a partial function to the other one.

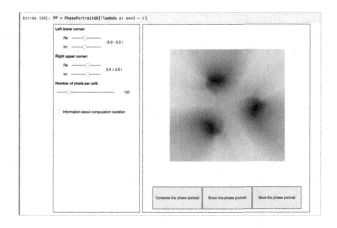

Fig. 4. Interactive Jupyter widget to construct phase portraits

3   Computation of the 1D-Zeta Value

In this section, we present a poorly known formula due to Lindelöf (see [8], chapter 3) which explains how to compute the sum of the values at integers of holomorphic functions. Then, we apply it to the Riemann Zeta Function and compare its performances to the Euler-Mac Laurin summation.

However, we warn the reader that Lindelöf formula application to the Zeta function could not compete computation technics dedicated to the Zeta function, such that the Riemann-Siegel formula or the Odlyzko-Schönhage algorithm (see [5] for example). However, Eqs. (7) and (9) will prove that Lindelöf formula can be extended to the numerical computation of double sums, and so to the numerical computation of 2D-Zeta Values, while it seems nowadays impossible to adapt specific ζ computational technics to 2D-Zeta Values.

3.1   On Lindelöf Formula

Before giving the Lindelöf formula, we first emphasize a technical definition from Chapter 3, p. 54, of [8].

Definition 1. *Let $m_0 > \dfrac{1}{2}$ be a real number. Let also Ω_s be the set defined for all real number s by $\Omega_s = \{z \in \mathbb{C} \, , \, Re\, z \geq s\}$.*
A holomorphic function $f : \Omega_{m_0 - \frac{1}{2}} \longrightarrow \mathbb{C}$ satisfies the 1D-Lindelöf hypothesis if:

1. for all $s \geq m_0 - \dfrac{1}{2}$, $\displaystyle\lim_{t \longrightarrow \pm\infty} e^{-2\pi|t|} f(\tau + it) = 0$ uniformly for $\tau \in \left[m_0 - \dfrac{1}{2}; s\right]$;

2. for all $s \geq m_0 - \dfrac{1}{2}$, $t \longmapsto e^{-2\pi|t|} f(s + it) \in \mathcal{L}^1(\mathbb{R})$ and

$$\lim_{s \longrightarrow +\infty} \int_{-\infty}^{+\infty} e^{-2\pi|t|} |f(s + it)| \, dt = 0. \tag{4}$$

Now, exploring Chapter 3 of [8], one can reconstruct the following theorem which gives an integral representation of sums of values of holomorphic functions. The proof is based on a clever use of the residuum principle. Nevertheless, the result is quite unknown.

Theorem 1. *(Lindelöf, 1905)*

Let $m_0 > \dfrac{1}{2}$ be a real number, m a positive integer such that $m \geq m_0$ and $f : \Omega_{m_0 - \frac{1}{2}} \longrightarrow \mathbb{C}$ be an holomorphic function over $\Omega_{m_0 - \frac{1}{2}}$ satisfying

- *the 1D-Lindelöf hypothesis ;*   • $\displaystyle\sum_{\nu \geq m} f(\nu)$ *is a convergent series.*

  Then, $f \in \mathcal{L}^1([m_0 - \frac{1}{2}; +\infty[)$ and

$$\sum_{\nu \geq m} f(\nu) = \int_{m-\frac{1}{2}}^{+\infty} f(t)\ dt - i \int_0^{+\infty} \frac{f(m - \frac{1}{2} + it) - f(m - \frac{1}{2} - it)}{e^{2\pi t} + 1}\ dt. \quad (5)$$

The Lindelöf formula (5) can be used to numerically compute sums, expanding its second integral using Taylor expansion with integral remainder: this gives us a Lindelöf's Euler-Maclaurin like formula.

Theorem 2. *(Lindelöf, 1905)*
Let m_0, m and f be like in Theorem 1. Then:

$$\forall K \in \mathbb{N}, \ \sum_{\nu \geq m} f(\nu) \approx \int_{m-\frac{1}{2}}^{+\infty} f(t)\ dt + \sum_{k=1}^{K} \left(1 - \frac{1}{2^{2k-1}}\right) \frac{B_{2k}}{(2k)!} f^{(2k-1)} \left(m - \frac{1}{2}\right). \quad (6)$$

Coefficients in (6) are nothing else than the Taylor coefficients of $z \longmapsto \dfrac{x}{2\sinh\left(\dfrac{z}{2}\right)}$. So, in a certain sens, Eq. (6) means that:

$$\sum_{\nu \geq m} f(\nu) \approx \int_{m-\frac{1}{2}}^{+\infty} \frac{\dfrac{d}{dz}}{2\sinh\left(\dfrac{1}{2}\dfrac{d}{dz}\right)}(f)(t)\ dt. \quad (7)$$

While in Lindelöf's days, it was not necessary to have a precise estimate of the approximation error, we now need it due to the existence of computers and their immense associated computing power.

Theorem 3. *Let m_0, m, K and f be like in Theorem 1. Let also assume that, for all $u \geq m_0 - \dfrac{1}{2}$, the quantity $M_u(f) = \sup\limits_{\substack{\zeta \in \mathbb{C} \\ Re\ \zeta \geq u}} |f(\zeta)|$ is well-defined.*

If we denote by $\mathcal{R}_{K,m}(f)$ the Lindelöf remainder in Eq. (6), i.e. the difference between the right hand side and the left hand side of Eq. (6), then we have the following upper bound:

$$|\mathcal{R}_{K,m}(f)| \leq \frac{M_{m_0-\frac{1}{2}}(f)}{(m - m_0)^{2K+1}} \cdot \frac{(2K+1)!}{(2\pi)^{2K+1}}. \quad (8)$$

3.2   Application to the Riemann Zeta Function

To produce a phase portrait of the Riemann Zeta function ζ, we will compute the values $\zeta(s)$ for s in a grid up to 10^{-4}.

First, the Euler-Maclaurin summation process applies (see [2]).

Moreover, the functions $f_s : z \longmapsto e^{i\pi z} z^{-s}$ satisfy the $1D$-Lindelöf hypothesis for all complex numbers s such that $\Re es > 0$. So, Eq. (5) can be used, its second integral being numerically computed with the mpmath Python library.

Equation (6) can also be used. We just have to find the best choice of triplet (m_0, m, K). Note that the bigger K is, the longest the computation, due to the computation of the successive derivative which concentrates most of the multiplications. However, the size of m also intervenes.

This allows us to compute $\eta(s) = \sum_{n>0} (-1)^n n^{-s}$ and $\zeta(s)$, for all $s \in \Omega_{\frac{1}{2}}$.

Now, classical tools can be used to compute ζ on the whole punctured complex plane $\mathbb{C} - \{1\}$ (reflection formula, analytic continuation process; see [4]). Using the Jupyter widget [1], this generates the phase portraits available in Figs. 5a–5c.

3.3   Comparaison of the Three Methods

Figures 6a to 6d show some comparisons between the three previous methods (*i.e.* the Euler-Maclaurin summation process, the Lindelöf formula (5) and the Lindelöf Euler-Maclaurin like formula given by Eq. (6)).

First, we can see that the integral Lindelöf method is the most stable method, even if there are some discontinuities (see Fig. 6c and 6d). This is explained by the different numerical integration methods used by the mpmath Python library.

According to these Figures, we see that to have numerous exact digits, the bigger $|s|$ is, the more efficient the Lindelöf Euler-Maclaurin-like method is.

Finally, for the three methods, we observe that the computation gets longer as we want more exact digits. This increase is correlated to the modulus of $|s|$. It turns out that Lindelöf Euler-Maclaurin-like formula is the method with the slower computation time increase relatively to the required number of digits in the computation of $\zeta(s)$.

 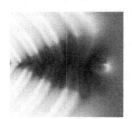 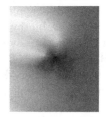

(a) On $[-40; 20] + i[-40; 40]$ (b) On $[-18; 14] + i[-10; 10]$ (c) Near the first non trivial zero $z_1 \approx 0.5 + 14.135$

Fig. 5. Three phase portraits of the Rieman Zeta function ζ

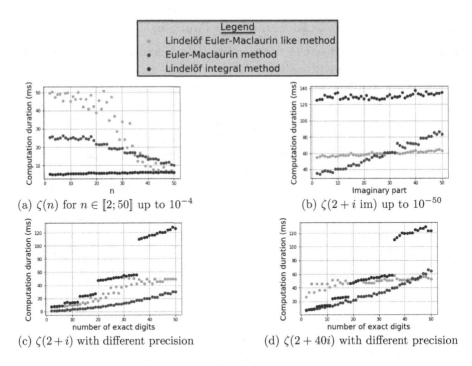

(a) $\zeta(n)$ for $n \in [\![2;50]\!]$ up to 10^{-4}

(b) $\zeta(2 + i\text{ im})$ up to 10^{-50}

(c) $\zeta(2+i)$ with different precision

(d) $\zeta(2 + 40i)$ with different precision

Fig. 6. Comparaison between three methods to compute Zeta Values

4   Computation of Convergent Two Dimentional Multiple Zeta Values by a 2D-Lindelöf Formula

To the best of the author's knowledge, no one was able to numerically compute a convergent Multiple Zeta Values (MZV) with complex exponents so far. The only known algorithm was dedicated to MZV with integers parameters: it uses binomial expansions where exponents are related to the MZV parameters (see [3]).

To produce and explore phase portraits of 2D-Zeta Values, we nevertheless need, first, to *be able to compute* these numbers, then to compute them *efficently* because producing phase portraits of 2D-Zeta Values could easily require millions of different evaluations. This difficulty is, of course, the cornerstone of our quest for a 2D-Riemann hypothesis.

Hopefully, one of the most important advantages of Lindelöf formulas (5) and (6) over Euler-Maclaurin formula is that it could be extended to higher dimensions. Consequently, in this section, we will present the first general method to compute double sums and then apply it to produce the first bi-dimensional Zeta Values with complex exponents computing algorithm.

Efficientness will not be discussed here for technical reasons, as well as the upper bound of the error term in the computation of a double sum.

4.1   On the 2D-Lindelöf Formula

First, we extend the 1D-Lindelöf hypothesis to the bi-dimensional case:

Definition 2. *We say that a function* $f : \Omega_{m_{01}} \times \Omega_{m_{02}} \longrightarrow \mathbb{C}$ *satisfies the 2D-Lindelöf hypothesis when:*

1. $\forall z_1 \in \Omega_{m_{01}}, f(z_1, \cdot)$ *satisfies the 1D-Lindelöf hypothesis over* $\Omega_{m_{02}}$;
2. $S = \sum\limits_{l \geq m_2} f(\cdot, l)$ *is a well-defined function over* $\Omega_{m_{01}}$ *satisfying the 1D-Lindelöf hypothesis;*

Then, using an iteration of (7), we can prove the following:

Theorem 4. *Let* m_{01} *and* m_{02} *be two real number greater than* $\frac{1}{2}$, m_1 *and* m_2 *be two positive integers such that* $m_i \geq m_{0i}$, $i \in \{1; 2\}$ *and* $f : \Omega_{m_{01}} \times \Omega_{m_{02}} \longrightarrow \mathbb{C}$ *a fonction satisfying the 2D-Lindelöf hypothesis.*
Let us also suppose that $\sum\limits_{\substack{k \geq m_1 \\ l \geq m_2}} f(k, l)$ *is a convergent double series. Then:*

$$\bullet \sum_{\substack{k \geq m_1 \\ l \geq m_2}} f(k, l) \approx \int_{m_1 - \frac{1}{2}}^{+\infty} \int_{m_2 - \frac{1}{2}}^{+\infty} \frac{\dfrac{d}{dz_1}}{2\sinh\left(\dfrac{1}{2}\dfrac{d}{dz_1}\right)} \frac{\dfrac{d}{dz_2}}{2\sinh\left(\dfrac{1}{2}\dfrac{d}{dz_2}\right)}(f)(u, v)\, du dv \qquad (9)$$

$$\bullet \sum_{\substack{k \geq m_1 \\ l \geq m_2}} f(k, l) \approx \sum_{\substack{p, q \in \mathbb{N} \\ p+q \leq K}} \left(1 - \frac{1}{2^{2p-1}}\right)\left(1 - \frac{1}{2^{2q-1}}\right)\frac{B_{2p}}{(2p)!}\frac{B_{2q}}{(2q)!}$$

$$\times \left(\int_{m_1-\frac{1}{2}}^{+\infty}\int_{m_2-\frac{1}{2}}^{+\infty}\frac{\partial^{2p+2q}f}{\partial_{z_1}^{2p}\partial_{z_2}^{2q}}(u, v)\, du\, dv\right) \qquad (10)$$

for all integers K.

Some of the integrals in Eq. (10) could be explicitly computed; some could not. So, to obtain numerical approximation, we use the mpmath Python library for arbitrary-precision floating-point arithmetic (see [6]).

4.2   Computation of a Double Sum over $\mathbb{N}^2$

Now, we are able to compute a double sum over $\mathbb{N}^2$ by cutting $\mathbb{N}^2$ onto six parts (see Fig. 7):

$$\sum_{k\geq 0}\sum_{l\geq 0} f(k,l) = \sum_{k=0}^{m_1-1}\sum_{l=0}^{m_2-1} f(k,l) + \sum_{k=0}^{m_1-1}\sum_{l=m_2}^{q-1} f(k,l) + \sum_{k=0}^{m_1-1}\sum_{l\geq q} f(k,l)$$

$$+ \sum_{k=m_1}^{p-1}\sum_{l=0}^{m_2-1} f(k,l) + \sum_{k\geq p}\sum_{l=0}^{m_2-1} f(k,l) + \sum_{k\geq m_1}\sum_{l\geq m_2} f(k,l). \quad (11)$$

For "good" functions f, we can compute $\displaystyle\sum_{k=0}^{m_1-1}\sum_{l\geq q} f(k,l)$ and $\displaystyle\sum_{k\geq p}\sum_{l=0}^{m_2-1} f(k,l)$ using the 1D-Lindelöf Euler-Maclaurin-like formula, while $\displaystyle\sum_{k\geq m_1}\sum_{l\geq m_2} f(k,l)$ is computed using the 2D-Lindelöf Euler-Maclaurin-like formula.

Let us consider the function $f_{s_1,s_2}:(z_1,z_2)\longmapsto (z_1+1)^{-s_2}(z_1+z_2+2)^{-s_1}$. f_{s_1,s_2} satisfies the $2D$ Lindelöf hypothesis, while its partial functions satisfy the $1D$-Lindelöf hypothesis. Therefore, we can use the decomposition (11) to compute the double Zeta Value (see Fig. 8a and 8b).

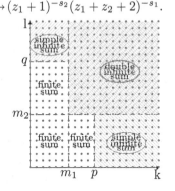

According to Fig. 8a, we could conjecture that there is a pole near $(s_1, s_2) = (1, 0)$.

According to [9], we know exactly the localization of the poles of double Zeta values and $(1, 0)$ is actualy a pole. Up to long computations, we now are able to find out zeros of convergent double Zeta Values.

Fig. 7. Decomposition of $\mathbb{N}^2$ onto 6 parts to compute a double sum

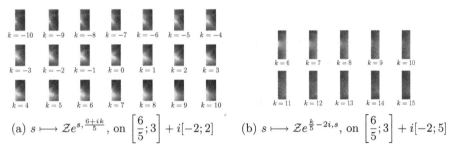

(a) $s\longmapsto \mathcal{Z}e^{s,\frac{6+ik}{5}}$, on $\left[\frac{6}{5};3\right]+i[-2;2]$ (b) $s\longmapsto \mathcal{Z}e^{\frac{k}{5}-2i,s}$, on $\left[\frac{6}{5};3\right]+i[-2;5]$

Fig. 8. Expansion of phase portrait of partial functions of double Zeta values

Acknowledgment. The author would like to warmly thank the reviewers for their time and effort devoted to improving the quality of this work.

References

1. Bouillot, O.: Phaseportrait, a github repository (2020). https://github.com/tolliob/PhasePortrait
2. Cohen, H., Olivier, M.: Calcul des valeurs de la fonction zêta de riemann en multiprécision. C.R. Acad. Sci. Paris Sér I Math. **314**, 427–430 (1992)
3. Crandall, R.E.: Fast evaluation of multiple zeta sums. Math. Comp. **67**(223), 1163–1172 (1998)

4. Edwards, H.E.: Riemann's Zeta Function. Academic Press, New York (1974)
5. Gourdon, X., Sebah, P.: Numerical evaluation of the riemann ζ-function. http:// numbers.computation.free.fr/Constants/Miscellaneous/zetaevaluations.pdf
6. Johansson, F., et al.: mpmath: a Python library for arbitrary-precision floating-point arithmetic (version 0.18), December 2013. http://mpmath.org/
7. Kluyver, T., et al.: Jupyter notebooks - a publishing format for reproducible computational workflows. In: Loizides, F., Schmidt, B. (eds.) Positioning and Power in Academic Publishing: Players, Agents and Agendas, pp. 87–90. IOS Press, Amsterdam (2016)
8. Le Lindelöf, E.L.: calcul des résidus et ses applications à la théorie des fonctions. Gauthier-Villars, Paris (1905)
9. Mehta, J., Saha, B., Viswanadham, G.K.: Analytic properties of multiple zeta functions and certain weighted variants, an elementary approach. J. Number Theor. **168**, 487–508 (2016)
10. Riemann, B.: Über die anzahl der primzahlen unter einer gegebenen grösse. Mon. Not. Berlin Akad. **2**, 671–680 (1859)
11. Semmler, G., Wegert, E.: Phase plots of complex functions: a journey in illustration. Notices Am. Math. Soc. **58**, 768–780 (2011)
12. Wegert, E.: Complex function explorer. https://www.mathworks.com/ matlabcentral/fileexchange/45464-complex-function-explorer. Accessed 08 May 2020. MATLAB Central File Exchange
13. Wegert, E.: Visual Complex Functions. An Introduction with Phase Portraits. Springer, Basel (2012). https://doi.org/10.1007/978-3-0348-0180-5
14. Jupyter widgets community: ipywidgets, a github repository (2015). https:// github.com/jupyter-widgets/ipywidgets
15. Wilkinson, L., Friendly, M.: The history of the cluster heat map. Am. Stat. **63**, 179–184 (2009)

Prototyping Controlled Mathematical Languages in Jupyter Notebooks

Jan Frederik Schaefer$^{(\boxtimes)}$, Kai Amann, and Michael Kohlhase$^{(\boxtimes)}$

Computer Science, FAU Erlangen-Nürnberg, Erlangen, Germany
{jan.frederik.schaefer,michael.kohlhase}@fau.de

Abstract. The Grammatical Logical Framework (GLF) is a framework for prototyping the translation of natural language sentences into logic. The motivation behind GLF was to apply it to mathematical language, as the classical compositional approach to semantics construction seemed most suitable for a domain where high precision was mandatory—even at the price of limited coverage. In particular, software for formal mathematics (such as proof checkers) require formal input languages. These are typically difficult to understand and learn, raising the entry barrier for potential users. A solution is to design input languages that closely resemble natural language. Early results indicate that GLF can be a useful tool for quickly prototyping such languages. In this paper, we will explore how GLF can be used to prototype such languages and present a new Jupyter kernel that4 adds visual support for the development of GLF-based syntax/semantics interfaces.

1 Introduction

The work of mathematicians is increasingly supported by computer software ranging from computer algebra systems to proof checkers and automated theorem provers. Such software typically requires a specialized input language, which users have to learn to use the software themselves and in order to understand how it was used by other people. The latter is of particular interest when it comes to computer supported theorem proving. In mathematics, proofs are much more than mere correctness certificates: they give insights into *why a theorem is true*. A computer proof cannot fulfil this duty if the reader cannot understand it in the first place. The obvious consequence is that input languages should be designed to be as intuitive as possible. In some cases, this could be in the form of a **controlled natural language**—a formal language with well-defined semantics that closely resembles natural language. There are some general-purpose controlled natural languages, most notably Attempto Controlled English (ACE) [FSS98]. But for input languages for mathematical software, we need **controlled mathematical languages**.

State of the Art. A **controlled natural language** (CNL) consists of *i)* a natural language fragment e.g. defined by a grammar, *ii)* a formal target language,

© Springer Nature Switzerland AG 2020
A. M. Bigatti et al. (Eds.): ICMS 2020, LNCS 12097, pp. 406–415, 2020.
https://doi.org/10.1007/978-3-030-52200-1_40

and *iii)* a program that translates from *i*) to *ii*). Different controlled mathematical languages (CML) have been developed in the past, especially for automatic proof checkers. An example for this is ForTheL [Pas07], the language of the System for Automated Deduction (SAD). It appears that ForTheL has reached a sweet spot between expressivity and parseability. Implemented with hand-crafted parser combinators in Haskell, however, it is hard to maintain and even harder to extend. More recently, SAD was extended by some of the people behind the Naproche system [Cra13], which has a controlled mathematical language that supports a "controlled" LATEX input. This resulted in the Naproche-SAD project [FK19]. Over the last few years, research into controlled mathematical languages has gained momentum with Thomas Hales' Formal Abstracts project (e.g. [Hal19]). Its goal is the creation of a controlled mathematical language that translates into the language of the lean theorem prover – a type theory based on the calculus of inductive constructions.

Overview. In this paper, we present a setup for prototyping controlled mathematical language. Section 2 describes the underlying technology: the Grammatical Logical Framework (GLF) [KS19]. In Sect. 3 we introduce a new Jupyter kernel for GLF that makes GLF much more accessible and supports the development and testing of controlled mathematical languages with a variety of features. All listings in this paper are screenshots of Jupyter notebooks. In Sect. 4, we will discuss some of our insights from our attempts to re-implement ForTheL with GLF. Section 5 concludes the paper.

2   Grammatical Logical Framework

The **Grammatical Logical Framework** (GLF) [KS19] is a tool for prototyping translation pipelines from natural language to logic. As a running example, we will develop a pipeline that translates sentences like *"the derivative of any holomorphic function is holomorphic"* into expressions in first-order logic: $\forall f(\text{holomorphic}(f) \Rightarrow \text{holomorphic}(\text{derivative}(f)))$. This translation pipeline consists of two steps: parsing and semantics construction.

```
abstract Grammar = {
  cat
    Stmt; Term; Notion; Prop;
  fun
    state : Term->Prop->Stmt;
    every : Notion -> Term;
    -- ...
    integer : Notion;
    even : Prop;
    derivative : Term -> Term;
}
```

```
concrete GrammarEng of Grammar = {
  lincat Stmt=Str;Term=Str; -- ...
  lin
    state t p = t ++ "is" ++ p;
    every n = ("every"|"any")++n;
    -- ...
    integer = "integer";
    even = "even";
    derivative t =
      "the derivative of" ++ t;
}
```

Listing 1.1. Sketch of a very simple GF grammar to talk about mathematics.

Parsing is done with the **Grammatical Framework** (GF) [Ran11], which is a powerful tool for the development of natural-language grammars. A GF grammar consists of an **abstract syntax** that describes the parse trees and (possibly multiple) **concrete syntaxes** that describe how these parse trees correspond to strings in a particular language. Listing 1.1 sketches an example GF grammar that can parse sentences like *"every integer is even"*[1]. The abstract syntax introduces categories (node types) and function constants that describe how nodes can be combined. E.g. `state` combines a term and a property into a statement. The sentence *"every integer is even"* thus corresponds to the expression `state (every integer) even`. For our simple example, the concrete syntax is very straight-forward, but in general the concrete syntax has to handle the complex morphology and syntax of natural language. GF supports this with a powerful type system and various mechanisms for modularity and reusability. GF also supplies the *Resource Grammar Library*, which provides re-usable implementations of the morphology and basic syntax for many (≥ 35) languages.

```
theory FOL : ur:?LF =                    view GrammarSemantics :
  propositions : type | # o |              ?Grammar -> ?DomainTheory =
  individuals : type | # ι |               Stmt = o |
  not : o → o | # ¬ 1 |                     Term = (ι → o) → o |
  and : o → o → o | # 1 ∧ 2 |              Notion = ι → o |
  // ... |                                  Prop = ι → o |
  forall : (ι→o) → o | # ∀ 1 |
  exists : (ι→o) → o | # ∃ 1 |             state = [term,pr] term pr |
                                           every = [notion] [p]
                                                 ∀ [x] notion x ⇒ p x |
theory DomainTheory : ?FOL =               // ... |
  integer : ι → o |                        integer = integer |
  even : ι → o |                           even = even |
  derivative : ι → ι |                     derivative = [term] [p]
  // ... |                                        term ([x]p(derivative x))|
```

Listing 1.2. Example logic, domain theory, and semantics construction in MMT.

The **semantics construction** describes how the parse trees are translated into logical expressions. GLF uses the **Meta Meta Tool** (MMT) for the logic development and semantics construction. MMT is a foundation-independent framework for knowledge representation [Uni]. In MMT, knowledge is represented as **theories**, which contain sequences of constant declarations of the form

---

[1] Note that neither parsing nor the semantics construction are concerned with the validity of a statement.

```
CONSTANT [: TYPE] [ | = DEFINITION] [ | # NOTATION] |
```

While MMT itself is foundation independent, MMT theories are typically based on the Edinburgh Logical Framework (LF) [HHP93] and extensions of that in practice. Listing 1.2 contains a theory FOL that defines the syntax of first-order logic. First, we need types for propositions and individuals, denoted by o and ι respectively. Afterwards, we can declare logical connectives as binary/ternary operators on propositions with the expected prefix/infix notations. For quantifiers, we use higher-order abstract syntax. Together with a domain theory, this allows us to express the meaning of our example sentence as $\forall x(\text{integer}(x) \Rightarrow \text{even}(x))$.

Before we can define the semantics construction, we need to be able to represent GF parse trees as MMT terms. For this, GLF creates a **language theory** from the abstract syntax, i.e. an MMT theory that contains the GF categories as type constants and the GF functions as function constants. Then, we can define the semantics construction as an MMT **view** from the language theory into the domain theory (see Listing 1.2). A view maps every constant in the source theory to a term in the target theory – e.g. statements are mapped to propositions (o) and properties to unary predicates ($\iota \to o$). A term like *"every integer"* should have the meaning $\lambda p.\forall x(\text{integer}(x) \Rightarrow p(x))$, i.e. we apply properties to terms, not the other way around. Note that $\lambda x.M$ is denoted in MMT by [x] M.

With all this in place, we can parse the sentence *"every integer is even"* to obtain the parse tree state (every integer) even, and then apply the semantics construction to obtain the MMT expression

([term,prop] term prop) (([n] [p] $\forall$ [x] n x $\Rightarrow$ p x) integer) even, which β-reduces to the desired $\forall$ [x] integer x $\Rightarrow$ even x.

Given the declarative treatment of semantics construction and target logic, GLF can serve as the basis for a rapid prototyping system for the development and implementation of controlled (mathematical) languages. To complete that, we need a good user interface (the extended GF shell that GLF comes with does not qualify).

3   Jupyter Integration

Jupyter [Jup] provides a user environment for working with notebooks, which can contain code cells, explanatory text and interactive widgets. The code cells can be executed in-document, resulting in a very interactive experience. We developed a new Jupyter kernel to bring these features to GLF.

The code cells in a GLF notebook either enrich the language context or contain executable commands. The **language context** consists of GF grammars and MMT theories and views. A user can explore and test the language context with **commands** for e.g. parsing a sentence with the specified grammar and applying the semantics construction.

When a code cell is executed, the first step is to identify its content type. We use simple pattern matching for this. If the code cell extends the language context, we write its content to a file. The file name is simply the name of the grammar/theory/view, which is also extracted during the pattern matching. Afterwards, grammars are imported into GF and MMT (for the language theory) and theories and views are imported into MMT.

For this, GF and MMT are running as subprocesses in the background and the GLF kernel communicates with them via pipes and HTTP respectively. The user gets feedback whether the imports succeeded along with possible error messages.

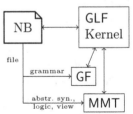

If a code cell contains commands, on the other hand, they are executed and the output is returned to the user. As GF is an integral part of the system, the GLF kernel supports all of the GF shell commands by simply passing them on to the GF shell. On top of that, we have added a number of kernel commands peculiar to GLF. Some of them are stand-alone commands such as for specifying where the GF and MMT files should be stored. Other commands are intended to be used in combination with GF commands. In the GF shell, commands can be combined with the pipe operator |. For example, one might want to use the parse command to obtain the parse tree of an English sentence and then use the linearize command to transform the parse tree into e.g. an Italian sentence. The Jupyter kernel imitates this behaviour, i.e. if the user enters a command of the form a | b, the output of command a is used as input for command b. By imitating this behaviour, rather than letting the GF shell handle the piping, we can add kernel commands that can also be used in combination with pipes, which lets them blend in more naturally. The construct command takes a parse tree as argument and sends a semantics construction request to MMT. Therefore, it is commonly used in combination with the parse command. The show command can be used for in-notebook visualization of parse trees (see Fig. 1). It is usually used in combination with certain GF commands that generate graph descriptions in the .dot format. The show command then uses GraphViz to generate images that are displayed in a widget. If there are multiple parse trees (e.g. due to ambiguity), a drop-down menu is created, where the user can select which parse tree to display.

The GLF kernel provides a number of convenience features. Syntax highlighting is based on CodeMirror. In JupyterLab, which has been used for the screenshots in this paper, syntax highlighting is provided by an extension. The syntax highlighting also depends on the content type of the cell, i.e. different rules are used for GF content, MMT content and commands.

Other features are built around tab-completion. MMT theories often use unicode characters for notations. The Jupyter kernel has a list of (currently 426) character sequences that can be tab-completed into unicode characters This is based on a similar list used in other MMT services. The character sequences are inspired by LATEX macros, which most users should be familiar with. For example, \subseteq is completed to ⊆.

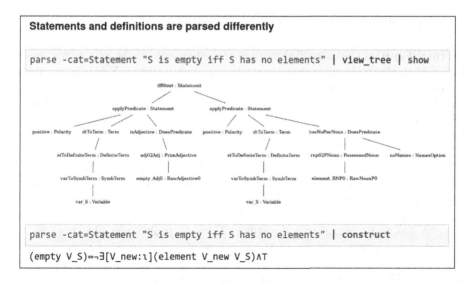

Fig. 1. Fragment of a GLForTheL notebook.

Tab-completion is also used for stub generation. A common workflow when writing a GF grammar is to first write the abstract syntax and then implement (possibly multiple) concrete syntaxes. A concrete syntax simply defines a linearization for all symbols introduced in the abstract syntax. The GLF kernel comes with a small script that can parse GF's abstract syntaxes—as long as they only contain commonly used features—and then generate a stub for the concrete syntaxes. This allows the user to fill out the concrete syntax without having to repeatedly scroll back to the abstract syntax to copy all the symbol names. Similarly, views for the semantics construction have to map every symbol in the abstract syntax to a logical expression, so stubs are generated for that as well (Fig. 2). Experience has shown that it is beneficial to add the types of function constants as comments in the generated stubs.

Smaller GLF pipelines like the example in Sect. 2 can be conveniently implemented and tested in Jupyter notebooks. Of course, larger projects (like the GLForTheL project discussed in the next section) are usually not implemented inside notebooks, but rather in text editors and IDEs. GLF pipelines inherit the modularity of GF grammars and MMT pipelines, and the GLF kernel can use grammars/theories/ views defined elsewhere. This way, notebooks can be used for testing and documenting the pipeline, as well as for interactively exploring specific problems during the course of the project.

```
view GrammarSemantics : http://mathhi
  Stmt = _ |
  Term = _ |
  Notion = _ |
  Prop = _ |

  // state : Term → Prop → Stmt |
  state = _ |
  // every : Notion → Stmt |
  every = _ |
  // integer : Notion |
  integer = _ |
  // even : Prop |
```

Fig. 2. Generated stub.

4   Case Study: **GLForTheL**

ForTheL [Pas07] is the controlled mathematical language of the System for Automated Deduction. We have recently started to experiment with re-implementing ForTheL in GLF (we call the result GLForTheL). This can serve as a case study that highlights both the capabilities of GLF and the role of Jupyter notebooks during development. In the following paragraphs, we will discuss some of the challenges we encountered during the implementation of GLForTheL.

Binding Variables. Our running example already covers quantification in natural language (*"every integer"*). However, in mathematical language this is further complicated by the use of variables, as exemplified in this ForTheL statement:
    "there is an integer" N *"that is even"*
Here, N has to be bound to a quantifier. This is a problem in GLF, because N is treated as a constant during the semantics construction. Our current workaround in GLForTheL is a special λ binder that turns the bound constant into a variable. For example, the semantics construction might map the statement above to $\exists(\lambda' N. (\text{int}(N) \wedge \text{even}(N)))$, where λ' and N are simple (function) constants. A post-processing step transforms the "bound constant" N into a variable (including all its occurrences in the body) and replaces λ' by a real λ.

Redundancy in Logical Expressions. The handling of variable sequences in GLForTheL results in some artefacts in the logical expressions. An example are trailing $\wedge true$ in some statements. While simple artefacts like this could be removed by MMT, there are also more complicated redundancies: GLForTheL translates the statement *"there are sets X, Y such that every element of X is an element of Y"* into the expression

$$\exists Y(\exists X(\text{set}(Y) \wedge \forall n(n \in X \Rightarrow n \in Y) \quad \wedge \quad \text{set}(X) \wedge \forall n(n \in X \Rightarrow n \in Y)))$$

because both X and Y are *"sets such that every. . . "*. We are currently working on an extension of GLF that adds an inference step to the pipeline (see [SK20]), which could—among other things—be used to implement advanced simplification algorithms.

Lexicon Management. Adding another word to the grammar usually requires new entries to the abstract syntax, concrete syntax, domain theory, and semantics construction. To simplify this, we have developed a tool that automatically creates these entries from a custom lexicon file. Especially in mathematics, though, there is another problem: new words and notations are introduced whenever needed, so the lexicon is growing while a document is processed. This problem is currently unsolved in GLF, which relies on a pre-defined lexicon. In the context of controlled languages, it may be argued that a document should have a preamble defining the necessary lexicon. Another solution could be a two-pass process, where a document-specific lexicon could be generated in a pre-processing step that harvests the definienda.

Better Target Logic. DRT Discourse Representation Theory (DRT) [KR93] solves various problems that arise from using first-order logic as the target representation in compositional natural-language semantics by introducing discourse representation structures as an intermediate representation, which can be compiled into first-order logic. Variants of DRT have been repeatedly used in the context of mathematical language (e.g. [Cra+10]).

Neither ForTheL nor GLForTheL use DRT. One of the consequences is that statements like

$$\textit{if the square of some integer } N \textit{ is even then } N \textit{ is even}$$

get translated into

$$(\exists v(\mathrm{int}(v) \wedge \mathrm{even}(\mathrm{square}(v)))) \Rightarrow \mathrm{even}(N),$$

because *"some"* usually means that the expression is existentially quantified (think e.g. of "H is contained in some ball $B \subseteq U$"). Of course, one would expect to get

$$\forall v((\mathrm{int}(v) \wedge \mathrm{even}(\mathrm{square}(v))) \Rightarrow \mathrm{even}(v)).$$

To remedy this in the future, we are currently looking into different ways of representing DRT in MMT.

From Sentences to Discourse. GLF operates on the sentence level. This becomes a problem when variables are introduced in one sentence (*"let G be a group"*) and then used in another sentence. Our implementation extracts the restrictions ($G : group$) and keeps G as a free variable in the following sentences. In general, our goal is to extend GLF with an inference step (as mentioned above), which could be used to combine this information. It would also allow experimentation with other discourse-level challenges such as anaphor resolution.

5   Conclusion and Evaluation

We have introduced a new Jupyter front-end for GLF; together they form a rapid prototyping system for controlled mathematical languages.

So far, it has been mostly used in a one-semester course on logic-based natural language processing [LBS20] at FAU Erlangen-Nürnberg. In previous years, we connected GF and MMT via some rather fragile Scala code, which was so inconvenient that we mostly used GF and MMT independently. Last semester we had the first iteration of the lecture using GLF and Jupyter. In the lab sessions (half of the course), we explored different natural-language semantics phenomena by implementing the language-to-logic pipeline in GLF. Since the class was rather small, we basically asked the students to tell us what to enter into the notebook and could immediately test the ideas. After some cleanup, the notebooks could be shared with students—much more easily than the messy file collections we had in previous years. The modularity of GLF also allowed us to try different semantics constructions for the same grammar. We also used Jupyter notebooks for homework assignments, providing partial implementations where the students had to implement the critical parts. The students also had the option to implement the homework without Jupyter notebooks (using command line tools for testing), but almost all students chose to use Jupyter notebooks. One of the biggest challenges was the installation on students' computers. This was further complicated by the need for updates throughout the semester as we improved the implementation. Going forward, we are planning to provide Docker images and online notebooks that can be used instead.

The newly reached maturity of GLF allowed us to return to our original motivation to apply it to mathematical language. As a larger case study, we have discussed our attempts to implement GLForTheL, a variant of ForTheL, which was our first larger project. Since the goal of GLForTheL is to imitate ForTheL, our experimentation was primarily focused on the ways technical challenges can be handled. While Jupyter notebooks were the go-to tool for these experiments, they were less useful for implementing the actual GLForTheL project, since it was much larger (currently 39 different node types and over 50 production rules). GLForTheL imposes tighter restrictions on the input language than ForTheL, rejecting ungrammatical statements like *"S are a sets"*, which are accepted by ForTheL. For a long time we supported both a German and an English concrete syntax. All this was possible without much effort, due to GF's powerful grammar mechanisms.

At its current state, our GLForTheL re-implementation can translate two example files from the SAD repository (excluding proofs). Since GLForTheL requires well-formed English sentences, it will never have the same coverage as ForTheL, but this was not our goal after all. One of the more complex sentences GLForTheL can currently handle is the definition *"a subset of S is a set T such that every element of T belongs to S"*, which results (after some α-renaming for readability) in

$$\forall T.(subsetof\ T\ S) \Leftrightarrow (set\ T) \wedge \forall x.(elementof\ x\ T) \wedge \top \Rightarrow (belongto\ x\ S) \wedge \top.$$

Expanding the coverage to more examples mostly boils down to extending the lexicon and adding the occasional grammatical rule. However, more work is needed to handle binary relations imposed on variable sequences as in *"let x, y, z be pairwise linearly independent vectors"*.

Our GLForTheL case study also indicated the need for a processing step after the semantics construction. [SK20] describes an extension of GLF that adds an inference component, which can be used for e.g. simplification, ambiguity resolution or theorem proving. Additional work is needed for lexicon management and regression testing.

Overall, we believe our experiences with GLForTheL confirm our hypothesis that GLF+Jupyter provide a flexible framework for the quick prototyping of controlled mathematical languages. The Jupyter kernel along with a link to an online version can be found at [GLFa] and the GLForTheL code at [GLFb].

References

[Cra+10] Cramer, M., Fisseni, B., Koepke, P., Kühlwein, D., Schröder, B., Veldman, J.: The naproche project controlled natural language proof checking of mathematical texts. In: Fuchs, N.E. (ed.) CNL 2009. LNCS (LNAI), vol. 5972, pp. 170–186. Springer, Heidelberg (2010). https://doi.org/10.1007/978-3-642-14418-9_11

[Cra13] Cramer, M.: Proof-checking mathematical texts in controlled natural language. eng. Ph.D. thesis. Rheinische Friedrich-Wilhelms-Universität Bonn (2013). http://hss.ulb.uni-bonn.de/2013/3390/3390.pdf

[FK19] Frerix, S., Koepke, P.: Making set theory great again. Talk at the Artificial Intelligence and Theorem Proving Conference (2019). http://aitp-conference.org/2019/slides/PK.pdf

[FSS98] Fuchs, N.E., Schwertel, U., Schwitter, R.: Attempto controlled english—not just another logic specification language. In: Flener, P. (ed.) LOPSTR 1998. LNCS, vol. 1559, pp. 1–20. Springer, Heidelberg (1999). https://doi.org/10.1007/3-540-48958-4_1

[GLFa] GLF Kernel. https://github.com/kaiamann/glf_kernel. Visited on 27 Mar 2020

[GLFb] GLForTheL – implementing ForTheL in GLF. https://gl.mathhub.info/comma/glforthel. Visited on 27 Mar 2020

[Hal19] Hales, T.: An argument for controlled natural languages in mathematics (2019). https://jiggerwit.wordpress.com/2019/06/20/an-argument-for-controlled-natural-languages-in-mathematics/. Visited on 9 May 2020

[HHP93] Harper, R., Honsell, F., Plotkin, G.: A framework for defining logics. J. Assoc. Comput. Mach. **40**(1), 143–184 (1993)

[Jup] Project Jupyter. http://www.jupyter.org. Visited on 22 Aug 2017

[KR93] Kamp, H., Reyle, U.: From Discourse to Logic: Introduction to Model-Theoretic Semantics of Natural Language, Formal Logic and Dis-course Representation Theory. Kluwer, Dordrecht (1993)

[KS19] Kohlhase, M., Schaefer, J.F.: GF + MMT = GLF – from language to semantics through LF. In: Miller, D., Scagnetto, I. (eds.) Proceedings of the Fourteenth Workshop on Logical Frameworks and Meta-Languages: Theory and Practice, LFMTP 2019, vol. 307. Electronic Proceedings in Theoretical Computer Science (EPTCS), pp. 24–39 (2019). https://doi.org/10.4204/EPTCS.307.4

[LBS20] Kohlhase, M.: Logic-Based Natural Language Processing: Lecture Notes (2020). http://kwarc.info/teaching/LBS/notes.pdf. Visited on 7 May 2020

[Pas07] Paskevich, A.: The syntax and semantics of the ForTheL language (2007)

[Ran11] Ranta, A.: Grammatical framework: programming with multilingual grammars. CSLI Publications, Stanford (2011). ISBN-10: 1-57586-626-9 (Paper), 1-57586-627-7 (Cloth)

[SK20] Schaefer, J.F., Kohlhase, M.: The GLIF system: a framework for inference-based natural-language understanding (2020, submitted). http://kwarc.info/kohlhase/papers/cicm20-glif.pdf

[Uni] MMT – Language and System for the Uniform Representation of Knowledge. https://uniformal.github.io/

General Session

Method to Create Multiple Choice Exercises for Computer Algebra System

Tatsuyoshi Hamada[1]([⊠])[iD], Yoshiyuki Nakagawa[2][iD], and Makoto Tamura[3][iD]

[1] Nihon University, Kameino, Fujisawa, Kanagawa 252-0880, Japan
`hamada.tatsuyoshi@nihon-u.ac.jp`
[2] Ryukoku University, Fukakusa, Fushimi-ku, Kyoto 612-8577, Japan
`nakagawa@mail.ryukoku.ac.jp`
[3] Osaka Sangyo University, Nakagaito, Daito, Osaka 574-8530, Japan
`mtamura@edo.osaka-sandai.ac.jp`

Abstract. When studying mathematics, it is important to solve exercises of an appropriate level. Recently, web-based assessment systems with a computer algebra system (CAS), e.g., Moodle with Stack and the Möbius platform with Maple, have become popular. Such web-based systems are convenient; however, they have some problems relative to inputting and evaluating mathematical formulas. In addition, when considering and solving mathematical problems, handwriting mathematics is important. We want management system of paper-oriented exercises. Auto multiple choice (AMC), which was developed by Alexis Bienvenüe, is open source software for creating and managing multiple choice questionnaires with automated marking. LaTeX is the native AMC language for questionnaire descriptions. We propose to combine AMC and CAS using LuaTeX, which is a TeX-based computer typesetting system with an embedded Lua scripting engine. We can embed CAS scripts into LuaTeX source, and, by creating exercises with CAS, we can generate various problems with random coefficients or terms. By providing various patterns of practice problems and facilitating discussions with each student, we expect sufficient educational benefits of providing opportunities to communicate about mathematical concepts and algorithms among students.

Keywords: CAS · Multiple choice exercises · Calculus

1 Introduction

Exercises are important when learning mathematics, especially if we really want to understand something. We can find many exercises of various levels in textbooks. However, solving exercises in textbooks can be difficult for some students who are not good at calculations. Teachers must prepare and mark various levels of exercises for students. Feedback to students after evaluation should be given quickly; however, for many students, evaluation is very hard work for teachers.

This work was supported by KAKENHI 16K05139. The authors would like to thank Enago (www.enago.jp) for the English language review.

Recently, web-based assessment systems that use a computer algebra system (CAS), e.g., Moodle with Stack [1], which uses the Maxima [2] open source computer algebra system, and the Möbius platform [3] with Maple [4], have become popular. Creating exercises for web-based educational system has been investigated by Yoshitomi [5], where he attempted to generate question data for linear algebra. However, with web-based systems, there are problems related to inputting and evaluating mathematical formulas. Some studies have investigated mathematical input methods, e.g., [6–9]. However, there is no standard method for typical usage. In addition, handwritten work is important when considering and solving mathematical problems.

We would like to exploit the power of CAS to create mathematical exercises with auto multiple choice (AMC), which was developed by Alexis Bienvenüe [10]. AMC is open source software for creating and managing multiple choice questionnaires with automated marking. LaTeX is the native AMC language for questionnaire descriptions. The potential of AMC has been examined by many researchers and educators, e.g., [11] and [12]. Milana Lima dos Santos et al. presented results from the application of examination papers for engineering courses developed using Matlab/Octave scripts and the AMC, [13]. The first and second author presented the applications of AMC at the workshop in Japan, [19].

We use LuaTeX [14] for AMC. LuaTeX is an extended version of pdfTeX that uses Lua as an embedded scripting language. Lua [15] is a powerful and lightweight scripting language, that supports procedural programming and data descriptions. Using Lua's "table" data structure, we can implement arrays, records, lists, queues and sets efficiently.

In addition, we can embed Maxima scripts into LuaTeX source. Maxima is the most famous general purpose open source computer algebra system (cf. [16]). By creating exercises using Maxima, we can generate various calculus problems with random coefficients or terms. By offering various practice problem pattern and facilitating discussions for each student, we expect sufficient educational effects of providing opportunities to communicate about mathematical concepts and algorithms among students.

2   AMC and CAS

AMC is a set of utilities that use of multiple choice questionnaires written in plain text or LaTeX, as well automated correction and grading from scans of answer sheets using optical mark recognition. The following software packages are required to use AMC: LaTeX, the ImageMagick image processing libraries, OpenCV, and Perl with Gtk2-Perl and Glade::XML for the graphical user interface. AMC has effective documentation, the developers of AMC operate a community support website and French and English user forums.

For users who are not ready to use LaTeX, AMC includes a filter to process simple plain text files in AMC-TXT format, which is a type of markdown script. However, we embed CAS script in LaTeX code (we do not use AMC-TXT). AMC can create questions and answers in random order for each sheet, and we can distribute unique exercises for each student. Mathematical questions with randomized statements are supported by the original AMC with the fp

LaTeX package or LuaTeX. However, these are for random coefficients of fixed or floating point numbers, (not rational numbers), and they do not support mathematical symbolic formulations. Thus we must create naturally symbolic formulated exercises with rational numbers. Initially, it may a little time to create problems, because of the incomprehensible error messages of LaTeX and CAS. However, we can recycle sets of questions easily. Using automated grading and creating answers, teacher's time is much saved. We found that students are less likely to give up when using the proposed approach, in fact, even after the lecture, many students didn't give up for solving problems.

In the following, we describe a simple example. In this source code, we changed the mark box to an oval using the `\AMCboxDimensions` LaTeX command, and the script of code acquisition was omitted. Code acquisition can be performed easily using the `\AMCcodeGridInt[options]{key}{n}` LaTeX command, e.g., to allow each student to enter her/his student id on an answer sheet.

The `\onecopy` LaTeX command produces as many distinct realizations of the test as desired (10 in our case).

```
\documentclass[a4paper]{article}
\usepackage[box,completemulti,lang=EN]{automultiplechoice}
\usepackage{luacode, tikz}
\newcommand*{\var}[1]{\luaexec{tex.print(#1)}}
\begin{document}
\AMCboxDimensions{shape=oval,width=1.8ex,height=2.5ex}
\AMCcodeVspace=0em
\luaexec{math.randomseed(20200713)}
\begin{luacode*}
function execMaxima(cmd)
   local texcmd="echo 'tex1("..cmd..");'|maxima --very-quiet"
   local hdl=io.popen(texcmd, "r")
   local content=string.gsub(hdl:read("*all"), "\n", "")
   hdl:close()
   return content
end
\end{luacode*}
\onecopy{10}{
   %%% start of the header
   % Code acquisition script
   %%% end of the header
   %%% start of the questions
   \begin{question}{1st_question}
   ...
   \end{question}
   \begin{question}{2nd_question}
   ...
   \end{question}
   ...
   %%% end of the questions
}
\end{document}
```

We used Maxima in LuaTeX to create exercises. LuaTeX supports Lua as embedded scripting. The io.popen() function of Lua is used to execute command line arguments. Note that this function is system dependent and is not available on all platforms. We used MathLibre [17] or the Debian GNU/Linux 10.3 "buster" release. One of the authors (Y. Nakagawa) is currently checking Lua scripting under Microsoft Windows (with some modifications). LuaTeX supports the \directlua command, which can sometimes be tricky. When executing Lua code within TeX using the \directlua command, there is no easy way to use the percent character, and counting backslashes can be difficult. We used luacode environments and the \luaexec command from the luacode package [18]. We set the default LaTeX engine using "lualatex --shell-escape" in AMC preferences to execute CAS script[1].

First, we call the environment luacode in our TeX source code.

```
\usepackage{luacode}
```

The following TeX command is used to evaluate variables as TeX output.

```
\newcommand*{\var}[1]{\luaexec{tex.print(#1)}}
```

To fix the random seed to obtain the same results across different typesetting, we must set random seed after the \begin{document}.

```
\luaexec{math.randomseed(20200713)}
```

We define the new execMaxima() function in luacode*. In this script, cmd is the argument for an arbitrary Maxima script.

```
\begin{luacode*}
function execMaxima(cmd)
  local texcmd="echo 'tex1("..cmd..");'|maxima --very-quiet"
  local hdl=io.popen(texcmd, "r")
  local content=string.gsub(hdl:read("*all"), "\n", "")
  hdl:close()
  return content
end
\end{luacode*}
```

We used Maxima in the article. However many of CASs support LaTeX formatting conventions; thus you can use your preferred CAS, e.g., Sage, Maple, or Mathematica.

```
texcmd="sage -c 'print(latex("..cmd.."))'"
```

```
texcmd="echo 'latex("..cmd..");' | maple -q"
```

```
texcmd="wolframscript -code 'TeXForm["..cmd.."]'"
```

---

[1] Enabling "shell-escape" by default is dangerous, because it makes the LaTeX binary execute arbitrary shell commands in LaTeX files.

3   Example

The following example is a partial differentiation exercise. Here, coefficients a, b, s, t in this script are random numbers created using the Lua `math.random()` function. The string concatenation operator in Lua is denoted by two dots ('..'), the `exp` function represents an exponential function and `diff(expr, x)` returns the derivative of *expr* relative to variable x in Maxima.

```
\begin{question}{pdiff01}
\luaexec{
  a=math.random(2, 9);
  b=math.random(2, 9);
  s=(-1)^(math.random(0, 1));
  t=(-1)^(math.random(0, 1));
  g=s*a..'*x+'..t*b..'*y';
  g1=s*(a-1)..'*x+'..t*b..'*y';
  g2=s*a..'*x+'..t*(b-1)..'*y';
  f='exp('..g..')';
  f1='exp('..g1..')';
  f2='exp('..g2..')';
  formula=execMaxima(f);
  correct1=execMaxima('diff('..f..', x)');
  wrong1=execMaxima('diff('..f1..', x)');
  wrong2=execMaxima('diff('..f2..', x)');
  wrong3=execMaxima('diff('..f..', x)/'..a*s);
  wrong4=execMaxima('diff('..f1..', x)/'..a*s);
}

Find \(f_{x}\) where \(f(x, y)=\var{formula}\).

  \begin{choiceshoriz}
    \correctchoice{\(\var{correct1}\)}
    \wrongchoice{\(\var{wrong1}\)}
    \wrongchoice{\(\var{wrong2}\)}
    \wrongchoice{\(\var{wrong3}\)}
    \wrongchoice{\(\var{wrong4}\)}
  \end{choiceshoriz}
\end{question}
```

The result of upper script is here (Fig. 1).

1   Find f_x where $f(x, y) = e^{7x-4y}$.

○ $\frac{6e^{6x-4y}}{7}$ ○ $7e^{7x-4y}$ ○ e^{7x-4y} ○ $6e^{6x-4y}$ ○ $7e^{7x-3y}$

Fig. 1. A question for partial differentiation

4   Conclusions

This paper has proposed an automated approach to create several patterns of calculus exercises for different students at the undergraduate level. The primary advantages of the proposed approach are summarized as follows:

- Initially, it may take a little time to create problems using this system. However with random coefficients and the CAS, we can recycle sets of questions, easily.
- The proposed approach provides different types of problems for different student.
- The proposed approach facilitates discussions about different problems for each student, and it provides opportunities to communicate about mathematical concepts and algorithms.
- We found that students are less likely to give up when using the proposed approach.

We believe that automatically creating mathematical exercises at appropriate levels is significant challenge in mathematical software research.

The Scripts demonstrated in this paper are available at the GitHub public repository [20].

References

1. STACK—The University of Edinburgh. https://www.ed.ac.uk/maths/stack/. Accessed 27 Mar 2020
2. Maxima, A Computer Algebra System. http://maxima.sourceforge.net/. Accessed 27 Mar 2020
3. Möbius platform. https://www.digitaled.com/platform. Accessed 27 Mar 2020
4. Maple. https://www.maplesoft.com/. Accessed 27 Mar 2020
5. Yoshitomi, K.: Generation of abundant multi-choice or STACK type questions using CAS for random assignments. In: Davenport, J.H., Kauers, M., Labahn, G., Urban, J. (eds.) ICMS 2018. LNCS, vol. 10931, pp. 492–497. Springer, Cham (2018). https://doi.org/10.1007/978-3-319-96418-8_58
6. Shirai, S., et al.: Intelligent editor for authoring educational materials in mathematics e-learning systems. In: Davenport, J.H., Kauers, M., Labahn, G., Urban, J. (eds.) ICMS 2018. LNCS, vol. 10931, pp. 431–437. Springer, Cham (2018). https://doi.org/10.1007/978-3-319-96418-8_51
7. Fujimoto, M.: An implementation method of a CAS with a handwriting interface on tablet devices. In: Hong, H., Yap, C. (eds.) ICMS 2014. LNCS, vol. 8592, pp. 545–548. Springer, Heidelberg (2014). https://doi.org/10.1007/978-3-662-44199-2_82
8. Su, W., Wang, P.S., Li, L.: A touch-based mathematical expression editor. In: Hong, H., Yap, C. (eds.) ICMS 2014. LNCS, vol. 8592, pp. 635–640. Springer, Heidelberg (2014). https://doi.org/10.1007/978-3-662-44199-2_95
9. EquateIO. https://equatio.texthelp.com/. Accessed 27 Mar 2020
10. Auto Multiple Choice. https://www.auto-multiple-choice.net/. Accessed 27 Mar 2020

11. Siclet, C.: Teaching of mathematics thanks to multiple choice questionnaires. In: 27th EAEEIE Annual Conference (EAEEIE) (2017)
12. Clancy, I., Marcus-Quinn, A.: Exploring the possibilities of automated feedback for third level students/Esplorare le possibilita di feedback automatizzato per gli studenti universitari. Form@re **19**(3), 247–256 (2019). Gale Academic OneFile. Accessed 6 May 2020
13. dos Santos, M.L., Schmidt, H.P., Manassero, G., Pellini, E.L.: Automated design for engineering student examinations using Matlab/Octave scripts and the auto multiple choice package. In: 2019 IEEE World Conference on Engineering Education (EDUNINE), Lima, Peru, pp. 1–5 (2019)
14. LuaTEX. http://www.luatex.org/. Accessed 27 Mar 2020
15. Lua. https://www.lua.org/. Accessed 27 Mar 2020
16. Joyner, D.: OSCAS: maxima. ACM Commun. Comput. Algebra **40**(3/4), 108–111 (2006)
17. MathLibre. http://www.mathlibre.org/. Accessed 27 Mar 2020
18. Luacode. https://ctan.org/pkg/luacode. Accessed 27 Mar 2020
19. Hamada, T., Nakagawa, Y.: An application of CAS for creating multiple choice exercises. Suuriken-Kokyuroku **2142**, 108–116 (2019)
20. Hamada, T.: mc-calculus. https://github.com/knxm/mc-calculus. Accessed 8 May 2020

A Flow-Based Programming Environment for Geometrical Construction

Kento Nakamura[✉][iD] and Kazushi Ahara[iD]

Meiji University, Tokyo, Japan
`somaarcr@gmail.com, ahara@meiji.ac.jp`
`https://soma-arc.net`
`https://aharalab.sakura.ne.jp`

Abstract. In this article, we show a flow-based programming environment for interactive geometry software. Flow-based programming is one of the programming paradigms. All of the processes and data are represented as nodes, and we connect processes and data with edges. We call the figure with nodes and edges *graph* because the figure looks like a planar graph.

There is a lot of software implementing flow-based programming. However, there are few mathematical software based on a flow-based programming environment. So, we develop experimental interactive geometry software to generate kaleidoscope patterns based on flow-based programming.

The software shows us some advantages of flow-based programming. First, it is easy to understand the procedure of construction. Second, flow-based programming is flexible. Third, flow-based programming has high extensibility. We seek possibilities of practical use of the geometrical construction software with flow-based programming.

Keywords: Flow-based programming · Interactive geometry software · Kaleidoscope patterns

1 Introduction

In this article, we show interactive geometry software with flow-based programming. Interactive geometry software is software that creates geometrical constructions, and we can manipulate geometrical objects by hand keeping relationships between the geometrical objects. For example, see Fig. 1. There are three points, and they form three lines. When we move one of the points, the lines are also moved according to the positions of points. One of the most famous interactive geometry software is GeoGebra.

Flow-based programming is one of the programming paradigms. See Fig. 2. All processes and data are represented as nodes. We connect sockets of nodes with edges. We also call the resulting figure *graph* because the figure looks like a planar graph. In Fig. 2, there are three Point nodes, and they are connected to the three LineTwoPoints nodes by edges. The data is sent left to right through the edges. The geometrical objects are rendered as in Fig. 1.

© Springer Nature Switzerland AG 2020
A. M. Bigatti et al. (Eds.): ICMS 2020, LNCS 12097, pp. 426–431, 2020.
https://doi.org/10.1007/978-3-030-52200-1_42

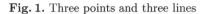

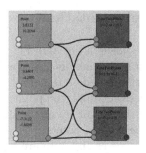

Fig. 1. Three points and three lines   **Fig. 2.** Nodes and edges

Flow-based programming is often used for creative coding or algorithmic design environments. For example, Max/MSP, PureData, vvvv, TouchDesigner, Grasshopper for Rhinoceros, and shader node graph editor implemented in three-dimensional modeling software like Blender or Maya. However, there is few mathematical software adopting a flow-based programming environment. We guess flow-based programming is suited for geometrical construction, that is, interactive geometry software. There are three advantages of flow-based programming. They are shown in the summary part.

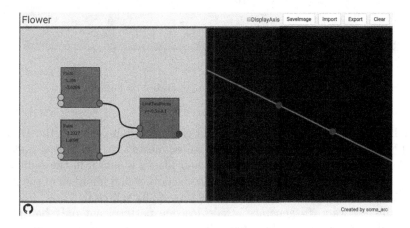

Fig. 3. Point and LineTwoPoints nodes generating a line

The first author is developing an experimental interactive geometry software based on flow-based programming. The software also can draw kaleidoscope patterns. The name of the software is *Flower*. Flower is web application developed by JavaScript. It is published on the web site of the first author[1]. Also, we have an introductory video on YouTube[2]. It introduces basic operations of Flower.

---

[1] https://soma-arc.net/Flower/.
[2] https://youtu.be/FWp-eF5gz5o.

2   Implementation

Flower is shown in Fig. 3. In Fig. 3, the left panel is a node graph editor. There are two Point nodes and one LineTwoPoints node. The right panel shows geometrical construction based on the graph. In Fig. 3, the Point nodes represent points, and a line passes through the points. They are rendered in the right panel. Also, we can move the point by mouse dragging. Then, the line is also moved according to the positions of points.

All processes and data are represented as nodes and edges. The flows of data are represented by edges. We can find a procedure of construction from the node graph.

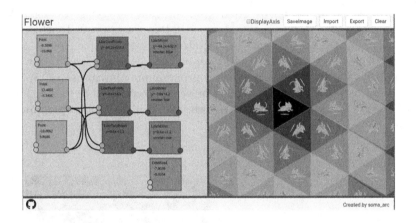

Fig. 4. Kaleidoscope pattern generated by half-planes

A LineMirror node receives a line and generates half-plane. In the graph of Fig. 4, all of the LineTwoPoints nodes are connected to LineMirror nodes. A kaleidoscope pattern generated by three half-planes is shown in the right panel of Fig. 4. Three half-planes are reflected in each other and filling all of the planes with triangles. Also, the image of a cat is repeatedly reflected.

In this software, to generate a kaleidoscope pattern, not only half-planes, but we also use reflections by circles. It is also called inversion in circles. The equation is as follows. Let C be the center of the circle and let R be the radius of the circle. Inversion map of the circle is $f(z) = \frac{R^2}{z-C} + C$ in the complex coordinate. Figure 5 shows the image of the cat and its inverted image. Circle inversion transforms a circle as a circle, but other images are inverted with distortion.

Figure 6 shows a kaleidoscope pattern generated by inversions in four circles. The cats and circles are transformed infinitely repeated reflections by circles. So, it may take much time to render Fig. 6 because the number of circles and cats exponentially increases in the process of the reflections. So, we use our original algorithm called *Iterated Inversion System (IIS)* to render the right panel in Fig. 6.

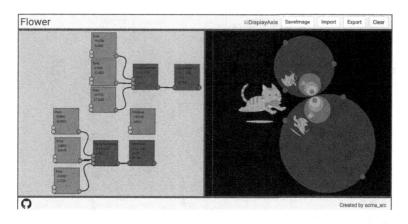

Fig. 5. Reflections of circles

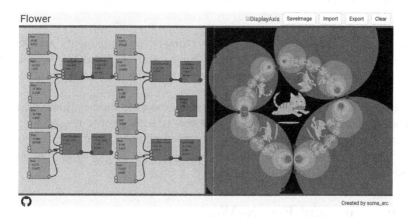

Fig. 6. Kaleidoscope pattern generated by circles

IIS is simple. For each point in the canvas, if the point is in the circle or half-plane, we apply a reflection of the geometrical object. We continue iterating the reflections until the point transformed outside of all of the objects. Finally, we determine color according to the number of reflections or refer to the pixel value of the iterated point. We can parallelize this algorithm easily. So, the figures are rendered in real-time when we use parallel processing. For more details, read [1] or [2].

We obtain this kind of fractal patterns in Fig. 7. It is generated by reflections of three half-planes and one circle. It is also rendered with IIS in real-time.

Next, we show technical details about Flower. The left panel is rendered by *Canvas 2D Context* of html5. On the other hand, We have to use parallel processing to render images of the right panel in real-time with IIS. So, Flower uses fragment shader of *OpenGL Shading Language (GLSL)* on WebGL. From graph

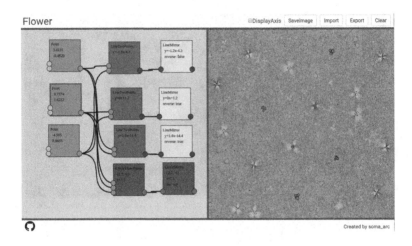

Fig. 7. Fractal patterns generated by inversions in half-planes and a circle

of the left panel, shader code to render geometrical components and kaleidoscope patterns is generated by template engine and compiled by WebGL. The positions of geometrical components are operated by uniform variables of the shader in the every frame.

Originally fragment shader is used to shade polygon model. However, in this case, we prepare rectangle composed of two triangles cover the screen. Then fragment shader determines the colors of screen pixel by pixel. For more details about this kind of shader graphics technique, read *The Book of Shaders*[3] by Patricio Gonzalez Vivo and Jen Lowe.

3   Summary and Future Work

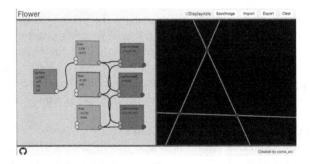

Fig. 8. Sin wave node connected to the y-coordinate of the Point node

---

[3] https://thebookofshaders.com/.

There are some advantages to adopting flow-based programming. Firstly, we can understand the procedure of constructions easily. We understand the relationships between mathematical components from the graph. Secondly, flow-based programming is flexible. We can insert processes wherever we like. For example, see Fig. 8. We can use a time-series data node such as SinWave node. It is connected to the y-coordinate socket of the Point node, and the point moves according to sin wave. Thirdly, flow-based programming has high extensibility. The users of the software can make their own node or script by their ideas, and add or extend functions of the software easily.

In this system, we only construct kaleidoscope patterns. Software with flow-based has prominent features. We aim to develop not only kaleidoscope pattern editor but also general-purpose geometrical construction software. We pursue the possibility of flow-based programming and interactive geometry software.

References

1. Nakamura, K., Ahara, K.: A new algorithm for rendering kissing Schottky groups. In: Proceedings of Bridges 2016: Mathematics, Music, Art, Architecture, Education, Culture, pp. 367–370. Tessellations Publishing, Phoenix (2016)
2. Nakamura, K., Ahara, K.: A geometrical representation and visualization of Möbius transformation groups. In: Proceedings of Bridges 2017: Mathematics, Music, Art, Architecture, Education, Culture, pp. 159–166. Tessellations Publishing, Phoenix (2017)

MORLAB – A Model Order Reduction Framework in MATLAB and Octave

Peter Benner[1,2] and Steffen W. R. Werner[1(✉)]

[1] Max Planck Institute for Dynamics of Complex Technical Systems,
Sandtorstr. 1, 39106 Magdeburg, Germany
{benner,werner}@mpi-magdeburg.mpg.de
[2] Faculty of Mathematics, Otto von Guericke University,
Universitätsplatz 2, 39106 Magdeburg, Germany
peter.benner@ovgu.de

Abstract. When synthesizing feedback controllers for large-scale dynamical systems, often a reduction of the plant model by model order reduction is required. This is a typical task in computer-aided control system design environments. Therefore, in the last years, model order reduction became an essential tool for the practical use of mathematical models in engineering processes. For the integration of established model order reduction methods into those processes, software solutions are needed. In this paper, we describe the MORLAB (Model Order Reduction LABoratory) toolbox as such a software solution in MathWorks MATLAB® and GNU Octave, and its featured integration into established software tools used in simulations and controller design. We give benchmark examples for two important extensions of the toolbox.

Keywords: Model order reduction · Dynamical systems · MATLAB · Octave

1 Introduction

Dynamical input-output systems are a usual way of modeling natural phenomena as, e.g., fluid dynamics, mechanical systems or the behavior of electrical circuits. In general, they are described by differential and algebraic equations

$$G : \begin{cases} 0 = f(x(t), Dx(t), \ldots, D^k x(t), u(t)), \\ y(t) = h(x(t), Dx(t), \ldots, D^k x(t), u(t)), \end{cases} \tag{1}$$

This work was supported by the German Research Foundation (DFG) Research Training Group 2297 "MathCoRe", Magdeburg.

where the inputs, $u(t) \in \mathbb{R}^m$, are used to influence the internal states, $x(t) \in \mathbb{R}^n$, to get the desired outputs, $y(t) \in \mathbb{R}^p$. Nowadays, dynamical systems (1) are an essential tool in simulation and the design of controllers. But due to the use of highly accurate models, involving a large number of differential equations in (1), the demand for computational resources, e.g., time and memory, is often too high for a practical usage. The aim of model order reduction is the construction of a surrogate model $\widehat{G}$, with a much smaller number of internal states, $\hat{x}(t) \in \mathbb{R}^r$, and differential equations, $r \ll n$, which approximates the input-to-output behavior of (1), i.e.,

$$\|y - \hat{y}\| \leq \text{tolerance} \cdot \|u\|,$$

for an appropriately defined norm and all admissible inputs u, where $\hat{y}$ denotes the outputs of the reduced-order model.

For the integration into established engineering processes, software solutions for the model reduction problem are needed. One such software solution, compatible with MathWorks MATLAB® and GNU Octave, is the **MORLAB** (Model Order Reduction **LABoratory**) toolbox [6,8]. As a free and open source software, the main aim of the toolbox is the model order reduction of linear, medium-scale dynamical systems.

In Sect. 2, we will briefly describe the fundamentals of the MORLAB toolbox and afterwards, in Sect. 3, provide the ideas behind the integration of MORLAB in other MATLAB and Octave software, also presenting some numerical examples.

2   The MORLAB Toolbox

The MORLAB toolbox originated in [2] as a repository of MATLAB codes for model order reduction of linear standard systems. The current version [6] comes with 10 different model reduction techniques for continuous- and discrete-time standard, descriptor and second-order systems. The implementation of the toolbox is based on spectral projection methods, like the matrix sign function [11] and the right matrix pencil disk function [3], which are used for the solution of underlying matrix equations. Therefore, the toolbox comes with many different solvers for different types of matrix equations, e.g., for continuous- and discrete-time algebraic Riccati or Lyapunov equations. To support the work with linear dynamical systems, the toolbox also implements a number of system-theoretic subroutines, e.g., the additive decomposition of dynamical systems, and evaluation tools for the time and frequency domain with support for the different implemented system structures.

An overview of the basic principles of the MORLAB toolbox can be found in [8] and is also given in the following:

Open source and free The toolbox is licensed under the GNU Affero General Public License v3.0 and is freely available on the project website and Zenodo.

Fast and accurate Based on the spectral projection methods, the toolbox can outperform other available model reduction and system-theoretic software.

Unified framework All model reduction routines share the same interface and allow for quick exchange and easy comparison between the methods.

Configurable All routines can be configured separately using option structs.

Modular Each subroutine can be called on its own by the user to be used and combined in various ways.

Portable No binary extensions are required, which allows for running the toolbox with bare MATLAB or Octave installations.

Documentation The toolbox comes with an extensive documentation for every function in HTML and MATLAB inline format.

Dependencies MATLAB ($\geq$ 2012b), Octave ($\geq$ 4.0.0).

See [8] for a more detailed discussion of the toolbox structure, implementational details and overviews about the currently supported system classes.

3   Toolbox Integration in MATLAB and Octave

An important point for the design decisions in MORLAB was the compatibility with other highly-used system-theoretic software tools. Two of the most important ones are Simulink® and the Control System Toolbox™ in MATLAB. In the following, we describe first the integration of MORLAB into those toolboxes using the state-space object and give then an overview of the extensions that MORLAB introduces itself.

The numerical experiments, used here for illustration, have been executed on a machine with 2 Intel(R) Xeon(R) Silver 4110 CPU processors running at 2.10 GHz and equipped with 192 GB total main memory. The computer runs on CentOS Linux release 7.5.1804 (Core) with MATLAB 9.7.0.1190202 (R2019b).

3.1   Standard First-Order Systems and the State-Space Object

The system class corresponding to state-space models as (1), which is fully supported in MATLAB, are standard linear time-invariant systems, e.g., in the continuous-time case

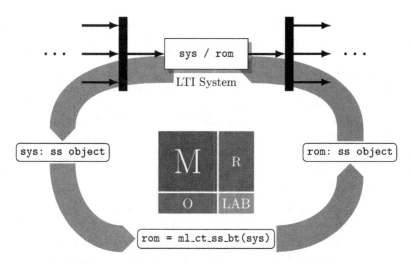

Fig. 1. Idea of integrating MORLAB into Simulink and the Control System Toolbox in MATLAB.

$$\dot{x}(t) = Ax(t) + Bu(t),$$
$$y(t) = Cx(t) + Du(t), \tag{2}$$

with $A \in \mathbb{R}^{n \times n}$, $B \in \mathbb{R}^{n \times m}$, $C \in \mathbb{R}^{p \times n}$ and $D \in \mathbb{R}^{p \times m}$. The way in MATLAB to efficiently deal with (2) are so-called *state-space model objects* (ss objects). Those objects are, in principle, a collection of the most important properties of (2), e.g., the matrices A, B, C, D, corresponding measured data, units of the state variables and so forth.

The Control System Toolbox is the main toolbox for dealing with dynamical systems in MATLAB. It provides a lot of important system-theoretic routines for the analysis and simulation of (2). The toolbox uses the ss object as its main interface. On the other hand, Simulink is a modeling toolkit for hierarchical system design and numerical experiments using graphical blocks. Complete dynamical processes are usually modeled here, especially embedding LTI system blocks, which correspond to (2) and use the ss object in its backend.

Now, the task is the integration of model order reduction into the computer-aided control system design workflow. Concerning the model order reduction of (2), the reduced-order surrogate model reads as

$$\hat{x}(t) = \widehat{A}\hat{x}(t) + \widehat{B}u(t),$$
$$\hat{y}(t) = \widehat{C}\hat{x}(t) + \widehat{D}u(t),$$

with $\widehat{A} \in \mathbb{R}^{r \times r}$, $\widehat{B} \in \mathbb{R}^{r \times m}$, $\widehat{C} \in \mathbb{R}^{p \times r}$ and $\widehat{D} \in \mathbb{R}^{p \times m}$, which belongs to the same system class as (2). Therefore, MORLAB provides consistent interfaces for the model order reduction routines, i.e., the reduced-order models will be of the same data type as the original models. Besides an interface for the single matrices of (2) and one using `structs`, MORLAB fully supports `ss` objects. An example of the resulting workflow is given in Fig. 1. Let `sys` be the original `ss` object, e.g., in a Simulink `LTI system` block. This object can be directly put into a MORLAB routine (here `ml_ct_ss_bt`), which produces a new `ss` object, namely `rom`. Due to the consistent data type of `rom`, the reduced-order model can now directly replace the original `sys` object for the acceleration of the surrounding process. The important point here is that the model reduction with MORLAB only concerns the `ss` object to be exchanged but the surrounding process is untouched. The example function used in Fig. 1 can be replaced by any model order reduction routine of choice from MORLAB. Overviews about supported methods for (2) can be found in [4, 8].

Remark 1. Parts of the Control System Toolbox are also available in Octave implemented in the 'control' package. There, the same interfaces and integration options are provided as for the MATLAB software. Due to the lack of system-theoretic subroutines in Octave, MORLAB additionally provides analysis and simulation tools that are compatible with MATLAB and Octave, which highly extend the functionality of Octave for dynamical systems in its current state.

Remark 2. Besides the continuous-time version (2), the `ss` object supports discrete-time linear time-invariant standard systems, which are also implemented in the MORLAB toolbox with appropriate model reduction routines.

It should be noted that also the default MATLAB toolboxes provide some model reduction routines for small-scale dense standard systems in the Control System Toolbox and the Robust Control Toolbox^{TM}. In comparison, MORLAB can be applied to way larger systems due to its efficient underlying spectral projection routines, which also often allow MORLAB to outperform the routines from the default MATLAB toolboxes by far. Also, the variety of the MOR-LAB model order reduction routines is so far not known to be matched by any other comparable model order reduction software for any of the provided system classes. Especially, the following two sections describe significant extensions of the MORLAB toolbox, which, in large parts, cannot be found in any other published model reduction software.

3.2   Extension for Descriptor Systems

The first extension of the standard MATLAB functionality for linear dynamical systems that was done in MORLAB are descriptor systems, e.g., in the continuous-time case

$$
\begin{aligned}
E\dot{x}(t) &= Ax(t) + Bu(t), \\
y(t) &= Cx(t) + Du(t),
\end{aligned}
\tag{3}
$$

with $E \in \mathbb{R}^{n \times n}$ and A, B, C, D as in (2). In principle, the `ss` object supports general E matrices but most of the routines in the system-theoretic toolboxes reformulate (3) as (2) by explicitly inverting the E matrix to the right. Beside

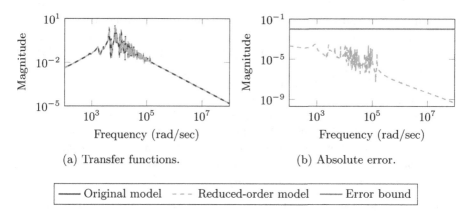

(a) Transfer functions.   (b) Absolute error.

———— Original model   - - - Reduced-order model   ———— Error bound

Fig. 2. Frequency response results for the PI-circuit example.

explicit inversion often has a negative influence on numerical accuracy, in case of differential-algebraic equations describing the system dynamics, the E matrix might not be invertible anymore. In that case, a lot of the routines in the Control System Toolbox and Simulink are unusable. But MORLAB supports this case in corresponding model reduction and system-theoretic routines by proper handling of possible algebraic system parts. The idea of model order reduction of descriptor systems with MORLAB can be found in [5] and the underlying additive decomposition of descriptor systems also in [8].

In many cases, the use of the MORLAB toolbox makes the integration of descriptor system models into other MATLAB software possible as we will see in the following example. We consider the model of an RLC circuit described by cascaded PI-circuits from [9]. The chosen S40PI_n benchmark system has the form (3) with $n = 2\,182$, $m = 1$ and $p = 1$ and we apply an α-shift with $\alpha = 10^{-4}$ as in [10] to stabilize the system. By construction, the E matrix is not invertible and the overall system has a complicated nested structure of differential-algebraic equations. For this system, we are going to use the balanced truncation method for descriptor systems as it is implemented in MORLAB by

[rom, info] = ml_ct_dss_bt(sys, opts),

where sys contains the original model as ss object, opts is an option struct that could be used to adjust parameters of the subroutines, rom is again the reduced-order model and info contains information about the subroutines and the full- and reduced-order systems.

First, we take a closer look at the resulting info struct. Here, we can see the resulting sizes of additive decomposed system parts. The matrix pencil $\lambda E - A$ of the original model had $n_f = 2\,028$ finite eigenvalues in the left open half-plane and $n_\infty = 154$ infinite eigenvalues, which correspond to the algebraic equations of the system. The reduced-order model instead has $r = 262$ states, where the matrix pencil $\lambda \hat{E} - \hat{A}$ has only finite stable eigenvalues left, i.e., the $\hat{E}$ matrix is now invertible and the reduced-order system can be treated as (2). In other words, the reduced-order model generated by MORLAB can now be used in all the system-theoretic routines in MATLAB, which was not possible for the original descriptor system.

To analyze the quality of the resulting reduced-order model, we plotted the frequency response and the absolute error of the original and reduced-order system using the ml_sigmaplot function from MORLAB. The results can be seen in Fig. 2. The two transfer functions are indistinguishable in the eye ball norm and the absolute error shows the good approximation behavior of the reduced-order model. Additionally, we added the absolute error bound to the plot that was provided in the info struct.

Remark 3. As for the standard system case, MORLAB supports discrete-time descriptor systems with appropriate model reduction methods.

3.3   Extension for Second-Order Systems

A very important but currently unsupported system class in the Control System Toolbox and Simulink are linear systems involving second-order time derivatives

$$\begin{aligned} M\ddot{x}(t) + E\dot{x}(t) + Kx(t) &= B_{\mathrm{u}}u(t), \\ y(t) &= C_{\mathrm{p}}x(t) + C_{\mathrm{v}}\dot{x}(t) + Du(t), \end{aligned} \tag{4}$$

with $M, E, K \in \mathbb{R}^{n \times n}$, $B_{\mathrm{u}} \in \mathbb{R}^{n \times m}$, $C_{\mathrm{p}}, C_{\mathrm{v}} \in \mathbb{R}^{p \times n}$, $D \in \mathbb{R}^{p \times m}$. Those systems usually arise in the modeling process of mechanical or electro-mechanical systems. In principle, every second-order system (4) can be rewritten as a first-order system (3), e.g., by introducing an extended state vector $q(t) = \left[x(t)^{\mathsf{T}}, \dot{x}(t)^{\mathsf{T}} \right]^{\mathsf{T}}$. The drawback of this process is the resulting first-order system of order $2n$, for which most computational methods ignore the internal second-order structure. This makes computations with second-order systems potentially more expensive than they need to be. This can be avoided by directly working on the second-order system structure (4). In MORLAB, the struct data type "soss" is used to handle second-order system objects. Appropriate evaluation tools for the time and frequency evaluation that work directly with the second-order structure are implemented in the toolbox to give, as addition to established software, the opportunity to efficiently work with second-order systems.

Corresponding to the direct use of the second-order structure in analysis and evaluation of dynamical systems, the application of model order reduction should preserve the system structure, i.e., it is beneficial if a reduced-order model of (4) has the same structure

$$\begin{aligned} \widehat{M}\ddot{\hat{x}}(t) + \widehat{E}\dot{\hat{x}}(t) + \widehat{K}\hat{x}(t) &= \widehat{B}_{\mathrm{u}}u(t), \\ y(t) &= \widehat{C}_{\mathrm{p}}\hat{x}(t) + \widehat{C}_{\mathrm{v}}\dot{\hat{x}}(t) + \widehat{D}u(t), \end{aligned} \tag{5}$$

with $\widehat{M}, \widehat{E}, \widehat{K} \in \mathbb{R}^{r \times r}$, $\widehat{B}_{\mathrm{u}} \in \mathbb{R}^{r \times m}$, $\widehat{C}_{\mathrm{p}}, \widehat{C}_{\mathrm{v}} \in \mathbb{R}^{p \times r}$, $\widehat{D} \in \mathbb{R}^{p \times m}$. This not only allows to use (5) directly as a surrogate model that can be handled by the same

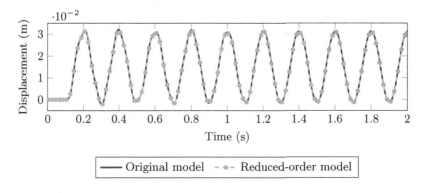

Fig. 3. Time simulation of the artificial fishtail example.

tools as (4) but usually a structure-preserving approximation is more accurate than an unstructured one and also might allow for interesting re-interpretations of the reduced-order system quantities. MORLAB provides several structure-preserving model reduction methods for (4).

As example, we take a look at the benchmark model [13], describing the movement of an artificial fishtail designed for autonomous underwater vehicles. The model has a mechanical second-order structure (4) with $C_v = 0$, $D = 0$ and $n = 779\,232$, $m = 1$, $p = 3$. It is crucial to employ the second-order structure since the corresponding first-order realization would be of order $\approx 1.5 \cdot 10^6$. Due to the size of the model, the dense model reduction methods from MORLAB cannot be applied directly to the system. Therefore, we use the same approach as in [7,12] by applying a structure-preserving pre-reduction following the theory from [1] with the same setup as in [7]. The second-order frequency-limited balanced truncation method (ml_ct_soss_flbt) was then applied to the pre-reduced model of order 100 for the frequency interval $[0, 20]$ Hz. As this MORLAB routine allows for computing several reduced-order models at once, we created reduced-order models for the 8 different balancing formulas from [7]. Independent of the balancing formula, the reduced-order models were of order 1 and stable.

To test the results, the original system and one of the reduced-order models, namely the one created by the so formula (see [7] for details), were simulated with the input signal

$$u(t) = 2500(\sin(10\pi(t - 1.35)) + 1)$$

using the MORLAB function ml_ct_soss_simulate_ss22. Figure 3 shows the second output entry with the displacement of the fishtail tip in y-direction. In the simulation, no difference between the full- and reduced-order models is visible. To compare, the simulation of the full-order model took around 4 h.

The main computational work for the model order reduction was the intermediate model, which took 21 h followed by the computation of the final reduced-order models with 0.45 s. The simulation of the reduced-order model took then 0.06 s using the same time discretization scheme as for the original model, i.e., the reduced-order model allows now for real time simulations, which could be even implemented on an onboard chip in a fish drone.

References

1. Beattie, C.A., Gugercin, S.: Interpolatory projection methods for structure-preserving model reduction. Syst. Control Lett. **58**(3), 225–232 (2009). https://doi.org/10.1016/j.sysconle.2008.10.016
2. Benner, P.: A MATLAB repository for model reduction based on spectral projection. In: 2006 IEEE Conference on Computer Aided Control System Design, 2006 IEEE International Conference on Control Applications, 2006 IEEE International Symposium on Intelligent Control, pp. 19–24, October 2006. https://doi.org/10.1109/CACSD-CCA-ISIC.2006.4776618
3. Benner, P., Byers, R.: Disk functions and their relationship to the matrix sign function. In: Proceedings of European Control Conference, ECC 97, p. 936. BELWARE Information Technology, Waterloo, Belgium (1997). CD-ROM
4. Benner, P., Werner, S.W.R.: Balancing related model reduction with the MORLAB toolbox. Proc. Appl. Math. Mech. **18**(1), e201800083 (2018). https://doi.org/10.1002/pamm.201800083
5. Benner, P., Werner, S.W.R.: Model reduction of descriptor systems with the MORLAB toolbox. IFAC-PapersOnLine **51**(2), 547–552 (2018). https://doi.org/10.1016/j.ifacol.2018.03.092. 9th Vienna International Conference on Mathematical Modelling MATHMOD 2018, Vienna, Austria, 21–23 February 2018
6. Benner, P., Werner, S.W.R.: MORLAB - Model Order Reduction LABoratory (version 5.0) (2019). https://doi.org/10.5281/zenodo.3332716. https://www.mpi-magdeburg.mpg.de/projects/morlab
7. Benner, P., Werner, S.W.R.: Frequency- and time-limited balanced truncation for large-scale second-order systems. e-print 2001.06185, arXiv (2020). http://arxiv.org/abs/2001.06185. math.OC
8. Benner, P., Werner, S.W.R.: MORLAB - The Model Order Reduction LABoratory. e-print 2002.12682, arXiv (2020). https://arxiv.org/abs/2002.12682. cs.MS
9. Freitas, F., Martins, N., Varricchio, S.L., Rommes, J., Veliz, F.C.: Reduced-order transfer matrices from RLC network descriptor models of electric power grids. IEEE Trans. Power Syst. **26**(4), 1905–1916 (2011). https://doi.org/10.1109/TPWRS.2011.2136442
10. Freitas, F., Rommes, J., Martins, N.: Gramian-based reduction method applied to large sparse power system descriptor models. IEEE Trans. Power Syst. **23**(3), 1258–1270 (2008). https://doi.org/10.1109/TPWRS.2008.926693
11. Roberts, J.D.: Linear model reduction and solution of the algebraic Riccati equation by use of the sign function. Int. J. Control **32**(4), 677–687 (1980). https://doi.org/10.1080/0020717800892288. (Reprint of Technical Report No. TR-13, CUED/B-Control, Cambridge University, Engineering Department, 1971)

12. Saak, J., Siebelts, D., Werner, S.W.R.: A comparison of second-order model order reduction methods for an artificial fishtail. at-Automatisierungstechnik **67**(8), 648–667 (2019). https://doi.org/10.1515/auto-2019-0027

13. Siebelts, D., Kater, A., Meurer, T., Andrej, J.: Matrices for an artificial fishtail. Hosted at MORwiki - Model Order Reduction Wiki (2019). https://doi.org/10.5281/zenodo.2558728

FlexRiLoG—A SageMath Package
for Motions of Graphs

Georg Grasegger[1]($\boxtimes$) and Jan Legerský[2,3]

[1] Johann Radon Institute for Computational and Applied Mathematics (RICAM),
Austrian Academy of Sciences, Linz, Austria
georg.grasegger@ricam.oeaw.ac.at
[2] Research Institute for Symbolic Computation, Johannes Kepler University Linz,
Linz, Austria
[3] Department of Applied Mathematics, Faculty of Information Technology,
Czech Technical University in Prague, Prague, Czech Republic

Abstract. In this paper we present the SAGEMATH package FLEX-RiLoG (short for flexible and rigid labelings of graphs). Based on recent results the software generates motions of graphs using special edge colorings. The package computes and illustrates the colorings and the motions. We present the structure and usage of the package.

Keywords: Motion · Flexible labeling · Flexible graph · NAC-coloring

1 Introduction

A graph with a placement of its vertices in the plane is considered to be flexible if the placement can be continuously deformed by an edge length preserving motion into a non-congruent placement. The study of such graphs and their motions has a long history (see for instance [1,4,11,12,14–17]). Recently we provided a series of results [7,8] with a deeper analysis of the existence of flexible placements. This is done via special edge colorings called NAC-colorings ("No Almost Cycles", see [7]). These colorings classify the existence of a flexible placement in the plane and give a construction of the motion.

Basic Definitions. We briefly give a precise definition of flexibility of a graph. A *framework* is a pair (G, p) where $G = (V_G, E_G)$ is a graph and $p : V_G \to \mathbb{R}^2$ is a *placement* of G in $\mathbb{R}^2$. The placement might be possibly non-injective but for all edges $uv \in E_G$ we require $p(u) \neq p(v)$.

Two frameworks (G, p) and (G, q) are *equivalent* if for all $uv \in E_G$,

$$\|p(u) - p(v)\| = \|q(u) - q(v)\|. \tag{1}$$

Two placements p, q of G are said to be *congruent* if (1) holds for all pairs of vertices $u, v \in V_G$. Equivalently, p and q are congruent if there exists a Euclidean isometry M of $\mathbb{R}^2$ such that $Mq(v) = p(v)$ for all $v \in V_G$.

© Springer Nature Switzerland AG 2020
A. M. Bigatti et al. (Eds.): ICMS 2020, LNCS 12097, pp. 442–450, 2020.
https://doi.org/10.1007/978-3-030-52200-1_44

A *flex* of the framework (G, p) is a continuous path $t \mapsto p_t$, $t \in [0, 1]$, in the space of placements of G such that $p_0 = p$ and each (G, p_t) is equivalent to (G, p). The flex is called trivial if p_t is congruent to p for all $t \in [0, 1]$. We define a framework to be *flexible* if there is a non-trivial flex in $\mathbb{R}^2$. Otherwise is is called *rigid* (see Fig. 1). We say that a labeling $\lambda : E_G \to \mathbb{R}_{>0}$ of a graph G is *flexible* if there is a flexible framework (G, p) such that p induces λ, namely, $\|p(u) - p(v)\| = \lambda(uv)$ for all $uv \in E_G$. On the other hand, λ is *rigid* if (G, p) is rigid for all placements p inducing λ. A flexible labeling λ of a graph is *proper* if there exists a framework (G, p) such that p induces λ and it has a non-trivial flex with all but finitely many placements being injective. We call a graph *movable* if it has a proper flexible labeling.

Fig. 1. A rigid and a flexible framework for the three-prism graph.

Outline of the Paper. We have given the necessary definitions. Section 2 describes the main functionality of the FLEXRILOG dealing with colorings and motions. In this paper we do not provide the algorithms themselves but refer to the respective theorems and literature. In Sect. 3 we describe how to use the package to ask for movable graphs.

2 The Package

FLEXRILOG [5] is a package for SAGEMATH running in versions 8.9 and 9.0 [13]. The latest release of the package can be installed by executing:

```
sage -pip install --upgrade flexrilog
```

The development version of FLEXRILOG can be found in the repository [6], where also other options of installation are described.

A convenient way of using the package instead of the **sage** console is a Jupyter notebook (coming with SAGEMATH, launch by **sage -n jupyter**). The file **examples/flexrilog_Motions_of_Graphs.ipynb** in [6] provides a Jupyter notebook version of this paper[1].

The package allows to check whether a graph has a NAC-coloring, in particular to list all of them. A motion or flex obtained from a NAC-coloring can be constructed and displayed. Moreover, it implements the results of [8] regarding

---

[1] See also https://jan.legersky.cz/flexrilogICMS2020 redirecting to a version of the notebook executable on-line using Binder.

the existence of proper flexible labelings, namely, the check of a necessary condition and construction of a proper flex from a pair of NAC-colorings. There is also functionality providing tools for classification of all proper flexible labeling, which is out of the scope of this paper (see [9] for details).

2.1   Data Types

The package provides data types in different classes for dealing with graphs, colorings and motions. In order to use the data types provided by the package, they have to be loaded.

```
sage: from flexrilog import FlexRiGraph, GraphMotion
```

The main object will always be of type **FlexRiGraph**. This class inherits properties of the standard **Graph** from SAGEMATH and adds specific properties for investigations of flexibility and rigidity. In this paper we focus on the flexibility part. A **FlexRiGraph** can be constructed by the following command.

```
sage: FlexRiGraph([[0,1],[1,2],[0,2]])
```

FlexRiGraph with the vertices [0, 1, 2] and edges [(0, 1), (0, 2), (1, 2)]

Further constructions can be made via integer encoding described in [2] and via objects of the standard **Graph** class.

```
sage: FlexRiGraph(graphs.CompleteBipartiteGraph(2,3))
```

Complete bipartite graph of order 2+3: FlexRiGraph with the vertices [0, 1, 2, 3, 4] and edges [(0, 2), (0, 3), (0, 4), (1, 2), (1, 3), (1, 4)]

Besides the class for graphs there is a class for colorings (**NACcoloring**). We do not discuss the class itself here but rather show how to compute colorings of graphs (see Sect. 2.2). Furthermore, motions are stored in a third class, **GraphMotion**. They are discussed in Sect. 2.3. The **GraphGenerator** class stores the code for some important graphs from the area of rigidity and flexibility theory. We do not go into detail but some of the graphs are used in the paper.

2.2   NAC-Colorings

NAC-colorings are a special type of edge colorings using two colors. Unlike proper edge colorings in Graph Theory we do not require incident edges to have different colors.

Definition 1. *Let G be a graph. A coloring of edges $\delta\colon E_G \to \{blue, red\}$ is called a* NAC-coloring, *if it is surjective and for every cycle in G, either all edges have the same color, or there are at least 2 edges in each color.*

FLEXRILOG contains functionality for computing and showing NAC-colorings of a given graph. The standard output is a textual list but the colorings can be shown in figures as well.

```
sage: C4 = FlexRiGraph([[0,1],[1,2],[2,3],[0,3]])
sage: C4.NAC_colorings()
```
[NAC-coloring with red edges [[0, 1], [0, 3]] and blue edges [[1, 2], [2, 3]],
NAC-coloring with red edges [[0, 1], [1, 2]] and blue edges [[0, 3], [2, 3]],
NAC-coloring with red edges [[0, 1], [2, 3]] and blue edges [[0, 3], [1, 2]]]
```
sage: C4.show_all_NAC_colorings()
```

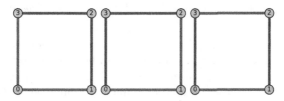

It can be checked in polynomial time whether a coloring is a NAC-coloring using [7, Lemma 2.4]. Hence, the question if a graph has a NAC-coloring is in NP, but it is subject to further investigations whether there is a polynomial time algorithm for a general graph. For instance every graph that is not generically rigid has a NAC-coloring due to Theorem 1. It can be checked in polynomial time whether a graph is generically rigid [10].

In order to compute all NAC-colorings, lists of edges that necessarily have the same color due to being in 3-cycles (so called $\triangle$-*connected components*, see [7]) are determined. So far, we then test all possible combinations how to color them by red and blue. The edges colored the same in the following picture must have the same color in any NAC-coloring, but no combination satisfies the conditions of NAC-coloring.

```
sage: from flexrilog import GraphGenerator
sage: N = GraphGenerator.NoNACGraph()
sage: N.has_NAC_coloring()
```
False
```
sage: N.plot(show_triangle_components=True)
```

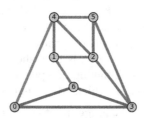

For graphs with symmetries we get many similar colorings, in a sense that after applying the symmetry one NAC-coloring yields the other. We call such NAC-colorings isomorphic. In order to visualize this, NAC-colorings can be named so that isomorphic ones have the same Greek letter but differ by their index.

```
sage: C4.set_NAC_colorings_names()
sage: C4.NAC_colorings_isomorphism_classes()
```
[[alpha1: NAC-coloring with red edges [[0, 1], [0, 3]] and blue edges [[1, 2], [2, 3]],
```
```

```
alpha2: NAC-coloring with red edges [[0, 1], [1, 2]] and blue edges [[0, 3], [2, 3]]],
[beta: NAC-coloring with red edges [[0, 1], [2, 3]] and blue edges [[0, 3], [1, 2]]]]
```

2.3 Constructing Motions

Given a NAC-coloring we are able to construct a motion. The following result from [7] describes the relation.

Theorem 1. *A connected non-trivial graph has a flexible labeling if and only if it has a NAC-coloring.*

The main idea to construct a flex is to place the vertices on a grid in such a way that all the edges lie on grid lines. This can be achieved by placing vertices according to the color component of the graph. For color components we remove all edges of the other color and then take connected components of the remaining graph. Then all vertices which lie in the same red component are placed in the same column of the grid and all vertices from the same blue component are placed in the same row of the grid. By this procedure each vertex is assigned a unique grid point and all edges of the graph lie on the grid lines. In FLEXRILOG this can be done with the classmethod GraphMotion.GridConstruction.

```
sage: from flexrilog import GraphMotion, GraphGenerator
sage: P = GraphGenerator.ThreePrismGraph()
sage: delta = P.NAC_colorings()[0]
sage: motion_P = GraphMotion.GridConstruction(P, delta)
sage: motion_P.parametrization()
{0: (0, 0),
 1: (sin(alpha) + 1, cos(alpha)),
 2: (2*sin(alpha) + 1, 2*cos(alpha)),
 3: (2*sin(alpha), 2*cos(alpha)),
 4: (sin(alpha), cos(alpha)),
 5: (1, 0)}
```

There is also the option to generate an animated SVG showing the NAC-coloring, which is automatically displayed when used in a Jupyter notebook (the picture below is a screenshot). If the fileName is specified, the SVG animation is stored and a web browser can be used to view it. Note that not all web browsers support SVG animations. It can be chosen, whether the edges are colored according to the NAC-coloring in use. The package also distinguishes the vertex layout depending on whether it is drawing a graph having no specific placement properties (dark vertices), or drawing a motion, in which edge lengths are fixed (light vertices).

```
sage: motion_P.animation_SVG(edge_partition="NAC",
....:                        fileName="3-prism_grid")
```

More generally the base points of the grid can be chosen arbitrarily to get a zig-zag grid. This can be used to avoid degenerate subgraphs. Base points consist of two lists. The standard values consists of lists with points $(i, 0)$ and $(0, i)$ respectively. Using them we get a rectangular initial grid. A zig-zag grid in general does not need to be initially rectangular. It is uniquely determined by the base points and drawing parallel lines. Doing so the grid consists of parallelograms. Usually the grid itself is not easily visible from the output motion.

```
sage: motion_P = GraphMotion.GridConstruction(P, delta,
....:    zigzag=[[[0,0], [3/4,1/2], [2,0]],
....:           [[0,0], [1,0]]])
sage: motion_P.animation_SVG(edge_partition="NAC")
```

3   Movable Graphs

Using the grid construction, non-adjacent vertices might overlap, i.e., the constructed framework is not proper. Note, that this cannot be avoided by zig-zag constructions either but depends solely on the NAC-coloring in use. For some graphs all NAC-colorings result in overlapping vertices. In FLEXRILOG it can be checked whether this is the case.

```
sage: P.has_injective_grid_construction()
True
sage: Q1 = GraphGenerator.Q1Graph()        # see the picture below
sage: Q1.has_injective_grid_construction()
False
```

For some graphs, a proper flexible labeling exists due to the following lemma [8], which relates movability to spatial embeddings.

Lemma 1. *Let G be a graph with an injective embedding $\omega : V_G \to \mathbb{R}^3$ such that for every edge $uv \in E_G$, the vector $\omega(u) - \omega(v)$ is parallel to one of the four vectors $(1, 0, 0)$, $(0, 1, 0)$, $(0, 0, 1)$, $(-1, -1, -1)$, and all four directions are present. Then G is movable. Moreover, there exist two NAC-colorings of G such that two edges are parallel in the embedding ω if and only if they receive the same pair of colors.*

The package tries to construct such a spatial embedding for all pairs of NAC-colorings.

```
sage: inj, nacs = Q1.has_injective_spatial_embedding(
....:                        certificate=True); inj
True
sage: graphics_array([d.plot() for d in nacs])
```

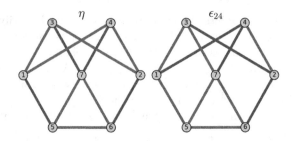

From the spatial embedding we can construct a motion of the graph. The motion can be transformed in such a way that a particular edge is fixed.

```
sage: motion_Q1 = GraphMotion.SpatialEmbeddingConstruction(Q1, nacs)
sage: motion_Q1.fix_edge([5,6])
sage: motion_Q1.parametrization()
{1: ((3*t^2 - 3)/(t^2 + 1), -6*t/(t^2 + 1)),
 2: ((t^4 + 23*t^2 + 4)/(t^4 + 5*t^2 + 4), (6*t^3 - 12*t)/(t^4 + 5*t^2 + 4)),
 3: ((4*t^2 - 2)/(t^2 + 1), -6*t/(t^2 + 1)),
 4: (18*t^2/(t^4 + 5*t^2 + 4), (6*t^3 - 12*t)/(t^4 + 5*t^2 + 4)),
 5: (0, 0),
 6: (2, 0),
 7: (1, 0)}
sage: motion_Q1.animation_SVG()
```

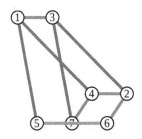

Besides the sufficient condition on movability above, there is also a necessary condition given in [8]. For this condition we consider all NAC-colorings and look for monochromatic paths. Adding certain edges according to these paths we get a bigger graph with similar movability properties.

For a graph G, let $U(G)$ denote the set of all pairs $\{u, v\} \subset V_G$ such that $uv \notin E_G$ and there exists a path from u to v which is monochromatic for all NAC-colorings δ of G. If there exists a sequence of graphs $G_0, \ldots, G_n$ such that $G = G_0$, $G_i = (V_{G_{i-1}}, E_{G_{i-1}} \cup U(G_{i-1}))$ for $i \in \{1, \ldots, n\}$, and $U(G_n) = \emptyset$, then the graph G_n is called *the constant distance closure* of G, denoted by $\mathrm{CDC}(G)$.

Theorem 2. *A graph G is movable if and only if* $\mathrm{CDC}(G)$ *is movable. Particularly, if* $\mathrm{CDC}(G)$ *is the complete graph, then G is not movable.*

We can see that the following graph G is not movable (G_1 has no NAC-coloring since $\{3, 4\}, \{5, 6\} \in U(G)$, hence, $U(G_1)$ are all non-edges of G_1).

```
sage: G = GraphGenerator.MaxEmbeddingsLamanGraph(7)
sage: G.show_all_NAC_colorings()
```

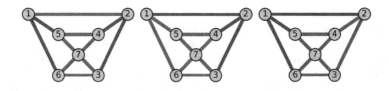

```
sage: G.constant_distance_closure().is_complete()
True
```

4 Conclusion

We gave a brief overview of the package and therefore did not cover all functionality. The package contains a documentation. As research in the field of flexible and movable graphs is going on the package is further developed, both regarding improvements as well as new functionality (for instance n-fold rotationally symmetric frameworks, see [3]). The most current version of FLEXRILOG can be found in [6].

Acknowledgments. This project was supported by the Austrian Science Fund (FWF): P31061, P31888 and W1214-N15, and by the Ministry of Education, Youth and Sports of the Czech Republic, project no. CZ.02.1.01/0.0/0.0/16_019/0000778. The project has received funding from the European Union's Horizon 2020 research and innovation programme under the Marie Skłodowska-Curie grant agreement No 675789.

References

1. Burmester, L.: Die Brennpunktmechanismen. Zeitschrift für Mathematik und Physik **38**, 193–223 (1893)
2. Capco, J., Gallet, M., Grasegger, G., Koutschan, C., Lubbes, N., Schicho, J.: An algorithm for computing the number of realizations of a Laman graph. Zenodo (2018). https://doi.org/10.5281/zenodo.1245506
3. Dewar, S., Grasegger, G., Legerský, J.: Flexible placements of graphs with rotational symmetry. arXiv:2003.09328 (2020)
4. Dixon, A.: On certain deformable frameworks. Messenger **29**(2), 1–21 (1899)
5. Grasegger, G., Legerský, J.: FlexRiLoG – SAGEMATH package for flexible and rigid labelings of graphs. Zenodo (2020). https://doi.org/10.5281/zenodo.3078757
6. Grasegger, G., Legerský, J.: FlexRiLoG – SAGEMATH package for Flexible and Rigid Labelings of Graphs, repository (2020). https://github.com/Legersky/flexrilog/
7. Grasegger, G., Legerský, J., Schicho, J.: Graphs with flexible labelings. Discrete Comput. Geometry **62**(2), 461–480 (2018). https://doi.org/10.1007/s00454-018-0026-9
8. Grasegger, G., Legerský, J., Schicho, J.: Graphs with flexible labelings allowing injective realizations. Discrete Math. **343**(6), Art. 111713 (2020). https://doi.org/10.1016/j.disc.2019.111713

9. Grasegger, G., Legerský, J., Schicho, J.: On the classification of motions of paradoxically movable graphs. arXiv:2003.11416 (2020)
10. Jacobs, D.J., Hendrickson, B.: An algorithm for two-dimensional rigidity percolation: the pebble game. J. Comput. Phys. **137**(2), 346–365 (1997). https://doi.org/10.1006/jcph.1997.5809
11. Kempe, A.: On conjugate four-piece linkages. Proc. Lond. Math. Soc. **s1–9**(1), 133–149 (1877). https://doi.org/10.1112/plms/s1-9.1.133
12. Stachel, H.: On the flexibility and symmetry of overconstrained mechanisms. Philos. Trans. R. Soc. Lond. A Math. Phys. Eng. Sci. **372** (2013). https://doi.org/10.1098/rsta.2012.0040
13. The Sage Developers: SageMath, the Sage Mathematics Software System (Version 9.0) (2020). https://www.sagemath.org
14. Walter, D., Husty, M.: On a nine-bar linkage, its possible configurations and conditions for paradoxical mobility. In: 12th World Congress on Mechanism and Machine Science, IFToMM (2007)
15. Wunderlich, W.: Ein merkwürdiges Zwölfstabgetriebe. Österreichisches. Ingenieur-Archiv **8**, 224–228 (1954)
16. Wunderlich, W.: On deformable nine-bar linkages with six triple joints. Indagationes Mathematicae (Proceedings) **79**(3), 257–262 (1976). https://doi.org/10.1016/1385-7258(76)90052-4
17. Wunderlich, W.: Mechanisms related to Poncelet's closure theorem. Mech. Mach. Theory **16**, 611–620 (1981). https://doi.org/10.1016/0094-114X(81)90067-7

Markov Transition Matrix Analysis of Mathematical Expression Input Models

Francis Quinby[1], Seyeon Kim[1], Sohee Kang[3], Marco Pollanen[1],
Michael G. Reynolds[2], and Wesley S. Burr[1(✉)]

[1] Department of Mathematics, Trent University, Peterborough, ON, Canada
wesleyburr@trentu.ca
[2] Department of Psychology, Trent University, Peterborough, ON, Canada
[3] Department of Computer and Mathematical Sciences, University of Toronto
Scarborough, Toronto, ON, Canada

Abstract. Computer software interfaces for mathematics collaboration and problem solving rely, as all interfaces do, on user identification and recognition of symbols (via icons and other contextual widgets). In this paper we examine the results of a short study which examined users interacting with mathematics software (Mathematics Classroom Communicator, MC^2) designed for education, real-time communication and collaboration. Videos were recorded of 14 users working through seven comprehensive problems in the MC^2 interface. Extensive second-by-second coding was completed of the user's actions and status throughout their work, and a set of transition matrices were tabulated, estimating transition probabilities between symbols, operators and other aspects of mathematical expressions. We discuss the results of these matrices, and their implications in the translation of abstract mathematical concepts into software interfaces, and further conclude with a brief discussion of suggestions for mathematical software interface design. This study also has applications in mathematical software usability and accessibility.

Keywords: Mathematical notation · Eye-tracking · Mathematical software interfaces · Transitions

1 Introduction

Computer software applications which support the transcription of mathematics make use of a number of models to represent mathematical structure. For most experts, the 1-dimensional model used in LaTeX is preferred once time is taken to learn the syntax due to the power and expressiveness of the system. In LaTeX the transcription proceeds as a single string of characters, transformed into possibly complex two-dimensional mathematical structures upon compilation. Due to the lack of immediate connection between this 1-dimensional LaTeX code and the compiled output, and the accompanying learning curve, this approach can present significant challenges for novice users. Due to the extreme market penetration of Microsoft Word, many novice users simply do any required electronic

© Springer Nature Switzerland AG 2020
A. M. Bigatti et al. (Eds.): ICMS 2020, LNCS 12097, pp. 451–461, 2020.
https://doi.org/10.1007/978-3-030-52200-1_45

mathematical transcription using the WYSIWYG (what you see is what you get) Microsoft Word Equation Editor or similar products. These products tend to be somewhat intuitive in their use, with extensive menu and icon systems that allow for eventual (if not efficient) implementation of marginally complex mathematical expressions. Further, as WYSIWYG editors provide users with the ability to directly edit the mathematical output within the user interface without a need to recompile the document, their use by novice level users has been shown to result in fewer output errors [4]. While LaTeX provides users with many positive features not available in WYSIWYG editors there are a large number of individuals who rely solely on WYSIWYG editors (and often are not even aware of the existence of LaTeX). Therefore, it is important that users are able to use these editors with the highest level of efficiency and effectiveness. To improve these editors we must have a thorough understanding of how individuals interact with their respective environments.

Within the subset of WYSIWYG editors, there are different representative models for mathematical structure utilized, and it remains unclear which model provides an overall more positive user experience in their respective environments [4–6]. It is likely the case that there is not one model which is superior to others for all elements of mathematical notation, but rather that each model handles some mathematical structures better than others. To examine this, the following study was designed to compare the execution of insertion and editing of different types of mathematical structures by novice users through examination of the probabilities of symbol entry order.

MC^2, an application created by Pollanen and Kang (co-authors on this paper), with colleagues Cater, Chen & Lee, allows users to construct expressions in a free-form manner through a 2-dimensional drag and drop interface in which symbols are inserted onto a 'blank canvas' workspace from a series of menus and can subsequently be moved and resized using the mouse. The technical details of MC^2 specifications can be found in Pollanen et al. [7]. As a typical default choice due to ubiquitous availability, Microsoft Equation Editor was used as a comparator. Equation Editor is a 2-dimensional structure-based editor, in which users generally insert the mathematical structures first (with imposed dimensionality, e.g., powers or subscripts), and then users interact with the input options of the structures to insert symbols in the form of 1-dimensional strings. Though users interact directly with the 2-dimensional mathematical structures in both programs, they must do so in different ways [1]. Through this study we aim to show that the model chosen within these typesetting applications changes the order in which users enter symbols into their workspace resulting in a change in their workflow, with some implications for future design.

2   Methods

The study consisted of 14 first-year undergraduate student participants (10 female and 4 male), who are not from quantitative fields, and none of whom had any previous experience using any mathematical equation software. They were instructed to transcribe a series of four different mathematical expressions using the two digital typesetting applications (MC^2 and Equation Editor). Each expression was first transcribed with MC^2 and subsequently with Microsoft Equation Editor (in Word). This resulted in 8 experimental trials per participant. All participants completed the experiment simultaneously and communication with the researcher occurred through a chat screen contained within the user interface of MC^2. All participants completed the 8 trials in identical order. The expressions provided to the participants are displayed in Table 1.

Table 1. Expressions provided to participants for transcription using the computational environment.

$$1. \quad s = \sqrt{\frac{1}{n-1}\sum_{i=1}^{n}(x-\bar{x})^2} \qquad 3. \quad \int_0^{\infty} e^{-y}dy$$

$$2. \quad \int \frac{1}{\sqrt{2\pi}}e^{-\frac{x^2}{2}}dx \qquad 4. \quad \frac{\sqrt{\frac{\sqrt{x^2+2x+2}}{x}}}{x^2+1}$$

The video recordings (screen captures) of the 14 participants' actions were analyzed by a research assistant, who carefully watched the videos and assigned a time-stamped code to each significant, discrete action performed by the participant while working towards completion of the transcription task. Coded actions consisted of: insertion of a symbol; deletion of a symbol; resizing of a symbol; and movement of a symbol. Each symbol in every expression was assigned a unique reference within the given expression. The assigned codes contained information indicating both the action and the unique symbol upon which the action was performed. For example, the action of inserting the symbol assigned the number 4 would have been assigned the code, "*i-4*".

From this analysis, we were then able to analyze the order in which individuals inserted the mathematical structures relative to other symbols in the expression through Markov Chain transition matrices. Markov Chains were applied in a similar fashion by Jansen, Marriott, and Yelland to model the relative order in which individuals read symbols in algebraic expressions [2]. As this proved to be an effective analysis, we applied this approach to model the relative order of symbol insertion when digitally transcribing mathematical expression.

The resulting transition matrices contain the conditional probabilities that a symbol in a particular part of the expression will be inserted, given that a symbol from another part of the expression was the previously inserted symbol. Therefore, in a transition matrix T, the entry in position $T_{i,j}$ represents the probability that the symbol from the part of the expression indicated in column j is inserted, given that the symbol from part of the expression in row i was the symbol previously inserted. We initially created these transition matrices to represent the probability of transition between all individual symbols within the four expressions. However, the experimental expressions contained numerous symbols (>20), and we therefore decided it was more clear and concise to create transition matrices which tabulated transition probabilities relative to the mathematical *structure* found in these expressions with a unique transition matrix. We then analyzed the transitions between structural components of the expression. To create these transition matrices, we first recorded all insertion events which occurred in each experiment video recording. We then categorized the transitions between insertions, increasing the count in $T_{i,j}$ by one when the symbol in column j was inserted following insertion of the symbol in row i. For these matrices we chose to ignore transitions occurring within the same region of an expression. Therefore, all counts of $T_{i,i}$ were 0. We then normalized each row, so that the sum of all entries in each respective row was 1 or 0 in the case where the symbol represented by that row was the last symbol inserted in all trials. The result was a transition probability matrix for each experimental trial for a given structure, for the 14 participants in the experiment. The transition matrices from the applicable individual trials within each participant were then normalized, summed, and re-normalized, resulting in our final transition probability matrices for each of the mathematical structures found in the experimental stimuli.

The transition probability matrices for the four types of mathematical structures included in the experimental stimuli are shown in the Results below (Tables 2, 3, 4 and 5), displaying transition probabilities related to all square roots, fractions, definite integrals and summations present in the four expressions used in the study. Though the exponent structure was featured in these expressions, we did not choose to model their insertion order due to significant inconsistencies in the manner of their insertion when using MC^2 and Microsoft Word Equation Editor.

3   Results

3.1   Markov Chain Transition Probability Matrices for Square Root Structures

The matrices in Table 2 highlight some key differences observed in how users inserted the square root structure into their workspace. Notably, users were about 16% more likely to insert a symbol inside of the square root immediately following insertion of the square root structure when using Equation Editor than when using MC^2. Users were also about 23% more likely to insert the square root structure immediately following insertion of a symbol inside of the square root when using MC^2 than they were when using Equation Editor.

From these transition probability matrices, we can compute the probability of the order of entry for an expression containing a radical when using both interfaces. One logical order of entry, if we assume the user starts from outside of the radical, for such an expression is *Outside* → *Radical* → *Radicand*. The probability of this symbol entry order after 2 transitions when using MC^2 was approximately 0.56 versus 0.70 when using Equation Editor. On the other hand, an alternative order of symbol entry is *Outside* → *Radicand* → *Radical*. The probability of symbols being inserted in this order after 2 transitions when using MC^2 was approximately 0.16 versus 0.09 in Equation Editor, indicating users were nearly twice as likely to insert the expression in this order when using the 2-dimensional free-form model found in the MC^2 user interface.

Table 2. Square root transition probability matrix for expressions transcribed with MC^2 (left) and Microsoft Word Equation Editor (right)

From	To			From	To		
	Outside	Radical	Radicand		Outside	Radical	Radicand
Outside	0.000	0.726	0.274	Outside	0.000	0.758	0.242
Radical	0.232	0.000	0.768	Radical	0.071	0.000	0.929
Radicand	0.399	0.601	0.000	Radicand	0.633	0.367	0.000

3.2   Markov Chain Transition Probability Matrices for Fraction Structures

Table 3. Fraction transition probability matrix for expressions transcribed with MC^2 (top) and Microsoft Word Equation Editor (bottom)

From	To			
	Numerator	Denominator	Fraction bar	Outside
Numerator	0.000	0.387	0.472	0.141
Denominator	0.262	0.000	0.071	0.667
Fraction bar	0.507	0.332	0.000	0.161
Outside	0.506	0.018	0.476	0.000

From	To			
	Numerator	Denominator	Fraction bar	Outside
Numerator	0.000	0.386	0.391	0.223
Denominator	0.268	0.000	0.089	0.643
Fraction bar	0.691	0.214	0.000	0.095
Outside	0.281	0.018	0.701	0.000

Similar to the matrices exploring square roots in the previous section, the matrices in Table 3 show important differences in how users insert fractions when using the two models featured in the applications featured in this study. When using Microsoft Equation Editor, we can see from the transition matrices that users are about 18% more likely to enter the numerator following insertion of the fraction bar or structure than they are when using MC^2. Furthermore, users were about 23% more likely the insert the numerator of a fraction following insertion of a symbol outside the fraction when using MC^2 than when using Equation Editor, where they are more likely to insert the fraction structure following insertion of a symbol outside of the fraction. These results show evidence that users are more likely to enter a fraction in the order, *Outside* → *Numerator* → *Fraction Bar* → *Denominator*, when using MC^2, with a probability of 0.08 after 3 transitions, than they were when using Equation Editor with a probability of 0.02 after three transitions. On the other hand, the transition matrices also show evidence that users are more likely to enter a fraction in the order, *Outside* → *Fraction Bar* → *Numerator* → *Denominator* when using Microsoft Equation Editor than they are when using MC^2. Again, we propose that this is due in part to differences in the flexibility of symbol insertion order when using these two applications. Generally, when using the structure based model featured in Equation Editor, an attempt to insert the fraction structure following insertion of the numerator will not result in the intended fraction and will require the user to delete the numerator and re-insert it into the numerator position of the fraction structure.

3.3 Markov Chain Transition Probability Matrices for Definite Integral Structures

Table 4. Definite integral transition probability matrix for expressions transcribed with MC^2 (top) and Microsoft Word Equation Editor (bottom)

From	To			
	Integral	Upper bound	Lower bound	Integrand
Integral	0.000	0.536	0.428	0.036
Upper bound	0.000	0.000	0.571	0.429
Lower bound	0.000	0.462	0.000	0.538
Integrand	1.000	0.000	0.000	0.000

From	To			
	Integral	Upper bound	Lower bound	Integrand
Integral	0.000	1.000	0.000	0.000
Upper bound	0.000	0.000	0.893	0.107
Lower bound	0.036	0.036	0.000	0.928
Integrand	0.500	0.000	0.500	0.000

The transition matrices for the order of insertion of symbols relative to definite integral structures, shown in Table 4, also show interesting results. Of primary interest is the conditional probability of the subsequent symbol inserted, given that the integral was inserted. In Microsoft Equation Editor, the upper bound of the definite integral was always the symbol inserted following insertion of the integral. However, in MC^2, the upper bound was inserted about 54% of the time following the insertion of the integral, and the lower bound was inserted about 43% of the time. Furthermore, users of MC^2 never inserted the integral following insertion of the upper bound or lower bound. This may be a result of the different models used in the two applications. We suggest there may be an element of the 2-dimensional structural model used in Equation Editor which makes users more likely to enter in the upper bound of a definite integral following insertion of the integral itself. It should also be noted that there was only one definite integral present in the experimental stimuli and that the trials using MC^2 always occurred prior to those with Equation Editor. Therefore, these differences could be attributed to a learning or familiarity effect rather than a true difference in the two models.

3.4 Markov Chain Transition Probability Matrices for Summation Notation Structures

Table 5. Summation transition probability matrix for expressions transcribed with MC^2 (top) and Microsoft Word Equation Editor (bottom)

From	Sum	Upper index	To Lower index	Inside	Outside
Sum	0.000	0.769	0.077	0.154	0.000
Upper index	0.000	0.000	0.917	0.083	0.000
Lower index	0.000	0.167	0.000	0.833	0.000
Inside	0.000	0.111	0.333	0.000	0.566
Outside	0.929	0.000	0.000	0.071	0.000

From	Sum	Upper index	To Lower index	Inside	Outside
Sum	0.000	0.869	0.071	0.036	0.024
Upper index	0.000	0.000	0.823	0.107	0.000
Lower index	0.000	0.077	0.000	0.808	0.115
Inside	0.100	0.167	0.000	0.000	0.733
Outside	0.881	0.000	0.0595	0.0595	0.000

In Table 5 we see the transition matrices related to the summation notation for MC^2 and Microsoft Equation Editor respectively. These matrices show less drastic differences than those related to the other structures. One difference to note is the conditional probabilities of symbol insertion following insertion of the summation symbol. We can see that users are approximately 10% more likely to insert the upper index following the summation symbol when using Equation Editor than MC^2, similar to the integral approach explored above. Though not a large difference, this is interesting to note as there was only one summation notation present in the experimental stimuli. The relative similarity between these transition matrices gives more merit to the differences seen in the transition matrices for the definite integral structures, of which there was also only one present in the stimuli. Therefore, if there were learning effects which occurred, they were not necessarily consistent across all structures.

4   Limitations and Discussion

This study was designed to test and compare properties of the two featured models for representing mathematical structure in the applications used: a 2-dimensional structure-based model (Equation Editor) and a 2-dimensional free-form model (MC^2), rather than test which of the two programs performs better as a whole. Previously, Kang *et al.* [3] analyzed the completion time between these two software platforms, controlled for quality of expressions. It is worth noting that Microsoft Equation Editor has been adapted over the years to improve the user experience, and now includes many shortcuts which allow users to enter expressions in a more intuitive manner (including a limited subset of the LaTeX syntax). Participants were not given any instruction on the manner in which to use Equation Editor, and were therefore free to make use of these shortcuts if they wished. Most did not, most likely due to the fact that they were not aware that these existed, as the participants in this study were novices. To evaluate the overall effectiveness of the Equation Editor application in Microsoft Word, it might be more useful to make users aware that these shortcuts exist. However, the fact that most beginner level users were not able to discern that these shortcuts could be used may speak to the user experience of Equation Editor in itself.

There were a number of limitations present in this study. The most notable was the fact that all expressions were first transcribed using MC^2 followed by transcription with Microsoft Equation Editor. Therefore, for all 4 experimental stimuli, participants were more familiar with the expression to be transcribed during the Equation Editor trial than during the MC^2 trial. This could result in less interference with working memory caused by the Equation Editor user interface as the mathematical content would be internalized to a greater degree during these trials [8]. Furthermore, the mathematical expressions which were chosen as experimental stimuli were not chosen in a systematic fashion controlling for the complexity of each expression or the number of symbols or different types of mathematical structures found in each expression. This makes it difficult to fully attribute some of the effects found in this study to how the models handle certain structures without considering some sort of interaction which may have occurred. Therefore, future research should be done to examine how the two featured models in these applications handle each of the structures, such as square roots, fractions, and exponents, in less complex expressions in which these structures are not featured in combination. Finally, the sample size for this study was rather small and although we control for the technical levels of students, we did not consider covariates while we constructed the Markov Chains, many of which may provide insight if sample sizes were sufficient.

Despite these limitations, this study presents a number of notable findings related to the user experience of typesetting applications which feature two different structural models for representation of mathematical notation. The Markov Chain analysis provides us with interesting insight into the process users are more likely to follow regarding the order of symbol insertion while transcribing mathematical content. Perhaps the most important finding is simply the fact that there is a difference in the order in which symbols are entered in both interfaces. If we assume that there is a natural order which individuals are more likely to follow when transcribing these expressions, regardless of the environment in which they are doing so, any sort of difference found here shows that either one, or possibly both of the models used in these interfaces forces users to need to adapt and adjust this order of insertion. Given the tendency found for users of the structure-based editor to insert the symbols in the upper position of an inserted structure (such as the numerator of a fraction) immediately following insertion of that structure, we recommend that editors using structure-based models place the cursor in this upper location immediately following insertion of the structure. In the Microsoft Equation Editor interface the cursor is placed outside and to the right of an inserted fraction structure. If the user wishes to insert the numerator next, they must navigate to the respective location using the arrow keys (pressing the left arrow twice) or click in the location using the mouse. Placing the cursor in the numerator position immediately following insertion of the fraction would increase the efficiency of the program as well as reduce the cognitive load, which could in turn decrease the working memory interference which has been shown to occur when digitally typesetting mathematics [8].

A number of differences in the conditional probabilities can likely be attributed to the reduced flexibility allowed by the 2-dimensional structure-based model found in Equation Editor. For example, when using the traditional implementation of this model in Word, a user cannot insert the numerator of a fraction before inserting the fraction bar itself. However, it is quite interesting that in all 2-dimensional structures with notation in which a symbol is located in an upper position and another one located in a lower position (fractions, definite integrals, summations) users were more likely when using Equation Editor to insert the symbol in the upper location following insertion of the structure itself (as compared to when using MC^2). While the reasoning behind this effect is unclear, it merits further study. As all of the experimental stimuli featured expressions with multiple structures, it would be interesting to observe whether the transition matrices were similar for expressions containing only one of the featured structures, or if there is some level of interaction taking place.

5   Conclusion

In this paper we explored the user experience while engaged with two mathematical typesetting models and discovered several insights into the order symbols were entered when novice users transcribed mathematics. These insights suggest that further research in this direction could have important implications into the design of mathematical user interfaces, in particular interfaces for mathematical novices. Though findings should be interpreted with caution, the application of Markov transition matrices to symbol order entry in this research serves to guide future studies on the characteristics and usability of different models featured in WYSIWYG mathematical typesetting applications. Similar experiments should be conducted with more standardized experimental stimuli (mathematical expressions), varying the nested nature of the mathematical structure, to investigate how certain interactions might manifest themselves. Furthermore, for a more complete picture of how users see and interact with mathematical expression structure in software, future research might also include how novice users revise and fix input errors in an expression they have written. As more education moves online, developing and studying mathematical input models that are intuitive for novice users will likely become a more pressing issue. While this area has so far not received significant attention in the academic literature, we believe this paper makes an important first step in this direction.

References

1. Gozli, D.G., Pollanen, M., Reynolds, M.: The characteristics of writing environments for mathematics: behavioral consequences and implications for software design and usability. In: Carette, J., Dixon, L., Coen, C.S., Watt, S.M. (eds.) CICM 2009. LNCS (LNAI), vol. 5625, pp. 310–324. Springer, Heidelberg (2009). https://doi.org/10.1007/978-3-642-02614-0_26
2. Jansen, A.R., Marriott, K., Yelland, G.W.: Parsing of algebraic expressions by experienced users of mathematics. Eur. J. Cogn. Psychol. **19**(2), 286–320 (2007)
3. Kang, S., Pollanen, M., Damouras, S., Cater, B.: Mathematics classroom collaborator (MC2): technology for democratizing the classroom. In: Davenport, J.H., Kauers, M., Labahn, G., Urban, J. (eds.) ICMS 2018. LNCS, vol. 10931, pp. 280–288. Springer, Cham (2018). https://doi.org/10.1007/978-3-319-96418-8_33
4. Knauff, M., Nejasmic, J.: An efficiency comparison of document preparation systems used in academic research and development. PloS ONE **9**(12), e115069 (2014)
5. Loch, B., Lowe, T.W., Mestel, B.D.: Master's students' perceptions of Microsoft Word for mathematical typesetting. Teach. Math. Its Appl. Int. J. IMA **34**(2), 91–101 (2015)
6. Padovani, L., Solmi, R.: An investigation on the dynamics of direct-manipulation editors for mathematics. In: Asperti, A., Bancerek, G., Trybulec, A. (eds.) MKM 2004. LNCS, vol. 3119, pp. 302–316. Springer, Heidelberg (2004). https://doi.org/10.1007/978-3-540-27818-4_22

7. Pollanen, M., Kang, S., Cater, B., Chen, Y., Lee, K.: MC^2: mathematics classroom collaborator. In: Proceedings of the Workshop on Mathematical User Interfaces (2017)
8. Quinby, F., Pollanen, M., Reynolds, M.G., Burr, W.S.: Effects of digitally typesetting mathematics on working memory. In: Harris, D., Li, W.-C. (eds.) Engineering Psychology and Cognitive Ergonomics. Mental Workload, Human Physiology, and Human Energy (LCAI), pp. 1–20. Springer (2020, in press)

Certifying Irreducibility in $\mathbb{Z}[x]$

John Abbott[(✉)]

Fakultät für Informatik und Mathematik, Universität Passau, Passau, Germany
abbott@dima.unige.it

Abstract. We consider the question of certifying that a polynomial in $\mathbb{Z}[x]$ or $\mathbb{Q}[x]$ is irreducible. Knowing that a polynomial is irreducible lets us recognise that a quotient ring is actually a field extension (equiv. that a polynomial ideal is maximal). Checking that a polynomial is irreducible by factorizing it is unsatisfactory because it requires trusting a relatively large and complicated program (whose correctness cannot easily be verified). We present a practical method for generating certificates of irreducibility which can be verified by relatively simple computations; we assume that primes and irreducibles in $\mathbb{F}_p[x]$ are self-certifying.

Keywords: Certificate · Irreducibility

1 Introduction

1.1 What Is a "Certificate"?

A certificate that object X has property P is a "small" amount of extra information C such that some *quick and simple computations* with X and C suffice to confirm that X does have the property. We illustrate this vague definition with a well-known, concrete example.

Example 1. We can certify that a positive integer n is prime using a *Lucas-Pratt certificate* [9]. The idea is to find a witness w such that $w^{n-1} \equiv 1 \mod n$ and $w^{(n-1)/q} \not\equiv 1 \mod n$ for all prime factors q of $n-1$.

These certificates have a recursive structure, since in general we must certify each prime factor q of $n-1$. To avoid infinite recursion we say that all small primes up to some limit are "self-certifying" (*i.e.* they need no certificate).

Thus a Lucas-Pratt certificate comprises a witness w, and a list of prime factors $q_1, q_2, \ldots$ of $n-1$ (and certificates for each q_j). Verification involves:

- verify that $w^{n-1} \equiv 1 \mod n$;
- verify that each $w^{(n-1)/q_j} \not\equiv 1 \mod n$;
- verify that $n - 1 = \prod_j q_j^{e_j}$ for positive exponents e_j;
- recursively verify that each q_j is prime.

The operations required to verify such a certificate are: iteration over a list, exponentiation modulo an integer, comparison with 1, division of integers, and divisibility testing of integers. These are all simple operations, and the entire function to verify a Lucas-Pratt certificate is small enough to be fully verifiable itself.

© Springer Nature Switzerland AG 2020
A. M. Bigatti et al. (Eds.): ICMS 2020, LNCS 12097, pp. 462–472, 2020.
https://doi.org/10.1007/978-3-030-52200-1_46

An important point in this example is that the certificate actually involves several cases: namely, if the prime is small enough, the certificate just says that it is a "small prime" (*e.g.* we can verify by table-lookup); otherwise the certificate contains a non-trivial body. In this instance there are just two possible cases.

We note that *generating* a Lucas-Pratt certificate could be costly because the prime factorization of $n - 1$ must be computed.

1.2 Costs of a Certificate

The total cost of a certificate comprises several components:

- computational cost of generating the certificate;
- size of the certificate (*e.g.* cost of storage or transmission);
- computational cost of verification given the certificate;
- size and code complexity of the verifier.

In the case of certifying the irreducibility of a polynomial in $\mathbb{Z}[x]$ we could issue trivial certificates for all polynomials, and say that the verifier simply has to be an implementation of a polynomial factorizer. We regard this as unsatisfactory because the size and code complexity of the verifier are too high.

2 Irreducibility Criteria for $\mathbb{Z}[x]$ and $\mathbb{Q}[x]$

We can immediately reduce from $\mathbb{Q}[x]$ to $\mathbb{Z}[x]$ thanks to Gauss's Lemma (for polynomials): let $f \in \mathbb{Q}[x]$ be non-constant then f is irreducible if and only if $\mathrm{prim}(f) \in \mathbb{Z}[x]$ is irreducible, where $\mathrm{prim}(f) = \alpha f$ and the uniquely defined, non-zero factor $\alpha \in \mathbb{Q}$ is such that all coefficients of $\mathrm{prim}(f)$ are integers with common factor 1, and the leading coefficient is positive,

The problem of certifying irreducibility in $\mathbb{Z}[x]$ has a long history, and has already been considered by several people. Here is a list of some approaches:

- give a "large" evaluation point n such that $f(n)$ has a large prime factor;
- degree analysis (from factorizations over one or more finite fields)[1];
- a linear polynomial is obviously irreducible;
- Newton polygon methods (*e.g.* Schönemann, Eisenstein, and Dumas [4]);
- Vahlen-Capelli lemma [10] for binomials
- Perron's Criterion [8];
- the coefficients are (non-negative) digits of a prime to some base b (*e.g.* [8]).

The first technique in the list was inspired by ideas from [3]; it seems to be new.

In this presentation, we shall assume that the degree is at least 2, and shall concentrate on the first two methods as they are far more widely applicable than others listed.

---

[1] *Degree analysis* has likely been known for a long time.

2.1   Factor Degree Analysis

Factor degree analysis is a well-known, behind-the-scenes technique in polynomial factorization. It involves using degrees of modular factors to obtain a list of *excluded degrees* for factors in $\mathbb{Z}[x]$.

We define a **factor degree lower bound** for $f \in \mathbb{Z}[x]$ to be $\Delta \in \mathbb{N}$ such that we have excluded all degrees less than Δ, *e.g.* through factor degree analysis. We can certify this lower bound by accompanying it with the modular factorizations used. Clearly, if degree analysis excludes all degrees up to $\frac{1}{2} \deg f$ then *we have proved that f is irreducible*. Finally, we may always take $\Delta = 1$ without any degree analysis.

In many cases we can indeed prove/certify irreducibility via degree analysis. However, there are some (infinite) families of polynomials where one must use "larger" primes, and there are also (infinite) families where irreducibility cannot be proved via factor degree analysis (*e.g.* resultants, in particular Swinnerton-Dyer polynomials, see also [6]).

Example 2. The well-known, classical example of a polynomial which cannot be proved irreducible by degree analysis is $x^4 + 1$: every modular factorization is into either 4 linears or 2 quadratics, so this does not let us exclude the possible existence of a degree 2 factor.

There are also many polynomials which can be proved irreducible by degree analysis, but are not irreducible modulo any prime; this property depends on the Galois group of the polynomial. For instance, $f = x^4 + x^3 + 3x + 4$ is one such polynomial: modulo 2 the irreducible factors have degrees 1 and 3, and modulo 5 both factors have degree 2; but it is never irreducible modulo p.

Degree Analysis Certificate. A degree analysis certificate comprises

- a subset $D \subseteq \{1, 2, \ldots, \frac{1}{2} \deg f\}$ of "not excluded" factor degrees
- a list, L, of pairs: a prime p, and the irreducible factors of f modulo p

If $D = \emptyset$, we have a certificate of ireducibility; otherwise the smallest element of the set is a *factor degree lower bound*.

Verification of the certificate involves the following steps:

- for each entry in L, check that the product of the modular factors is f;
- for each entry in L, compute the set of degrees of all possible products of the modular factors; verify that their intersection is D;
- check that each modular factor is irreducible (*e.g.* use gaussian reduction to compute the rank of $B - I$ where B is the Berlekamp matrix).

The main cost of the verification is the computation of B and the rank of $B - I$; the cost of computing B is greater for larger primes, so we prefer to generate certificates which use smaller primes if possible.

Practical Matters. We would like to know, in practice, how costly it is to produce a useful degree analysis certificate, and how large the resulting certificate could be. More specifically:

- How many different primes should we consider? And how large?
- How to find a minimal set of primes yielding the factor degree subset?
- How many primes are typically in the minimal set?

In our experience, a minimal length list very rarely contains more than 3 entries, but we should expect to consider many more primes during generation of the certificate. We can construct irreducible polynomials which require considering "large" primes to obtain useful degree information (*e.g.* $x^2 + Nx + N$ where $N = 1000000!$) but in many cases "small" primes up to around $\deg f$ suffice.

2.2 Irreducibility Certificates for $\mathbb{Z}[x]$ via Evaluation

Bunyakowski's conjecture (*e.g.* see page 323 of [7]) states that if $f \in \mathbb{Z}[x]$ is irreducible (and has trivial fixed divisor) then $|f(n)|$ is prime for infinitely many $n \in \mathbb{Z}$. Assuming the conjecture is true, we can get a certificate of irreducibility by finding a suitable evaluation point n (and perhaps including a certificate that $|f(n)|$ is prime).

Applying Bunyakowski's conjecture directly is inconvenient for two reasons:

- we want to handle polynomials with non-trivial fixed divisor;
- finding a suitable n may be costly, and the resulting $|f(n)|$ may be large.

The first point is solved by an easy generalization of the conjecture: let $f \in \mathbb{Z}[x]$ be irreducible and δ be its fixed divisor, then there are infinitely many $n \in \mathbb{Z}$ such that $|f(n)|/\delta$ is prime. The second point is a genuine inconvenience: for some polynomials, it can be costly to find a "Bunyakowski prime," and the prime itself will be large (and thus costly to verify). For example, let $f = x^{16} + 4x^{14} + 6x^2 + 4$ then the smallest good evaluation point is $n = 6615$, and $|f(n)| \approx 1.3 \times 10^{61}$.

A Large Prime Factor Suffices. Here we present a much more practical way of certifying irreducibility by evaluation: we require just a sufficiently large prime factor. Let $f \in \mathbb{Z}[x]$ be non-constant, and let $\rho \in \mathbb{Q}$ be a **root bound** for f: that is, for every $\alpha \in \mathbb{C}$ such that $f(\alpha) = 0$ we have $|\alpha| \leq \rho$. We note that it is relatively easy to compute root bounds (*e.g.* see [2]). The following proposition was partly inspired by Theorem 2 in [3], but appears to be new.

Proposition 1. *Let $f \in \mathbb{Z}[x]$ be non-constant, and let $\rho \in \mathbb{Q}$ be a root bound for f. Let $\Delta \in \mathbb{N}$ be a factor degree lower bound for f. If we have $n \in \mathbb{Z}$ with $|n| > 1 + \rho$ such that $|f(n)| = sp$ where $s < (|n| - \rho)^{\Delta}$ and p is prime then f is irreducible.*

Proof. For a contradiction, suppose that $f = gh \in \mathbb{Z}[x]$ is a non-trivial factorization. We may assume that $\Delta \leq \deg g \leq \deg h$. We have $f = C_f \prod_{j=1}^{d}(x - \alpha_j)$

where $d = \deg f$, $C_f \in \mathbb{Z}$ is the leading coefficient, and the α_j are the roots of f in $\mathbb{C}$. We may assume that the α_j are indexed so that the roots of g are $\alpha_1, \ldots, \alpha_{d_g}$ where $d_g = \deg g$.

By evaluation we have $f(n) = g(n) h(n)$ with all values in $\mathbb{Z}$. Also $f(n) \neq 0$ since $|n| > \rho$. We now estimate $|g(n)|$:

$$g(n) = C_g \prod_{j=1}^{d_g} (n - \alpha_j)$$

where $C_g \in \mathbb{Z}$ is the leading coefficient. Each factor in the product has magnitude greater than 1, so $|g(n)| \geq (|n| - \rho)^{\Delta} > s$. Similarly, $|h(n)| > s$. This contradicts the given factorization $f(n) = sp$. $\qquad\square$

When we have an evaluation point to which Proposition 1 applies we call it a **large prime factor witness** (abbr. **LPFW**) for f, ρ and Δ. We conjecture that every irreducible polynomial has infinitely many LPFWs; note that Bunyakowski's conjecture implies this.

Example 3. This example shows that it can be beneficial to look for large prime factor witnesses rather than Bunyakowski prime witnesses.

Let $f = x^{12} + 12x^4 + 92$ and take $\Delta = 1$. We compute $\rho = \frac{7}{4}$ as root bound, and then we obtain a LPFW at $n = 5$ with prime factor $p = 81382739$. In contrast, the smallest Bunyakowski prime is $\approx 3.06 \times 10^{41}$ at $n = 2865$.

In the light of this example we exclude consideration of a certificate based on Bunyakowski's conjecture, and consider only LPFWs.

We prefer to issue an LPFW certificate where the prime p is as small as "reasonably possible". Our implementation searches for suitable n in an incremental way, since smaller values of $|n|$ produce smaller values of $|f(n)|$, and we expect smaller values of $|f(n)|$ to be more likely to lead to an "sp" factorization with small prime factor p—this is only a heuristic, and does not guarantee to find the smallest such p. We look for the factorization $|f(n)| = sp$ by trial division by the first few small primes (and GMP's probabilistic prime test for p).

LPFW Certificate. An LPFW certificate comprises the following information:

- a root bound ρ,
- a factor degree lower bound Δ ←— with degree analysis certificate,
- the evaluation point $n > 1 + \rho$,
- the large prime factor p of $|f(n)|$ ←— (opt.) with certificate of primality.

Verification of an LPFW certificate entails:

- evaluating $f(n)$ and verifying that p is a factor;
- verifying that the discarded factor $s = |f(n)|/p$ satisfies $s < (|n| - \rho)^{\Delta}$;
- verifying that ρ is a root bound for f ←— see comment below;
- (if $\Delta > 1$) verifying that Δ is a factor degree lower bound;
- verifying that p is (probably) prime.

In many cases the root bound can be verified simply by evaluation of a modified polynomial: let $f(x) = \sum_{j=0}^{d} a_j x^j$ and set $f^*(x) = |a_d| x^d - \sum_{j=0}^{d-1} |a_j| x^j$, then if $f^*(\rho) > 0$ then ρ is a root bound for f. Some tighter root bounds may require applying an (iterated) Gräffe transform to f first (e.g. see [2]).

Example 4. This example shows how degree information can be useful in finding a small LPFW. Let $f = x^4 - 1036x^2 + 7744$. We find that $\rho = 33$ is a root bound. Without degree information (*i.e.* taking $\Delta = 1$) we obtain the first LPFW at $n = 65$ with corresponding prime $p = 13481269$. In contrast, from the factorization of f modulo 3 we can certify that $\Delta = 2$ is a factor degree lower bound for f. This information lets us obtain an LPFW at $n = 47$ with far smaller corresponding prime $p = 14519$.

3   Möbius Transformations

We define a (minor generalization of) a *Möbius transformation* for $\mathbb{Z}[x]$. The crucial property for us is that these transformations preserve irreducibility (except for some polynomials of degree 1).

Definition 1. *Let $M = \begin{pmatrix} a & b \\ c & d \end{pmatrix}$ be a 2×2 matrix. Let $f = \sum_{j=0}^{\deg(f)} c_j x^j$ be a polynomial in $\mathbb{Z}[x]$. We define the **Möbius transform** of f induced by M to be the polynomial $\mu_M(f) = \sum_{j=0}^{\deg f} c_j (ax + b)^j (cx + d)^{\deg(f)-j}$.*

In our applications the matrix entries will be integers, and we shall suppose that at least one of a and c is non-zero.

Definition 2. *A Möbius transformation μ_M is **degenerate** if $\det M = 0$.*

Definition 3. *Let μ_M be a Möbius transform. We define the **pseudo-inverse** of μ_M to be the Möbius transformation corresponding to the classical adjoint $M^{adj} = \begin{pmatrix} d & -b \\ -c & a \end{pmatrix}$. We write μ_M^* to denote the pseudo-inverse.*

Here is a summary of useful properties of a Möbius transformation μ_M.

Proposition 2. *Let $M = \begin{pmatrix} a & b \\ c & d \end{pmatrix}$ be non-singular, so μ_M is non-degenerate.*

(a) *Let $f = \alpha x + \beta$ be a linear polynomial. If $f(\frac{a}{c}) \neq 0$ then $\mu_M(f)$ is linear; otherwise $\mu_M(f) = \alpha b + \beta d$ is a non-zero constant.*

(b) *μ_M respects multiplication: $\mu_M(gh) = \mu_M(g)\,\mu_M(h)$.*

(c) *$\deg(\mu_M(f)) = \deg(f) \iff f(\frac{a}{c}) \neq 0$.*

(d) *If $\deg(\mu_M(f)) = \deg(f)$ then $\mu_M^*(\mu_M(f)) = D^{\deg(f)} f(x)$ where $D = \det M$.*

(e) *If $\deg(\mu_M^*(f)) = \deg(f)$ then $\mu_M(\mu_M^*(f)) = D^{\deg(f)} f(x)$ where $D = \det M$.*

(f) *If $a, b, c, d \in \mathbb{Z}$ and $f \in \mathbb{Z}[x]$ is irreducible and $\deg(\mu_M(f)) = \deg(f)$ then $\mathrm{prim}(\mu_M(f))$ is irreducible.*

Proof. Parts (a) and (b) are elementary algebra. Part (c) follows from (a) and (b) by considering the factorization of f over a splitting field. Parts (d) and (e) are elementary for linear f; the general case follows by repeated application of part (b).

For part (f), suppose we have a counter-example $f \in \mathbb{Z}[x]$, then we have a non-trivial factorization $\mu_M(f) = gh$, but by (b) and (d) we deduce that $D^{\deg(f)} f = \mu_M^*(g)\,\mu_M^*(h)$ which is a non-trivial factorization, contradicting the assumption that f was irreducible.

Our interest in Möbius transformations is that they offer the possibility of finding a better LPFW certificate. Unfortunately we do not yet have a good way of determining which Möbius transformations are helpful.

Example 5. Let $f = 97x^4 + 76x^3 + 78x^2 + 4x + 2$. We obtain a LPFW certificate with $\rho = 7/5$, $\Delta = 1$, $n = -4$ with corresponding prime factor $p = 10601$.

Let $M = \left(\begin{smallmatrix} 1 & 1 \\ -3 & 2 \end{smallmatrix}\right)$. Let $g = \mathrm{prim}(\mu_M(f)) = (x^4 + 1)$; by Proposition 2.(f) since $\deg g = \deg f$ a LPFW certificate for g also certifies that f is irreducible. For g we obtain a certificate with $\rho = 1$, $\Delta = 1$, $n = 2$ with *much smaller* corresponding prime factor $p = 17$.

Unsolved Problem: How to find a good Möbius matrix M given just f?

3.1   Certifying a Transformed Polynomial

Naturally, if we generate a LPFW certificate for a transformed polynomial $\mu_M(f)$ then we must indicate which Möbius transformation was used. Given two polynomials $f, g \in \mathbb{Z}[x]$ of the same degree d, and $M \in \mathrm{Mat}_{2\times 2}(\mathbb{Z})$, one can easily verify that $g = \mathrm{prim}(\mu_M(f))$ by evaluating f at $\deg(f)$ distinct rational points, and g at the (rational) transforms of these points, and then checking that the ratios of the values are all equal. So the extra information needed is M and $\mu_M(f)$.

3.2   Fixed Divisors

Definition 4. *Let $f \in \mathbb{Z}[x]$ be non-zero. The* **fixed divisor** *of f is defined to be* $\mathrm{FD}(f) = \gcd\{f(n) \mid n \in \mathbb{Z}\}$.

Some content-free polynomials have non-trivial fixed divisors: an example is $f = x^2 + x + 2$ which is content-free but has fixed divisor 2.

Proposition 3. *Let $f \in \mathbb{Z}[x]$ be non-zero. Its fixed divisor is equal to:*

$$\mathrm{FD}(f) = \gcd(f(1), f(2), \ldots, f(\deg f))$$

Proof. The standard proof follows easily from representating of f with respect to the "binomial basis" for $\mathbb{Z}[x]$, namely $\{\binom{x}{k} \mid k \in \mathbb{N}\}$.

Polynomials having large fixed divisor δ cannot have small LPFW certificates because we are forced to choose large evaluation points since we must have $(|n| - \rho)^\Delta > \delta$. This problem becomes more severe for higher degree polynomials since the fixed divisor can be as large as $d!$ where d is the degree.

We can reduce the size of the fixed divisor by scaling the indeterminate (*i.e.* a Möbius transformation for a diagonal matrix), or perhaps reversing the polynomial and scaling the indeterminate (*i.e.* a Möbius transformation for an anti-diagonal matrix). We have not yet investigated the use of more general Möbius transformations.

Let $f \in \mathbb{Z}[x]$ be content-free, irreducible with fixed divisor δ. Let q be a prime factor of δ, and let k be the multiplicity of q in $|f(0)|$. Then $g(x) = q^{-k} f(q^k x) \in \mathbb{Z}[x]$ has fixed divisor δ/q^k. In practice, we consider several polynomials obtained by scaling x by $q^1, q^2, \ldots, q^k$; in fact scaling by $q^{-1}, q^{-2}, \ldots$ can also be beneficial.

4   Implementation and Experimentation

Our prototype implementation runs degree analysis and LPFW search "in parallel": *i.e.* it repeatedly alternates a few iterations of degree analysis with a few iterations of LPFW search. If degree analysis finds a new factor degree lower bound, Δ, this information is passed to the LPFW search.

4.1   Degree Analysis

We adopted the following strategy for choosing primes during degree analysis: initially we create a list of "preferential primes" (*e.g.* including the first few primes greater than the degree), then we pick primes alternately from this list or from a random generator. The range for randomly generated primes is gradually increased to favour finding quickly a certificate involving smaller primes (since these are computationally cheaper to verify).

This strategy was inspired by some experimentation. There exist polynomials whose degree analysis certificates must involve "large" primes: *e.g.* a good set of primes for $x^4 + 16x^3 + 5x^2 - 14x - 18$ must contain at least one prime greater than 101. Also, empirically we find that a degree analysis certificate for an (even) Hermite polynomial must use primes greater than the degree.

To issue a certificate, we look for a minimal cardinality subset of the primes used which suffices. This subset search is potentially exponential, but in our experiments it is very rare for a minimal subset to need more than 3 primes.

4.2   Large Prime Factor Witness

As already mentioned, not all polynomials can be certified irreducible by degree analysis. A well-known class of polynomials for which irreducibility cannot be shown by degree analysis are the *Swinnerton-Dyer polynomials*: they are the minimal polynomials for sums of square-roots of "independent" integers. A more general class of such polynomials was presented in [6].

We saw in Example 5, it can be better to issue a LPFW certificate for a transformed polynomial, but we do not yet have a good way of finding a good Möbius transformation. Our current prototype implementation considers only indeterminate scaling and possibly reversal: *i.e.* the Möbius matrix must be diagonal or anti-diagonal. A list of all scaling and reverse-scaling transforms by "simple" rationals is maintained, and the resulting polynomials are considered "in parallel". For each transformed polynomial we keep track of two evaluation points (one positive, one negative) and the corresponding evaluations. The evaluations are then considered in order of increasing absolute value; once an evaluation has been processed the corresponding evaluation point is incremented (or decremented, if it is negative).

The LPFW search depends on a factor degree lower bound, Δ, which is initially 1. The degree analysis "thread" may at any time furnish a better value for Δ. So that this asynchrony can work well the LPFW search records, *for each possible factor degree lower bound,* any certificates it finds. When a higher Δ is received, the search first checks whether a corresponding LPFW certificate has already been recorded; if so, that certificate is produced as output. Otherwise searching proceeds using the new Δ.

4.3   Examples

Here are a few examples as computed by the current prototype, since degree analysis picks primes in a pseudo-random order different certificates may be issued for the same polynomial.

- $x^{16} + 4x^{14} + 6x^2 + 4$: degree analysis with prime list $L = [13, 127]$
- $x^4 + 16x^3 + 5x^2 - 14x - 18$: degree analysis with prime list $L = [107]$
- 21-st cyclotomic polynomial: LPFW with $\rho = 2$, $\Delta = 1$, $n = 3$, and prime factor $p = 368089$
- Swinnerton-Dyer polynomial for $[71, 113, 163]$: LPFW with $\rho = 43$, $\Delta = 2$ (with $L = [3]$), $n = 82$ and prime factor $p = 2367715751029$
- $97x^4 + 76x^3 + 78x^2 + 4x + 2$: **transform** $x \mapsto \frac{2}{x}$, LPFW $\rho = 67/5$, $\Delta = 2$ (with $L = [3]$), $n = -29$ and prime factor $p = 3041$

A quick comment about run-times: our *interpreted prototype* favours producing certificates which are cheap to verify (rather than cheap to generate); the degree analysis certificates took ~ 0.25 s each to generate, the others ~ 0.5 s each. We did not measure verification run-time, but fully expect it to be less than 0.01 s in each case. In comparison, the polynomial factorizer in CoCoA took less than 0.01 s for all of these polynomials.

As a larger example: the prototype took ~ 20 s (we expect the final implementation to be significantly faster) to produce a certificate for the degree 64 (Swinnerton-Dyer) minimal polynomial of

$$\sqrt{61} + \sqrt{79} + \sqrt{139} + \sqrt{181} + \sqrt{199} + \sqrt{211}$$

This polynomial has fixed divisor $\delta = 2^{29}\,5^{14}\,13^4 \approx 1.2 \times 10^{28}$. Our prototype found and applied the transformation $x \mapsto \frac{52}{15}x$, then produced an LPFW certificate for the transformed polynomial: $\rho = 451/16$, $\Delta = 2$ (with $L = [19]$), $n = 46$ and $p \approx 7.5 \times 10^{180}$ which was confirmed to be "probably prime" (according to GMP [5]). The classical Berlekamp-Zassenhaus factorizer in CoCoA [1] took about 300 s to recognize irreducibility.

4.4 A Comment About Run-Time

An anonymous referee reasonably asked about expected run-time or a (possibly heuristic) complexity analysis. The answer is *"It depends . . . "*. For "almost all" polynomials, degree analysis suffices and is quick. In our setting, the LPFW search effectively happens only if a degree analysis certificate cannot be quickly found. In our experiments, the number of iterations in LPFW search before producing a certificate was quite irregular.

5 Conclusion

As mentioned in the introduction there are many different criterions for certifying the irreducibility of a polynomial in $\mathbb{Z}[x]$. Here we have concentrated on just two of them, and have pointed out how they can "collaborate".

We have built a prototype implementation in CoCoA [1], and plan to integrate it into CoCoALib, the underlying C++ library (where we expect significant performance gains).

An interesting future possibility is for the requester of the certificate to state which criterions may be used (dictated by the implemented verifiers that the requester has available). But, a too restrictive choice of criterions may make it impossible to generate a certificate: *e.g.* there is no "Eisenstein" certificate for most polynomials.

References

1. Abbott, J., Bigatti, A.M., Robbiano, L.: CoCoA: a system for doing Computations in Commutative Algebra. http://cocoa.dima.unige.it/
2. Abbott, J.: Bounds on factors in $\mathbb{Z}[x]$. J. Symb. Comput. **50**, 532–563 (2013)
3. Davenport, J., Padget, J.: Heugcd: how elementary upperbounds generate cheaper data. In: Caviness, B.F. (ed.) EUROCAL 1985. LNCS, vol. 204, pp. 18–28. Springer, Heidelberg (1985). https://doi.org/10.1007/3-540-15984-3_231
4. Dumas, G.: Sur quelques cas d'irréductibilité des polynomes à coefficients rationnels. Journ. de Math. **6**(2), 191–258 (1906)
5. Granlund, T., et al.: GNU multiprecision library. http://www.gmplib.org/
6. Kaltofen, E., Musser, D.R., Saunders, B.D.: A generalized class of polynomials that are hard to factor. SIAM J. Comput. **12**, 473–483 (1983)
7. Lang, S.: Algebra, 3rd edn. Addison Wesley, Reading (1993)
8. Perron, O.: Neue Kriterien für die Irreduzibilität algebraischer Gleichungen. J. Reine Angew. Math. **132**, 288–307 (1907)

9. Pratt, V.R.: Every prime has a succinct certificate. SIAM J. Comput. **4**, 214–220 (1975)

10. Rowlinson, E.: New proofs for two theorems of Capelli. Can. Math. Bull. **7**, 431–433 (1964)

A Content Dictionary for In-Object Comments

Lars Hellström[(⊠)] [iD]

Division of Applied Mathematics, The School of Education, Culture and
Communication, Mälardalen University, Box 883, 721 23 Västerås, Sweden
lars.hellstrom@mdh.se

Abstract. It is observed that some OpenMath objects may benefit from
containing comments. A content dictionary with suitable attribution
symbols is proposed. This content dictionary also provides application
symbols for constructing comments that are somewhat more than just
plain text strings.

1 Introduction

OpenMath [1], of which is Content MathML [6, Ch. 4] is one encoding, is the
open standard for machine-readable formalised mathematics. Though an early
motivating application was the exchange of formulae between different Computer
Algebra Systems, the standard is by means restricted to that context; other
applications include the formalised mathematics aspect of documents written in
the Lurch [2] word processor and semantic export of material [5] from the Digital
Library of Mathematical Functions (DLMF) [3].

One key ingredient in the standard is that the many symbols needed to
express mathematical claims as object are not defined in the standard itself,
but by separate documents called Content Dictionaries (CDs) that may be created by anyone; ultimately each symbol is uniquely identified by a URI, and a
web browser might use this to fetch the defining CD if that is made available
by an appropriately located and configured server. The content dictionaries are
also where OpenMath (OM) gets a bit recursive, since CDs typically contain
OM objects (formulae) stating Formal Mathematical Properties (FMPs) of the
symbols being defined. Though not *necessarily* the kind of rigorous axiomati-
sations one would expect in an automated theorem prover system, FMPs have
the potential of fulfilling that role, and as such they may sometimes need to be
rather large. Sheer size can then make it difficult for a reader to understand what
is being stated (e.g., what would happen in a particular edge case), even when
being well versed in the mathematics as such. This phenomenon of unavoidable
complexity making things difficult to understand is utterly familiar from the
realm of programming, and there it has a time-tested solution: document your
code, and in particular *include appropriate comments* in it!

Being an application of XML, the OpenMath CD format of course allows
XML comments, but those would only benefit a person editing the CD, which is

A. M. Bigatti et al. (Eds.): ICMS 2020, LNCS 12097, pp. 473–481, 2020.
https://doi.org/10.1007/978-3-030-52200-1_47

a minor activity. More common is to use the CD as a reference or even authority on the meanings of symbols, and what that kind of document needs in order to be more easily digested are *annotations*—comments that are visible to the reader of a presentation of the document. Since the normative form of the CD will go through XML processing when generating such a presentation, XML comments are no good. Luckily the OpenMath standard already contains a feature that suits this end perfectly: attributions. One may to any OpenMath element (subformula) attach anything, and as long as the attribution symbol is not of the semantic variety, any OpenMath phrasebook (processor of OM data) is free to ignore it.

What prevents authors of CDs, and other OM objects of a more persistent nature, from already including annotations in them is that there has not existed any CD with appropriate symbols (they need to be declared as having the Role of 'attribution') to use as keys in these attributions. That is of course easily remedied by creating such a CD, and the primary contribution in this work. Secondarily, this work also provides some additional symbols that may be used to provide simple markup of the comments; the rationale behind this is discussed below, in conjunction with the actual symbol definitions.

One example below shows the OM-XML encoding of a comment attribution, for maximal clarity, but most of them employ a more compact "semiformula" format. In those examples, @ denotes application (OMA), **attr** denotes attribution (OMATTR), **bind** denotes binding (OMBIND), and symbols are written as *cd.name* (e.g. 'relation1.eq' for the standard equality relation =). Outside of semiformulae, references to symbols are rather written on the URI form *cd#name*, e.g. relation1#eq.

2   The comment1 Content Dictionary

Description: Symbols to attach comments as attributions of OpenMath objects.
Standard OpenMath licence terms apply.

Note 1. Due to the size constraints of the ICMS proceedings, some material (including entire symbol definitions) have been omitted from this presentation. On the other hand, this section is based on the literate source for the content dictionary.

2.1   Basics

The purpose of attaching comments to objects may need an explanation, since not all kinds of comments make sense in all OpenMath usage scenarios. One factor of importance is the lifetime of the objects in question; objects that only stick around for a short time will at most have comments that are generated automatically (perhaps inserted by tools examining OpenMath objects, as a means of reporting on their findings), but objects of a very long duration (such as those in Formal Mathematical Properties) could well acquire manually written comments as part of content dictionary maintenance.

Symbol `remark` (attribution)

> The generic (neutral) comment attribution symbol.

Commented Mathematical Property. As an attribution, the value may be data in some foreign format, but expressing comments as OpenMath objects can be more to the point as well as easier on implementors (since any tool working with comment attributions must already be able to read and/or write OM). In particular, a comment that is a piece of plain text can be succintly encoded as an OpenMath string.

Example.

```
<OMOBJ>
  <OMA> <OMS cd="arith1" name="divide"/>
    <OMATTR>
      <OMATP>
        <OMS cd="comment1" name="remark"/>
        <OMSTR>This cancellation may degrade numeric precision.</OMSTR>
      </OMATP>
      <OMA> <OMS cd="arith1" name="minus"/>
        <OMI> 1 </OMI>
        <OMA> <OMS cd="transc1" name="cos"/>
          <OMV name="x"/>
        </OMA>
      </OMA>
    </OMATTR>
    <OMA> <OMS cd="transc1" name="sin"/>
      <OMV name="x"/>
    </OMA>
  </OMA>
</OMOBJ>
```

In particular, using instead e.g. XHTML+MathML as format for comments, one would create an "industry standard" situation rather than a "math in the middle" situation: it would be easy on tools that aim to output HTML or the like, but very demanding for tools that do not target HTML.

Comments are primarily aimed at human readers, so the tools that take notice of comments would normally be those that generate a presentation. Even in that context, comments need not have much visible effect; in an interactive medium (e.g. a web page) it might be most appropriate to leave comments invisible until the user undertakes some action to display them (such as placing the cursor on top of the commented object, for displaying the comment as a tooltip).

One type of comment that might require visibility by default are headings, i.e., comments that aim to help a reader discern the essential structure of an OM object. In normal text, headings present little trouble as they apply at distinct vertical positions and may be inserted between paragraphs or in the margin, but an OM object presented as a standard typeset formula may well warrant several headings on a single line, and then the problem of what falls under which heading becomes more troublesome. (One approach worth exploring could be to tint the background according to the heading which applies.) Generating presentations is not the issue here, but providing the information required to do so is, and just saying "heading" risks creating an ambiguity regarding the commenter's intents.

The problem is that text headings are normally understood as applying to all text up to the next (same level) heading, and someone editing an OM object in e.g. OM-XML format is likely to apply that interpretation, but OM objects grammatically rather have a tree structure, so phrasebook authors are likely to instead interpret a heading as applying only to the explicitly attributed element. In order to curtail that ambiguity, this CD has two heading symbols that explicitly state which kind of scope is intended.

Symbol `s-heading` (attribution)

> This symbol attaches a comment to an object, classifying that comment as a heading. The scope of that heading consists of the attributed element and all its later siblings (up to the next sibling with an s-heading attribution, if any).

Example. While giving a long list of conditions, an author will often spontaneously organise these into blocks, but is perhaps not prepared to let those blocks be explicit in the structure of the formula; authors may be more willing to write

@(logic1.and, **attr**(comment1.s-heading "`First block`", $c1$), $c2$, $c3$, $c4$, $c5$,
  attr(comment1.s-heading "`Second block`", $c6$), $c7$, $c8$)

than the logically equivalent

@(logic1.and, **attr**(comment1.t-heading "`First block`",
    @(logic1.and, $c1$, $c2$, $c3$, $c4$, $c5$)),
  attr(comment1.t-heading "`Second block`", @(logic1.and, $c6$, $c7$, $c8$)))

Symbol `t-heading` (attribution)

> This symbol attaches a comment to an object, classifying that comment as a heading. The scope of that heading is exactly the element to which the comment is attached, i.e., it adheres strictly to the tree structure of the OM object.

Example. The distinction between left and right hand side of an implication is explicit in the tree structure of the formula, but inserting explicit headings can still improve the overall human readability.

bind(quant1.forall; *desc, p, epsilon*;

    @(logic1.implies; **attr**(comment1.t-heading "`Conditions`",

        @(logic1.and, @(Riemann-surface.is_description, *desc*),

          @(set1.in, *p*, @(comment1.set, *desc*)),

            @(set1.in, *epsilon*, setname1.R), @(relation1.gt, *epsilon*; 0))),

      attr(comment1.t-heading "`Claims`";

        @(logic1.and, @(set1.prsubset, @(set1.set, *p*),

          @(Riemann-surface.neighbourhood, *desc, p, epsilon*)),

        @(set1.subset, @(Riemann-surface.neighbourhood, *desc, p, epsilon*),

          @(Riemann-surface.set, *desc*))))))

Symbol `question` (attribution)

> This symbol attaches a comment to an object, classifying that comment as a question.

Example

attr(comment1.question

    "`Is this formalisation what we want also under intuitionistic logic?`",

    @(logic1.implies, @(logic1.not, Q), @(logic1.not, P)))

Commented Mathematical Property. The extra signal sent by classifying a comment as a question is that an answer to it is sought. Presentation-wise, it might not be much different from a plain remark, but if one has a large collection of OM objects then the ability to easily search for those with pending questions is obviously valuable.

Symbol `warning` (attribution)

> This symbol attaches a comment to an object, classifying that comment as a warning.

Symbol `diagnostic` (attribution)

> This symbol attaches a comment to an object, classifying that comment as containing diagnostic information.

2.2   Comment Format Functions

A human reader can usually guess what format a piece of text is, but automated processing is much aided by allowing material of a specific form to be marked up as such.

Symbol `uri` (application)

> This function expects a string as argument, and asserts that this string is a Uniform (or more generally Internationalized) Resource Identifier, producing a corresponding comment object as result.

Example. Does fns1#range denote the image of a function or the codomain of a function? There is a tracker issue on this.

> @(**attr**(comment1.question
>
> @(comment1.uri, "`https://github.com/OpenMath/CDs/issues/4`"),
>
> fns1.range), f)

Symbol `formula` (application)

> This function can take an arbitrary OpenMath object as argument, and asserts that the comment is this object as a formula, rather than the interpretation of that object. To the author, this is essentially to enter "math mode".

Example. An expression ln(-x) strikes most readers as strange, but in the case that x < 0 there is of course nothing strange or problematic about it at all.

> **attr**(comment1.remark@(comment1.formula, @(relation1.lt, x, 0)),
>
> @(transc1.ln, @(arith1.unary_minus, x)))

A content dictionary only specifies what an OM object means, not how it is to be processed, but a reasonable mode of processing comments is to see each function as returning a presentation (in some suitable form) of the comment in question; this is feasible because the vocabulary of "comment-valued" applications is fairly small. However if a general formula is to be used as a comment then this may obviously contain arbitrary symbols and constructions, so the formula symbol serves as a sign to any processing phrasebook that a different mode of processing should be applied for this subobject.

The `meta#Example` symbol arguably provides an imperfect encoding of OM CD Examples because it lacks this kind of symbol: it cannot distinguish between a string that is text and a string that is an embedded OMOBJ that is just a string, whereas the standard encoding of that content dictionary document would make this distinction.

Commented Mathematical Property. It is a feature of formulae in comments that they may contain variables whose scopes extend outside the comment. Renaming a variable through alpha-conversion then causes not only occurrences of that variable in the main OMOBJ to be renamed, but also occurrences in comment attributions in that OMOBJ, and this happens as a consequence of the general principles in the OM standard. Foreign encodings of comments would not be so lucky.

Symbol mixed (application)

> This function concatenates an arbitrary number of arbitrary comment pieces into one comment.

This is named for the `multipart/mixed` content-type in MIME.

Commented Mathematical Property. Heuristics may be applied when joining different pieces together, but expect there to be interword spaces between the pieces when concatenated.

Example

attr(comment1.remark

@(comment1.mixed, "`This comment is`", "`split over two strings.`"),

@(relation1.eq, @(arith1.plus, 1, 1), 2))

2.3   Comment Structuring Functions

It can sometimes be useful to attach metadata to comments. The following are some function symbols which take a comment as their first argument, producing an embellished comment as result.

Symbol when (application)

> This function attaches a timestamp to a comment.

Commented Mathematical Property. The first argument is the comment to act upon, and the second argument is the timestamp to attach. If the second argument is a number, then that is interpreted as the number of seconds since the new year 1970 CE (the UNIX epoch). The second argument may alternatively be a string expressing the date; the format of that string is then not mandated, but ISO 8601 is recommended.

Symbol language (application)

> This function specifies the language used in a comment.

Commented Mathematical Property. The first argument is the comment. The second argument is an IETF language tag.

Symbol `alternatives` (application)

> This function takes an arbitrary number of comments as arguments, making the assertion that they all say the same thing (although probably in different ways) and return a combined comment where the individual comments are distinct pieces.

Example. This can be combined with the language symbol to collect different language versions of the same comment.

> **attr**(comment1.remark
>
> @(comment1.alternatives, @(comment1.language, "`continuous`", "en"),
>
> @(comment1.language, "`stetig`", "de")), f)

Symbol `signed` (application)

> This function attaches a signature to a comment.

Commented Mathematical Property. The first argument is the comment. The second argument identifies the entity (not necessarily a person) signing the comment.

Commented Mathematical Property. If there are additional arguments, then the third argument names a digital signature scheme and the fourth is a signature under that scheme. (This CD does not define any signature schemes.)

On the other hand, for example RFC5652 [4] does define signature schemes. What those sign are octet-sequences, so one need only define a way of encoding the OpenMath objects being signed as such. One obvious possibility is to use the binary encoding of the material signed, normalised to make this encoding canonical (expanding all references and using the shortest possible encoding form should suffice quite far).

References

1. Buswell, S., et al.: The openmath standard. Technical report, The OpenMath Society, July 2019. https://www.openmath.org/standard/om20-2019-07-01/
2. Carter, N.C., Monks, K.G.: Lurch: a word processor that can grade students' proofs. In: Lange, C., et al. (eds.) MathUI, OpenMath, PLMMS and ThEdu Workshops and Work in Progress at the Conference on Intelligent Computer Mathematics. No. 1010 in CEUR Workshop Proceedings, Aachen (2013). http://ceur-ws.org/Vol-1010/paper-04.pdf
3. Olver, F.W.J., et al.: NIST Digital Library of Mathematical Functions. Release 1.0.26 of 2020-03-15. http://dlmf.nist.gov/
4. Housley, R.: Cryptographic Message Syntax (CMS). RFC 5652 (Internet Standard), September 2009. https://doi.org/10.17487/RFC5652. https://www.rfc-editor.org/rfc/rfc5652.txt

5. Rabe, F., Farmer, W.M., Passmore, G.O., Youssef, A. (eds.): CICM 2018. LNCS (LNAI), vol. 11006. Springer, Cham (2018). https://doi.org/10.1007/978-3-319-96812-4
6. Miner, R.R., Ion, P.D.F., Carlisle, D.: Mathematical markup language (MathML) version 3.0, 2nd edn. W3C recommendation, W3C, April 2014. http://www.w3.org/TR/2014/REC-MathML3-20140410/

Implementing the Tangent Graeffe Root Finding Method

Joris van der Hoeven[1,2] and Michael Monagan[1(✉)]

[1] Department of Mathematics, Simon Fraser University, Burnaby, Canada
mmonagan@sfu.ca
[2] CNRS, LIX, École polytechnique, Palaiseau, France

Abstract. The tangent Graeffe method has been developed for the efficient computation of single roots of polynomials over finite fields with multiplicative groups of smooth order. It is a key ingredient of sparse interpolation using geometric progressions, in the case when blackbox evaluations are comparatively cheap. In this paper, we improve the complexity of the method by a constant factor and we report on a new implementation of the method and a first parallel implementation.

1 Introduction

Consider a polynomial function $f : \mathbb{K}^n \to \mathbb{K}$ over a field $\mathbb{K}$ given through a black box capable of evaluating f at points in $\mathbb{K}^n$. The problem of *sparse interpolation* is to recover the representation of $f \in \mathbb{K}[x_1, \ldots, x_n]$ in its usual form, as a linear combination

$$f = \sum_{1 \leqslant i \leqslant t} c_i \boldsymbol{x}^{e_i} \qquad (1)$$

of monomials $\boldsymbol{x}^{e_i} = x_1^{e_{1,1}} \cdots x_n^{e_{1,n}}$. One popular approach to sparse interpolation is to evaluate f at points in a geometric progression. This approach goes back to work of Prony in the eighteen's century [15] and became well known after Ben-Or and Tiwari's seminal paper [2]. It has widely been used in computer algebra, both in theory and in practice; see [16] for a nice survey.

More precisely, if a bound T for the number of terms t is known, then we first evaluate f at $2T - 1$ pairwise distinct points $\boldsymbol{\alpha}^0, \boldsymbol{\alpha}^1, \ldots, \boldsymbol{\alpha}^{2T-2}$, where $\boldsymbol{\alpha} = (\alpha_1, \ldots, \alpha_n) \in \mathbb{K}^n$ and $\boldsymbol{\alpha}^k := (\alpha_1^k, \ldots, \alpha_n^k)$ for all $k \in \mathbb{N}$. The generating function of the evaluations at $\boldsymbol{\alpha}^k$ satisfies the identity

$$\sum_{k \in \mathbb{N}} f(\boldsymbol{\alpha}^k) z^k = \sum_{1 \leqslant i \leqslant t} \sum_{k \in \mathbb{N}} c_i \boldsymbol{\alpha}^{e_i k} z^k = \sum_{1 \leqslant i \leqslant t} \frac{c_i}{1 - \boldsymbol{\alpha}^{e_i} z} = \frac{N(z)}{\Lambda(z)},$$

where $\Lambda = (1 - \boldsymbol{\alpha}^{e_1} z) \cdots (1 - \boldsymbol{\alpha}^{e_t} z)$ and $N \in \mathbb{K}[z]$ is of degree $< t$. The rational function N/Λ can be recovered from $f(\boldsymbol{\alpha}^0), f(\boldsymbol{\alpha}^1), \ldots, f(\boldsymbol{\alpha}^{2T-2})$ using fast Padé

Note: This paper received funding from NSERC (Canada) and "Agence de l'innovation de défense" (France).

Note: This document has been written using GNU TEX_MACS [13].

A. M. Bigatti et al. (Eds.): ICMS 2020, LNCS 12097, pp. 482–492, 2020.
https://doi.org/10.1007/978-3-030-52200-1_48

approximation [4]. For well chosen points $\boldsymbol{\alpha}$, it is often possible to recover the exponents e_i from the values $\boldsymbol{\alpha}^{e_i} \in \mathbb{K}$. If the exponents e_i are known, then the coefficients c_i can also be recovered using fast structured linear algebra [5]. This leaves us with the question how to compute the roots $\boldsymbol{\alpha}^{-e_i}$ of Λ in an efficient way.

For practical applications in computer algebra, we usually have $\mathbb{K} = \mathbb{Q}$, in which case it is most efficient to use a multi-modular strategy, and reduce to coefficients in a finite field $\mathbb{K} = \mathbb{F}_p$, where p is a prime number that we are free to choose. It is well known that polynomial arithmetic over $\mathbb{F}_p$ can be implemented most efficiently using FFTs when the order $p - 1$ of the multiplicative group is smooth. In practice, this prompts us to choose p of the form $s2^l + 1$ for some small s and such that p fits into a machine word.

The traditional way to compute roots of polynomials over finite fields is using Cantor and Zassenhaus' method [6]. In [10,11], alternative algorithms were proposed for our case of interest when $p - 1$ is smooth. The fastest algorithm was based on the *tangent Graeffe transform* and it gains a factor $\log t$ with respect to Cantor–Zassenhaus' method. The aim of the present paper is to report on a parallel implementation of this new algorithm and on a few improvements that allow for a further constant speed-up.

In Sect. 2, we recall the Graeffe transform and the heuristic root finding method based on the tangent Graeffe transform from [10]. In Sect. 3, we present the main new theoretical improvements, which all rely on optimizations in the FFT-model for fast polynomial arithmetic. Our contributions are twofold. In the FFT-model, one backward transform out of four can be saved for Graeffe transforms of order two (see Sect. 3.2). When composing a large number of Graeffe transforms of order two, FFT caching can be used to gain another factor of 3/2 (see Sect. 3.3). In the longer preprint version of the paper [12], we also show how to generalize our methods to Graeffe transforms of general orders and how to use it in combination with the truncated Fourier transform.

Section 4 is devoted to our new sequential and parallel implementations of the algorithm in C and Cilk C. Our sequential implementation confirms the gain of a new factor of two when using the new optimizations. So far, we have achieved a parallel speed-up by a factor of 4.6 on an 8-core machine. Our implementation is freely available at http://www.cecm.sfu.ca/CAG/code/TangentGraeffe.

2   Root Finding Using the Tangent Graeffe Transform

2.1   Graeffe Transforms

The traditional *Graeffe transform* of a monic polynomial $P \in \mathbb{K}[z]$ of degree d is the unique monic polynomial $G(P) \in \mathbb{K}[z]$ of degree d such that

$$G(P)(z^2) = P(z)P(-z). \qquad (2)$$

If P splits over $\mathbb{K}$ into linear factors $P = (z - \beta_1) \cdots (z - \beta_d)$, then one has

$$G(P) = (z - \beta_1^2) \cdots (z - \beta_d^2).$$

More generally, given $r \geqslant 2$, we define the *Graeffe transform of order r* to be the unique monic polynomial $G_r(P) \in \mathbb{K}[z]$ of degree d such that $G_r(P)(z) = (-1)^{rd} \operatorname{Res}_u(P(u), u^r - z)$. If $P = (z - \beta_1) \cdots (z - \beta_d)$, then

$$G_r(P) = (z - \beta_1^r) \cdots (z - \beta_d^r).$$

If $r, s \geqslant 2$, then we have

$$G_{rs} = G_r \circ G_s = G_s \circ G_r. \tag{3}$$

2.2   Root Finding Using Tangent Graeffe Transforms

Let ϵ be a formal indeterminate with $\epsilon^2 = 0$. Elements in $\mathbb{K}[\epsilon]/(\epsilon^2)$ are called *tangent numbers*. Now let $P \in \mathbb{K}[z]$ be of the form $P = (z - \alpha_1) \cdots (z - \alpha_d)$ where $\alpha_1, \ldots, \alpha_d \in \mathbb{K}$ are pairwise distinct. Then the *tangent deformation* $\tilde{P}(z) := P(z + \varepsilon)$ satisfies

$$\tilde{P} = P + P'\epsilon = (z - (\alpha_1 - \epsilon)) \cdots (z - (\alpha_d - \epsilon)).$$

The definitions from the previous subsection readily extend to coefficients in $\mathbb{K}[\epsilon]$ instead of $\mathbb{K}$. Given $r \geqslant 2$, we call $G_r(\tilde{P})$ the *tangent Graeffe transform* of P of order r. We have

$$G_r(\tilde{P}) = (z - (\alpha_1 - \epsilon)^r) \cdots (z - (\alpha_d - \epsilon)^r),$$

where

$$(\alpha_k - \epsilon)^r = \alpha_k^r - r\alpha_k^{r-1}\epsilon, \qquad k = 1, \ldots, d.$$

Now assume that we have an efficient way to determine the roots $\alpha_1^r, \ldots, \alpha_d^r$ of $G_r(P)$. For some polynomial $T \in \mathbb{K}[z]$, we may decompose $G_r(\tilde{P}) = G_r(P) + T\epsilon$ For any root α_k^r of $G_r(P)$, we then have

$$\begin{aligned} G_r(\tilde{P})(\alpha_k^r - r\alpha_k^{r-1}\epsilon) &= G_r(P)(\alpha_k^r) + (T(\alpha_k^r) - G_r(P)'(\alpha_k^r)r\alpha_k^{r-1})\epsilon \\ &= (T(\alpha_k^r) - G_r(P)'(\alpha_k^r)r\alpha_k^{r-1})\epsilon = 0. \end{aligned}$$

Whenever α_k^r happens to be a single root of $G_r(P)$, it follows that

$$r\alpha_k^{r-1} = \frac{T(\alpha_k^r)}{G_r(P)'(\alpha_k^r)}.$$

If $\alpha_k^r \neq 0$, this finally allows us to recover α_k as $\alpha_k = r\dfrac{\alpha_k^r}{r\alpha_k^{r-1}}$.

2.3   Heuristic Root Finding over Smooth Finite Fields

Assume now that $\mathbb{K} = \mathbb{F}_p$ is a finite field, where p is a prime number of the form $p = \sigma 2^m + 1$ for some small σ. Assume also that $\omega \in \mathbb{F}_p$ be a primitive element of order $p - 1$ for the multiplicative group of $\mathbb{F}_p$.

Let $P = (z - \alpha_1) \cdots (z - \alpha_d) \in \mathbb{F}_p[z]$ be as in the previous subsection. The tangent Graeffe method can be used to efficiently compute those α_k of P for which α_k^r is a single root of $G_r(P)$. In order to guarantee that there are a sufficient number of such roots, we first replace $P(z)$ by $P(z + \tau)$ for a random shift $\tau \in \mathbb{F}_p$, and use the following heuristic:

H For any subset $\{\alpha_1, \dots, \alpha_d\} \subseteq \mathbb{F}_p$ of cardinality d and any $r \leqslant (p-1)/(4d)$, there exist at least $p/2$ elements $\tau \in \mathbb{F}_p$ such that $\{(\alpha_1 - \tau)^r, \dots, (\alpha_d - \tau)^r\}$ contains at least $2d/3$ elements.

For a random shift $\tau \in \mathbb{F}_p$ and any $r \leqslant (p-1)/(4d)$, the assumption ensures with probability at least $1/2$ that $G_r(P(z + \tau))$ has at least $d/3$ single roots.

Now take r to be the largest power of two such that $r \leqslant (p-1)/(4d)$ and let $s = (p-1)/r$. By construction, note that $s = O(d)$. The roots $\alpha_1^r, \dots, \alpha_d^r$ of $G_r(P)$ are all s-th roots of unity in the set $\{1, \omega^r, \dots, \omega^{(s-1)r}\}$. We may thus determine them by evaluating $G_r(P)$ at ω^i for $i = 0, \dots, s - 1$. Since $s = O(d)$, this can be done efficiently using a discrete Fourier transform. Combined with the tangent Graeffe method from the previous subsection, this leads to the following probabilistic algorithm for root finding:

Algorithm 1
Input: $P \in \mathbb{F}_p[z]$ of degree d and only order one factors, $p = \sigma 2^m + 1$
Output: the set $\{\alpha_1, \dots, \alpha_d\}$ of roots of P

---

1. If $d = 0$ then return $\varnothing$
2. Let $r = 2^N \in 2^{\mathbb{N}}$ be largest such that $r \leqslant (p-1)/(4d)$ and let $s := (p-1)/r$
3. Pick $\tau \in \mathbb{F}_p$ at random and compute $P^* := P(z + \tau) \in \mathbb{F}_p[z]$
4. Compute $\tilde{P}(z) := P^*(z + \epsilon) = P^*(z) + P^*(z)'\epsilon \in (\mathbb{F}_p[\epsilon]/(\epsilon^2))[z]$
5. For $i = 1, \dots, N$, set $\tilde{P} := G_2(\tilde{P}) \in (\mathbb{F}_p[\epsilon]/(\epsilon^2))[z]$
6. Let ω have order $p-1$ in $\mathbb{F}_p$. Write $\tilde{P} = A + B\epsilon$ and compute $A(\omega^{ir})$, $A'(\omega^{ir})$, and $B(\omega^{ir})$ for $0 \leqslant i < s$
7. If $P(\tau) = 0$, then set $S := \{\tau\}$, else set $S := \varnothing$
8. For $\beta \in \{1, \omega^r, \dots, \omega^{(s-1)r}\}$ if $A(\beta) = 0$ and $A'(\beta) \neq 0$, set $S := S \cup \{r\beta A'(\beta)/B(\beta) + \tau\}$
9. Compute $Q := \prod_{\alpha \in S}(z - \alpha)$
10. Recursively determine the set of roots S' of P/Q
11. Return $S \cup S'$

Remark 1. To compute $G_2(\tilde{P}) = G_2(A + B\epsilon)$ we may use $G_2(\tilde{P}(z^2)) = A(z)A(-z) + (A(z)B(-z) + B(z)A(-z))\epsilon$, which requires three polynomial multiplications in $\mathbb{F}_p[z]$ of degree d. In total, step 5 thus performs $O(\log(p/s))$ such multiplications. We discuss how to perform step 5 efficiently in the FFT model in Sect. 3.

Remark 2. For practical implementations, one may vary the threshold $r \leqslant (p-1)/(4d)$ for r and the resulting threshold $s \geqslant 4d$ for s. For larger values of s, the computations of the DFTs in step 6 get more expensive, but the proportion of single roots goes up, so more roots are determined at each iteration. From an asymptotic complexity perspective, it would be best to take $s \asymp d\sqrt{\log p}$. In practice, we actually preferred to take the *lower* threshold $s \geqslant 2d$, because the constant factor of our implementation of step 6 (based on Bluestein's algorithm [3]) is significant with respect to our highly optimized implementation of the tangent Graeffe method. A second reason we prefer s of size $O(d)$ instead of $O(d\sqrt{\log p})$ is that the total space used by the algorithm is linear in s. In the future, it would be interesting to further speed up step 6 by investing more time in the implementation of high performance DFTs of general orders s.

3   Computing Graeffe Transforms

3.1   Reminders About Discrete Fourier Transforms

Assume $n \in \mathbb{N}$ is invertible in $\mathbb{K}$ and let $\omega \in \mathbb{K}$ be a primitive n-th root of unity. Consider a polynomial $A = a_0 + a_1 z + \cdots + a_{n-1} z^{n-1} \in \mathbb{K}[z]$. Then the discrete Fourier transform (DFT) of order n of the sequence $(a_i)_{0 \leqslant i < n}$ is defined by

$$\mathrm{DFT}_\omega((a_i)_{0 \leqslant i < n}) := (\hat{a}_k)_{0 \leqslant k < n}, \qquad \hat{a}_k := A(\omega^k).$$

We will write $\mathsf{F}_\mathbb{K}(n)$ for the cost of one discrete Fourier transform in terms of the number of operations in $\mathbb{K}$ and assume that $n = o(\mathsf{F}_\mathbb{K}(n))$. For any $i \in \{0, \ldots, n-1\}$, we have

$$\mathrm{DFT}_{\omega^{-1}}((\hat{a}_k)_{0 \leqslant k < n})_i = \sum_{0 \leqslant k < n} \hat{a}_k \omega^{-ik} = \sum_{0 \leqslant j < n} a_j \sum_{0 \leqslant k < n} \omega^{(j-i)k} = n a_i. \qquad (4)$$

If n is invertible in $\mathbb{K}$, then it follows that $\mathrm{DFT}_\omega^{-1} = n^{-1} \mathrm{DFT}_{\omega^{-1}}$. The costs of direct and inverse transforms therefore coincide up to a factor $O(n)$.

If $n = n_1 n_2$ is composite, $0 \leqslant k_1 < n_1$, and $0 \leqslant k_2 < n_2$, then it is well known [7] that

$$\hat{a}_{k_2 n_1 + k_1} = \mathrm{DFT}_{\omega^{n_1}} \left(\left(\omega^{i_2 k_1} \mathrm{DFT}_{\omega^{n_2}}((a_{i_1 n_2 + i_2})_{0 \leqslant i_1 < n_1})_{k_1} \right)_{0 \leqslant i_2 < n_2} \right)_{k_2}. \qquad (5)$$

This means that a DFT of length n reduces to n_1 transforms of length n_2 plus n_2 transforms of length n_1 plus n multiplications in $\mathbb{K}$:

$$\mathsf{F}_\mathbb{K}(n_1 n_2) \leqslant n_1 \mathsf{F}_\mathbb{K}(n_2) + n_2 \mathsf{F}_\mathbb{K}(n_1) + O(n).$$

In particular, if $r = O(1)$, then $\mathsf{F}_\mathbb{K}(rn) \sim r\mathsf{F}_\mathbb{K}(n)$.

It is sometimes convenient to apply DFTs directly to polynomials as well; for this reason, we also define $\mathrm{DFT}_\omega(A) := (\hat{a}_k)_{0 \leqslant k < n}$. Given two polynomials $A, B \in \mathbb{K}[z]$ with $\deg(AB) < n$, we may then compute the product AB using

$$AB = \mathrm{DFT}_\omega^{-1}(\mathrm{DFT}_\omega(A) \, \mathrm{DFT}_\omega(B)).$$

In particular, if $\mathsf{M}_{\mathbb{K}}(n)$ denotes the cost of multiplying two polynomials of degree $< n$, then we obtain $\mathsf{M}_{\mathbb{K}}(n) \sim 3\mathsf{F}_{\mathbb{K}}(2n) \sim 6\mathsf{F}_{\mathbb{K}}(n)$.

Remark 3. In Algorithm 1, we note that step 6 comes down to the computation of three DFTs of length s. Since r is a power of two, this length is of the form $s = \sigma 2^k$ for some $k \in \mathbb{N}$. In view of (5), we may therefore reduce step 6 to 3σ DFTs of length 2^k plus $3 \cdot 2^k$ DFTs of length σ. If σ is very small, then we may use a naive implementation for DFTs of length σ. In general, one may use Bluestein's algorithm [3] to reduce the computation of a DFT of length σ into the computation of a product in $\mathbb{K}[z]/(z^\sigma - 1)$, which can in turn be computed using FFT-multiplication and three DFTs of length a larger power of two.

3.2   Graeffe Transforms of Order Two

Let $\mathbb{K}$ be a field with a primitive $(2n)$-th root of unity ω. Let $P \in \mathbb{K}[z]$ be a polynomial of degree $d = \deg P < n$. Then the relation (2) yields

$$G(P)(z^2) = \mathrm{DFT}_\omega^{-1}(\mathrm{DFT}_\omega(P(z)) \, \mathrm{DFT}_\omega(P(-z))). \qquad (6)$$

For any $k \in \{0, \dots, 2n - 1\}$, we further note that

$$\mathrm{DFT}_\omega(P(-z))_k = P(-\omega^k) = P(\omega^{(k+n) \,\mathrm{rem}\, 2n}) = \mathrm{DFT}_\omega(P(z))_{(k+n) \,\mathrm{rem}\, 2n}, \qquad (7)$$

so $\mathrm{DFT}_\omega(P(-z))$ can be obtained from $\mathrm{DFT}_\omega(P)$ using n transpositions of elements in $\mathbb{K}$. Concerning the inverse transform, we also note that

$$\mathrm{DFT}_\omega(G(P)(z^2))_k = G(P)(\omega^{2k}) = \mathrm{DFT}_{\omega^2}(G(P))_k,$$

for $k = 0, \dots, n - 1$. Plugging this into (6), we conclude that

$$G(P) = \mathrm{DFT}_{\omega^2}^{-1}((\mathrm{DFT}_\omega(P)_k \, \mathrm{DFT}_\omega(P)_{k+n})_{0 \leqslant k < n}).$$

This leads to the following algorithm for the computation of $G(P)$:

Algorithm 2
Input: $P \in \mathbb{K}[z]$ with $\deg P < n$ and a primitive $(2n)$-th root of unity $\omega \in \mathbb{K}$
Output: $G(P)$

---

1. Compute $(\hat{P}_k)_{0 \leqslant k < 2n} := \mathrm{DFT}_\omega(P)$
2. For $k = 0, \dots, n - 1$, compute $\hat{G}_k := \hat{P}_k \hat{P}_{k+n}$
3. Return $\mathrm{DFT}_{\omega^2}^{-1}((\hat{G}_k)_{0 \leqslant k < n})$

Proposition 1. *Let $\omega \in \mathbb{K}$ be a primitive $2n$-th root of unity in $\mathbb{K}$ and assume that 2 is invertible in $\mathbb{K}$. Given a monic polynomial $P \in \mathbb{K}[z]$ with $\deg P < n$, we can compute $G(P)$ in time $\mathsf{G}_{2,\mathbb{K}}(n) \sim 3\mathsf{F}_{\mathbb{K}}(n)$.*

Proof. We have already explained the correctness of Algorithm 2. Step 1 requires one forward DFT of length $2n$ and cost $F_{\mathbb{K}}(2n) = 2F_{\mathbb{K}}(n) + O(n)$. Step 2 can be done in $O(n)$. Step 3 requires one inverse DFT of length n and cost $F_{\mathbb{K}}(n) + O(n)$. The total cost of Algorithm 2 is therefore $3F_{\mathbb{K}}(n) + O(n) \sim 3F_{\mathbb{K}}(n)$.

Remark 4. In terms of the complexity of multiplication, we obtain $G_{2,\mathbb{K}}(n) \sim (1/2)M_{\mathbb{K}}(n)$. This gives a 33.3% improvement over the previously best known bound $G_{2,\mathbb{K}}(n) \sim (2/3)M_{\mathbb{K}}(n)$ that was used in [10]. Note that the best known algorithm for squaring polynomials of degree $< n$ is $\sim (2/3)M_{\mathbb{K}}(n)$. It would be interesting to know whether squares can also be computed in time $\sim (1/2)M_{\mathbb{K}}(n)$.

3.3   Graeffe Transforms of Power of Two Orders

In view of (3), Graeffe transforms of power of two orders 2^m can be computed using

$$G_{2^m}(P) = \left(G \circ \overset{m\times}{\cdots} \circ G\right)(P). \tag{8}$$

Now assume that we computed the first Graeffe transform $G(P)$ using Algorithm 2 and that we wish to apply a second Graeffe transform to the result. Then we note that

$$\mathrm{DFT}_{\omega}(G(P))_{2k} = \mathrm{DFT}_{\omega^2}(G(P))_k = \hat{G}_{2k} \tag{9}$$

is already known for $k = 0, \ldots, n-1$. We can use this to accelerate step 1 of the second application of Algorithm 2. Indeed, in view of (5) for $n_1 = 2$ and $n_2 = n$, we have

$$\mathrm{DFT}_{\omega}(G(P))_{2k+1} = \mathrm{DFT}_{\omega^2}((\omega^i G(P)_i)_{0 \leqslant i < n})_k \tag{10}$$

for $k = 0, \ldots, n-1$. In order to exploit this idea in a recursive fashion, it is useful to modify Algorithm 2 so as to include $\mathrm{DFT}_{\omega^2}(P)$ in the input and $\mathrm{DFT}_{\omega^2}(G(P))$ in the output. This leads to the following algorithm:

Algorithm 3
Input: $P \in \mathbb{K}[z]$ with $\deg P < n$, a primitive $(2n)$-th root of unity $\omega \in \mathbb{K}$,
          and $(\hat{Q}_k)_{0 \leqslant k < n} = \mathrm{DFT}_{\omega^2}(P)$
Output: $G(P)$ and $\mathrm{DFT}_{\omega^2}(G(P))$

---

1. Set $(\hat{P}_{2k})_{0 \leqslant k < n} := (\hat{Q}_k)_{0 \leqslant k < n}$
2. Set $(\hat{P}_{2k+1})_{0 \leqslant k < n} := \mathrm{DFT}_{\omega^2}((\omega^i P_i)_{0 \leqslant i < n})$
3. For $k = 0, \ldots, n-1$, compute $\hat{G}_k := \hat{P}_k \hat{P}_{k+n}$
4. Return $\mathrm{DFT}_{\omega^2}^{-1}((\hat{G}_k)_{0 \leqslant k < n})$ and $(\hat{G}_k)_{0 \leqslant k < n}$

Proposition 2. *Let $\omega \in \mathbb{K}$ be a primitive $2n$-th root of unity in $\mathbb{K}$ and assume that 2 is invertible in $\mathbb{K}$. Given a monic polynomial $P \in \mathbb{K}[z]$ with $\deg P < n$ and $m \geqslant 1$, we can compute $G_{2^m}(P)$ in time $G_{2^m,\mathbb{K}}(n) \sim (2m+1)F_{\mathbb{K}}(n)$.*

Proof. It suffices to compute $\mathrm{DFT}_{\omega^2}(P)$ and then to apply Algorithm 3 recursively, m times. Every application of Algorithm 3 now takes $2\mathsf{F}_{\mathbb{K}}(n) + O(n) \sim 2\mathsf{F}_{\mathbb{K}}(n)$ operations in $\mathbb{K}$, whence the claimed complexity bound.

Remark 5. In [10], Graeffe transforms of order 2^m were directly computed using the formula (8), using $\sim 4m\mathsf{F}_{\mathbb{K}}(n)$ operations in $\mathbb{K}$, which is twice as slow as the new algorithm.

4   Implementation and Benchmarks

We have implemented the tangent Graeffe root finding algorithm (Algorithm 1) in C with the optimizations presented in Sect. 3. Our C implementation supports primes of size up to 63 bits. In what follows all complexities count arithmetic operations in $\mathbb{F}_p$.

In Tables 1 and 2 the input polynomial $P(z)$ of degree d is constructed by choosing d distinct values $\alpha_i \in \mathbb{F}_p$ for $1 \leqslant i \leqslant d$ at random and creating $P(z) = \prod_{i=1}^{d}(z - \alpha_i)$. We will use $p = 3 \times 29 \times 2^{56} + 1$, a smooth 63 bit prime. For this prime $\mathsf{M}(d)$ is $O(d \log d)$.

One goal we have is to determine how much faster the Tangent Graeffe (TG) root finding algorithm is in practice when compared with the Cantor-Zassenhaus (CZ) algorithm which is implemented in many computer algebra systems. In Table 1 we present timings comparing our sequential implementation of the TG algorithm with Magma's implementation of the CZ algorithm. For polynomials in $\mathbb{F}_p[z]$, Magma uses Shoup's factorization algorithm from [17]. For our input $P(z)$, with d distinct linear factors, Shoup uses the Cantor–Zassenhaus equal degree factorization method. The average complexity of TG is $O(\mathsf{M}(d)(\log(p/s)+\log d))$ and of CZ is $O(\mathsf{M}(d) \log p \log d)$.

Table 1. Sequential timings in CPU seconds for $p = 3 \cdot 29 \cdot 2^{56} + 1$ and using $s \in [2d, 4d]$.

d	Our sequential TG implementation in C						Magma CZ timings	
	Total	First	%roots	Step 5	Step 6	Step 9	V2.25-3	V2.25-5
$2^{12} - 1$	0.11 s	0.07 s	69.8%	0.04 s	0.02 s	0.01 s	23.22 s	8.43
$2^{13} - 1$	0.22 s	0.14 s	69.8%	0.09 s	0.03 s	0.01 s	56.58 s	18.94
$2^{14} - 1$	0.48 s	0.31 s	68.8%	0.18 s	0.07 s	0.02 s	140.76 s	44.07
$2^{15} - 1$	1.00 s	0.64 s	69.2%	0.38 s	0.16 s	0.04 s	372.22 s	103.5
$2^{16} - 1$	2.11 s	1.36 s	68.9%	0.78 s	0.35 s	0.10 s	1494.0 s	234.2
$2^{17} - 1$	4.40 s	2.85 s	69.2%	1.62 s	0.74 s	0.23 s	6108.8 s	534.5
$2^{18} - 1$	9.16 s	5.91 s	69.2%	3.33 s	1.53 s	0.51 s	NA	1219
$2^{19} - 1$	19.2 s	12.4 s	69.2%	6.86 s	3.25 s	1.13 s	NA	2809

The timings in Table 1 are sequential timings obtained on a Linux server with an Intel Xeon E5-2660 CPU with 8 cores. In Table 1 the time in column "first" is for the first application of the TG algorithm (steps 1–9 of Algorithm 1), which obtains about 69% of the roots. The time in column "total" is the total time for the TG algorithm. Columns step 5, step 6, and step 9 report the time spent in steps 5, 6, and 9 in Algorithm 1 and do not count time in the recursive call in step 10.

The Magma timings are for Magma's `Factorization` command. The timings for Magma version V2.25-3 suggest that Magma's CZ implementation involves a subalgorithm with quadratic asymptotic complexity. Indeed it turns out that the author of the code implemented all of the sub-quadratic polynomial arithmetic correctly, as demonstrated by the second set of timings for Magma in column V2.25-5, but inserted the d linear factors found into a list using linear insertion! Allan Steel of the Magma group identified and fixed the offending subroutine for Magma version V2.25-5. The timings show that TG is faster than CZ by a factor of 76.6 (=8.43/0.11) to 146.3 (=2809/19.2).

We also wanted to attempt a parallel implementation. To do this we used the MIT Cilk C compiler from [8]. Cilk provides a simple fork-join model of parallelism. Unlike the CZ algorithm, TG has no gcd computations that are hard to parallelize. We present some initial parallel timing data in Table 2. The timings in parentheses are parallel timings for 8 cores.

Table 2. Real times in seconds for 1 core (8 cores) and $p = 3 \cdot 29 \cdot 2^{56} + 1$.

d	Our parallel tangent Graeffe implementation in Cilk C				
	Total	First	Step 5	Step 6	Step 9
$2^{19} - 1$	18.30 s(9.616 s)	11.98 s(2.938 s)	6.64 s(1.56 s)	3.13 s(0.49 s)	1.09 s(0.29 s)
$2^{20} - 1$	38.69 s(12.40 s)	25.02 s(5.638 s)	13.7 s(3.03 s)	6.62 s(1.04 s)	2.40 s(0.36 s)
$2^{21} - 1$	79.63 s(20.16 s)	52.00 s(11.52 s)	28.1 s(5.99 s)	13.9 s(2.15 s)	5.32 s(0.85 s)
$2^{22} - 1$	166.9 s(41.62 s)	107.8 s(23.25 s)	57.6 s(11.8 s)	28.9 s(4.57 s)	11.7 s(1.71 s)
$2^{23} - 1$	346.0 s(76.64 s)	223.4 s(46.94 s)	117 s(23.2 s)	60.3 s(9.45 s)	25.6 s(3.54 s)
$2^{24} - 1$	712.7 s(155.0 s)	459.8 s(95.93 s)	238 s(46.7 s)	125 s(19.17)	55.8 s(7.88 s)
$2^{25} - 1$	1465 s(307.7 s)	945.0 s(194.6 s)	481 s(92.9 s)	259 s(39.2 s)	121 s(16.9 s)

4.1   Implementation Notes

To implement the Taylor shift $P(z + \tau)$ in step 3, we used the $O(M(d))$ method from [1, Lemma 3]. For step 5 we use Algorithm 3. It has complexity $O(M(d) \log \frac{p}{s})$. To evaluate $A(z), A'(z)$ and $B(z)$ in step 6 in $O(M(s))$ we used the Bluestein transformation [3]. In step 9 to compute the product $Q(z) = \Pi_{\alpha \in S}(z - \alpha)$, for $t = |S|$ roots, we used the $O(M(t) \log t)$ product tree multiplication algorithm [9]. The division in step 10 is done in $O(M(d))$ with the fast division.

The sequential timings in Tables 1 and 2 show that steps 5, 6 and 9 account for about 90% of the total time. We parallelized these three steps as follows.

For step 5, the two forward and two inverse FFTs are done in parallel. We also parallelized our radix 2 FFT by parallelizing recursive calls for size $n \geqslant 2^{17}$ and the main loop in blocks of size $m \geqslant 2^{18}$ as done in [14]. For step 6 there are three applications of Bluestein to compute $A(\omega^{ir})$, $A'(\omega^{ir})$ and $B(\omega^{ir})$. We parallelized these (thereby doubling the overall space used by our implementation). The main computation in the Bluestein transformation is a polynomial multiplication of two polynomials of degree s. The two forward FFTs are done in parallel and the FFTs themselves are parallelized as for step 5. For the product in step 9 we parallelize the two recursive calls in the tree multiplication for large sizes and again, the FFTs are parallelized as for step 5.

To improve parallel speedup we also parallelized the polynomial multiplication in step 3 and the computation of the roots in step 8. Although step 8 is $O(|S|)$, it is relatively expensive because of two inverse computations in $\mathbb{F}_p$. Because we have not parallelized about 5% of the computation the maximum parallel speedup we can obtain is a factor of $1/(0.05 + 0.95/8) = 5.9$. The best overall parallel speedup we obtained is a factor of $4.6 = 1465/307.7$ for $d = 2^{25}-1$.

References

1. Aho, A.V., Steiglitz, K., Ullman, J.D.: Evaluating polynomials on a fixed set of points. SIAM J. Comput. **4**, 533–539 (1975)
2. Ben-Or, M., Tiwari, P.: A deterministic algorithm for sparse multivariate polynomial interpolation. In: STOC 1988: Proceedings of the Twentieth Annual ACM Symposium on Theory of Computing, pp. 301–309. ACM Press (1988)
3. Bluestein, L.I.: A linear filtering approach to the computation of discrete Fourier transform. IEEE Trans. Audio Electroacoust. **18**(4), 451–455 (1970)
4. Brent, R.P., Gustavson, F.G., Yun, D.Y.Y.: Fast solution of Toeplitz systems of equations and computation of Padé approximants. J. Algorithms **1**(3), 259–295 (1980)
5. Canny, J., Kaltofen, E., Lakshman, Y.: Solving systems of non-linear polynomial equations faster. In: Proceedings of the ACM-SIGSAM 1989 International Symposium on Symbolic and Algebraic Computation, pp. 121–128. ACM Press (1989)
6. Cantor, D.G., Zassenhaus, H.: A new algorithm for factoring polynomials over finite fields. Math. Comput. **36**(154), 587–592 (1981)
7. Cooley, J.W., Tukey, J.W.: An algorithm for the machine calculation of complex Fourier series. Math. Comput. **19**, 297–301 (1965)
8. Frigo, M., Leisorson, C.E., Randall, R.K.: The implementation of the Cilk-5 multithreaded language. In: Proceedings of PLDI 1998, pp. 212–223. ACM (1998)
9. von zur Gathen, J., Gerhard, J.: Modern Computer Algebra, 3rd edn. Cambridge University Press, New York (2013)
10. Grenet, B., van der Hoeven, J., Lecerf, G.: Randomized root finding over finite fields using tangent Graeffe transforms. In: Proceedings of the ISSAC 2015, pp. 197–204. ACM, New York (2015)
11. Grenet, B., van der Hoeven, J., Lecerf, G.: Deterministic root finding over finite fields using Graeffe transforms. Appl. Algebra Eng. Commun. Comput. **27**(3), 237–257 (2015). https://doi.org/10.1007/s00200-015-0280-5
12. van der Hoeven, J., Monagan, M.: Implementing the tangent Graeffe root finding method. Technical report, HAL (2020). http://hal.archives-ouvertes.fr/hal-02525408

13. van der Hoeven, J., et al.: GNU TeXmacs (1998). http://www.texmacs.org

14. Law, M., Monagan, M.: A parallel implementation for polynomial multiplication modulo a prime. In: Proceedings of PASCO 2015, pp. 78–86. ACM (2015)

15. Prony, R.: Essai expérimental et analytique sur les lois de la dilatabilité des fluides élastiques et sur celles de la force expansive de la vapeur de l'eau et de la vapeur de l'alkool, à différentes températures. J. de l'École Polytechnique Floréal et Plairial, an III 1(cahier 22), 24–76 (1795)

16. Roche, D.S.: What can (and can't) we do with sparse polynomials? In: Arreche, C. (ed.) ISSAC 2018: Proceedings of the 2018 ACM International Symposium on Symbolic and Algebraic Computation, pp. 25–30. ACM Press (2018)

17. Shoup, V.: A new polynomial factorization and its implementation. J. Symb. Comput. **20**(4), 363–397 (1995)

Author Index

Printed in the United States
By Bookmasters